Stochastic Analysis, Random Fields and Integrable Probability
— Fukuoka 2019

ADVANCED STUDIES
IN PURE MATHEMATICS 87

Chief Editor of the Series: Shigeki Aida (University of Tokyo)

Stochastic Analysis, Random Fields and Integrable Probability — Fukuoka 2019

Edited by

Yuzuru Inahama (Kyushu University, Chief Editor)
Hirofumi Osada (Kyushu University)
Tomoyuki Shirai (Kyushu University)

 Mathematical Society of Japan

This book was typeset by $\mathcal{A}_\mathcal{M}\mathcal{S}$-TEX and $\mathcal{A}_\mathcal{M}\mathcal{S}$-LATEX, the TEX macro systems of the American Mathematical Society, together with the style files aspm.sty *and* aspmfm.sty *for $\mathcal{A}_\mathcal{M}\mathcal{S}$-TEX written by Dr. Chiaki Tsukamoto and* aspmproc.sty *for $\mathcal{A}_\mathcal{M}\mathcal{S}$-LATEX written by Dr. Akihiro Munemasa and* aspm.cls *for $\mathcal{A}_\mathcal{M}\mathcal{S}$-LATEX provided by Livretech Co., Ltd.*

TEX is a trademark of the American Mathematical Society.

Edited by the Mathematical Society of Japan.

Published by the Mathematical Society of Japan.

Distributed by the Mathematical Society of Japan, the American Mathematical Society and the World Scientific Publishing Co., Ltd.
Distributed exclusively in North America by the American Mathematical Society.
Partially supported by Grant-in-Aid for Publication of Scientific Research Results, Japan Society for the Promotion of Science, Grant Number 17HP1002.

Advanced Studies in Pure Mathematics 87
ISBN 978-4-86497-094-5

PRINTED IN JAPAN
by Livretech Co., Ltd.

2010 Mathematics Subject Classification.
Primary 60Hxx, 60G60, 60K35.
Secondary 60C05, 82B41, 60D05, 60H15, 60J67, 60F17, 60G55.

Preface

This volume is a collection of eighteen research and survey papers written by speakers of the conference of the 12th Mathematical Society of Japan, Seasonal Institute (MSJ-SI), titled as

"Stochastic Analysis, Random Fields and Integrable Probability,"
31st July–9th August 2019
Kyushu University, Fukuoka, Japan.

MSJ-SI is an official workshop of the Mathematical Society of Japan (MSJ), which has been held annually since 2008. Each year, the workshop's topic is chosen by MSJ, and leading mathematicians worldwide are invited accordingly.

The conference focuses on recent progress in stochastic analysis, random fields, and integrable probability. Six course lectures by

Louigi Addario-Berry (McGill University)
Alexander I. Bufetov (Aix-Marseille Université)
Ivan Corwin (Columbia University)
Frank den Hollander (Universiteit Leiden)
Gregory F. Lawler (The University of Chicago)
Grégory Miermont (École Normale Supérieure de Lyon),

Eighteen invited talks, seven contributed talks, and sixteen poster presentations were presented along with active and fruitful discussion by 118 participants.

The conference was organized by the Society's committee. The member of the committee consists of Yuzuru Inahama, Takashi Kumagai, Hirofumi Osada, Tomohiro Sasamoto, Tomoyuki Shirai, and Hideki Tanemura—in collaboration with the local organizing committee—Syota Esaki, Yuu Hariya, Masato Hoshino, Yuzuru Inahama, Masato Shinoda, and Tomoyuki Shirai.

The conference was supported by the Mathematical Society of Japan, Faculty of Mathematics (Kyushu University), Institute of Mathematics for Industry (Kyushu University), and Japan Society for the Promotion of Science, Grand-in-Aid for Scientific Research (S) JP16H06338 (PI: H. Osada), Scientific Research (A) JP17H01093 (PI: T. Kumagai), Scientific Research (A) JP19H00643 (PI: M. Hino).

We thank all the speakers and the participants for their invaluable contributions, making the conference most successful. We also thank all the referees for their help with this volume by carrying out painful but excellent jobs. Besides, we thank Tsubura Imabayashi and Seiko Sasaguri for their contributions to the workshop. Finally, we thank Ryuichi Honda for his editing work on the proceedings.

March 2021
Yuzuru Inahama
Hirofumi Osada
Tomoyuki Shirai

All papers in this volume have been refereed and are in final form. No version of any of them will be submitted for publication elsewhere.

Program of the Conference

July 31st (Wed.)

9:00–9:45	Registration
9:45–10:00	**President of the Mathematical Society of Japan** Opening Speech
10:00–11:00	**Frank den Hollander** (Universiteit Leiden) Complex Networks: Structure and Functionality. I. Spectra
11:20–12:20	**Ivan Corwin** (Columbia University) Stochastic Vertex Models I
12:20–13:40	Lunch
13:40–14:20	**Tadahisa Funaki** (Waseda University) Large Deviation for Lozenge Tiling Dynamics
14:30–15:10	**Rongchan Zhu** (Beijing Institute of Technology) Stochastic Cahn-Hilliard Equation
15:20–15:40	**Masato Hoshino** (Kyushu University) Paracontrolled Calculus and Regularity Structures
15:40–16:10	Coffee break
16:10–16:50	**Subhroshekhar Ghosh** (National University of Singapore) Two Manifestations of Rigidity in Point Sets: Forbidden Regions and Maximal Degeneracy
17:00–17:20	**Kenkichi Tsunoda** (Osaka University) Derivation of Viscous Burgers Equations from Weakly Asymmetric Exclusion Processes
18:00–20:00	Welcome Party

August 1st (Thu.)

10:00–11:00	**Ivan Corwin** (Columbia University) Stochastic Vertex Models II
11:20–12:20	**Ivan Corwin** (Columbia University) Stochastic Vertex Models III
12:20–13:40	Lunch
13:40–14:20	**Takashi Imamura** (Chiba University) Determinantal Structures in the q-Whittaker Measure
14:30–15:10	**Nina Holden** (ETH-ITS Zurich) Cardy Embedding of Random Planar Maps

15:20–15:40 **Yoshihiro Abe** (Chiba University)
Exceptional Points of Two-Dimensional Random Walks at Multiples of the Cover Time

15:40–16:10 Coffee break

16:10–16:50 **Makoto Katori** (Chuo University)
Gaussian Free Fields with Boundary Points, Multiple SLEs, and Log-Gases

17:00–17:20 **Jacek Małecki** (Wrocław University of Science and Technology)
Universality Classes for General Random Matrix Flows

August 2nd (Fri.)

10:00–11:00 **Frank den Hollander** (Universiteit Leiden)
Complex Networks: Structure and Functionality. II. Equivalence

11:20–12:20 **Frank den Hollander** (Universiteit Leiden)
Complex Networks: Structure and Functionality. III. Exploration

12:20–13:40 Lunch

13:40–14:20 **Pierre-François Rodriguez** (Institut des Hautes Études Scientifiques)
Isomorphism Theorems and the Sign Cluster Geometry of the Gaussian Free Field

14:30–15:10 **Julien Berestycki** (The University of Oxford)
Branching Particle Systems with Selection and Free Boundary Problems

15:20–15:40 **Izumi Okada** (Kyushu University)
Exponents for High Points of Simple Random Walks in Two Dimensions

15:40–16:10 Coffee break

16:10–16:50 **Li-Cheng Tsai** (Rutgers University)
Lower-Tail Large Deviations of the KPZ Equation

17:00–17:20 **Xinyi Li** (The University of Chicago)
Natural Parametrization for the Scaling Limit of Loop-Erased Random Walk in Three Dimensions

August 5th (Mon.)

10:00–11:00 **Gregory F. Lawler** (The University of Chicago)
Loop Measures and Loop-Erased Random Walk I

11:20–12:20	**Louigi Addario-Berry** (McGill University) Algorithms and Random Trees I
12:20–13:40	Lunch
13:40–14:40	**Gregory F. Lawler** (The University of Chicago) Loop Measures and Loop-Erased Random Walk II
14:50–15:30	**Daisuke Shiraishi** (Kyoto University) Scaling Limit of Uniform Spanning Tree in Three Dimensions
15:40–16:40	Poster Presentations
16:40–17:40	Poster Session and Coffee break

August 6th (Tue.)

10:00–11:00	**Louigi Addario-Berry** (McGill University) Algorithms and Random Trees II
11:20–12:20	**Gregory F. Lawler** (The University of Chicago) Loop Measures and Loop-Erased Random Walk III
12:20–13:40	Lunch
13:40–14:20	**Makiko Sasada** (The University of Tokyo) Generalized Pitman's Transform and Discrete Integrable Systems
14:30–15:10	**Riddhipratim Basu** (Tata Institute of Fundamental Research) Bigeodesics and Polymer Watermelons in Last Passage Percolation
15:20–15:40	**Shuta Nakajima** (Nagoya University) Divergence of Non-Random Fluctuation in First-Passage Percolation
15:40–16:10	Coffee break
16:10–16:50	**Insuk Seo** (Seoul National University) The Cut-Off Phenomenon for the Random-Cluster Model
18:30–20:30	Banquet (Hyakunen-Gura)

August 7th (Wed.)

10:00–11:00	**Grégory Miermont**, (École Normale Supérieure de Lyon) Brownian Surfaces I
11:20–12:20	**Louigi Addario-Berry** (McGill University) Algorithms and Random Trees III

12:20–19:00 Excursion (Karatsu)

August 8th (Thu.)

10:00–11:00 **Alexander I. Bufetov** (Aix-Marseille Université)
Determinantal Point Processes and Extrapolation I

11:20–12:20 **Grégory Miermont** (École Normale Supérieure de Lyon)
Brownian Surfaces II

12:20–13:40 Lunch

13:40–14:40 **Grégory Miermont** (École Normale Supérieure de Lyon)
Brownian Surfaces III

14:50–15:30 **Akira Sakai** (Hokkaido University)
Critical Two-Point Function for Long-Range Models with Power-Law Couplings: The Marginal Case for $d \geq d_c$

15:30–16:10 Coffee break

16:10–16:50 **Seiichiro Kusuoka** (Kyoto University)
Approach to the Quantum Field with Exponential Interactions by Singular SPDEs

17:00–17:40 **Yanqi Qiu** (Chinese Academy of Sciences)
A Class of Positive Functions on the Circle and Their Applications on Hankel Operators

August 9th (Fri.)

10:00–11:00 **Alexander I. Bufetov** (Aix-Marseille Université)
Determinantal Point Processes and Extrapolation II

11:20–12:20 **Alexander I. Bufetov** (Aix-Marseille Université)
Determinantal Point Processes and Extrapolation III

12:20–13:40 Lunch

13:40–14:20 **Naotaka Kajino** (Kobe University)
The Laplacian on Some Self-Conformal Fractals and Weyl's Asymptotics for Its Eigenvalues

14:30–15:10 **Ryoki Fukushima** (Kyoto University)
Zero Temperature Limits for Directed Polymers in Random Environment

15:15–15:30 **Chair of the Scientific Committee**
Closing Speech

Posters

Eleanor Archer (The University of Warwick)
Brownian Motion on Stable Looptrees
Sung-Soo Byun (Seoul National University)
Annulus SLE Partition Functions and Martingale-Observables
Taiki Endo (Chuo University)
Three-Parametric Marcenko–Pastur Density
Xiang Fang (National Central University (Taiwan))
Random Weighted Shifts
Yueyuan Gao (AIST, Tohoku University)
Uniqueness of the Entropy Solution of a Stochastic Conservation Law with a Q-Brownian Motion
Masayuki Kageyama (Nagoya City University)
A New Criteria of Risk in Markov Decision Processes
Yoshinori Kamijima (Hokkaido University)
Mean-Field Behavior for the Quantum Ising Model
Shinji Koshida (Chuo University)
Free Field Approach to the Macdonald Process
Matteo Mucciconi (Tokyo Institute of Technology)
Yang Baxter Field and Stochastic Vertex Mode
Takuya Murayama (Kyoto University)
Loewner Chains and Evolution Families on Parallel Slit Half-Planes
Takuya Nakagawa (Ritsumeikan University)
$L^{\alpha-1}$ Distance between Two One-Dimensional Stochastic Differential Equations Driven by a Symmetric α-Stable Process
Shota Osada (Kyushu University)
Isomorphisms between DPPs and PPPs
Yuki Tokushige (Kyoto University)
The Differentiability of the Speed of Biased RWs on Galton-Watson Trees
Aurélien Velleret (Aix-Marseille Université)
Large Deviations on the Profile of Mutation Effects for a Model of Populations Adapting to a Changing Environment
Linjie Zhao (Peking University)
Scaling Limits for the Exclusion Process with a Slow Site

CONTENTS

Advanced Studies in Pure Mathematics 87, 2021
Stochastic Analysis, Random Fields and Integrable Probability — Fukuoka 2019
pp. 1–57

Random tree-weighted graphs

Louigi Addario-Berry and Jordan Barrett

Abstract.

For each $n \geq 1$, let $\mathrm{d}^n = (d^n(i), 1 \leq i \leq n)$ be a sequence of positive integers with even sum $\sum_{i=1}^n d^n(i) \geq 2n$. Let (G_n, T_n, Γ_n) be uniformly distributed over the set of simple graphs G_n with degree sequence d^n, endowed with a spanning tree T_n and rooted along an oriented edge Γ_n of G_n which is not an edge of T_n. Under a finite variance assumption on degrees in G_n, we show that, after rescaling, T_n converges in distribution to the Brownian continuum random tree as $n \to \infty$. Our main tool is a new version of Pitman's additive coalescent [P], which can be used to build both random trees with a fixed degree sequence, and random tree-weighted graphs with a fixed degree sequence. As an input to the proof, we also derive a Poisson approximation theorem for the number of loops and multiple edges in the superposition of a fixed graph and a random graph with a given degree sequence sampled according to the configuration model; we find this to be of independent interest.

§1. Introduction

By a *rooted tree* we mean a labeled tree $t = (\mathrm{v}(t), \mathrm{e}(t))$, with a distinguished root node denoted $r(t)$. A *tree-rooted graph* is a pair (g, t, γ) where $g = (\mathrm{v}(g), \mathrm{e}(g))$ is a labeled graph, $t = (\mathrm{v}(t), \mathrm{e}(t))$ is a spanning tree of g, and $\gamma = uv$ is a distinguished oriented edge with $\{u, v\} \in \mathrm{e}(g) \setminus \mathrm{e}(t)$. We view t as a rooted tree by setting $r(t) = u$.

Throughout this work, we allow our graphs to have multiple edges and loops; in tree-rooted graphs, the root edge is allowed to be a loop. We say (g, t, γ) is *simple* if g is simple, i.e., if g contains no multiple edges or loops.

Received August 28, 2020.
Revised January 23, 2021.
2010 *Mathematics Subject Classification.* Primary 60C05, 05C80; Secondary 05C05, 60F05.
Key words and phrases. Random graphs, tree-weighted graphs, continuum random tree, branching processes, configuration model.

For a node u of a rooted tree t, we write $c_t(u)$ for the number of children of u in t. Given a rooted tree t with $\mathrm{v}(t) = [n] := \{1, 2, \ldots, n\}$, the *child sequence* of t is the sequence $\mathrm{c}_t = (c_t(i), 1 \le i \le n)$. Similarly, given a tree-rooted graph (g, t, γ) with vertex set $\mathrm{v}(g) = [n]$, the *degree sequence* of (g, t, γ) is the sequence $(d_g(i), 1 \le i \le n)$, where $d_g(i)$ is the number of endpoints of edges incident to i in g; here loops are counted twice.

For any sequence $\mathrm{d} = (d(1), \ldots, d(n))$ of non-negative integers, we define the *degree distribution* $p_{\mathrm{d}} = (p_{\mathrm{d}}(k), k \ge 1)$ of d by letting $p_{\mathrm{d}}(k) = \#\{i \in [n] : d(i) = k\}/n$.

The following theorem contains our main result, which is an invariance principle for the spanning trees in random tree-rooted graphs with a fixed degree sequence. To state it, two further pieces of notation are needed. Given a finite graph $g = (v, e)$ and a constant $c > 0$, we write cg for the measured metric space (v, dist, π) whose points are the elements of v, with $\mathrm{dist}(x, y) := c \cdot \mathrm{dist}_g(x, y)$, where $\mathrm{dist}_g(x, y)$ denotes graph distance in g, and with π the uniform probability measure on v. Also, for a sequence $p = (p(k), k \ge 1)$ of real numbers, we write $\mu_1(p) := \sum_{k \ge 1} kp(k)$ and $\mu_2(p) := \sum_{k \ge 1} k^2 p(k)$.

Theorem 1.1. *For each $n \ge 1$ let $\mathrm{d}^n = (d^n(i), 1 \le i \le n)$ be a degree sequence with $\min_{1 \le i \le n} d^n(i) \ge 1$, with $\sum_{i \in [n]} d^n(i) \ge 2n$ and with $\sum_{i \in [n]} d^n(i)$ even. Let p^n be the degree distribution of d^n. Suppose that there exists a probability distribution $p = (p(k), k \ge 1)$ such that* (a) *$p^n \to p$ pointwise and $p(2) < 1$, and* (b) *$\mu_2(p^n) \to \mu_2(p) \in (0, \infty)$. Then there exists $\sigma = \sigma(p) \in (0, \infty)$ such that the following holds.*

For $n \ge 1$ let (G_n, T_n, Γ_n) be chosen uniformly at random among all simple tree-rooted graphs with vertex set $[n]$ and degree sequence d^n. Then

$$\frac{\sigma}{n^{1/2}} T_n \xrightarrow{\mathrm{d}} \mathcal{T}$$

as $n \to \infty$ with respect to the Gromov-Hausdorff-Prokhorov topology, where $\mathcal{T}$ is the Brownian continuum random tree.

We refer the reader to [ADH] for a good discussion of the Gromov-Hausdorff-Prokhorov topology aimed at probabilists. The technical insight underlying the proof of Theorem 1.1 is the fact that Pitman's additive coalescent [P] can be modified to yield a simple construction procedure for random tree-weighted graphs with a given degree sequence. We anticipate that this procedure has further interesting features to be explored.

1.1. Related work

The enumerative combinatorics of tree-rooted maps was developed in the 1960's and 1970's [M3, WL]. The area has seen renewed attention over the last decade or so [BMC, B1, B2]. Random tree-rooted maps can be interpreted as samples from a Fortuin-Kastelyn model at zero temperature, and are an active object of study in the planar probability community (see, e.g., [GHS, HS, MS, GKMW, BLR, GS, LSW]).

There has also been some work on the typical number of spanning trees in uniformly random graphs [M, GIKM, GKW] with given degree sequences. (In such models, the underlying graph is sampled uniformly at random from some set of allowed graphs; in our model, it is the tree-weighted graph which is uniformly random, which means the underlying measure on graphs is biased in favour of graphs with a greater number of spanning trees.)

Except in the setting of graphs on surfaces, we have not found any previous work on tree-weighted graphs, random or otherwise.

1.2. Overview of the proof

We begin with a small number of facts and definitions that are required for the overview. We say a sequence $\mathrm{c} = (c(i), 1 \le i \le n)$ of non-negative integers is a child sequence if it is the child sequence of some tree. Note that c is a child sequence if and only if $\sum_{1 \le i \le n} c(i) = n - 1$, in which case

$$(1) \qquad \#\{\text{rooted trees } t : \mathrm{c}_t = \mathrm{c}\} = \binom{n-1}{c(1), \ldots, c(n)} = \frac{1}{n} \frac{n!}{\prod_{i=1}^{n} c(i)!};$$

see [M2], Section 3.3.

Given any sequence $\mathrm{c} = (c(i), 1 \le i \le n)$ of non-negative integers, for $k \ge 1$ we write $Q_\mathrm{c}(k) = \#\{i \in [n] : c(i) = k\}$. We call $Q_\mathrm{c} = (Q_\mathrm{c}(k), k \ge 0)$ the *child statistics vector* of c. For a tree t with child sequence c_t, we will sometimes write $Q_t = Q_{\mathrm{c}_t}$ for succinctness.

Given a graph g, for an edge $e \in \mathrm{e}(g)$ we write $m_g(e)$ for the multiplicity of edge e in g. Given a degree sequence $\mathrm{d} = (d(1), \ldots, d(n))$, the classical *configuration model* [vdH, Chapter 7] produces a random graph G such that for any fixed graph g with degree sequence d,

$$(2) \qquad \mathbf{P}\{G = g\} \propto \frac{1}{\prod_{i=1}^{n} 2^{m_g(ii)} \prod_{e \in \mathrm{e}(g)} m_g(e)!}.$$

In Section 2, we define a sampling procedure, inspired by the configuration model and by Pitman's additive coalescent [P], which produces a random tree-weighted graph (G, T, Γ) with the property that for any fixed tree-weighted graph (g, t, γ) with degree sequence d,

$$\mathbf{P}\{(G, T, \Gamma) = (g, t, \gamma)\} \propto \frac{2^{\mathbf{1}_{[\gamma \text{ is a loop}]}} \cdot m_{g-t}(\gamma)}{\prod_{i=1}^{n} 2^{m_{g-t}(ii)} \cdot \prod_{e \in \mathrm{e}(g)} m_{g-t}(e)!}, \tag{3}$$

where $g - t$ is the graph with the same vertex set as g and with edge multiplicities given by

$$m_{g-t}(e) = \begin{cases} m_g(e) & \text{if } e \notin \mathrm{e}(t) \\ m_g(e) - 1 & \text{if } e \in \mathrm{e}(t). \end{cases}$$

We call a random tree-weighted graph (G, T, Γ) with distribution given by (3) a *random tree-weighted graph with degree sequence* d. Note that in this case, conditionally given that G is simple, (G, T, Γ) is uniformly distributed over simple tree-rooted graphs with degree sequence d.

The sampling procedure we use has enough exchangeability that, conditional on its child sequence, the resulting spanning tree T is uniformly distributed; that is, for any fixed child sequence $\mathrm{c} = (c(i), 1 \leq i \leq n)$, and any tree t with $\mathrm{c}_t = \mathrm{c}$,

$$\mathbf{P}\{T = t \mid \mathrm{c}_T = \mathrm{c}\} = \binom{n-1}{c(1), \ldots, c(n)}^{-1}.$$

Now let $(\mathrm{d}^n, n \geq 1)$ be a sequence of degree sequences satisfying the conditions of Theorem 1.1; for each n let $(G(\mathrm{d}^n), T(\mathrm{d}^n), \Gamma(\mathrm{d}^n))$ be a random tree-weighted graph with degree sequence d^n. We prove (see Proposition 3.1) that there is a probability distribution $q = (q(i), i \geq 0)$ with $\mu_2(q) < \infty$ such that the child statistics vector $Q_{T(\mathrm{d}^n)}$ satisfies that $n^{-1} Q_{T(\mathrm{d}^n)}(a) \to q(a)$ in probability for all $a \geq 0$, and moreover that $\mu_2(n^{-1} Q_{T(\mathrm{d}^n)}) \to \mu_2(q)$ in probability. It then follows from a result of Broutin and Marckert [BM] that

$$\frac{\sigma}{n^{1/2}} T(\mathrm{d}^n) \xrightarrow{\mathrm{d}} \mathcal{T} \tag{4}$$

in the Gromov-Hausdorff-Prokhorov sense, where $\sigma^2 = \mu_2(q) - 1$ and $\mathcal{T}$ is the Brownian continuum random tree.

This is not quite the convergence claimed in Theorem 1.1, because $(G(\mathrm{d}^n), T(\mathrm{d}^n), \Gamma(\mathrm{d}^n))$ is a random tree-weighted graph with degree sequence d^n, whereas Theorem 1.1 concerns random *simple* tree-weighted graphs. To obtain Theorem 1.1 from (4), we show that there is $\alpha \in (0, 1]$ such that as $n \to \infty$,

$$\mathbf{P}\{G(\mathrm{d}^n) \text{ is simple} \mid T(\mathrm{d}^n)\} \to \alpha \tag{5}$$

in probability. The value of (5), informally, is that it implies that conditioning $G(\mathrm{d}^n)$ to be simple has an asymptotically negligible effect on the law of $T(\mathrm{d}^n)$. Since the law of $(G(\mathrm{d}^n), T(\mathrm{d}^n), \Gamma(\mathrm{d}^n))$, conditional on the simplicity of $G(\mathrm{d}^n)$, is uniform over simple tree-weighted graphs with degree sequence d^n, we can then conclude straightforwardly.

We finish the overview with a brief discussion of how we prove (5). Our procedure for constructing random tree-weighted graphs with a given degree sequence first constructs the tree $T(\mathrm{d}^n)$, then randomly pairs the remaining half-edges as in the standard configuration model. Viewing $T(\mathrm{d}^n)$ as fixed, this leads us to the following more general question. Let $G = (V, E)$ be a random graph with a given degree sequence generated according to the configuration model, and let $T = (V, E')$ be a fixed, simple graph with the same vertex set. What is the probability that the union of G and T forms a simple graph (i.e. that G is a simple graph and that E and E' are disjoint)? We provide a partial answer to this question by proving a fairly general Poisson approximation theorem for the number of loops and multiple edges in the superposition of a fixed graph and a random graph drawn from the configuration model (see Theorem 4.1). In order to apply Theorem 4.1, we need that the joint degree statistics in $G(\mathrm{d}^n)$ and in $T(\mathrm{d}^n)$) are sufficiently well-behaved; proving this is the task of Section 3. We then state and prove Theorem 4.1 in Section 4, and finally put all the pieces together to prove Theorem 1.1 in Section 5.

§2. Pitman's additive coalescent with a fixed degree sequence

2.1. The sampling process

Let $\mathrm{d} = (d(1), \ldots, d(n))$ be a degree sequence, which is to say that $d(1), \ldots, d(n)$ are non-negative integers. To be well-defined, the next process requires that $\sum_{1 \le i \le n} d(i) \ge 2n - 1$ and that $d(i) \ge 1$ for all $1 \le i \le n$.

Pitman's additive coalescent. The process has $n - 1$ steps, and at the start of step k consists of a rooted forest $F_k(\mathrm{d}) = \{T_1^k(\mathrm{d}), \ldots, T_{n+1-k}^k(\mathrm{d})\}$ with $n + 1 - k$ trees. At the start of step 1, these trees are isolated vertices with labels $1, \ldots, n$. Vertex i has $d(i)$ half-edges $(i1, i2, \ldots, id(i))$ attached to it, and $id(i)$ is distinguished as the *root half-edge*.

Step k:

Choose a uniformly random pair (r_k, s_k), where r_k is a root half-edge which is not paired in $F_k(\mathrm{d})$ and s_k is a non-root half-edge which is not paired in $F_k(\mathrm{d})$ and additionally belongs to a different tree of $F_k(\mathrm{d})$ from r_k.

Pair the half-edges r_k and s_k to create an edge e_k connecting their endpoints; this merges two trees of $F_k(\mathrm{d})$. The root of the new tree is the same as the root of the tree of $F_k(\mathrm{d})$ containing s_k. In the new tree, the vertex incident to r_k is the child of the vertex incident to s_k.

Define $F_{k+1}(\mathrm{d})$ to be the forest consisting of the new tree thus created, together with the remaining $n-k-1$ unaltered trees of $F_k(\mathrm{d})$.

An example is shown in Figure 1. Write $T(\mathrm{d}) = T_1^n(\mathrm{d})$ for the single tree in the random forest $F_n(\mathrm{d})$. Attached to the tree $T(\mathrm{d})$ there is a single pendant (unpaired) root half-edge which is incident to the root of $T(\mathrm{d})$, and if $\sum_{i=1}^n d(i) > 2n-1$ then there are also other pendant half-edges. By ignoring pendant half-edges, we may view $T(\mathrm{d})$ as a random rooted tree with vertex set $[n]$.

It will be useful to additionally define two edge labellings of $T(\mathrm{d})$, denoted K and H. We define $\mathrm{K}(e)$ to be the step at which edge e was added; so $\mathrm{K}(e_k) := k$. We define $\mathrm{H}(e)$ to be the non-root half-edge used in creating e; so $\mathrm{H}(e_k) = s_k$. Note that K is a bijection between $\mathrm{e}(T(\mathrm{d}))$ and $[n-1]$. Also, if $i \in [n]$ has $c_{T(\mathrm{d})}(i) = c$, then H assigns c distinct half-edges from the set $\{i1, \ldots, i(d(i)-1)\}$ to the edges between i and its children in $T(\mathrm{d})$.

We use the phrase "execution path" to mean a sequence of pairs $(r_1, s_1), \ldots, (r_{n-1}, s_{n-1})$ which may concievably appear as the ordered sequence of pairs of half-edges added during the course of Pitman's coalescent.

The next proposition fully describes the joint distribution of $T(\mathrm{d})$, K, and H. In its proof, and in what follows, for a rooted tree t and a node $u \in \mathrm{v}(t) \setminus \{r(t)\}$ we write $\mathrm{par}(u)$ for the parent of u in t. Also, we use the falling factorial notation $(k)_\ell := k(k-1) \cdot \ldots \cdot (k-\ell+1) = k!/(k-\ell)!$.

Proposition 2.1. *Let* $\mathrm{d} = (d(1), \ldots, d(n))$ *be a degree sequence with* $d(i) \geq 1$ *for all* $i \in [n]$ *and with* $\sum_{i=1}^n d(i) \geq 2(n-1)$, *and write* $m = \frac{1}{2}\sum_{i=1}^n d(i)$. *(Note: we allow that* $\sum_{i=1}^n d(i)$ *is odd.) Then the following properties all hold.*

(1) *For any fixed rooted tree* t *with vertex set* $[n]$,

$$\mathbf{P}\{T(\mathrm{d}) = t\} = \frac{1}{(2m-n)_{n-1}} \prod_{i=1}^n (d(i)-1)_{c_t(i)}.$$

(2) *Fix any set* $\mathcal{H} \subset \bigcup_{i=1}^n \{i1, \ldots, i(d(i)-1)\}$ *with* $|\mathcal{H}| = n-1$. *Conditionally given that* $\{s_1, \ldots, s_{n-1}\} = \mathcal{H}$, *the triple* $(T(\mathrm{d}),$

K, H) *is uniformly distributed over the* $((n-1)!)^2$ *triples which are consistent with the event* $\{s_1, \ldots, s_{n-1}\} = \mathcal{H}$.

(3) *The sequence* $(s_1, \ldots, s_{n-1})$ *of non-root half-edges, added by Pitman's coalescent, is uniformly distributed over the set of sequences of* $(n-1)$ *distinct elements of* $\bigcup_{1\le i\le n}\{i1, \ldots, i(d(i)-1)\}$. *Consequently,* $\{s_1, \ldots, s_{n-1}\}$ *is a uniformly random size-*$(n-1)$ *subset of* $\bigcup_{1\le i\le n}\{i1, \ldots, i(d(i)-1)\}$.

(4) *Finally, conditionally given that* $T(\mathrm{d}) = t$ *and given the set* $\{s_1, \ldots, s_{n-1}\}$ *of non-root half-edges added by Pitman's coalescent, the ordering* $(e_1, \ldots, e_{n-1})$ *of* $\mathrm{e}(t)$ *is uniformly distributed over the* $(n-1)!$ *possible orderings of* $\mathrm{e}(t)$.

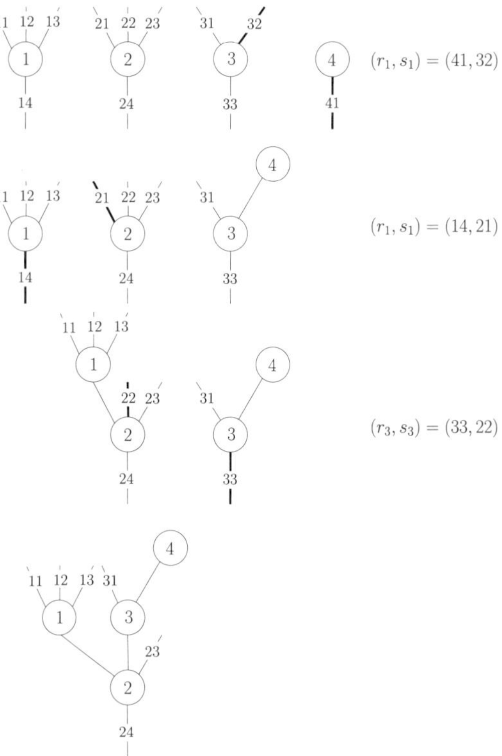

Fig. 1. An example of an execution path of Pitman's additive coalescent. The forests F_1, F_2, F_3 and F_4 are displayed in successive rows.

Proof. At step i of the process, there are $n+1-i$ components and $2m-n+1-i$ unpaired non-root half-edges. We may specify the pair (r_i, s_i) by first revealing the non-root half-edge s_i, then revealing r_i. Whatever the choice of s_i, there are $n-i$ possibilities for r_i, so the number of distinct choices for the pair (r_i, s_i) is $(2m-n+1-i)(n-i)$. Thus, the total number of possible execution paths for the process is

$$\prod_{i=1}^{n-1}(2m-n+1-i)(n-i) = (n-1)!(2m-n)_{n-1}. \tag{6}$$

The execution path followed by the process is uniquely determined by the tree $T(\mathrm{d})$ and the functions $\mathrm{K} : \mathrm{e}(T(\mathrm{d})) \to [n-1]$ and $\mathrm{H} : \mathrm{e}(T(\mathrm{d})) \to \bigcup_{i=1}^{n}\{i1, \ldots, i(\mathrm{d}_i - 1)\}$. To see this, fix any $k \in [n-1]$. Then the edge e_k created at step k of Pitman's coalescent may be recovered as $e_k = \mathrm{K}^{-1}(k)$; and, if $e_k = uv$ with $v = \mathrm{par}(u)$ then the half-edges paired to create e_k are the root half-edge $vd(v)$ incident to v and the half-edge $\mathrm{H}^{-1}(e_k)$.

Now, fix any tree t with degree sequence d, any bijection $\mathrm{k} : \mathrm{e}(t) \to [n-1]$, and any function $\mathrm{h} : \mathrm{e}(t) \to \mathbb{N}$ which, for all $i \in [n]$, assigns $c_t(i)$ distinct values from the set $\{1, \ldots, (d(i)-1)\}$ to the edges between i and its children in t. Together with (6), the observation of the preceding paragraph implies that

$$\mathbf{P}\{T(\mathrm{d}) = t, \mathrm{K} = \mathrm{k}, \mathrm{H} = \mathrm{h}\} = \frac{1}{(n-1)!(2m-n)_{n-1}}.$$

Having fixed the tree t, the number of possible values for K is $(n-1)!$ and the number of possible values for H is $\prod_{i\in[n]}(d(i)-1)_{c_t(i)}$. It follows that

$$\mathbf{P}\{T(\mathrm{d}) = t\} = \frac{(n-1)! \cdot \prod_{i\in[n]}(d(i)-1)_{c_t(i)}}{(n-1)!(2m-n)_{n-1}} = \frac{\prod_{i\in[n]}(d(i)-1)_{c_t(i)}}{(2m-n)_{n-1}},$$

which proves the first claim of the proposition.

Next, fix $\mathcal{H}$ as in the second assertion of the proposition, and any ordering of $\mathcal{H}$ as $(h_1, \ldots, h_{n-1})$. Then the number of execution paths which yield that $s_k = h_k$ for $k \in [n-1]$ is precisely $(n-1)!$. To see this, note that if $s_j = h_j$ for $1 \le j \le k$ then, whatever the choices of the root half-edges $(r_j, 1 \le j \le k)$, the forest F_k^n has $n+1-k$ component trees so there are $n-k$ unpaired root half-edges in components different from that of s_k; any such root half-edge may be chosen as r_k. Since there are also $(n-1)!$ possible orderings of $\mathcal{H}$, the second assertion of the proposition follows.

To prove the third statement, fix a set $\mathcal{H}$ and an ordering $(h_1, \ldots, h_{n-1})$ of its elements, as in the previous paragraph. For each $1 \le k < n-1$, given that $s_j = h_j$ for $1 \le j < k$, whatever the choices of $(r_j, 1 \le j < k)$ may be, there are $n-k$ ways to choose r_k in a distinct tree from h_k. It follows that there are $\prod_{k=1}^{n-2}(n-k) = (n-1)!$ execution paths with the property that $s_k = h_k$ for each $1 \le k \le n-1$. Since this number does not depend on $\mathcal{H}$, it follows that each size-$(n-1)$ subset of $\bigcup_{1 \le i \le n}\{i1, \ldots, i(d(i)-1)\}$ is equally likely.

Finally, fix both the tree t and an unordered set $\mathcal{H}$ of non-root half-edges with $|\mathcal{H} \cap \{i1, \ldots, i(d(i)-1)\}| = c_t(i)$ for all $i \in [n]$. We consider the number of execution paths which yield $T(\mathrm{d}) = t$ and $\{s_1, \ldots, s_{n-1}\} = \mathcal{H}$. The number of choices of an ordering function $\mathrm{k} : \mathrm{e}(t) \to [n-1]$ consistent with these constraints is still $(n-1)!$. Moreover, whatever the choice of k, under the further constraint $\mathrm{K} = \mathrm{k}$, the number of possibilities for H is $\prod_{i \in [n]} c_t(i)!$. To see this, note that for each $i \in [n]$, the constraints precisely imply that $\mathcal{H} \cap \{i1, \ldots, i(d(i)-1)\} = \{s_1, \ldots, s_{n-1}\} \cap \{i1, \ldots, i(d(i)-1)\}$, and H is fixed once we additionally specify which of these $c_t(i)$ half-edges is matched to which child of i, for each $i \in [n]$. It follows that the number of execution paths which yield that $T(\mathrm{d}) = t$, that $\mathrm{K} = \mathrm{k}$ and that $\{s_1, \ldots, s_n\} = \mathcal{H}$ is

$$\prod_{i=1}^{n} c_t(i)!\,.$$

As this quantity doesn't depend on the choice of the ordering function k, the final assertion of the proposition follows. Q.E.D.

We state a corollary of the above proposition, for later use.

Corollary 2.2. $T(\mathrm{d})$ *is a uniformly random rooted tree with child sequence* c_T.

The corollary follows since the formula for $\mathbf{P}\{T(\mathrm{d}) = t\}$ from Proposition 2.1 only depends on t through c_t.

We now assume that $\sum_{i=1}^{n} d(i) \ge 2n$ and that $\sum_{i=1}^{n} d(i)$ is even, and define a random tree-rooted graph $(G, T, \Gamma) = (G(\mathrm{d}), T(\mathrm{d}), \Gamma(\mathrm{d}))$ as follows: First, let $T = T(\mathrm{d})$ be the random tree built by Pitman's coalescent, and let $\Gamma^+ = \Gamma^+(\mathrm{d})$ be its root half-edge. We refer to T as the *spanning tree-elect* of a to-be-constructed tree-rooted graph. Next, choose a uniformly random matching of the $2m - 2(n-1)$ pendant half-edges attached to T, and pair the half-edges according to this matching to create $G = G(\mathrm{d})$. Then let Γ be the edge containing Γ^+, oriented so that Γ^+ is at the head; for later use, let $\Gamma^- = \Gamma^-(\mathrm{d})$ be the other half-edge of Γ. We call $(G(\mathrm{d}), T(\mathrm{d}), \Gamma(\mathrm{d}))$, or any other graph with the same distribution, a *random tree-rooted graph with degree sequence* d. The tree T has now taken office.

The next proposition describes the distribution of $(G(\mathrm{d}), T(\mathrm{d}), \Gamma(\mathrm{d}))$. For a tree-rooted graph (g, t, γ),

Proposition 2.3. *Let* $\mathrm{d} = (d(1), \ldots, d(n))$ *be a degree sequence with* $d(i) \geq 1$ *for all* $i \in [n]$, *and write* $m = \frac{1}{2}\sum_{i=1}^{n} d(i)$. *Fix a tree-rooted graph* (g, t, γ) *where* g *is a graph with degree sequence* d. *Then*

$$\mathbf{P}\{(G(\mathrm{d}), T(\mathrm{d}), \Gamma(\mathrm{d})) = (g, t, \gamma)\} \propto \frac{2^{\mathbf{1}_{[\gamma \text{ is a loop}]}} \cdot m_{g-t}(\gamma)}{\prod_{i=1}^{n} 2^{m_{g-t}(ii)} \cdot \prod_{e \in \mathrm{e}(g)} m_{g-t}(e)!}.$$

Proof. Proposition 2.1 gives us a formula for $\mathbf{P}\{T(\mathrm{d}) = t\}$. We next focus on computing

$$\mathbf{P}\{G(\mathrm{d}) = g \mid T(\mathrm{d}) = t\}.$$

Write r for the root of t, and $\gamma = qr$ for the oriented root edge of g. Given that $T(\mathrm{d}) = t$, each $i \in [n]$ with $i \neq r(t)$ has $d'(i) := d(i) - c_t(i) - 1$ pendant half-edges attached to it, and r has $d'(r) := d(r) - c_t(r)$ half-edges attached to it. Conditionally given that $T(\mathrm{d}) = t$, the graph $G(\mathrm{d}) - T(\mathrm{d})$ is distributed as $CM(\mathrm{d}')$, a random graph with degree sequence $\mathrm{d}' = (d'(1), \ldots, d'(n))$ sampled according to the configuration model, so with distribution as in (2). Writing $m' := m - (n-1) = \frac{1}{2}\sum_{i=1}^{n} d'(i)$ and $g' = (\mathrm{v}(g), \mathrm{e}(g) \setminus \mathrm{e}(t))$, it follows that

$$\begin{aligned}
\mathbf{P}\{G(\mathrm{d}) = g \mid T(\mathrm{d}) = t\} &= \mathbf{P}\{CM(\mathrm{d}') = g'\} \\
&= \frac{2^{m'}(m')!}{(2m')!} \frac{\prod_{i=1}^{n} d'(i)!}{\prod_{i=1}^{n} 2^{m_{g'}(ii)} \cdot \prod_{e \in \mathrm{e}(g')} m_{g'}(e)!} \\
&= \frac{2^{m'}(m')!}{(2m')!} \frac{\prod_{i=1}^{n} d'(i)!}{\prod_{i=1}^{n} 2^{m_{g-t}(ii)} \cdot \prod_{e \in \mathrm{e}(g)} m_{g-t}(e)!}.
\end{aligned}$$

For the second equality we have used the exact expression for the distribution of $CM(\mathrm{d}')$, which can be found in, e.g., [vdH], equation (7.2.6). For the last equality, we use that $m_{g'}(ii) = m_{g-t}(ii)$ since t is a tree so contains no loops, and that $m_{g'}(e) = m_{g-t}(e)$ by definition when $e \in \mathrm{e}(g')$.

Given that $T(\mathrm{d}) = t$ and that $G(\mathrm{d}) = g$, in order to have $(G(\mathrm{d}), T(\mathrm{d}), \Gamma(\mathrm{d})) = (g, t, \gamma)$ it is necessary and sufficient that $\Gamma(\mathrm{d}) = \gamma$. This occurs precisely if γ^+, the half-edge of γ incident to r, was matched with some half-edge incident to q. Since the matching of half-edges in

$G(\mathrm{d})-T(\mathrm{d})$ is chosen uniformly at random, by symmetry the conditional probability that this occurred is $m_{g-t}(\gamma)/d'(r)$ if γ is not a loop, and is $2m_{g-t}(\gamma)/d'(r)$ if γ is a loop. We may unify these two formulas by writing

$$\mathbf{P}\{\Gamma(\mathrm{d})=\gamma \mid T(\mathrm{d})=t, G(\mathrm{d})=g\} = \frac{2^{\mathbf{1}_{[\gamma \text{ is a loop}]}} m_{g-t}(\gamma)}{d'(r)}.$$

Combined with the formula for $\mathbf{P}\{T(\mathrm{d})=t\}$ from Proposition 2.1, this gives

$$\begin{aligned}
&\mathbf{P}\{(G(\mathrm{d}),T(\mathrm{d}),\Gamma(\mathrm{d}))=(g,t,\gamma)\} \\
&= \frac{1}{(2m-n)_{n-1}}\prod_{i=1}^{n}(d(i)-1)_{c_t(i)} \\
&\quad \cdot \frac{2^{m'}(m')!}{(2m')!}\frac{\prod_{i=1}^{n} d'(i)!}{\prod_{i=1}^{n} 2^{m_{g-t}(ii)}\cdot \prod_{e\in e(g)} m_{g-t}(e)!} \\
&\quad \cdot \frac{2^{\mathbf{1}_{[\gamma \text{ is a loop}]}} m_{g-t}(\gamma)}{d'(r)} \\
&= \prod_{i=1}^{n}(d(i)-1)! \cdot \frac{2^{m-(n-1)}(m-(n-1))!}{2m'(2m-n)!} \\
&\quad \cdot \frac{2^{\mathbf{1}_{[\gamma \text{ is a loop}]}} m_{g-t}(\gamma)}{\prod_{i=1}^{n} 2^{m_{g-t}(ii)}\cdot \prod_{e\in e(g)} m_{g-t}(e)!}.
\end{aligned}$$

In the second equality we have used that $(2m-n)_{n-1}(2m')! = 2m'(2m-n)!$, that $(d(i)-1)_{c_t(i)}d'(i)! = (d(i)-1)!$ for $i \neq r$, and that $(d(r)-1)_{c_t(r)}d'(r)! = d'(r)(d(r)-1)!$. The first two terms on the final line do not depend on the triple (g,t,γ), so the result follows. Q.E.D.

§3. Concentration of degrees

Throughout this section, let $(\mathrm{d}^n, n \geq 1)$ be a sequence of degree sequences satisfying the conditions of Theorem 1.1, and also let p^n and p be as in Theorem 1.1. Next, for $n \geq 1$ let $T(\mathrm{d}^n)$ be the tree built by Pitman's additive coalescent applied to the degree sequence $\mathrm{d}^n = (d^n(i), 1 \leq i \leq n)$. Let $\mathrm{c}^n = (c^n(i), 1 \leq i \leq n)$ be the child sequence of $T(\mathrm{d}^n)$, and recall that $Q_{\mathrm{c}^n} = (Q_{\mathrm{c}^n}(a), a \geq 0)$ is the child statistics vector of c^n. Also, for $0 \leq a < b$, let $P^n_{b,a} = \#\{1 \leq i \leq n : d^n(i) = b, c^n(i) = a\}$. Finally, let $\rho := 1/(\mu_1(p) - 1)$. Note that since $\sum_{i\in[n]} d^n(i) \geq 2n$, necessarily $\mu_1(p^n) \geq 2$; since $p^n \to p$ pointwise and $\mu_2(p^n) \to \mu_2(p)$, it follows that $\mu_1(p^n) \to \mu_1(p)$, so $\mu_1(p) \geq 2$ and hence $\rho \in (0,1]$.

Proposition 3.1. *For $a \geq 0$ let*

$$q(a) := \sum_{b=a+1}^{\infty} p(b) \cdot \mathbf{P}\{\mathrm{Bin}(b-1, \rho) = a\}.$$

Then $\mu_2(q) < \infty$ and $\mu_2(n^{-1}Q_{c^n}) \to \mu_2(q)$ in probability as $n \to \infty$. Moreover, for all $0 \leq a < b$, $n^{-1}P^n_{b,a} \xrightarrow{\text{prob}} p(b) \cdot \mathbf{P}\{\mathrm{Bin}(b-1,\rho) = a\}$, and $n^{-1}Q_{c^n}(a) \xrightarrow{\text{prob}} q(a)$, in both cases as $n \to \infty$.

Let $(G(\mathrm{d}^n), T(\mathrm{d}^n), \Gamma(\mathrm{d}^n))$ be a random tree-weighted graph with degree sequence d^n. Using Proposition 3.1, together with existing results from the literature, it is fairly straightforward to establish that $(\sigma n^{-1/2})T(\mathrm{d}^n) \xrightarrow{\mathrm{d}} \mathcal{T}$, with $\sigma = \mu_2(q) - 1 \in (0, \infty)$, where $\mathcal{T}$ is the Brownian continuum random tree. However, in order to show that such convergence holds for the corresponding random *simple* tree-weighted graphs, we additionally need the next proposition, which establishes that the number of pairs of tree-adjacent vertices in $T(\mathrm{d}^n)$ with given fixed degrees is well-concentrated around its expected values. This will be used in order to show that the probability of $G(\mathrm{d}^n)$ being simple given $T(\mathrm{d}^n)$ asymptotically behaves like a constant.

Write $G_-(\mathrm{d}^n) = G(\mathrm{d}^n) - T(\mathrm{d}^n)$ and let $\mathrm{d}^n_- = (d^n_-(i), 1 \leq i \leq n)$ be the degree sequence of $G_-(\mathrm{d}^n)$. For integers $k, \ell \geq 0$, let

$$\alpha(k,\ell) = \sum_{a_1,a_2 \geq 0} a_2 p(\ell + a_2 + 1)\mathbf{P}\{\mathrm{Bin}(\ell + a_2, \rho) = a_2\} \cdot p(k + a_1 + 1)\mathbf{P}\{\mathrm{Bin}(k + a_1, \rho) = a_1\}. \tag{7}$$

Proposition 3.2. *For integers $k, \ell \geq 0$ let*

$$A^n(k,\ell) = \left|\{uv \in \mathrm{e}(T(\mathrm{d}^n)) : \mathrm{d}^n_-(u) = k, \mathrm{d}^n_-(v) = \ell\}\right|.$$

Then for all $k, \ell \geq 0$,

$$\frac{1}{n} A^n(k,\ell) \xrightarrow{\text{prob}} \alpha(k,\ell)$$

as $n \to \infty$, and also

$$\frac{1}{n} \sum_{k,\ell \geq 0} k\ell A^n(k,\ell) \xrightarrow{\text{prob}} \sum_{k,\ell \geq 0} k\ell\alpha(k,\ell).$$

The proofs of Propositions 3.1 and 3.2 appear in Appendix A.

To conclude the section, we observe that $\alpha(k,\ell)$ defines a probability distribution on pairs of non-negative integers. Indeed,

$$\begin{aligned}
&\sum_{k\geq 0}\sum_{a_1\geq 0} p(k+a_1+1)\,\mathbf{P}\left\{\mathrm{Bin}(k+a_1,\rho)=a_1\right\} \\
&= \sum_{m\geq 0}\sum_{a=0}^{m} p(m+1)\,\mathbf{P}\left\{\mathrm{Bin}(m,\rho)=a\right\} = \sum_{m\geq 0} p(m+1) = 1-p(0) = 1,
\end{aligned}$$

and

$$\begin{aligned}
&\sum_{\ell\geq 0}\sum_{a_2\geq 0} a_2 p(\ell+a_2+1)\,\mathbf{P}\left\{\mathrm{Bin}(\ell+a_2,\rho)=a_2\right\} \\
&= \sum_{m\geq 0}\sum_{a=0}^{m} a p(m+1)\,\mathbf{P}\left\{\mathrm{Bin}(m,\rho)=a\right\} \\
&= \sum_{m\geq 0} p(m+1)\cdot m\rho = \big(\mu_1(p)-(1-p(0))\big)\rho = (\mu_1(p)-1)\rho = 1,
\end{aligned}$$

so by factorizing $\sum_{k,\ell\geq 0}\alpha(k,\ell)$ we obtain

$$\sum_{k,\ell\geq 0}\alpha(k,\ell) = \left(\sum_{m\geq 0} p(m+1)\right)\cdot\left(\sum_{m\geq 0}\sum_{m\geq 0} p(m+1)\cdot m\rho\right) = 1;$$

the fact that $\sum_{k,\ell\geq 0}\alpha(k,\ell) = 1$ will be used in the proof of Proposition 3.2. A similar computation shows that

$$\begin{aligned}
\text{(8)}\qquad \sum_{k,\ell\geq 0} k\ell\cdot\alpha(k,\ell) &\leq \left(\sum_{m\geq 0} mp(m+1)\right)\cdot\left(\sum_{m\geq 0} m^2 p(m+1)\rho\right) \\
&= \mu_2(p) - 2\mu_1(p) + 1 < \infty,
\end{aligned}$$

a fact we will use in bounding the probability of simplicity of $G(\mathrm{d}^n)$.

§4. Poisson approximation for graph superpositions

In this section we state a Poisson approximation theorem for the number of loops and multiple edges in the superposition of a fixed simple graph and a random graph with a fixed degree sequence; this in particular allows us to control the probability that such a superposition yields a simple graph.

Let H be a simple graph with vertex set $v(H)=[n]$. Fix a degree sequence $\mathrm{d}=(d(1),\ldots,d(n))$ whose sum of degrees is even, and let G

be a random graph with degree sequence d sampled according to the configuration model. For vertices $u, v \in [n]$ and $i \in [d(u)], j \in [d(v)]$, let $\mathbf{1}_{[ui,vj]}$ be the indicator of the event that half-edge ui is matched with half-edge vj in G. Now write

$$\begin{aligned}\mathcal{L} &= \mathcal{L}(G) = \{(ui, uj) : u \in [n], i, j \in [d(u)], i < j\},\\ \mathcal{M} &= \mathcal{M}(G, H) = \big\{((ui_1, vj_1), (ui_2, vj_2)) : u, v \in [n], uv \notin e(H),\\ &\qquad\qquad i_1, i_2 \in [d(u)], j_1, j_2 \in [d(v)], u < v, i_1 < i_2, j_1 \neq j_2\big\},\end{aligned}$$

and

$$\mathcal{N} = \mathcal{N}(G, H) = \{(ui, vj) : uv \in e(H), i \in [d(u)], j \in [d(v)]\},$$

and let

$$\begin{aligned}L = L(G) &= \sum_{(ui,uj)\in\mathcal{L}} \mathbf{1}_{[ui,uj]},\\ M = M(G, H) &= \sum_{((ui_1,vj_1),(ui_2 vj_2))\in\mathcal{M}} \mathbf{1}_{[(ui_1,vj_1)]}\mathbf{1}_{[(ui_2 vj_2)]}, \text{ and}\\ N = N(G, H) &\sum_{(ui,vj)\in\mathcal{N}} \mathbf{1}_{[ui,vj]}.\end{aligned}$$

Note that the graph with edge set $e(G) \cup e(H)$ is simple precisely if $L + M + N = 0$.

Theorem 4.1. *Fix a sequence of simple graphs $(h_n, n \geq 1)$ with $v(h_n) = [n]$ for all $n \geq 1$ and $\max_{v\in[n]}\{\deg_{h_n}(v)\} = o(n)$. For each $n \geq 1$ let $\mathrm{d}^n = (d^n(v), 1 \leq v \leq n)$ be a degree sequence and let p^n be the degree distribution of d^n. Suppose that there exists a probability distribution $\mathrm{p} = (p(k), k \geq 0)$ with $\mu_2(\mathrm{p}) \in [0, \infty)$ and $p(0) < 1$ such that the following holds.*

First, $\mathrm{p}^n \to \mathrm{p}$ pointwise and $\mu_2(\mathrm{p}^n) \to \mu_2(\mathrm{p})$. Second, there are non-negative numbers $(\alpha(a, b), a, b \geq 0)$ such that for any $a, b \geq 0$

$$\alpha^n(a, b) := \frac{1}{n}\big|\{uv \in e(h_n) : d^n(u) = a, d^n(v) = b\}\big| \to \alpha(a, b),$$

and

$$\sum_{k,\ell\geq 0} k\ell\alpha^n(k, \ell) \to \sum_{k,\ell\geq 0} k\ell\alpha(k, \ell) < \infty. \tag{9}$$

For $n \geq 1$ let G_n be distributed according to the configuration model on graphs with vertex set $[n]$ and degree sequence d^n. Then with $L_n =$

$L(G_n)$, $M_n = M(G_n, h_n)$ and $N_n = N(G_n, h_n)$, we have

$$\left\|\mathrm{Dist}(L_n, M_n, N_n) - \mathrm{Poi}(\nu/2) \otimes \mathrm{Poi}(\nu^2/4) \otimes \mathrm{Poi}(\eta)\right\|_{\mathrm{TV}} \to 0$$

as $n \to \infty$, where $\nu = (\mu_2(p)/\mu_1(p)) - 1$ and $\eta = \frac{1}{\mu_1(p)} \sum_{i,j \geq 1} ij\alpha(i,j)$.

In the statement of Theorem 4.1 we have introduced the notation $\deg_{h_n}(v)$ for the degree of vertex v in h_n, and the notation $\|\mu - \nu\|_{\mathrm{TV}}$ for the total variation distance between probability measures. The proof of Theorem 4.1 appears in Appendix B. This theorem has the following consequence for random tree-weighted graphs, which we will use in the next section.

Corollary 4.2. *Let $(\mathrm{d}^n, n \geq 1)$ and $(p^n, n \geq 1)$ be as in Theorem 1.1, and for $n \geq 1$ let $(G(\mathrm{d}^n), T(\mathrm{d}^n), \Gamma(\mathrm{d}^n))$ be a random tree-weighted graph with degree sequence* d. *Then*

$$\mathbf{P}\left\{G(\mathrm{d}^n) \text{ simple} \mid T(\mathrm{d}^n)\right\} \xrightarrow{\text{prob}} \exp(-\nu/2 - \nu^2/4 - \eta),$$

as $n \to \infty$.

This corollary follows straightforwardly from Theorem 4.1 when $\mu_2(p^n) > 2$, in which case $G_-(\mathrm{d}^n) = G(\mathrm{d}^n) - T(\mathrm{d}^n)$ has a linear number of edges. However, when $\mu_2(p^n) = 2$, and the graph $G_-(\mathrm{d}^n)$ has a sub-linear number of edges, a separate argument is needed. The proof of Corollary 4.2 also appears in Appendix B.

§5. Proof of Theorem 1.1

Let $(\mathrm{d}^n, n \geq 1)$ be a sequence of degree sequences satisfying the conditions of Theorem 1.1. For $n \geq 1$ let $T(\mathrm{d}^n)$ be the tree built by Pitman's additive coalescent applied to degree sequence d^n, and let c^n be the child sequence of $T(\mathrm{d}^n)$. By Proposition 2.1 (1), conditionally given c^n, the tree $T(\mathrm{d}^n)$ is uniformly distributed over the set of trees with child sequence c^n.

By Proposition 3.1, the child statistics vectors $(Q_{\mathrm{c}^n}, n \geq 1)$ satisfy that, as $n \to \infty$, for all $a \geq 0$,

$$n^{-1} Q_{\mathrm{c}^n}(a) \xrightarrow{\text{prob}} q(a), \tag{10}$$

and moreover that $\mu_2(n^{-1}Q_{\mathrm{c}^n}(a)) \to \mu_2(q)$. Here $q = (q(a), a \geq 0)$ is as in Proposition 3.1, and in particular satisfies $\mu_2(q) < \infty$. We will also need that $\mu_2(q) > 1$, and we now justify this.

The convergence (10) and the fact that $\mu_2(n^{-1}Q_{\mathrm{c}^n}(a)) \to \mu_2(q)$ together imply that $\mu_1(n^{-1}Q_{\mathrm{c}^n}(a)) \to \mu_1(q)$. But $\mu_1(n^{-1}Q_{\mathrm{c}^n}(a)) =$

$(n-1)/n$ since Q_{c^n} is a child sequence, so necessarily $\mu_1(q) = 1$. By the definition of q, if $\rho = 1$ then $q(1) = p(2)$, and $p(2) < 1$ by assumption. If $\rho > 1$ then

$$q(0) := \sum_{b=1}^{\infty} p(b) \cdot \mathbf{P}\{\mathrm{Bin}(b-1, \rho) = 0\} > 0,$$

so again $q(1) \leq (1 - q(0)) < 1$. Thus, we always have $q(1) < 1$, which together with the fact that $\mu_1(q) = 1$ implies that $\mu_2(q) > 1$.

Writing $\sigma = \mu_2(q) - 1 \in (0, \infty)$, it then follows by Theorem 1 of [BM] that

$$\overline{T}(\mathrm{d}^n) := \frac{\sigma}{n^{1/2}} T(\mathrm{d}^n) \xrightarrow{\mathrm{d}} \mathcal{T},$$

in the Gromov-Hausdorff-Prokhorov sense.[1]

We aim to prove the same statement with $\overline{T}(\mathrm{d}^n)$ replaced by $\overline{T}_n := (\sigma/n^{1/2})T_n$, where (G_n, T_n, Γ_n) is is a uniformly random simple tree-rooted graph with degree sequence d^n. To accomplish this, we use that the law of (G_n, T_n, Γ_n) is precisely the conditional law of $(G(\mathrm{d}^n), T(\mathrm{d}^n), \Gamma(\mathrm{d}^n))$ given that $G(\mathrm{d}^n)$ is a simple graph.

Writing $\mathbb{K}$ for Gromov-Hausdorff-Prokhorov space as in [ADH], for any bounded continuous function $f : \mathbb{K} \to \mathbb{R}$ we have

$$\begin{aligned}
&\mathbf{E}\left(f(\overline{T}(\mathrm{d}^n)) \cdot \mathbf{1}_{[G(\mathrm{d}^n)\ \mathrm{simple}]}\right) \\
&= \mathbf{E}\Big(\mathbf{E}\big(f(\overline{T}(\mathrm{d}^n)) \cdot \mathbf{1}_{[G(\mathrm{d}^n)\ \mathrm{simple}]} \mid T(\mathrm{d}^n)\big)\Big) \\
&= \mathbf{E}\left(f(\overline{T}(\mathrm{d}^n)) \cdot \mathbf{P}\{G(\mathrm{d}^n) \text{ simple} \mid T(\mathrm{d}^n)\}\right).
\end{aligned}$$

Since $\mathbf{E}f(\overline{T}(\mathrm{d}^n)) \to \mathbf{E}f(\mathcal{T})$ and

$$\mathbf{P}\{G(\mathrm{d}^n) \text{ simple} \mid T(\mathrm{d}^n)\} \xrightarrow{\mathrm{prob}} \exp(-\nu/2 - \nu^2/4 - \eta)$$

by Corollary 4.2, it follows that

$$\mathbf{E}\left(f(\overline{T}(\mathrm{d}^n)) \cdot \mathbf{1}_{[G(\mathrm{d}^n)\ \mathrm{simple}]}\right) \to \exp(-\nu/2 - \nu^2/4 - \eta)\, \mathbf{E}\left(f(\mathcal{T})\right).$$

[1] Theorem 1 of [BM] is stated for plane trees with a fixed degree sequence, rather than labeled trees with a fixed degree sequence. However, as noted by Broutin and Marckert [BM, page 295], a straightforward combinatorial argument shows that the same result holds for labeled trees. Also, as stated, the theorem only yields convergence in the Gromov-Hausdorff sense; but the proof proceeds by establishing convergence distributional of coding functions. As explained in [ADH, Section 3], such proofs immediately yield the stronger Gromov-Hausdorff-Prokhorov convergence.

Furthermore,

$$\begin{aligned}\mathbf{P}\{G(\mathrm{d}^n)\text{ simple}\} &= \mathbf{E}\big(\mathbf{P}\{G(\mathrm{d}^n)\text{ simple} \mid T(\mathrm{d}^n)\}\big)\\ &\to \exp(-\nu/2-\nu^2/4-\eta),\end{aligned}$$

and therefore

$$\begin{aligned}\mathbf{E}\left(f(\overline{T}(\mathrm{d}^n)) \mid G(\mathrm{d}^n)\text{ simple}\right) &= \frac{\mathbf{E}\left(f(\overline{T}(\mathrm{d}^n))\cdot \mathbf{1}_{[G(\mathrm{d}^n)\text{ simple}]}\right)}{\mathbf{P}\{G(\mathrm{d}^n)\text{ simple}\}}\\ &\to \mathbf{E}(f(\mathcal{T})).\end{aligned}$$

Since

$$\mathbf{E}\left(f(\overline{T}_n)\right) = \mathbf{E}\left(f(\overline{T}(\mathrm{d}^n)) \mid G(\mathrm{d}^n)\text{ simple}\right),$$

the fact that $\overline{T}_n \xrightarrow{\mathrm{d}} \mathcal{T}$ now follows by the Portmanteau theorem.

Q.E.D.

§Appendix A. Proofs of Propositions 3.1 and 3.2

Before beginning the proofs in earnest, we state and prove a simple bound on the asymptotic behaviour of maximum degrees and sums of small sets of degrees, for sequences of degree sequences as in Theorems 1.1 and 4.1, which will be used multiple times below.

Fact A.1. *For each $n \geq 1$ let $\mathrm{d}^n = (d^n(v), 1 \leq v \leq n)$ be a degree sequence and let p^n be the degree distribution of d^n. Suppose that there exists a probability distribution $p = (p(k), k \geq 0)$ such that $p^n \to p$ pointwise and $\mu_2(p^n) \to \mu_2(p) \in [0, \infty)$. Then $\max_{1\leq i\leq n} d^n(i) = o(n^{1/2})$. Also, for any sets $(A_n, n \geq 1)$ with $A_n \subset [n]$ and $|A_n| = o(n)$, it holds that $\sum_{i\in A_n} d^n(i) = o(n)$.*

Proof. If $p^n \to p$ pointwise and $\mu_2(p^n) \to \mu_2(p) \in [0, \infty)$, then for all $\epsilon > 0$ there is M such that

$$\liminf_{n\to\infty} \sum_{k=1}^{M} k^2 p^n(k) \geq \mu_2(p) - \epsilon,$$

so $\sup_{M\geq 1} \liminf_{n\to\infty} \sum_{k=1}^{M} k^2 p^n(k) \geq \mu_2(p)$. If additionally there is $\delta > 0$ such that $\max_{1\leq i\leq n} d^n(i) \geq \delta n^{1/2}$ for infinitely many n, then

$$\limsup_{n\to\infty} \mu_2(p^n) \geq \delta^2 + \sup_{M\geq 1} \liminf_{n\to\infty} \sum_{k=1}^{M} k^2 p^n(k) > \mu_2(p),$$

so $\mu_2(p^n) \not\to \mu_2(p)$.

Similarly, for sets $(A_n, n \geq 1)$ as in the statement, since $|A_n| = o(n)$, for any $M \in \mathbb{N}$ we have $\sum_{i \in A_n} (d^n(i))^2 \mathbf{1}_{[d^n(i) \leq M]} = o(n)$, so for any $\epsilon > 0$ there is $M \in \mathbb{N}$ such that

$$\liminf_{n \to \infty} \frac{1}{n} \sum_{i=1}^{n} (d^n(i))^2 \mathbf{1}_{[d^n(i) \leq M]} \mathbf{1}_{[i \notin A_n]} \geq \mu_2(p) - \epsilon.$$

This implies that $\liminf_{n \to \infty} n^{-1} \sum_{i=1}^{n} (d^n(i))^2 \mathbf{1}_{[i \notin A_n]} \geq \mu_2(p)$. If also there is $\delta > 0$ such that $\sum_{i \in A_n} (d^n(i))^2 > \delta n$ for infinitely many n, then

$$\limsup_{n \to \infty} \mu_2(p^n) = \limsup_{n \to \infty} \frac{1}{n} \sum_{i=1}^{n} (d^n(i))^2 \geq \mu_2(p) + \delta,$$

so $\mu_2(p^n) \not\to \mu_2(p)$. Q.E.D.

Note that the conditions on the degree sequences in both Theorem 1.1 and Theorem 4.1 allow Fact A.1 to be applied.

To prove Proposition 3.1, we will make use of the following lemma, which uses the second moment method to control how subsampling affects degree distributions. The proof of the proposition immediately follows that of the lemma.

Lemma A.2. *For any integer $b \geq 1$ there exists n_0 such that for all $n \geq n_0$ the following holds. Let $\mathrm{d} = (d(1), \ldots, d(n))$ be a degree sequence with $d(i) \geq 1$ for all $i \in [n]$ and with $\sum_{i=1}^{n} d(i) \geq 2n - 1$, set $S = \bigcup_{i=1}^{n} \{i1, \ldots, i(d(i) - 1)\}$ and write $s = |S|$. Let U be a uniformly random subset of S with $|\mathrm{U}| = n - 1$, and for $1 \leq i \leq n$ write $U_i = \#\{1 \leq j < d(i) : (i, j) \in \mathrm{U}\}$. For $0 \leq a < b$, write $P_{b,a} = \#\{1 \leq i \leq n : (d(i), U_i) = (b, a)\}$. Then for all $\epsilon > 0$,*

$$\mathbf{P}\left\{|P_{b,a} - \mathbf{E}P_{b,a}| > \epsilon \mathbf{E}P_{b,a}\right\} < \frac{1}{\epsilon^2}\left(\frac{1}{\mathbf{E}P_{b,a}} + \frac{2b^2}{s}\right).$$

Proof. We fix $0 \leq a \leq b$ and compute the first and second moments of $P_{b+1,a}$; this makes the calculations slightly easier to read than they would be for $P_{b,a}$.

Fix indices k and ℓ with $k \neq \ell$ and $d(k) = d(\ell) = b + 1$. Since U is a uniformly random subset of S, by symmetry we have

$$\mathbf{P}\{|U_k| = a\} = \mathbf{P}\{|U_\ell| = a\} = \binom{b}{a} \cdot \binom{s-b}{n-1-a}\binom{s}{n-1}^{-1},$$

and

$$\mathbf{P}\{|U_k| = |U_\ell| = a\} = \binom{b}{a}^2 \cdot \binom{s-2b}{n-1-2a}\binom{s}{n-1}^{-1},$$

so writing $n_{b+1} = \#\{1 \le i \le n : d(i) = b+1\}$, we have

$$\mathbf{E}P_{b+1,a} = n_{b+1}\binom{b}{a}\cdot\binom{s-b}{n-1-a}\binom{s}{n-1}^{-1} \tag{11}$$

and

$$\begin{aligned}
&\mathbf{Var}\{P_{b+1,a}\} \\
&= n_{b+1}(n_{b+1}-1)\binom{b}{a}^2 \\
&\quad\cdot\left(\binom{s-2b}{n-1-2a}\binom{s}{n-1}^{-1} - \binom{s-b}{n-1-a}^2\binom{s}{n-1}^{-2}\right) \\
&\quad + n_{b+1}\left(\binom{b}{a}\binom{s-b}{n-1-1}\binom{s}{n-1}^{-1}\right. \\
&\qquad\qquad\left. - \binom{b}{a}^2\binom{s-b}{n-1-1}^2\binom{s}{n-1}^{-2}\right),
\end{aligned}$$

where the final line accounts for the diagonal terms. Bounding the final line from above by $\mathbf{E}P_{b+1,a}$ and cancelling terms in the parenthetical expression in the middle line gives

$$\begin{aligned}
&\mathbf{Var}\{P_{b+1,a}\} - \mathbf{E}P_{b+1,a} \\
&\le n_{b+1}(n_{b+1}-1)\binom{b}{a}^2 \\
&\quad\cdot\left(\frac{(n-1)_{2a}(s-(n-1))_{2(b-a)}}{(s)_{2b}} - \frac{(n-1)_a^2(s-(n-1))_{b-a}^2}{(s)_b^2}\right).
\end{aligned}$$

The ratio of the first and the second term in the final parentheses is

$$\frac{(n-1)_{2a}}{(n-1)_a^2}\frac{(s-(n-1))_{2(b-a)}}{(s-(n-1))_{b-a}^2}\frac{(s)_b^2}{(s)_{2b}} \le \frac{(s)_b^2}{(s)_{2b}} \le \left(1+\frac{b}{s-2b}\right)^b \le 1+\frac{2b^2}{s},$$

the last bound holding for b fixed and s large. This gives

$$\begin{aligned}
&\mathbf{Var}\{P_{b+1,a}\} \\
&\le \mathbf{E}P_{b+1,a} + n_{b+1}(n_{b+1}-1)\binom{b}{a}^2\frac{2b^2}{s}\binom{s-b}{n-1-a}^2\binom{s}{n-1}^{-2} \\
&< \mathbf{E}P_{b+1,a} + \frac{2b^2}{s}(\mathbf{E}P_{b+1,a})^2,
\end{aligned}$$

and the lemma follows by Chebyshev's inequality. Q.E.D.

Proof of Proposition 3.1. We first bound $\mu_2(q)$ by writing

$$\begin{aligned}\mu_2(q) &= \sum_{a\geq 0} a^2 q(a) \\ &= \sum_{a\geq 0} a^2 \sum_{b>a} p(b)\cdot \mathbf{P}\{\mathrm{Bin}(b-1,\rho)=a\} \\ &\leq \sum_{b>0} b^2 p(b)\cdot \sum_{0\leq a<b} \mathbf{P}\{\mathrm{Bin}(b-1,\rho)=a\} \\ &= \mu_2(p)^2 < \infty.\end{aligned}$$

Next, since $p^n \to p$ pointwise and $\mu_2(p^n)\to\mu_2(p)<\infty$, for any $\epsilon>0$ there is k such that $\sum_{d\geq k} d^2 p^n(d) < \epsilon$ and $\sum_{d\geq k} d^2 p(d) < \epsilon$. If node i has a children in $T(\mathrm{d}^n)$ then $d^n(i)\geq a+1$, so it follows that

$$\begin{aligned}\sum_{a=k}^{\infty} a^2 \frac{Q_{\mathrm{c}^n}(a)}{n} &= \sum_{a\geq k}\sum_{b>a} a^2 \frac{\#\{i\leq n : c^n(i)=a, d^n(i)=b\}}{n} \\ &\leq \sum_{b>k} b^2 \sum_{a<b} \frac{\#\{i\leq n : c^n(i)=a, d^n(i)=b\}}{n} \\ &= \sum_{b>k} b^2 p^n(b) < \epsilon.\end{aligned}$$

To complete the proof it thus suffices to show that $n^{-1}P^n_{b,a} \to p(b)\cdot \mathbf{P}\{\mathrm{Bin}(b-1,\rho)=a\}$ in probability for all $0\leq a<b$ and that $n^{-1}Q_{\mathrm{c}^n}\to q$ pointwise in probability; the fact that $\mu_2(n^{-1}Q_{\mathrm{c}^n})\to\mu_2(q)$ in probability then immediately follows.

By the third statement of Proposition 2.1, the set of non-root half-edges in $T(\mathrm{d}^n)$ is a uniformly random size-$(n-1)$ subset of the set $S^n := \bigcup_{1\leq i\leq n}\{i1,\ldots,i(d^n(i)-1)\}$. We will apply Lemma A.2 to control the numbers of nodes with a given number of children in $T(\mathrm{d}^n)$. To make the coming applications of that lemma transparent, we write $s^n := |S^n| = \sum_{1\leq i\leq n}(d^n(i)-1)$.

We handle the cases $\mu_1(p)=2$ and $\mu_1(p)>2$ separately. If $\mu_1(p)=2$ then $|S^n| = \sum_{1\leq i\leq n}(d^n(i)-1) = (1+o(1))n$ as $n\to\infty$. Note that in this case $\rho(p)=1/(\mu_1(p)-1)=1$ so $q(a)=p(a+1)$ for all $a\geq 0$. For any $a\geq 0$, by (11) we then have

$$\begin{aligned}\mathbf{E}P^n_{a+1,a} &= (1-o(1))np^n(a+1)\binom{a}{a}\binom{|S^n|-1-a}{n-1-a}\binom{|S^n|}{n-1}^{-1} \\ &= (1-o(1))np^n(a+1).\end{aligned}$$

If $p(a+1) > 0$ then $np^n(a+1) = \Theta(n)$, so

$$\frac{1}{\mathbf{E}\left(P^n_{a+1,a}\right)} + \frac{2(a+1)^2}{|S^n|} = o(1),$$

and hence by Lemma A.2,

$$\frac{P^n_{a+1,a}}{n} \xrightarrow{\text{prob}} p(a+1) = q(a).$$

If $p(a+1) = 0$ then $p^n(a+1) = o(1)$, so $\mathbf{E}(P^n_{a+1,a})/n \to 0$ and thus $P^n_{a+1,a}/n \xrightarrow{\text{prob}} 0 = q(a)$ by Markov's inequality. Since this holds for all $a \geq 0$, and $\sum_{a\geq 0} P^n_{a+1,a}/n \leq 1 = \sum_{a\geq 0} q(a)$, it follows that $\sum_{a\geq 0} P^n_{a+1,a}/n \to 1$ in probability. This implies that $\sum_{b>a+1} P^n_{b,a}/n \to 0$ in probability, so

$$\frac{Q_{c^n}(a)}{n} = \frac{1}{n}\sum_{b>a} P^n_{b,a} = \frac{P^n_{a+1,a}}{n} + \sum_{b>a+1} \frac{P^n_{b,a}}{n} \xrightarrow{\text{prob}} q(a),$$

and that for all $b > a+1$, $P^n_{b,a}/n \to 0 = p(b)\cdot \mathbf{P}\{\text{Bin}(b-1,\rho) = a\}$ in probability, as required.

We now assume $\mu_1(p) > 2$, so that $\rho(p) = 1/(\mu_1(p)-1) < 1$. Since $p = (p_k, k \geq 1)$ is supported on the positive integers,

$$\sum_{a\geq 0} q(a) = \sum_{a\geq 0}\sum_{b=a+1}^{\infty} p(b)\,\mathbf{P}\left\{\text{Bin}(b-1,\rho) = a\right\} = \sum_{b\geq 1} p(b) = 1.$$

Recalling that $Q_{c^n}(a) = \sum_{b>a} P^n_{b,a}$, to show that $n^{-1}Q_{c^n}(a) \to q(a)$ in probability, it therefore suffices to prove that $P^n_{b+1,a}/n \to p(b+1)\cdot \mathbf{P}\{\text{Bin}(b,\rho) = a\}$ for all $0 \leq a \leq b$, and we now turn to this.

Since $\mu_1(p^n) \to \mu_1(p)$, it follows that $|\sum_{i=1}^n d^n(i) - \mu_1(p)n| = n|\mu_1(p^n) - \mu_1(p)| = o(n)$ as $n \to \infty$, so $s^n = (1+o(1))n(\mu_1(p)-1)$. Thus, for any $b \geq 1$ and $0 \leq a \leq b$ we have

$$\begin{aligned}
\binom{b}{a}\cdot\binom{s^n-b}{n-1-a}\binom{s^n}{n-1}^{-1} &= \binom{b}{a}\frac{(n-1)_a(s^n-(n-1))_{b-a}}{(s^n)_b} \\
&= (1-o(1))\binom{b}{a}\frac{n^a((\mu_1(p)-2)n)^{b-a}}{((\mu_1(p)-1)n)^b} \\
&= (1-o(1))\binom{b}{a}\frac{(\mu_1(p)-2)^{b-a}}{(\mu_1(p)-1)^b} \\
&= (1-o(1))\,\mathbf{P}\left\{\text{Bin}(b,\rho) = a\right\}.
\end{aligned}$$

Using (11) we thus have

$$\begin{aligned}\mathbf{E}P^n_{b+1,a} &= (1-o(1))np^n(b+1)\,\mathbf{P}\left\{\mathrm{Bin}(b,\rho)=a\right\}\\ &= (1-o(1))np(b+1)\,\mathbf{P}\left\{\mathrm{Bin}(b,\rho)=a\right\},\end{aligned}$$

so again, applying Lemma A.2 in the case that $p(b+1)>0$, and applying Markov's inequality in the case that $p(b+1)=0$, we obtain that, as $n\to\infty$,

$$\frac{P^n_{b+1,a}}{n} \xrightarrow{\text{prob}} p(b+1)\,\mathbf{P}\left\{\mathrm{Bin}(b,\rho)=a\right\},$$

as required. Q.E.D.

We now turn to controlling the joint degrees of pairs of tree-adjacent vertices in tree-weighted graphs. Given a degree sequence $\mathrm{d}=(d(1),\ldots,d(n))$ and a tree t with $\mathrm{v}(t)=[n]$, for integers b_1,b_2,a_1,a_2 let

$$\begin{aligned}&R_{b_1,b_2,a_1,a_2}(t,\mathrm{d})\\ &= \#\{u\in\mathrm{v}(t)\setminus\\ &\qquad \{r(t)\}: d(u)=b_1, d(\mathrm{par}(u))=b_2, c_t(u)=a_1, c_t(\mathrm{par}(u))=a_2\}\\ &= \sum_{u\in\mathrm{v}(t)\setminus\{r(t)\}} \mathbf{1}_{[d(u)=b_1,c_t(u)=a_1]}\cdot\mathbf{1}_{[d(\mathrm{par}(u))=b_2,c_t(\mathrm{par}(u))=a_2]}.\end{aligned}$$

If (g,t,γ) is a tree-rooted graph and g has degree sequence d, then $R_{b_1,b_2,a_1,a_2}(t,\mathrm{d})$ counts the number of edges xy of t with $y=\mathrm{par}(x)$ such that $c_t(x)=a_1$, $c_t(y)=a_2$ and $d_g(x)=b_1$, $d_g(y)=b_2$.

Proposition A.3. *Under the assumptions of Theorem* 1.1, *for any integers* $0\le a_1<b_1$ *and* $0\le a_2<b_2$, *as* $n\to\infty$,

$$\begin{aligned}&\frac{R_{b_1,b_2,a_1,a_2}(T(\mathrm{d}^n),\mathrm{d}^n)}{n}\\ &\to a_2p(b_2)\,\mathbf{P}\left\{\mathrm{Bin}(b_2-1,\rho)=a_2\right\}\cdot p(b_1)\,\mathbf{P}\left\{\mathrm{Bin}(b_1-1,\rho)=a_1\right\}\end{aligned}$$

in probability, where $\rho=1/(\mu_1(p)-1)$.

We introduce two pieces of notation before beginning the proof. For a half-edge h we write $v(h)$ for the vertex incident to h. Also, for $r\in\mathbb{R}$ we write $r_+:=\max(r,0)$.

Proof. First, if $a_2=0$ then the right-hand side is zero, and also $R_{b_1,b_2,a_1,a_2}(T(\mathrm{d}^n),\mathrm{d}^n)=0$, since if $v=\mathrm{par}(u)\in T(\mathrm{d}^n)$ then $c_{T(\mathrm{d}^n)}(v)\ge 1$. The result thus holds trivially when $a_2=0$, and we assume hereafter that $a_2\ge 1$. For the remainder of the proof we write $R_{b_1,b_2,a_1,a_2}=R_{b_1,b_2,a_1,a_2}(T(\mathrm{d}^n),\mathrm{d}^n)$ for succinctness.

Let $\mathcal{H}$ be a fixed, size-$(n-1)$ subset of $S^n := \bigcup_{1\le i\le n}\{i1,\ldots,i(d^n(i)-1)\}$. Write $\mathcal{S}^n = \{s_1,\ldots,s_{n-1}\}$ for the (unordered) set of non-root half-edges of $T(\mathrm{d}^n)$. We now show that for any half edge $h \in \mathcal{H}$ and any root half-edge r with $v(r) \neq v(h)$, for all $1 \le i \le n-1$,

$$\mathbf{P}\{(r_i,s_i) = (r,h) \mid \mathcal{S}^n = \mathcal{H}\} = \mathbf{P}\{(r_1,s_1) = (r,h) \mid \mathcal{S}^n = \mathcal{H}\}. \tag{12}$$

To see this, note that by the second assertion of Proposition 2.1, the number of execution paths with $\mathcal{S}^n = \mathcal{H}$ is $((n-1)!)^2$. We claim that for any $i \in [n-1]$, the number of execution paths with $\mathcal{S}^n = \mathcal{H}$ which additionally satisfy that $(r_i,s_i) = (r,h)$ is $((n-2)!)^2$. As this number does not depend on $i \in [n-1]$, the displayed identity follows from this claim.

To prove the claim, simply note that there are $(n-2)!$ possible orderings of $\mathcal{H}$ consistent with the constraint that $s_i = h$. Having fixed such an ordering $(h_1,\ldots,h_{n-1})$, for each $j \in [n-1]$ with $j \neq i$, if $s_k = h_k$ for $1 \le k < j$ then, excluding r_i there are $n-j-\mathbf{1}_{[j<i]}$ unpaired root half-edges in components different from that of s_j, and any such root half-edge may be chosen as r_j. Thus the number of execution paths with $\mathcal{S}^n = \mathcal{H}$ and such that $(r_i,s_i) = (r,h)$ is $(n-2)! \cdot \prod_{j\in[n-1]\setminus\{i\}}(n-j-\mathbf{1}_{[j<i]}) = ((n-2)!)^2$.

Now fix a second non-root half-edge $h' \neq h$ and a second root half-edge $r' \neq r$ not incident to the same vertex as h'. Then a similar argument to the one leading to (12) shows that that for any $1 \le i < j \le n$,

$$\begin{aligned}\mathbf{P}\{(r_i,s_i) = (r,h),(r_j,s_j) = (r',s') \mid \mathcal{S}^n = \mathcal{H}\}\\ = \mathbf{P}\{(r_1,s_1) = (r,h),(r_2,s_2) = (r',s') \mid \mathcal{S}^n = \mathcal{H}\}.\end{aligned} \tag{13}$$

In the current case, the number of execution paths leading to the events in both the left- and right-hand probabilities is $((n-3)!)^2$.

We will next use the above identities in order to perform first and second moment computations. For any set $H \subset S^n$, for $0 \le a < b$ let $V^n_{b,a}(H) = \{i \in [n] : d^n(i) = b, |H \cap \{i1,\ldots,i(d^n(i)-1)| = a\}$. Note that $V^n_{b,a}(\mathcal{S}^n)$ is simply the set of nodes with degree b in $G(\mathrm{d}^n)$ and with a children in $T(\mathrm{d}^n)$; so $P^n_{b,a} = |V^n_{b,a}(\mathcal{S}^n)|$.

Fix a non-root node $u \in T(\mathrm{d}^n)$, and let $m \in [n-1]$ be such that $e_m = \{\mathrm{par}(u),u\}$. Then $v(r_m) = u$ and $v(s_m) = \mathrm{par}(u)$, so $u \in R_{b_1,b_2,a_1,a_2}$ if and only if $v(r_m) \in V^n_{b_1,a_1}(\mathcal{S}^n)$ and $v(s_m) \in V^n_{b_2,a_2}(\mathcal{S}^n)$. By (12), it follows that

$$\begin{aligned}&\mathbf{E}(R_{b_1,b_2,a_1,a_2} \mid \mathcal{S}^n = \mathcal{H})\\ &= (n-1)\mathbf{P}\left\{v(r_1) \in V^n_{b_1,a_1}(\mathcal{S}^n), v(s_1) \in V^n_{b_2,a_2}(\mathcal{S}^n) \mid \mathcal{S}^n = \mathcal{H}\right\}.\end{aligned} \tag{14}$$

Likewise, by (13) it follows that

(15)
$$\mathbf{E}\left(\binom{R_{b_1,b_2,a_1,a_2}}{2} \mid \mathcal{S}^n = \mathcal{H}\right)$$
$$= \binom{n-1}{2}$$
$$\cdot \mathbf{P}\left\{v(r_1), v(r_2) \in V^n_{b_1,a_1}(\mathcal{S}^n), v(s_1), v(s_2) \in V^n_{b_2,a_2}(\mathcal{S}^n) \mid \mathcal{S}^n = \mathcal{H}\right\}.$$

We develop the latter two identities in turn.

For integers $0 \le a < b$, the number of non-root half-edges $h \in \mathcal{S}^n$ with $v(h) \in V^n_{b,a}(\mathcal{S}^n)$ is $a \cdot |V^n_{b,a}(\mathcal{S}^n)|$, and the number of root half-edges h with $v(h) \in V^n_{b,a}(\mathcal{S}^n)$ is just $|V^n_{b,a}(\mathcal{S}^n)|$. Conditionally given that $\mathcal{S}^n = \mathcal{H}$, the half-edge s_1 is a uniformly random element of $\mathcal{H}$, so

$$\mathbf{P}\left\{v(s_1) \in V^n_{b_2,a_2}(\mathcal{S}^n) \mid \mathcal{S}^n = \mathcal{H}\right\} = \frac{a_2|V^n_{b_2,a_2}(\mathcal{H})|}{|\mathcal{H}|} = \frac{a_2|V^n_{b_2,a_2}(\mathcal{H})|}{n-1}.$$

Having chosen s_1, if $v(s_1) = v$ then $v(r_1)$ is a uniformly random element of $[n] \setminus \{v\}$, so

$$\mathbf{P}\left\{v(r_1) \in V^n_{b_1,a_1}(\mathcal{S}^n) \mid \mathcal{S}^n = \mathcal{H}, v(s_1) \in V^n_{b_2,a_2}(\mathcal{S}^n)\right\}$$
$$= \frac{(|V^n_{b_1,a_1}(\mathcal{H})| - \mathbf{1}_{[(b_1,a_1)=(b_2,a_2)]})_+}{n-1}.$$

Using these two identities in (14), it follows that

$$(n-1)\mathbf{E}\left(R_{b_1,b_2,a_1,a_2} \mid \mathcal{S}^n = \mathcal{H}\right)$$
$$= a_2|V^n_{b_2,a_2}(\mathcal{H})|(|V^n_{b_1,a_1}(\mathcal{H})| - \mathbf{1}_{[(b_1,a_1)=(b_2,a_2)]})_+,$$

so since $|V^n_{b,a}(\mathcal{S}^n)| = P^n_{b,a}$ for all $0 \le a < b$, by Proposition 3.1 we have

(16)
$$\mathbf{E}\left(\frac{R_{b_1,b_2,a_1,a_2}}{n} \mid \mathcal{S}^n\right) = \frac{1}{n(n-1)} a_2 P^n_{b_2,a_2}(P^n_{b_1,a_1} - \mathbf{1}_{[(b_1,a_1)=(b_2,a_2)]})$$
$$\xrightarrow{\text{prob}} a_2 p(b_2)\mathbf{P}\left\{\text{Bin}(b_2-1,\rho) = a_2\right\} \cdot p(b_1)\mathbf{P}\left\{\text{Bin}(b_1-1,\rho) = a_1\right\}.$$

For the second moment calculation, we need to additionally compute

(17)
$$\mathbf{P}\{v(r_2) \in V^n_{b_1,a_1}(\mathcal{S}^n), v(s_2) \in V^n_{b_2,a_2}(\mathcal{S}^n) \mid$$
$$\mathcal{S}^n = \mathcal{H}, v(r_1) \in V^n_{b_1,a_1}(\mathcal{S}^n), v(s_1) \in V^n_{b_2,a_2}(\mathcal{S}^n)\}.$$

Under the conditioning in (17), the number of non-root half-edges $h \in \mathcal{S}^n \setminus \{s_1\}$ with $v(h) \in V^n_{b_2,a_2}(\mathcal{H})$ is $(a_2 \cdot |V^n_{b_2,a_2}(\mathcal{H})| - 1)_+$, so

$$\begin{aligned}
&\mathbf{P}\left\{v(s_2) \in V^n_{b_2,a_2}(\mathcal{S}^n) \mid \mathcal{S}^n = \mathcal{H}, v(r_1) \in V^n_{b_1,a_1}(\mathcal{S}^n), v(s_1) \in V^n_{b_2,a_2}(\mathcal{S}^n)\right\} \\
&= \frac{(a_2 \cdot |V^n_{b_2,a_2}(\mathcal{H})| - 1)_+}{n-2}.
\end{aligned}$$

Now suppose that $\mathcal{S}^n = \mathcal{H}, v(r_1) \in V^n_{b_1,a_1}(\mathcal{S}^n), v(s_1) \in V^n_{b_2,a_2}(\mathcal{S}^n)$, and that $v(s_2) \in V^n_{b_2,a_2}(\mathcal{H})$, and consider the number of possible values for r_2. We claim that the number of unpaired root half-edges h with $v(h) \in V^n_{b_1,a_1}(\mathcal{S}^n)$ such that $v(h)$ is in a component different from $v(s_2)$ is

$$(|V^n_{b_1,a_1}(\mathcal{H})| - 1 - \mathbf{1}_{[(b_1,a_1)=(b_2,a_2)]})_+.$$

To see this, note that if $(b_1, a_1) = (b_2, a_2)$ and either $v(s_2) = v(s_1)$ or $v(s_2) = v(r_1)$, then we are precisely constrained constrained to choose h so that $v(h) \in V^n_{b_1,a_1}(\mathcal{H}) \setminus \{v(r_1), v(s_1)\}$. On the other hand, if $(b_1, a_1) = (b_2, a_2)$ and $v(s_2) \notin \{v(r_1), v(s_1)\}$ then we are constrained to choose h so that $v(h) \in V^n_{b_1,a_1}(\mathcal{H}) \setminus \{v(r_1), v(s_2)\}$. Both cases agree with the above formula. When $(b_1, a_1) = (b_2, a_2)$, the claim is straightforward, since in that case we are only constrained to choose h so that $v(h) \in V^n_{b_1,a_1}(\mathcal{H}) \setminus \{v(r_1)\}$. It follows that

$$\begin{aligned}
&\mathbf{P}\big\{v(r_2) \in V^n_{b_1,a_1}(\mathcal{S}^n) \mid \\
&\qquad \mathcal{S}^n = \mathcal{H}, v(s_2) \in V^n_{b_2,a_2}(\mathcal{S}^n), v(r_1) \in V^n_{b_1,a_1}(\mathcal{S}^n), v(s_1) \in V^n_{b_2,a_2}(\mathcal{S}^n)\big\} \\
&= \frac{(|V^n_{b_1,a_1}(\mathcal{H})| - 1 - \mathbf{1}_{[(b_1,a_1)=(b_2,a_2)]})_+}{n-2}.
\end{aligned}$$

Combining the above identities with (15) yields that

$$\begin{aligned}
&2(n-1)(n-2)\,\mathbf{E}\left(\binom{R_{b_1,b_2,a_1,a_2}}{2} \mid \mathcal{S}^n = \mathcal{H}\right) \\
&= a_2|V^n_{b_2,a_2}(\mathcal{H})|\big(a_2|V^n_{b_2,a_2}(\mathcal{H})| - 1\big)_+ \\
&\quad \cdot \big(|V^n_{b_1,a_1}(\mathcal{H})| - \mathbf{1}_{[(b_1,a_1)=(b_2,a_2)]}\big)_+ \big(|V^n_{b_1,a_1}(\mathcal{H})| - 1 - \mathbf{1}_{[(b_1,a_1)=(b_2,a_2)]}\big)_+,
\end{aligned}$$

so since $a_2 \geq 1$, Proposition 3.1 implies that

$$\mathbf{E}\left(\frac{R_{b_1,b_2,a_1,a_2}(R_{b_1,b_2,a_1,a_2}-1)}{n^2} \mid \mathcal{S}^n\right)$$
$$= \frac{a_2 P^n_{b_2,a_2}\left(a_2 P^n_{b_2,a_2}-1\right)_+}{n(n-1)}$$
$$\cdot \frac{\left(P^n_{b_1,a_1}-\mathbf{1}_{[(b_1,a_1)=(b_2,a_2)]}\right)_+\left(P^n_{b_1,a_1}-1-\mathbf{1}_{[(b_1,a_1)=(b_2,a_2)]}\right)_+}{n(n-2)}$$
$$\xrightarrow{\text{prob}} \left(a_2 p(b_2)\mathbf{P}\left\{\text{Bin}(b_2-1,\rho)=a_2\right\}\cdot p(b_1)\mathbf{P}\left\{\text{Bin}(b_1-1,\rho)=a_1\right\}\right)^2.$$

Also, (16) implies that $\mathbf{E}(n^{-2}R_{b_1,b_2,a_1,a_2} \mid \mathcal{S}^n) \xrightarrow{\text{prob}} 0$, which with the preceding asymptotic implies that

$$\mathbf{E}\left(\frac{R^2_{b_1,b_2,a_1,a_2}}{n^2} \mid \mathcal{S}^n\right)$$
$$\xrightarrow{\text{prob}} \left(a_2 p(b_2)\mathbf{P}\left\{\text{Bin}(b_2-1,\rho)=a_2\right\}\cdot p(b_1)\mathbf{P}\left\{\text{Bin}(b_1-1,\rho)=a_1\right\}\right)^2.$$

Combining this with (16) gives that

$$\mathbf{E}\left(\left(\frac{R_{b_1,b_2,a_1,a_2}}{n}\right)^2 \mid \mathcal{S}^n\right)-\left(\mathbf{E}\left(\frac{R_{b_1,b_2,a_1,a_2}}{n} \mid \mathcal{S}^n\right)\right)^2 \xrightarrow{\text{prob}} 0;$$

the conditional Chebyshev's inequality then gives that for all $\epsilon > 0$,

$$\mathbf{P}\left\{\left|\frac{R_{b_1,b_2,a_1,a_2}}{n}-\mathbf{E}\left(\frac{R_{b_1,b_2,a_1,a_2}}{n} \mid \mathcal{S}^n\right)\right| > \epsilon \mid \mathcal{S}^n\right\} \xrightarrow{\text{prob}} 0.$$

Taking expectations on the left of the previous inequality to remove the conditioning, and again using (16), this time to replace the term $\mathbf{E}(\frac{R_{b_1,b_2,a_1,a_2}}{n} \mid \mathcal{S}^n)$ in the probability by the constant

$$C := a_2 p(b_2)\mathbf{P}\big\{\text{Bin}(b_2-1,\rho)=a_2\big\}\cdot p(b_1)\mathbf{P}\left\{\text{Bin}(b_1-1,\rho)=a_1\right\},$$

we obtain that

$$\mathbf{P}\left\{\left|\frac{R_{b_1,b_2,a_1,a_2}}{n}-C\right| > \epsilon\right\} \to 0,$$

as required. Q.E.D.

Proof of Proposition 3.2. We may reexpress $A^n(k,\ell)$ as

$$\begin{aligned} A^n(k,\ell) = \sum_{a_1,a_2\geq 0} \sum_{u\in \mathrm{v}(T(\mathrm{d}^n))} & \mathbf{1}_{[r(T(\mathrm{d}^n))\notin\{u,\mathrm{par}(u)\}]} \\ & \cdot \mathbf{1}_{[\mathrm{d}^n(u)=k+a_1+1,\mathrm{c}^n(u)=a_1]} \\ & \cdot \mathbf{1}_{[\mathrm{d}^n(\mathrm{par}(u))=\ell+a_2+1,\mathrm{c}^n(\mathrm{par}(u))=a_2]} \\ + \sum_{a_1,a_2\geq 0} \sum_{u\in \mathrm{v}(T(\mathrm{d}^n))} & \mathbf{1}_{[\mathrm{par}(u)=r(T(\mathrm{d}^n))]} \\ & \cdot \mathbf{1}_{[\mathrm{d}^n(u)=k+a_1+1,\mathrm{c}^n(u)=a_1]} \\ & \cdot \mathbf{1}_{[\mathrm{d}^n(\mathrm{par}(u))=\ell+a_2,\mathrm{c}^n(\mathrm{par}(u))=a_2]}. \end{aligned}$$

For fixed $a_1, a_2 \geq 0$, if we replace $\mathbf{1}_{[r(T(\mathrm{d}^n))\notin\{u,\mathrm{par}(u)\}]}$ by $\mathbf{1}_{[u\neq r(T(\mathrm{d}^n))]}$ in the first double sum, then the inner sum is simply $R_{k+a_1+1,\ell+a_2+1,a_1,a_2}$. It follows that

$$\begin{aligned} A^n(k,\ell) = & \sum_{a_1,a_2\geq 0} R_{k+a_1+1,\ell+a_2+1,a_1,a_2} \\ & + \sum_{a_1,a_2\geq 0} \sum_{u\in \mathrm{v}(T(\mathrm{d}^n))} \mathbf{1}_{[\mathrm{par}(u)=r(T(\mathrm{d}^n))]} \\ & \qquad \cdot \mathbf{1}_{[\mathrm{d}^n(u)=k+a_1+1,\mathrm{c}^n(u)=a_1]} \\ & \qquad \cdot \mathbf{1}_{[\mathrm{d}^n(\mathrm{par}(u))=\ell+a_2,\mathrm{c}^n(\mathrm{par}(u))=a_2]} \\ & - \sum_{a_1,a_2\geq 0} \sum_{u\in \mathrm{v}(T(\mathrm{d}^n))} \mathbf{1}_{[\mathrm{par}(u)=r(T(\mathrm{d}^n))]} \\ & \qquad \cdot \mathbf{1}_{[\mathrm{d}^n(u)=k+a_1+1,\mathrm{c}^n(u)=a_1]} \\ & \qquad \cdot \mathbf{1}_{[\mathrm{d}^n(\mathrm{par}(u))=\ell+a_2+1,\mathrm{c}^n(\mathrm{par}(u))=a_2]}. \end{aligned}$$

But each of the last two double sums is bounded by $c^n(r(T(\mathrm{d}^n)))$, since they both count each child of the root at most once. Under the assumptions of Theorem 1.1, by Fact A.1 we have $c^n(r(T(\mathrm{d}^n))) \leq \max_{1\leq i\leq n} d^n(i) = o(n^{1/2})$, so the preceding identity gives

$$\left| A^n(k,\ell) - \sum_{a_1,a_2\geq 0} R_{k+a_1+1,\ell+a_2+1,a_1,a_2} \right| = o(n^{1/2}).$$

Since

$$\begin{aligned} & n^{-1} R_{k+a_1+1,\ell+a_2+1,a_1,a_2} \\ & \xrightarrow{\text{prob}} a_2 p(\ell+a_2+1)\mathbf{P}\left\{\mathrm{Bin}(\ell+a_2,\rho)=a_2\right\} \\ & \qquad \cdot p(k+a_1+1)\mathbf{P}\left\{\mathrm{Bin}(k+a_1,\rho)=a_1\right\} \end{aligned}$$

by Proposition A.3, and summing the right-hand side of the last expression over $a_1, a_2 \geq 0$ gives $\alpha(k,l)$, it follows that for any $\epsilon > 0$,

$$\mathbf{P}\left\{A^n(k,l)/n \geq \alpha(k,l) - \epsilon\right\} \to 1.$$

But also $n^{-1}\sum_{k,l\geq 0} A^n(k,l) = |\mathrm{e}(T(\mathrm{d}^n))| = (n-1)/n \to 1$; so since $\sum_{k,l\geq 0}\alpha(k,l) = 1$, we must in fact have that

$$\frac{A^n(k,l)}{n} \stackrel{\text{prob}}{\longrightarrow} \alpha(k,l)$$

for all $k, l \geq 0$, as required.

It remains to show that $n^{-1}\sum_{k,l\geq 0} klA^n(k,l) \stackrel{\text{prob}}{\longrightarrow} \sum_{k,l\geq 0} kl\alpha(k,l)$. For this we will exploit the exchangeability of Pitman's additive coalescent. Recall the notation $v(h)$ for the vertex incident to half-edge h. Note that for any $M \in \mathbb{N}$ we have

$$\begin{aligned}
&\sum_{k,l\geq 0} klA^n(k,l) - \sum_{0\leq k,l\leq M} klA^n(k,l) \\
&= \sum_{uv\in e(T^n)} d_-^n(u)d_-^n(v)\mathbf{1}_{[\max(d_-^n(u),d_-^n(v))>M]} \\
&\leq \sum_{uv\in e(T^n)} d^n(u)d^n(v)\mathbf{1}_{[\max(d^n(u),d^n(v))>M]} \\
&= \sum_{i=1}^{n-1} d^n(v(r_i))d^n(v(s_i))\mathbf{1}_{[\max(d^n(v(r_i)),d^n(v(s_i)))>M]}.
\end{aligned}$$

Now, by Proposition 2.1 (3) and the identity (12), for any $1 \leq i \leq n-1$ we have

$$\begin{aligned}
&\mathbf{E}\left(d^n(v(r_i))d^n(v(s_i))\mathbf{1}_{[\max(d^n(v(r_i)),d^n(v(s_i)))>M]}\right) \\
&\quad = \mathbf{E}\left(d^n(v(r_1))d^n(v(s_1))\mathbf{1}_{[\max(d^n(v(r_1)),d^n(v(s_1)))>M]}\right),
\end{aligned}$$

and by the definition of Pitman's additive coalescent we have

$$\begin{aligned}
&\mathbf{E}\left(d^n(v(r_1))d^n(v(s_1))\,\mathbf{1}_{[\max(d^n(v(r_1)),d^n(v(s_1)))>M]}\right) \\
&= \sum_{u\in[n]}\sum_{v\in[n]} d^n(u)d^n(v)\,\mathbf{1}_{[\max(d^n(u),d^n(v))>M]}\,\mathbf{P}\left\{v(s_1) = u, v(r_1) = v\right\} \\
&= \sum_{u\in[n]}\sum_{v\in[n]} d^n(u)d^n(v)\,\mathbf{1}_{[\max(d^n(u),d^n(v))>M]} \cdot \frac{d^n(u)}{n\mu_1(p^n)} \cdot \frac{1}{n-1},
\end{aligned}$$

where we have used that $\sum_{i\in[n]} d^n(i) = n\mu_1(p^n)$. Next,

$$\sum_{u\in[n]}\sum_{v\in[n]} (d^n(u))^2 d^n(v)\,\mathbf{1}_{[\max(d^n(u),d^n(v))>M]}$$
$$\leq \bigg(\Big(\sum_{u\in[n]:d^n(u)>M} (d^n(u))^2\Big)\cdot \sum_{v\in[n]} d^n(v)$$
$$\cdot\Big(\sum_{v\in[n]:d^n(v)>M} d^n(v)\Big)\cdot \sum_{u\in[n]} (d^n(u))^2\bigg),$$
$$= n\mu_1(p^n)\cdot \sum_{u\in[n]:d^n(u)>M} (d^n(u))^2 \;+\; n\mu_2(p^n)\cdot \sum_{v\in[n]:d^n(v)>M} d^n(v).$$

Since $p^n \to p$, $\mu_1(p^n) \to \mu_1(p)$ and $\mu_2(p_n) \to \mu_2(p)$, for any $\delta > 0$ we may choose $M = M(\delta)$ sufficiently large so that $\sum_{u\in[n]:d^n(u)>M}(d^n(u))^2 < \delta n$ and $\sum_{v\in[n]:d^n(v)>M} d^n(v) < \delta n$, for all $n \geq 1$. For such M, the previous bound and the two identities which precede it yield that

$$\mathbf{E}\left(\sum_{k,l\geq 0} klA^n(k,l) - \sum_{0\leq k,l\leq M} klA^n(k,l)\right)$$
$$\leq \frac{1}{\mu_1(p^n)}(\delta n\mu_1(p^n) + \delta n\mu_2(p^n)).$$

By Markov's inequality, it follows that for all $\epsilon > 0$ there is $M \in \mathbb{N}$ such that for all $n \in \mathbb{N}$,

$$\mathbf{P}\left\{\sum_{k,l\geq 0} klA^n(k,l) - \sum_{0\leq k,l\leq M} klA^n(k,l) > \epsilon n\right\} < \epsilon.$$

Finally, since $n^{-1}A^n(k,l) \stackrel{\text{prob}}{\longrightarrow} \alpha(k,l)$, it follows that for all $M \in \mathbb{N}$ we have

$$\frac{1}{n}\sum_{0\leq k,l\leq M} klA^n(k,l) \stackrel{\text{prob}}{\longrightarrow} \sum_{0\leq k,l\leq M} kl\alpha(k,l),$$

so the preceding probability bound implies that

$$\frac{1}{n}\sum_{k,l\geq 0} klA^n(k,l) \stackrel{\text{prob}}{\longrightarrow} \lim_{M\to\infty}\sum_{0\leq k,l\leq M} kl\alpha(k,l) = \sum_{k,l\geq 0} kl\alpha(k,l),$$

as required. Q.E.D.

§Appendix B. Proof of Theorem 4.1

Let H be a simple graph with vertex set $v(H) = [n]$, and let G be a random graph with degree sequence $\mathrm{d} = (d(1), \ldots, d(n))$ sampled according to the configuration model. Recall the definitions of $\mathcal{L}(G), \mathcal{M}(G,H), \mathcal{N}(G,H)$ and $L(G), M(G,H), N(G,H)$ from Section 4. The first subsection will provide a quantitative approximation result for mixed moments of L, M and N. In the second subsection, we will use this approximation to prove Theorem 4.1.

B.1. Deterministic bounds on loops and multi-edges

Our arguments in this section are based on and fairly closely parallel those from [vdH, Chapter 7]. We recall the falling factorial notation $(x)_\ell = x(x-1)\ldots(x-\ell+1)$. In what follows, it is convenient to define $(x)_\ell = 1$ if $\ell = 0$, and $(x)_\ell = 0$ if $\ell < 0$.

Proposition B.1. *Write* $m = \frac{1}{2}\sum_{i=1}^{n} d(i)$, *and write*

$$d_{\max} = \max\{d(1), \ldots, d(n)\}.$$

For any positive integers $q, r, s \in \mathbb{N}$,

$$\left| \mathbf{E}\left((L)_q (M)_r (N)_s\right) - \frac{(|\mathcal{L}|)_q (|\mathcal{M}|)_r (|\mathcal{N}|)_s}{\prod_{i=0}^{q+2r+s-1} 2m-1-2i} \right| \leq C(S_1 + S_2)$$

where C *is a constant depending only on* q, r *and* s, S_1 *is defined by the following identity,*

$$\begin{aligned}
S_1 &\prod_{i=0}^{q+2r+s-1} (2m-1-2i) \\
&= (|\mathcal{L}|)_{q-2}(|\mathcal{M}|)_r(|\mathcal{N}|)_s \sum_{1\leq u\leq n} d(u)^3 \\
&\quad + (|\mathcal{L}|)_{q-1}(|\mathcal{M}|)_{r-1}(|\mathcal{N}|)_s \sum_{1\leq u\neq v\leq n} d(u)^3 d(v)^2 \\
&\quad + (|\mathcal{L}|)_{q-1}(|\mathcal{M}|)_r(|\mathcal{N}|)_{s-1} \sum_{uv\in e(H)} d(u)^2 d(v) \\
&\quad + (|\mathcal{L}|)_q(|\mathcal{M}|)_{r-2}(|\mathcal{N}|)_s \sum_{\substack{1\leq u\leq n \\ u\notin\{v_1,v_2\}}} d(u)^3 d(v_1)^2 d(v_2)^2 \\
&\quad + (|\mathcal{L}|)_q(|\mathcal{M}|)_{r-1}(|\mathcal{N}|)_{s-1} \sum_{\substack{1\leq u,v_1,v_2\leq n \\ u,v_1,v_2\, distinct \\ uv_2\in e(H)}} d(u)^2 d(v_1)^2 d(v_2)
\end{aligned}$$

$$+ (|\mathcal{L}|)_q(|\mathcal{M}|)_r(|\mathcal{N}|)_{s-2} \sum_{\substack{uv_1, uv_2 \in e(H) \\ v_1 \neq v_2}} d(u)d(v_1)d(v_2),$$

and S_2 *is defined by*

$$S_2 = (|\mathcal{L}|)_q(|\mathcal{N}|)_s \sum_{k=1}^{r-1} (|\mathcal{M}|)_{r-k} \sum_{\ell=0}^{k} d_{\max}^{2\ell} \prod_{i=0}^{q+2r+s-1-(2k-\ell)} \frac{1}{2m-1-2i}.$$

Proof. Throughout the proof, write

$$(x, y, z) = \big((x_1, \ldots, x_q), (y_1, \ldots, y_r), (z_1, \ldots, z_s)\big) \in \mathcal{L}^q \times \mathcal{M}^r \times \mathcal{N}^s$$

to denote a generic element of $\mathcal{L}^q \times \mathcal{M}^r \times \mathcal{N}^s$. For $(x, y, z) \in \mathcal{L}^q \times \mathcal{M}^r \times \mathcal{N}^s$, write

$$\mathbf{1}_{[x]} = \prod_{i=1}^{q} \mathbf{1}_{[x_i]}, \qquad \mathbf{1}_{[y]} = \prod_{i=1}^{r} \mathbf{1}_{[y_i]}, \qquad \mathbf{1}_{[z]} = \prod_{i=1}^{s} \mathbf{1}_{[z_i]}.$$

We say (x, y, z) is *non-repeating* if $x_1, x_2, \ldots, x_q$ are pairwise distinct, $y_1, y_2, \ldots, y_r$ are pairwise distinct, and $z_1, z_2, \ldots, z_s$ are pairwise distinct. In what follows, write

$$\sum\nolimits^{\star} := \sum_{\substack{(x,y,z) \in \mathcal{L}^q \times \mathcal{M}^r \times \mathcal{N}^s \\ (x,y,z) \text{ is non-repeating}}},$$

and, for $S \subset \mathcal{L}^q \times \mathcal{M}^r \times \mathcal{N}^s$, write

$$\sum_{S}^{\star} := \sum_{\substack{(x,y,z) \in S \\ (x,y,z) \text{ is non-repeating}}}.$$

Note that $(L)_q(M)_r(N)_s = \sum^{\star} \mathbf{1}_{[x]}\mathbf{1}_{[y]}\mathbf{1}_{[z]}$, so

$$\mathbf{E}\left((L)_q(M)_r(N)_s\right) = \sum\nolimits^{\star} \mathbf{P}\left\{\mathbf{1}_{[x]}\mathbf{1}_{[y]}\mathbf{1}_{[z]} = 1\right\}. \tag{18}$$

We say (x, y, z) is *non-conflicting* if the $2q + 4r + 2s$ half-edges appearing in x, y and z are pairwise distinct, and otherwise we say (x, y, z) is conflicting. Since half-edges in G are paired uniformly at random, for non-conflicting (x, y, z) we have

$$\mathbf{P}\left\{\mathbf{1}_{[x]}\mathbf{1}_{[y]}\mathbf{1}_{[z]} = 1\right\} = \prod_{i=0}^{q+2r+s-1} \frac{1}{2m-1-2i}. \tag{19}$$

Now, for a given (x,y,z), let $er(x,y,z)$ be defined as follows:

$$er(x,y,z) := \mathbf{P}\left\{\mathbf{1}_{[x]}\mathbf{1}_{[y]}\mathbf{1}_{[z]} = 1\right\} - \prod_{i=0}^{q+2r+s-1} \frac{1}{2m-1-2i}.$$

By (19), if (x,y,z) is non-conflicting then $er(x,y,z)=0$, so

$$\begin{aligned}
&\mathbf{E}\left((L)_q(M)_r(N)_s\right) \\
&= \sum\nolimits^{\star} \prod_{i=0}^{q+2r+s-1} \frac{1}{2m-1-2i} + \sum\nolimits^{\star} er(x,y,z) \\
&= \frac{(|\mathcal{L}|)_q(|\mathcal{M}|)_r(|\mathcal{N}|)_s}{\prod_{i=0}^{q+2r+s-1} 2m-1-2i} + \sum_{(x,y,z) \text{ is conflicting}}^{\star} er(x,y,z).
\end{aligned}$$

By the triangle inequality this implies

$$\begin{aligned}
&\left| \mathbf{E}\left((L)_q(M)_r(N)_s\right) - \frac{(|\mathcal{L}|)_q(|\mathcal{M}|)_r(|\mathcal{N}|)_s}{\prod_{i=0}^{q+2r+s-1} 2m-1-2i} \right| \\
&\leq \sum_{(x,y,z) \text{ is conflicting}}^{\star} |er(x,y,z)|.
\end{aligned} \tag{20}$$

To bound the error terms, we must make a distinction between two types of conflicts. This distinction is most easily understood by way of an example. On the one hand, suppose $x_1 = (ui, uj_1)$ and $x_2 = (ui, uj_2)$ for $u \in V(G)$ and distinct $i, j_1, j_2 \in [d(u)]$. Then $\mathbf{1}_{[x_1]}\mathbf{1}_{[x_2]} = 1$ is the event that the half edge ui is joined to uj_1 and uj_2 simultaneously, and $\mathbf{P}\{\mathbf{1}_{[x_1]}\mathbf{1}_{[x_2]} = 1\} = 0$. On the other hand, suppose $y_1 = (ui_1, vj_1), (ui_2, vj_2)$ and $y_2 = (ui_1, vj_1), (ui_3, vj_3)$ for distinct $u, v \in V(G)$, distinct $i_1, i_2, i_3 \in [d(u)]$, and distinct $j_1, j_2, j_3 \in [d(v)]$. Then $\mathbf{1}_{[y_1]}\mathbf{1}_{[y_2]} = 1$ is the event that u and v are connected by a triple edge, and $\mathbf{P}\{\mathbf{1}_{[y_1]}\mathbf{1}_{[y_2]} = 1\} > 0$.

We say a conflicting triple (x,y,z) is a *bad conflict* if $\mathbf{P}\{\mathbf{1}_{[x]}\mathbf{1}_{[y]}\mathbf{1}_{[z]} = 1\} = 0$, and otherwise we say (x,y,z) is a *good conflict.* In the above examples, the first is a bad conflict and the second is a good conflict. Let $\mathcal{B} = \mathcal{B}(q,r,s) \subseteq \mathcal{L}^q \times \mathcal{M}^r \times \mathcal{N}^s$ and $\mathcal{G} = \mathcal{G}(q,r,s) \subseteq \mathcal{L}^q \times \mathcal{M}^r \times \mathcal{N}^s$ be the collections of bad conflicts and good conflicts respectively. The rest of this proof is dedicated to bounding $|\mathcal{B}|, |\mathcal{G}|$, and $er(x,y,z)$ for $(x,y,z) \in \mathcal{B} \cup \mathcal{G}$.

(*Bounding* $|\mathcal{B}|$): If (x,y,z) is a bad conflict then one of the following must hold (in reading the below descriptions, it may be useful to consult Figure 2):

(1) There exists $1 \le a < b \le q$, $u \in V(G)$ and distinct $i_1, i_2, i_3 \in [d(u)]$ such that $x_a = (ui_1, ui_2)$ and $x_b = (ui_1, ui_3)$. Write $\mathcal{B}_{xx}$ for the set of $(x, y, z) \in \mathcal{B}$ that contain a pair (x_a, x_b) of this form.

(2) There exists $1 \le a \le q$, $1 \le b \le r$, distinct $u, v \in V(G)$, distinct $i_1, i_2, i_3 \in [d(u)]$, and distinct $j_1, j_2 \in [d(v)]$ such that $x_a = (ui_1, ui_3)$ and $y_b = (ui_1, vj_1), (ui_2, vj_2)$. Write $\mathcal{B}_{xy}$ for the set of $(x, y, z) \in \mathcal{B}$ that contain a pair (x_a, y_b) of this form.

(3) There exists $1 \le a \le q$, $1 \le b \le s$, $uv \in e(H)$, distinct $i_1, i_2 \in [d(u)]$, and $j \in [d(v)]$ such that $x_a = (ui_1, ui_2)$ and $z_b = (ui_1, vj)$. Write $\mathcal{B}_{xz}$ for the set of $(x, y, z) \in \mathcal{B}$ that contain a pair (x_a, z_b) of this form.

(4) There exists $1 \le a < b \le r$, distinct $u, v_1, v_2 \in V(G)$, distinct $i_1, i_2, i_3 \in [d(u)]$, distinct $j_1, j_2 \in [d(v_1)]$, and distinct $k_1, k_2 \in [d(v_2)]$ such that $y_a = (ui_1, v_1j_1), (ui_2, v_1j_2)$ and $y_b = (ui_1, v_2k_1), (ui_3, v_2k_2)$. Write $\mathcal{B}_{yy}$ for the set of $(x, y, z) \in \mathcal{B}$ that contain a pair (y_a, y_b) of this form.

(5) There exists $1 \le a \le r$, $1 \le b \le s$ and distinct $u, v_1, v_2 \in V(G)$ such that $uv_2 \in e(H)$, distinct $i_1, i_2 \in [d(u)]$, distinct $j_1, j_2 \in [d(v_1)]$, and $k \in [d(v_2)]$ such that $y_a = (ui_1, v_1j_1), (ui_2, v_1j_2)$ and $z_b = (ui_1, v_2k)$. Write $\mathcal{B}_{yz}$ for the set of $(x, y, z) \in \mathcal{B}$ that contain a pair (y_a, z_b) of this form.

(6) There exists $1 \le a < b \le s$, distinct $uv_1, uv_2 \in e(H)$, $i \in [d(u)]$, $j \in [d(v_1)]$, and $k \in [d(v_2)]$ such that $z_a = (ui, v_1j)$ and $z_b = (ui, v_2k)$. Write $\mathcal{B}_{zz}$ for the set of $(x, y, z) \in \mathcal{B}$ that contain a pair (z_a, z_b) of this form.

We next turn to bounding the sizes of each set, starting with $\mathcal{B}_{xx}$. The number of choices for a and b is $q(q-1)/2$. Having chosen these, for each possible choice of $u \in V(G)$, there are less than $d(u)^3$ choices for i_1, i_2 and i_3. Then, having chosen these, we must choose one of i_1, i_2 and i_3 to be repeated in x_a and x_b. Lastly, we must choose the remaining $q-2$ entries for x, r entries for y, and s entries for z. Hence,

$$\begin{aligned} |\mathcal{B}_{xx}| &\le (|\mathcal{L}|)_{q-2}(|\mathcal{M}|)_r(|\mathcal{N}|)_s q(q-1)/2 \sum_{1 \le u \le n} 3d(u)^3 \\ &= C_1(|\mathcal{L}|)_{q-2}(|\mathcal{M}|)_r(|\mathcal{N}|)_s \sum_{1 \le u \le n} d(u)^3, \end{aligned}$$

where $C_1 = C_1(q) = 3q(q-1)/2$. Note that $\mathcal{B}_{xx}$ is empty if $q \le 1$, so $|\mathcal{B}_{xx}| = 0$. In this case the right hand side is also zero by our convention that $(k)_\ell = 0$ for $\ell < 0$. Therefore, the bound also holds for $q \le 1$. The

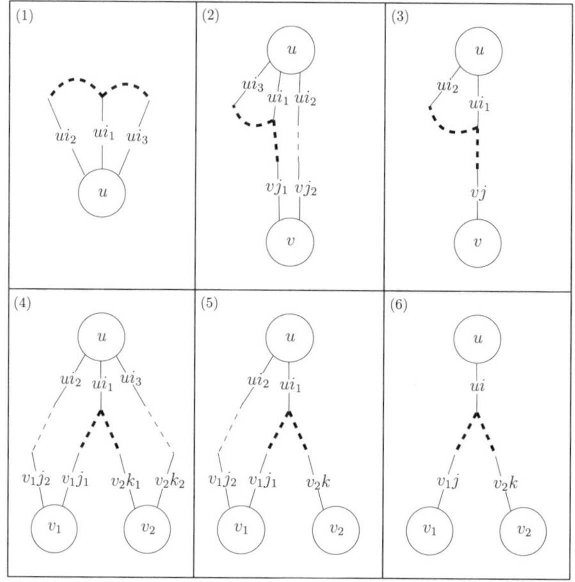

Fig. 2. An example of each of the six types of "bad" conflicts, depicted in the same order as they are described in the text. In the drawing, dashed lines represent the connections between half-edges required by the event. The bold dashed lines show the locations at which two half-edges must pair with a single half-edge, rendering the corresponding event impossible.

subsequent bounds can likewise be seen to hold when the right hand side is zero, though we do not explicitly verify this in every case.

When building an element of $\mathcal{B}_{xy}$, the number of ways to choose a and b is qr. Having chosen these, for each pair $u, v \in V(G)$, there are less than $d(u)^3 d(v)^2$ choices for $i_1, i_2, i_3 \in [d(u)]$ and $j_1, j_2 \in [d(v)]$. Then, there are a constant number of ways to arrange the half-edges in x_a and y_b. Lastly, we must choose the remaining entries for x, y and z. Hence,

$$|\mathcal{B}_{xy}| \leq C_2(|\mathcal{L}|)_{q-1}(|\mathcal{M}|)_{r-1}(|\mathcal{N}|)_s \sum_{1 \leq u \neq v \leq n} d(u)^3 d(v)^2,$$

where C_2 depends only on q and r.

For an element of $\mathcal{B}_{xz}$, for each $uv \in e(H)$, there are less than $d(u)^2 d(v)$ ways to choose $i_1, i_2 \in [d(u)]$ and $j \in [d(v)]$. Hence,

$$|\mathcal{B}_{xz}| \leq C_3(|\mathcal{L}|)_{q-1}(|\mathcal{M}|)_r(|\mathcal{N}|)_{s-1} \sum_{uv \in e(H)} d(u)^2 d(v).$$

For $\mathcal{B}_{yy}$, for each $u, v_1, v_2 \in V(G)$, there are less than $d(u)^3 d(v_1)^2 d(v_2)^2$ ways to choose $i_1, i_2, i_3 \in [d(u)]$, $j_1, j_2 \in [d(v_1)]$, and $k_1, k_2 \in [d(v_2)]$. Hence,

$$|\mathcal{B}_{yy}| \leq C_4(|\mathcal{L}|)_q(|\mathcal{M}|)_{r-2}(|\mathcal{N}|)_s \sum_{\substack{1 \leq u, v_1, v_2 \leq n \\ u, v_1, v_2 \text{ distinct}}} d(u)^3 d(v_1)^2 d(v_2)^2.$$

For $\mathcal{B}_{yz}$, for each $u, v_1, v_2 \in V(G)$ such that $uv_2 \in e(H)$, there are less than $d(u)^2 d(v_1)^2 d(v_2)$ ways to choose $i_1, i_2 \in [d(u)]$, $j_1, j_2 \in [d(v_1)]$, and $k \in [d(v_2)]$. Hence,

$$|\mathcal{B}_{yz}| \leq C_5(|\mathcal{L}|)_q(|\mathcal{M}|)_{r-1}(|\mathcal{N}|)_{s-1} \sum_{\substack{1 \leq u, v_1, v_2 \leq n \\ u, v_1, v_2 \text{ distinct} \\ uv_2 \in e(H)}} d(u)^2 d(v_1)^2 d(v_2).$$

Lastly, for $\mathcal{B}_{zz}$, for each $uv_1, uv_2 \in e(H)$, there are less than $d(u)d(v_1)$ $d(v_2)$ ways to choose $i \in [d(u)]$, $j \in [d(v_1)]$ and $k \in [d(v_2)]$. Hence,

$$|\mathcal{B}_{zz}| \leq C_6(|\mathcal{L}|)_q(|\mathcal{M}|)_r(|\mathcal{N}|)_{s-2} \sum_{\substack{uv_1, uv_2 \in e(H) \\ v_1 \neq v_2}} d(u)d(v_1)d(v_2).$$

Note that the values of C_1, C_2, C_3, C_4, C_5 and C_6 depend only on q, r and s.

(*Bounding* $|er(x, y, z)|$ *for* $(x, y, z) \in \mathcal{B}$): If $(x, y, z) \in \mathcal{B}$ then $\mathbf{P}\{\mathbf{1}_{[x]}\mathbf{1}_{[y]}$ $\mathbf{1}_{[z]} = 1\} = 0$, meaning

$$|er(x, y, z)| = \prod_{i=0}^{q+2r+s-1} \frac{1}{2m - 1 - 2i} \tag{21}$$

(*Bounding* $|\mathcal{G}|$): Suppose $(x, y, z) \in \mathcal{G}$. Then (x, y, z) is conflicting, meaning a half-edge appears more than once in $x \cup y \cup z$. However, since $\mathbf{P}\{\mathbf{1}_{[x]}\mathbf{1}_{[y]}\mathbf{1}_{[z]} = 1\} > 0$ for $(x, y, z) \in \mathcal{G}$, it cannot be the case where $\mathbf{1}_{[x]}\mathbf{1}_{[y]}\mathbf{1}_{[z]}$ contains the event that a half-edge is paired to two different half-edges simultaneously. Hence, there must be a half-edge pair that appears more than once in $x \cup y \cup z$. Furthermore, this half-edge pair must appear more than once in y, since the edges in y and z are disjoint. It follows that any $(x, y, z) \in \mathcal{G}$ can be constructed in the following way:

(1) Choose x and z arbitrarily. The number of choices here is $(|\mathcal{L}|)_q(|\mathcal{N}|)_s$.
(2) Choose a set of indices $1 \leq a_1 < a_2 < \cdots < a_k \leq r$ and arbitrarily choose the elements $y_a \in y$ such that $a \neq a_i$ for all $1 \leq i \leq k$. The number of choices here is $(|\mathcal{M}|)_{r-k}$ times a constant in terms of r.
(3) Choose $1 \leq b_1 \leq \cdots \leq b_\ell \leq r$ such that $\{b_1, \dots, b_\ell\} \subseteq \{a_1, \dots, a_k\}$. For each $1 \leq i \leq \ell$, build y_{b_i} by choosing a half-edge pair already in y, then choosing the other half-edge pair arbitrarily. The number of choices for each i is less than $d_{\max}^2$ times a constant in terms of r.
(4) For each $a \in \{a_1, \dots, a_k\} \setminus \{b_1 \dots, b_\ell\}$, build y_a by choosing two half-edge pairs already in y. The number of choices here is a constant in terms of r.

Since every element of $\mathcal{G}$ can be constructed in this way, we get

$$|\mathcal{G}| \leq C(r)(|\mathcal{L}|)_q(|\mathcal{N}|)_s \sum_{k=1}^{r-1}(|\mathcal{M}|)_{r-k} \sum_{\ell=0}^{k} d_{\max}^{2\ell}.$$

(*Bounding* $er(x,y,z)$ *for* $(x,y,z) \in \mathcal{G}$): Let $(x,y,z) \in \mathcal{G}$ and suppose we can construct (x,y,z) as above with a particular k and ℓ. Then there are $2k-\ell$ half-edge pairs that are redundant when calculating $\mathbf{P}\{\mathbf{1}_{[x]}\mathbf{1}_{[y]}\mathbf{1}_{[z]} = 1\}$. Hence,

$$\mathbf{P}\left\{\mathbf{1}_{[x]}\mathbf{1}_{[y]}\mathbf{1}_{[z]} = 1\right\} = \prod_{i=0}^{q+2r+s-1-(2k-\ell)} \frac{1}{2m-1-2i}.$$

Therefore, for such $(x,y,z) \in \mathcal{G}$,

(22)

$$\begin{aligned} |er(x,y,z)| &= \left| \prod_{i=0}^{q+2r+s-1} \frac{1}{2m-1-2i} - \prod_{i=0}^{q+2r+s-1-(2k-\ell)} \frac{1}{2m-1-2i} \right| \\ &\leq \prod_{i=0}^{q+2r+s-1-(2k-\ell)} \frac{1}{2m-1-2i}. \end{aligned}$$

Finally, by (20), we know that

$$\left| \mathbf{E}\left((L)_q (M)_r (N)_s\right) - \frac{(|\mathcal{L}|)_q (|\mathcal{M}|)_r (|\mathcal{N}|)_s}{\prod_{i=0}^{q+2r+s-1} 2m-1-2i} \right|$$
$$\leq \sum_{(x,y,z)\in\mathcal{B}}^{\star} |er(x,y,z)| + \sum_{(x,y,z)\in\mathcal{G}}^{\star} |er(x,y,z)|,$$

from which the result now follows by using the bounds on $|\mathcal{B}_{xx}|, |\mathcal{B}_{xy}|$, $|\mathcal{B}_{xz}|, |\mathcal{B}_{yy}|, |\mathcal{B}_{yz}|, |\mathcal{B}_{zz}|$ and on $|\mathcal{G}|$, together with (21) and (22). Q.E.D.

B.2. The probability of simplicity for a random superposition of graphs

Before proving Theorem 4.1, it will be useful to show some auxiliary bounds.

Lemma B.2. *Under the assumptions of Theorem* 4.1, *we have*

$$\sum_{v\in[n]} d^n(v) = O(n), \tag{23}$$

$$\sum_{v\in[n]} (d^n(v))^2 = O(n), \tag{24}$$

$$\sum_{uv\in e(h_n)} d^n(u) d^n(v) = O(n), \tag{25}$$

and

$$\sup_{u\in[n]} \sum_{v:uv\in e(h_n)} d^n(v) = o(n), \tag{26}$$

Proof. Equations (23) and (24) follow from the fact that $\mu_1(p^n) \to \mu_1(p) < \infty$ and $\mu_2(p^n) \to \mu_2(p) < \infty$. Indeed, we have

$$\sum_{v\in[n]} d^n(v) = n\sum_{k\geq 1} k p^n(k) = n\mu_1(p^n) = O(n),$$

and

$$\sum_{v\in[n]} (d^n(v))^2 = n\sum_{k\geq 1} k^2 p^n(k) = n\mu_2(p^n) = O(n).$$

Equation (25) follows from the convergence of $\sum_{i,j\geq 1} ij\alpha^n(i,j)$. Notice that, by the definition of α^n,

$$\sum_{uv\in e(h_n)} d^n(u)d^n(v) = \sum_{i,j\geq 1}\left(\sum_{uv\in e(h_n):d^n(u)=i,d^n(v)=j} ij\right)$$
$$= \sum_{i,j\geq 1} ij(n\alpha^n(i,j)).$$

Hence,

$$\frac{1}{n}\sum_{uv\in e(h_n)} d^n(u)d^n(v) \to \sum_{i,j\geq 1} ij\alpha(i,j) < \infty,$$

implying that

$$\sum_{uv\in e(h_n)} d^n(u)d^n(v) = O(n).$$

We will prove the fourth bound by contradiction. To this end, suppose (26) fails. Then we can find $c > 0$ and a sequence of vertices $(u_n, n \geq 1)$ with $u_n \in v(h_n)$ such that for all n sufficiently large,

$$\sum_{v:u_n v\in e(h_n)} d^n(v) \geq cn. \tag{27}$$

Write $\deg(u_n) = \deg_{h_n}(u_n)$. List the neighbours of u_n in h_n as $N_{h_n}(u_n) = \{v_1^n, \ldots, v_{\deg(u_n)}^n\}$ so that $d^n(v_i^n) \geq d^n(v_{i+1}^n)$ for all $1 \leq i < \deg(u_n)$.

Next, fix $D \in \mathbb{N}$ and let $k = k(n) = \max\{i : d^n(v_i^n) \geq D\}$. Then

$$\sum_{i=k+1}^{\deg(u_n)} d^n(v_i^n) \leq (\deg(u_n) - k)(D-1) = o(n),$$

the last bound holding since $\deg(u_n) = o(n)$ by assumption. Thus,

$$\sum_{v:u_n v\in e(h_n)} d^n(v)^2 = \sum_{i=1}^{\deg(u_n)} d^n(v_i^n)^2$$
$$\geq \sum_{i=1}^{k} d^n(v_i^n)^2$$
$$\geq D\sum_{i=1}^{k} d^n(v_i^n)$$

$$= D\left(\sum_{v:u_n v\in e(h_n)} d^n(v) - o(n)\right)$$
$$\geq D(c - o(1))n,$$

the last bound holding by (27). Since $D \in \mathbb{N}$ was arbitrary, it follows that

$$\sum_{v\in[n]} d^n(v)^2 \geq \sum_{v:u_n v\in e(h_n)} d^n(v)^2 = \omega(n),$$

contradicting (24). Q.E.D.

The next lemma is the last ingredient needed, and also assumes $p(0) + p(1) < 1$, i.e. an asymptotically non-zero proportion of the degrees in G_n are 2 or greater. We will show later that Theorem 4.1 is straightforward when $p(0) + p(1) = 1$.

Lemma B.3. *Under the assumptions of Theorem* 4.1, *suppose additionally that* $p(0) + p(1) < 1$. *Then*

$$|\mathcal{L}(G_n)| = \Theta(n), \tag{28}$$

and

$$|\mathcal{M}(G_n, h_n)| = \Theta(n^2). \tag{29}$$

Proof. Let $\mathcal{L}_n = \mathcal{L}(G_n), \mathcal{M}_n = \mathcal{M}(G_n, h_n)$, and $\mathcal{N}_n = \mathcal{N}(G_n, h_n)$. By their definitions, we have

$$|\mathcal{L}_n| = \sum_{v\in[n]} \frac{d^n(v)(d^n(v)-1)}{2}, \text{ and}$$
$$|\mathcal{M}_n| = \sum_{\{u<v:uv\notin e(h_n)\}} \frac{d^n(u)(d^n(u)-1)}{2} d^n(v)(d^n(v)-1).$$

For the upper bounds, by Lemma B.2 we have

$$|\mathcal{L}_n| = \sum_{v\in[n]} \frac{d^n(v)(d^n(v)-1)}{2} \leq \sum_{v\in[n]} d^n(v)^2 = O(n), \text{ and}$$
$$|\mathcal{M}_n| = \sum_{\{u<v:uv\notin e(h_n)\}} \frac{d^n(u)(d^n(u)-1)}{2} d^n(v)(d^n(v)-1)$$
$$\leq \left(\sum_{v\in[n]} d^n(v)^2\right)^2 = O(n^2).$$

For the lower bounds, first notice that

$$|\mathcal{L}_n| = \sum_{v\in[n]} \frac{d^n(v)(d^n(v)-1)}{2} = \sum_{v\in[n]:d^n(v)>1} \frac{d^n(v)(d^n(v)-1)}{2}$$
$$\geq |\{v\in[n] : d^n(v) > 1\}| = n(1-p^n(0)-p^n(1)).$$

Since $p^n(0)+p^n(1) \to p(0)+p(1) < 1$ by assumption, this implies $|\{v\in[n] : d^n(v)>1\}| = \Theta(n)$, and so

$$|\mathcal{L}_n| \geq \big|\{v\in[n] : d^n(v) > 1\}\big| = \Theta(n).$$

Similarly, we have

$$|\mathcal{M}_n| \geq \big|\{(u,v) : u < v, uv \notin e(h_n) \text{ and } d^n(u), d^n(v) > 1\}\big|.$$

From the conditions of Theorem 4.1 we know that

$$\max_{v\in[n]}\{\deg_{h_n}(v)\} = o(n),$$

which implies that $|e(h_n)| = o(n^2)$. Hence, we have

$$\big|\{(u,v) : u < v, uv \notin e(h_n) \text{ and } d^n(u), d^n(v) > 1\}\big|$$
$$= \big|\{(u,v) : u < v, d^n(u), d^n(v) > 1\}\big| - o(n^2),$$

and writing $k = k(n) = |\{v\in[n] : d^n(v) > 1\}|$, we have

$$\big|\{(u,v) : u < v, d^n(u), d^n(v) > 1\}\big| = \binom{k}{2} = \Theta(k^2) = \Theta(n^2),$$

and hence, $|\mathcal{M}_n| = \Theta(n^2)$. Q.E.D.

Proof of Theorem 4.1. Let $m = m(n) = \frac{1}{2}\sum_{v\in[n]} d^n(v)$, and let q, r and s be positive integers. Also, in what follows write $d^n_{\max} = \max_{1\leq i\leq n} d^n(i)$; by Fact A.1 we know that $d^n_{\max} = o(n^{1/2})$.

Assume for the time being that $p(0)+p(1) < 1$ and that $\eta > 0$. Notice that when $\eta > 0$ and $\mu_1(p) > 0$ we have

$$\frac{|\mathcal{N}_n|}{n} = \frac{1}{n}\sum_{uv\in e(h_n)} d^n(u)d^n(v) = \sum_{i,j\geq 1} ij\alpha^n(i,j)$$
$$\to \sum_{i,j\geq 1} ij\alpha(i,j) = \mu_1(p)\eta \in (0,\infty);$$

we have $\mu_1(p) > 0$ since $p(0) < 1$, and $\mu_1(p) < \infty$ since $\mu_2(p) < \infty$. It follows that $|\mathcal{N}_n| = \Theta(n)$.

We first claim that

$$\mathbf{E}\left((L_n)_q(M_n)_r(N_n)_s\right) = \frac{(|\mathcal{L}_n|)_q(|\mathcal{M}_n|)_r(|\mathcal{N}_n|)_s}{\prod_{i=0}^{q+2r+s-1} 2m-1-2i}(1+o(1)).$$

From Proposition B.1 we know that

$$\left|\mathbf{E}\left((L_n)_q(M_n)_r(N_n)_s\right) - \frac{(|\mathcal{L}_n|)_q(|\mathcal{M}_n|)_r(|\mathcal{N}_n|)_s}{\prod_{i=0}^{q+2r+s-1} 2m-1-2i}\right| \le C(S_1+S_2),$$

where S_1 is defined by the relationship

$$\begin{aligned}
S_1 &\cdot \prod_{i=0}^{q+2r+s-1} (2m-1-2i) \\
&= (|\mathcal{L}_n|)_{q-2}(|\mathcal{M}_n|)_r(|\mathcal{N}_n|)_s \sum_{v\in[n]} (d^n(v))^3 \\
&\quad + (|\mathcal{L}_n|)_{q-1}(|\mathcal{M}_n|)_{r-1}(|\mathcal{N}_n|)_s \sum_{1\le u\ne v\le n} (d^n(u))^3(d^n(v))^2 \\
&\quad + (|\mathcal{L}_n|)_{q-1}(|\mathcal{M}_n|)_r(|\mathcal{N}_n|)_{s-1} \sum_{uv\in e(h_n)} (d^n(u))^2 d^n(v) \\
&\quad + (|\mathcal{L}_n|)_q(|\mathcal{M}_n|)_{r-2}(|\mathcal{N}_n|)_s \sum_{\substack{1\le u,v_1,v_2\le n\\ u,v_1,v_2 \text{ distinct}}} (d^n(u))^3(d^n(v_1))^2(d^n(v_2))^2 \\
&\quad + (|\mathcal{L}_n|)_q(|\mathcal{M}_n|)_{r-1}(|\mathcal{N}_n|)_{s-1} \sum_{\substack{1\le u,v_1,v_2\le n\\ u,v_1,v_2 \text{ distinct}\\ uv_2\in e(h_n)}} (d^n(u))^2(d^n(v_1))^2 d^n(v_2) \\
&\quad + (|\mathcal{L}_n|)_q(|\mathcal{M}_n|)_r(|\mathcal{N}_n|)_{s-2} \sum_{\substack{uv_1,uv_2\in e(h_n)\\ v_1\ne v_2}} d^n(u)d^n(v_1)d^n(v_2),
\end{aligned}$$

and S_2 is defined by

$$\begin{aligned}
S_2 = (|\mathcal{L}_n|)_q(|\mathcal{N}_n|)_s \sum_{k=1}^{r-1} (|\mathcal{M}_n|)_{r-k} \sum_{\ell=0}^{k} (d^n_{\max})^{2\ell} \\
\cdot \prod_{i=0}^{q+2r+s-1-(2k-\ell)} \frac{1}{2m-1-2i};
\end{aligned}$$

recall that we set $(k)_\ell = 0$ if $\ell < 0$. We now show that

$$S_1 = o\left(\frac{(|\mathcal{L}_n|)_q (|\mathcal{M}_n|)_r (|\mathcal{N}_n|)_s}{\prod_{i=0}^{q+2r+s-1}(2m-1-2i)} \right) \text{ and} \tag{30}$$

$$S_2 = o\left(\frac{(|\mathcal{L}_n|)_q (|\mathcal{M}_n|)_r (|\mathcal{N}_n|)_s}{\prod_{i=0}^{q+2r+s-1}(2m-1-2i)} \right).$$

We will start by bounding S_1. Since $|\mathcal{L}_n|, |\mathcal{M}_n|, |\mathcal{N}_n| \to \infty$ as $n \to \infty$, to prove the first bound in (30) it suffices to establish the following bounds:

$$\begin{aligned}
&\sum_{v\in[n]} (d^n(v))^3 = o(|\mathcal{L}_n|^2), \\
&\sum_{1\le u\neq v\le n} (d^n(u))^3 (d^n(v))^2 = o(|\mathcal{L}_n||\mathcal{M}_n|), \\
&\sum_{uv\in e(h_n)} (d^n(u))^2 d^n(v) = o(|\mathcal{L}_n||\mathcal{N}_n|), \\
&\sum_{\substack{1\le u,v_1,v_2\le n \\ u,v_1,v_2 \text{ distinct}}} (d^n(u))^3 (d^n(v_1))^2 (d^n(v_2))^2 = o(|\mathcal{M}_n|^2), \\
&\sum_{\substack{1\le u,v_1,v_2\le n \\ u,v_1,v_2 \text{ distinct} \\ uv_2\in e(h_n)}} (d^n(u))^2 (d^n(v_1))^2 d^n(v_2) = o(|\mathcal{M}_n||\mathcal{N}_n|), \text{ and} \\
&\sum_{\substack{uv_1,uv_2\in e(h_n) \\ v_1\neq v_2}} d^n(u) d^n(v_1) d^n(v_2) = o(|\mathcal{N}_n|^2).
\end{aligned} \tag{31}$$

Using Lemmas B.2 and B.3, together with the fact that $d^n_{\max} = o(n^{1/2})$, we get the following results:

$$\sum_{v\in[n]} (d^n(v))^3 \le d^n_{\max} \sum_{v\in[n]} (d^n(v))^2 = o(n^{1/2}) \cdot O(n) = o(n^2) = o(|\mathcal{L}_n|^2),$$

$$\begin{aligned}
\sum_{1\le u\neq v\le n} (d^n(u))^3 (d^n(v))^2 &\le d^n_{\max} \sum_{1\le u\neq v\le n} (d^n(u))^2 (d^n(v))^2 \\
&\le d^n_{\max} \left(\sum_{v\in[n]} (d^n(v))^2 \right)^2 \\
&= o(n^3) = o(|\mathcal{L}_n||\mathcal{M}_n|),
\end{aligned}$$

$$\sum_{uv\in e(h_n)} (d^n(u))^2 d^n(v) \le d^n_{\max} \sum_{uv\in e(h_n)} d^n(u)d^n(v)$$
$$= d^n_{\max}|\mathcal{N}_n| = o(|\mathcal{L}_n||\mathcal{N}_n|),$$

$$\sum_{\substack{1\le u,v_1,v_2\le n\\ u,v_1,v_2 \text{ distinct}}} (d^n(u))^3(d^n(v_1))^2(d^n(v_2))^2$$
$$\le d^n_{\max} \sum_{\substack{1\le u,v_1,v_2\le n\\ u,v_1,v_2 \text{ distinct}}} (d^n(u))^2(d^n(v_1))^2(d^n(v_2))^2$$
$$\le d^n_{\max}\left(\sum_{v\in[n]}^{n}(d^n(v))^2\right)^3 = o(n^4) = o(|\mathcal{M}_n|^2),$$

$$\sum_{\substack{1\le u,v_1,v_2\le n\\ u,v_1,v_2 \text{ distinct}\\ uv_2\in e(h_n)}} (d^n(u))^2(d^n(v_1))^2 d^n(v_2) \le (d^n_{\max})^3 \sum_{uv\in e(h_n)} d^n(u)d^n(v)$$
$$= o(n^{3/2})|\mathcal{N}_n| = o(|\mathcal{M}_n||\mathcal{N}_n|),$$

and

$$\sum_{\substack{uv_1,uv_2\in e(h_n)\\ v_1\ne v_2}} d^n(u)d^n(v_1)d^n(v_2)$$
$$= \sum_{uv\in e(h_n)} d^n(u)d^n(v)\cdot \sum_{w:uw\in e(h_n)} d^n(w)$$
$$\le \sum_{uv\in e(h_n)} d^n(u)d^n(v)\cdot \left(\sup_{u\in[n]} \sum_{w:uw\in e(h_n)} d^n(w)\right)$$
$$= |\mathcal{N}_n|\cdot o(n)$$
$$= o(|\mathcal{N}_n|^2),$$

the last bound holding as $\mathcal{N}_n = \Theta(n)$.

To prove the bound on S_2 from (30), first notice that

$$S_2 = (|\mathcal{L}_n|)_q(|\mathcal{N}_n|)_s \sum_{k=1}^{r-1}(|\mathcal{M}_n|)_{r-k} \sum_{\ell=0}^{k}(d^n_{\max})^{2\ell} \cdot \prod_{i=0}^{q+2r+s-1-(2k-\ell)} \frac{1}{2m-1-2i}$$
$$= \frac{(|\mathcal{L}_n|)_q(|\mathcal{N}_n|)_s}{\prod_{i=0}^{q+s-1} 2m-1-2i} \sum_{k=1}^{r-1}(|\mathcal{M}_n|)_{r-k} \sum_{\ell=0}^{k}(d^n_{\max})^{2\ell} \cdot \prod_{i=q+s}^{q+2r+s-1-(2k-\ell)} \frac{1}{2m-1-2i}.$$

Hence, it suffices to show the following:

$$\sum_{k=1}^{r-1}(|\mathcal{M}_n|)_{r-k} \sum_{\ell=0}^{k}(d^n_{\max})^{2\ell} \prod_{i=q+s}^{q+2r+s-1-(2k-\ell)} \frac{1}{2m-1-2i}$$
$$= o\left(\frac{(|\mathcal{M}_n|)_r}{\prod_{i=q+s}^{q+2r+s-1} 2m-1-2i}\right).$$

Since r is fixed, we need only show that for arbitrary $k \in [1, r-1]$ and $\ell \in [0, k]$,

$$(|\mathcal{M}_n|)_{r-k}(d^n_{\max})^{2\ell} \prod_{i=q+s}^{q+2r+s-1-(2k-\ell)} \frac{1}{2m-1-2i}$$
$$= o\left(\frac{(|\mathcal{M}|)_r}{\prod_{i=q+s}^{q+2r+s-1} 2m-1-2i}\right).$$

By cancelling out some terms, this follows if we can show that

$$(d^n_{\max})^{2\ell} = o\left(\frac{(|\mathcal{M}_n| - (r-k))_k}{\prod_{i=q+2r+s-(2k-\ell)}^{q+2r+s-1} 2m-1-2i}\right).$$

Now since $m = \Theta(n)$, $|\mathcal{M}_n| = \Theta(n^2)$, and r is a constant, this in turn holds, provided that

$$(d^n_{\max})^{2\ell} = o\left(\frac{n^{2k}}{n^{2k-\ell}}\right)$$
$$= o(n^{\ell}),$$

which holds since $d^n_{\max} = o(n^{1/2})$.

Therefore,

$$S_1 + S_2 = o\left(\frac{(|\mathcal{L}_n|)_q(|\mathcal{M}_n|)_r(|\mathcal{N}_n|)_s}{\prod_{i=0}^{q+2r+s-1} 2m-1-2i}\right),$$

which proves that

$$\mathbf{E}\left((L_n)_q(M_n)_r(N_n)_s\right) = \frac{(|\mathcal{L}_n|)_q(|\mathcal{M}_n|)_r(|\mathcal{N}_n|)_s}{\prod_{i=0}^{q+2r+s-1} 2m-1-2i}(1+o(1)).$$

Furthermore, since q, r and s are fixed, we obtain that

$$\begin{aligned}
&\frac{(|\mathcal{L}_n|)_q(|\mathcal{M}_n|)_r(|\mathcal{N}_n|)_s}{\prod_{i=0}^{q+2r+s-1} 2m-1-2i} \\
&= \frac{|\mathcal{L}_n|^q|\mathcal{M}_n|^r|\mathcal{N}_n|^s}{(2m)^{q+2r+s}}(1+o(1)) \\
&= \left(\frac{|\mathcal{L}_n|}{\sum_{v\in[n]} d^n(v)}\right)^q \left(\frac{|\mathcal{M}_n|}{\left(\sum_{v\in[n]} d^n(v)\right)^2}\right)^r \left(\frac{|\mathcal{N}_n|}{\sum_{v\in[n]} d^n(v)}\right)^s (1+o(1))
\end{aligned}$$

Next, we claim that

$$\frac{|\mathcal{L}_n|}{\sum_{v\in[n]} d^n(v)} \to \nu/2, \quad \frac{|\mathcal{M}_n|}{\left(\sum_{v\in[n]} d^n(v)\right)^2} \to \nu^2/4, \text{ and } \frac{|\mathcal{N}_n|}{\sum_{v\in[n]} d^n(v)} \to \eta.$$

For the first of these three claims, we have

$$\begin{aligned}
\frac{|\mathcal{L}_n|}{\sum_{v\in[n]} d^n(v)} &= \frac{\frac{1}{2}\sum_{v\in[n]} d^n(v)(d^n(v)-1)}{\sum_{v\in[n]} d^n(v)} \\
&= \frac{1}{2}\frac{\sum_{v\in[n]}(d^n(v))^2 - \sum_{v\in[n]} d^n(v)}{\sum_{v\in[n]} d^n(v)} \\
&= \frac{1}{2}\left(\frac{\mu_2(p^n)-\mu_1(p^n)}{\mu_1(p^n)}\right) \to \nu/2.
\end{aligned}$$

For the second, we have

$$\frac{|\mathcal{M}_n|}{\left(\sum_{v\in[n]} d^n(v)\right)^2}$$
$$= \frac{\frac{1}{2}\sum_{\{u<v:uv\notin e(h_n)\}} d^n(u)(d^n(u)-1)d^n(v)(d^n(v)-1)}{\left(\sum_{v\in[n]} d^n(v)\right)^2}$$
$$= \frac{1}{4\left(\sum_{v\in[n]} d^n(v)\right)^2}\left(\left[\sum_{v\in[n]} d^n(v)(d^n(v)-1)\right]^2 - \sum_{v\in[n]} [d^n(v)(d^n(v)-1)]^2 - \sum_{uv\in e(h_n)} d^n(u)(d^n(u)-1)d^n(v)(d^n(v)-1)\right).$$

The second and third terms vanish in the limit since $\mu_1(p) > 0$ and so

$$\left(\sum_{v\in[n]} d^n(v)\right)^2 = \Omega(n^2),$$

and

$$\sum_{v\in[n]} \left[d^n(v)(d^n(v)-1)\right]^2 \le \left(d^n_{\max}\right)^2 \sum_{v\in[n]} (d^n(v))^2 = \left(d^n_{\max}\right)^2 n\mu_2(p^n) = o(n^2),$$

and

$$\sum_{uv\in e(h_n)} d^n(u)(d^n(u)-1)d^n(v)(d^n(v)-1) \le \left(d^n_{\max}\right)^2 \sum_{uv\in e(h_n)} d^n(u)d^n(v) = \left(d^n_{\max}\right)^2 |\mathcal{N}_n| = o(n^2).$$

Therefore,

$$\frac{|\mathcal{M}_n|}{\left(\sum_{v\in[n]} d^n(v)\right)^2} = \frac{1}{4}\frac{\left[\sum_{v\in[n]} d^n(v)(d^n(v)-1)\right]^2}{\left(\sum_{v\in[n]} d^n(v)\right)^2}(1+o(1))$$
$$\to \frac{1}{4}\frac{(\mu_2(p)-\mu_1(p))^2}{(\mu_1(p))^2}$$
$$= \nu^2/4.$$

For the third claim, we have

$$\frac{|\mathcal{N}_n|}{\sum_{v\in[n]} d^n(v)} = \frac{\sum_{uv\in e(h_n)} d^n(u)d^n(v)}{\sum_{v\in[n]} d^n(v)} = \frac{\sum_{i,j\geq 1} ij\alpha^n(i,j)}{\mu_1(p^n)}$$
$$\to \frac{\sum_{i,j\geq 1} ij\alpha(i,j)}{\mu_1(p)} = \eta.$$

Therefore,

$$\lim_{n\to\infty} \mathbf{E}\left((L_n)_q(M_n)_r(N_n)_s\right) \tag{32}$$
$$= \lim_{n\to\infty} \frac{(|\mathcal{L}_n|)_q(|\mathcal{M}_n|)_r(|\mathcal{N}_n|)_s}{\prod_{i=0}^{q+2r+s-1} 2m-1-2i}(1+o(1))$$
$$= \lim_{n\to\infty}\left(\frac{|\mathcal{L}_n|}{\sum_{v\in[n]} d^n(v)}\right)^q \left(\frac{|\mathcal{M}_n|}{\left(\sum_{v\in[n]} d^n(v)\right)^2}\right)^r \cdot \left(\frac{|\mathcal{N}_n|}{\sum_{v\in[n]} d^n(v)}\right)^s (1+o(1))$$
$$= (\nu/2)^q\,(\nu^2/4)^r\,(\eta)^s.$$

It then follows, by Theorem 2.6 of [vdH], that the random variables L_n, M_n and N_n converge to independent Poisson random variables with parameters $\nu/2, \nu^2/4$ and η respectively. This proves the theorem in the case that $p(0)+p(1)<1$ and $\eta>0$.

Lastly we will deal with the cases that arise if $p(0)+p(1)=1$ or if $\eta=0$. First, if $\eta=0$ then $\lim_{n\to\infty}\sum_{i,j\geq 1} ij\alpha^n(i,j)=\sum_{i,j\geq 1} ij\alpha(i,j)=0$. For any two half-edges ui and vj with $u,v\in[n]$, $i\in[d^n(u)]$ and $j\in[d^n(v)]$, we have $\mathbf{P}\{\mathbf{1}_{[ui,vj]}=1\}=\frac{1}{2m-1}$ from the definition of the configuration model. So by (18), since $m=m(n)=n\mu_1(p^n)/2=\Theta(n)$, we have that

$$\begin{aligned}\mathbf{E}(N_n) &= \sum_{uv\in e(h_n)} \sum_{i\in[d^n(u)]} \sum_{j\in[d^n(v)]} \mathbf{P}\left\{\mathbf{1}_{[ui,vj]} = 1\right\} \\ &= \frac{\sum_{uv\in e(h_n)} d^n(u)d^n(v)}{2m-1} \\ &= \frac{n}{2m-1}\sum_{i,j\geq 1} ij\alpha^n(i,j) = o(1).\end{aligned}$$

Hence, $\lim_{n\to\infty}\mathbf{E}(N_n) = 0$. In this case, if $p(0)+p(1) < 1$ then a reprise of the argument leading to (32) gives that for all $q, r \geq 1$,

$$\lim_{n\to\infty}\mathbf{E}\left((L_n)_q(M_n)_r\right) = (\nu/2)^q(\nu^2/4)^r,$$

and therefore that L_n and M_n converge to independent Poisson random variables with parameters $\nu/2$ and $\nu^2/4$ respectively.

Lastly, we deal with the case when $p(0) + p(1) = 1$. In this case, $\mu_2(p) = \mu_1(p)$, so $\nu = 0$. Furthermore,

$$\mathbf{E}(L_n) = \frac{\frac{1}{2}\sum_{v\in[n]} d^n(v)(d^n(v)-1)}{2m-1} = \frac{n}{4m-2}(\mu_2(p^n) - \mu_1(p^n)).$$

Since $\mu_2(p^n)-\mu_1(p^n) \to \mu_2(p)-\mu_1(p) = 0$, we get that $\lim_{n\to\infty}\mathbf{E}(L_n) = 0$, and an analogous argument shows that $\lim_{n\to\infty}\mathbf{E}(M_n) = 0$. In this case, another reprise of the argument leading to (32) gives that for all $s \geq 1$,

$$\lim_{n\to\infty}\mathbf{E}\left((N_n)_s\right) = \eta^s,$$

so N_n is asymptotically Poisson(η) distributed. Q.E.D.

Proof of Corollary 4.2. First, the fact that $\sum_{k,\ell\geq 0} k\ell\cdot\alpha(k,\ell) < \infty$ appears in (8), above, from which it is immediate that $\eta < \infty$. Next, let $G_-(\mathrm{d}^n)$ be the subgraph of $G(\mathrm{d}^n)$ with edge set $\mathrm{e}(G(\mathrm{d}^n))\setminus\mathrm{e}(T(\mathrm{d}^n))$, and let d^n_- be the degree sequence of $G_-(\mathrm{d}^n)$, as defined in Section 3 previous to Proposition 3.2. Finally, write $L_n = L(G_-(\mathrm{d}^n))$, $M_n = M(G_-(\mathrm{d}^n), T(\mathrm{d}^n))$, and $N_n = N(G_-(\mathrm{d}^n), T(\mathrm{d}^n))$. Our aim is to apply Theorem 4.1, with $h_n = T(\mathrm{d}^n)$ and $G_n = G_-(\mathrm{d}^n) = G(\mathrm{d}^n) - T(\mathrm{d}^n)$. Note that with these choices, we have $\alpha^n(k,l) = n^{-1}A^n(k,l)$, where $A^n(k,l)$ is as in Proposition 3.2. Moreover, we have

$$\sum_{l\geq 0} A^n(k,l) = \#\{u\in[n] : d^n_-(u) = k\}.$$

Conditionally given $T(\mathrm{d}^n)$, the graph $G_-(\mathrm{d}^n)$ is a random graph with degree sequence d^n_-. By Proposition 3.2 we know that $\alpha^n(k,l) \xrightarrow{\text{prob}} \alpha(k,l)$ for all $k,l \geq 0$, and that

$$\sum_{k,l\geq 0} kl\alpha^n(k,l) \xrightarrow{\text{prob}} \sum_{k,l\geq 0} kl\alpha(k,l).$$

Moreover, since α defines a probability distribution, it must be that for all k we have

$$\begin{aligned} (33) \qquad p^n_-(k) &:= \frac{1}{n}\#\{u \in [n] : d^n_-(u) = k\} \\ &= \frac{1}{n}\sum_{l\geq 0} A^n(k,l) = \sum_{l\geq 0} \alpha^n(k,l) \xrightarrow{\text{prob}} \sum_{l\geq 0} \alpha(k,l). \end{aligned}$$

Setting $p_-(k) = \sum_{l\geq 0} \alpha(k,l)$, then (33) states that $p^n_-(k) \xrightarrow{\text{prob}} p_-(k)$ for all $k \geq 0$. Moreover, since $p^n_-(k) \leq p^n(k)$, $p^n_-(k) \xrightarrow{\text{prob}} p_-(k)$, and $p^n(k) \xrightarrow{\text{prob}} p(k)$ for all $k \geq 0$, it follows that $\mu_2(p_-) \leq \mu_2(p) < \infty$ and $\mu_2(p^n_-) \xrightarrow{\text{prob}} \mu_2(p_-)$. From these observations, Fact A.1 then implies that

$$\max_{v\in[n]} \deg_{T(\mathrm{d}^n)}(v) \leq \max_{v\in[n]} \deg_{G_n}(v) = o(n^{1/2}).$$

Since $\mu_2(p^n_-) \xrightarrow{\text{prob}} \mu_2(p_-)$, we also have $\mu_1(p^n_-) \xrightarrow{\text{prob}} \mu_1(p_-)$; but

$$n\mu_1(p^n_-) = \sum_{i=1}^{n} d^n_-(u) = \left(\sum_{i=1}^{n} d^n(u)\right) - 2(n-1),$$

so $\mu_1(p_-) = \mu_1(p) - 2$. Thus, if $\mu_1(p) > 2$ then $\mu_1(p_-) > 0$, so $p_-(0) < 1$. In this case, applying Theorem 4.1, it follows that conditionally given $T(\mathrm{d}^n)$,

$$\left\| \mathrm{Dist}(L_n, M_n, N_n) - \mathrm{Poi}(\nu/2) \otimes \mathrm{Poi}(\nu^2/4) \otimes \mathrm{Poi}(\eta) \right\|_{\mathrm{TV}} \xrightarrow{\text{prob}} 0$$

as $n \to \infty$. If (L, M, N) is $\mathrm{Poi}(\nu/2) \otimes \mathrm{Poi}(\nu^2/4) \otimes \mathrm{Poi}(\eta)$-distributed, then we have $\mathbf{P}\{L = M = N = 0\} = \exp(-\nu/2 - \nu^2/4 - \eta)$; since $G(\mathrm{d}^n)$ is simple if and only if $L_n = M_n = N_n = 0$, it follows that

$$\begin{aligned} \mathbf{P}\left\{G(\mathrm{d}^n) \text{ simple} \mid T(\mathrm{d}^n)\right\} &= \mathbf{P}\left\{L_n = M_n = N_n = 0 \mid T(\mathrm{d}^n)\right\} \\ &\xrightarrow{\text{prob}} \mathbf{P}\left\{L = M = N = 0\right\} \\ &= \exp(-\nu/2 - \nu^2/4 - \eta), \end{aligned}$$

as required.

It remains to treat the case that $\mu_1(p) = 2$, which implies that $\mu_1(p_-) = 0$ and $p_-(0) = 1$. This case requires a separate argument, which is more involved than one might expect. We note immediately that in this situation, $m = (1 + o(1))n$ as $n \to \infty$.

Write G'_n for the graph obtained from $G_-(\mathrm{d}^n)$ by removing the edge $\Gamma(\mathrm{d}^n)$. Then let $L'_n = L(G'_n)$, $M'_n = M(G'_n, T(\mathrm{d}^n))$, and $N'_n = N(G'_n, T(\mathrm{d}^n))$. Then G'_n is simple precisely if $L'_n + M'_n + N'_n = 0$. We will first prove that $\mathbf{E}(L'_n + M'_n + N'_n) \stackrel{\text{prob}}{\longrightarrow} 0$, then explain how to deal with the root edge.

By Proposition 2.1, we know that the non-root half-edges chosen for $T(\mathrm{d}^n)$ form a uniformly random subset $\mathcal{S}^n$ of $\bigcup_{i=1}^n \{i1, \ldots, i(d^n(i) - 1)\}$ of size $n - 1$. The half-edges which are paired to form $G_-(\mathrm{d}^n)$ are precisely the edges of $\mathcal{U} := \bigcup_{i=1}^n \{i1, \ldots, i(d^n(i) - 1)\} \setminus \mathcal{S}^n$, together with the unique unpaired root half-edge of $T(\mathrm{d}^n)$.

Write $\mathcal{U}$ for the set of half-edges paired to form G'_n. By the observations of the preceding paragraph, $\mathcal{U}$ is a uniformly random subset of $\bigcup_{i=1}^n \{i1, \ldots, i(d^n(i) - 1)\}$ of size $2(m - n)$. Moreover, conditionally given $\mathcal{U}$, the pairing of half-edges in $\mathcal{S}$ is uniformly random and independent of $T(\mathrm{d}^n)$. Therefore, we can construct $(G(\mathrm{d}^n), T(\mathrm{d}^n), \Lambda(\mathrm{d}^n))$ as follows. First sample a sequence of $m-n$ disjoint pairs of half-edges from $\bigcup_{i=1}^n \{i1, \ldots, i(d^n(i) - 1)\}$ uniformly at random and join them to form edges; this determines the set $\mathcal{U}$, and all edges of G'_n. Next, build $T(\mathrm{d}^n)$ via Pitman's additive coalescent applied to $\bigcup_{i=1}^n \{i1, \ldots, i(d^n(i)-1)\} \setminus \mathcal{U}$. Finally, pair the root half-edge of T^n with the sole remaining unpaired non-root half-edge. We analyze this construction procedure in order to bound the expected number of loops and multiple edges in G'_n.

Let (h_1, h_2) be a half-edge pair chosen for G'_n in the construction procedure just above. Then h_1 and h_2 are uniform random half-edges chosen from $\bigcup_{i=1}^n \{i1, \ldots, i(d^n(i) - 1)\}$, and (h_1, h_2) is a loop if $v(h_1) = v(h_2)$. Hence,

$$\begin{aligned}
\mathbf{E}(L'_n) &= |e(G'_n)|\, \mathbf{P}\{v(h_1) = v(h_2)\} \\
&= (m - n) \sum_{i=1}^n \mathbf{P}\{v(h_1) = v(h_2) = i\} \\
&= (m - n) \sum_{i=1}^n \frac{(d^n(i) - 1)(d^n(i) - 2)}{(2m - (n - 1))(2m - n)}.
\end{aligned}$$

Similarly, edges (h_1, h_2) and (h_3, h_4) form a double edge if $v(h_1) = v(h_3)$ and $v(h_2) = v(h_4)$ or if $v(h_1) = v(h_4)$ and $v(h_2) = v(h_3)$. Hence,

$$\begin{aligned}
&\mathbf{E}\left(M'_n\right) \\
&= (|e(G'_n)|)(|e(G'_n)| - 1) \\
&\qquad \cdot \sum_{1 \le i < j \le n} \frac{4(d^n(i)-1)(d^n(i)-2)(d^n(j)-1)(d^n(j)-2)}{(2m-n+1)(2m-n)(2m-n-1)(2m-n-2)} \\
&= (m-n)(m-n-1) \\
&\qquad \cdot \sum_{1 \le i < j \le n} \frac{4(d^n(i)-1)(d^n(i)-2)(d^n(j)-1)(d^n(j)-2)}{(2m-n+1)(2m-n)(2m-n-1)(2m-n-2)} \\
&\le 2\left((m-n)\sum_{i=1}^{n} \frac{(d^n(i)-1)(d^n(i)-2)}{(2m-n-1)(2m-n-2)}\right)^2.
\end{aligned}$$

Since $m = n + o(n)$, we have $\frac{m-n-1}{2m-n-2} = o(1)$. Since also $\sum_{i=1}^{n}(d^n(i) - 1)(d^n(i)-2) \le \sum_{i=1}^{n} d^n(i)^2 = O(n) = O(2m-n-1)$, it follows from the two preceding displayed inequalities that $\mathbf{E}(L'_n + M'_n) \stackrel{\text{prob}}{\longrightarrow} 0$.

To show that $\mathbf{E}(N'_n) \stackrel{\text{prob}}{\longrightarrow} 0$, we will again use Proposition 2.1. Given a set $\mathcal{H} \subseteq \bigcup_{i=1}^{n}\{i1, \ldots, i(d^n(i)-1)\}$, write $\mathcal{H}_i = \mathcal{H} \cap \{i1, \ldots, i(d^n(i)-1)\}$. By (12) we know that for any set $\mathcal{H}$ as above with $|\mathcal{H}| = n-1$, for all $1 \le i \le n-1$, for any $h \in \mathcal{H}$ and any root half-edge r with $v(r) \ne v(h)$,

$$\mathbf{P}\left\{(r_i, s_i) = (r, h) \mid \mathcal{S}^n = \mathcal{H}\right\} = \mathbf{P}\left\{(r_1, s_1) = (r, h) \mid \mathcal{S}^n = \mathcal{H}\right\}.$$

Moreover, by construction, $T(d^n)$ and G'_n are conditionally independent given $\mathcal{S}^n$, so for any $1 \le i \le n-1$,

$$\begin{aligned}
&\mathbf{E}\left\{m_{G'_n}(v(r_i)v(s_i)) \;\middle|\; \mathcal{S}^n = \mathcal{H}\right\} \\
&= \mathbf{E}\left\{m_{G'_n}(v(r_1)v(s_1)) \;\middle|\; \mathcal{S}^n = \mathcal{H}\right\} \\
&= \sum_{j=1}^{n} \sum_{\substack{k=1 \\ k \ne j}}^{n} \mathbf{P}\left\{v(r_1) = j, v(s_1) = k \mid \mathcal{S}^n = \mathcal{H}\right\} \cdot \mathbf{E}\left\{m_{G'_n}(jk) \;\middle|\; \mathcal{S}^n = \mathcal{H}\right\}.
\end{aligned}$$

Now, by Proposition 2.1,

$$\mathbf{P}\left\{v(r_1) = j, v(s_1) = k \mid \mathcal{S}^n = \mathcal{H}\right\} = \frac{|\mathcal{H}_k|}{(n-1)^2}.$$

Also,

$$\begin{aligned}&\mathbf{E}\left\{m_{G'_n}(jk) \mid \mathcal{S}^n = \mathcal{H}\right\}\\&= (m-n)\cdot\frac{(d^n(j)-1-|\mathcal{H}_j|)(d^n(k)-1-|\mathcal{H}_k|)}{(2m-2(n-1))(2m-2(n-1)-1)}.\end{aligned}$$

The term $m-n$ above accounts for the number of edges of G'_n; the fraction is the probability that a uniformly random pair of half-edges from $(\bigcup_{l=1}^n\{l1,\ldots,l(d^n(i)-1)\})\setminus\mathcal{H}$ are incident to vertices j and k. Combining these formulas, we then have that for all $1\le i\le n-1$,

$$\begin{aligned}&\mathbf{E}\left\{m_{G'_n}(v(r_i)v(s_i)) \mid \mathcal{S}^n = \mathcal{H}\right\}\\&\le \sum_{j=1}^n\sum_{\substack{k=1\\k\ne j}}^n \frac{|\mathcal{H}_k|}{(n-1)^2}\cdot(m-n)\cdot\frac{(d^n(j)-1-|\mathcal{H}_j|)(d^n(k)-1-|\mathcal{H}_k|)}{(2m-2(n-1))(2m-2(n-1)-1)}\\&= O(1)\cdot\sum_{k=1}^n\frac{|\mathcal{H}_k|}{(n-1)^2}\cdot\frac{(d^n(k)-1-|\mathcal{H}_k|)}{(2(m-n)+1)}\sum_{\substack{j=1\\j\ne k}}^n(d^n(j)-1-|\mathcal{H}_j|).\end{aligned}$$

Now, since $\sum_{j=1}^n|\mathcal{H}_j| = |\mathcal{H}| = n-1$, we have

$$\begin{aligned}\sum_{\substack{j=1\\j\ne k}}^n(d^n(j)-1-|\mathcal{H}_j|) &= 2m-n-|\mathcal{H}|-(d^n(k)-1-|\mathcal{H}_k|)\\&\le 2(m-n)-1,\end{aligned}$$

and it follows that

$$\mathbf{E}\left\{m_{G'_n}(v(r_i)v(s_i)) \mid \mathcal{S}^n = \mathcal{H}\right\} = O(1)\cdot\sum_{k=1}^n\frac{|\mathcal{H}_k|\,(d^n(k)-1-|\mathcal{H}_k|)}{2(n-1)^2}.$$

Taking expectation over $\mathcal{S}^n$ in this bound, it follows that

$$\mathbf{E}\left(m_{G'_n}(v(r_i)v(s_i))\right) \le \frac{1}{(n-1)^2}\mathbf{E}\left(\sum_{k=1}^n|\mathcal{H}_k|\,(d^n(k)-1-|\mathcal{H}_k|)\right).$$

We now show that $\mathbf{E}(\sum_{k=1}^n|\mathcal{H}_k|(d^n(k)-1-|\mathcal{H}_k|)) = o(n)$. Notice that $|\mathcal{H}_k|\le d^n(k)$ and that $d^n(k)-1-|\mathcal{H}_k| = d^n_-(k)$, unless k is incident to the unpaired root half-edge, in which case $d^n(k)-1-|\mathcal{H}_k| = d^n_-(k)-1\ge 0$. Letting r be the unpaired root half-edge, it then follows that

$$\sum_{k=1}^n|\mathcal{H}_k|\,(d^n(k)-1-|\mathcal{H}_k|)\le\sum_{k=1}^n(d^n(k))^2\,\mathbf{1}_{[d^n_-(k)>0]}.$$

Since $|\{k : d_-^n(k) > 0\}| \leq m - n = o(n)$, by the second assertion of Fact A.1 it follows that $\sum_{k=1}^n d^n(k)^2 \mathbf{1}_{[d_-^n(k)>0]} = o(n)$, which combined with the two preceding inequalities yields that

$$\mathbf{E}\left(m_{G'_n}(v(r_i)v(s_i))\right) = o\left(\frac{1}{n}\right).$$

Summing over i, it follows that

$$\mathbf{E}(N'_n) = \sum_{i=1}^{n} \mathbf{E}\left(m_{G'_n}(v(r_i)v(s_i))\right) = o(1).$$

At this point we know that $\mathbf{E}(L'_n + M'_n + N'_n) \to 0$. We now show how to deduce that $L_n + M_n + N_n \to 0$ in probability. We provide full details only in the case that $m - n \to \infty$, i.e., that the number of edges of $G_-(\mathrm{d}^n)$ tends to infinity, and briefly explain the argument in the simpler case that $m - n = O(1)$.

By Proposition 2.3, for any pair (g, t) where g is a graph with degree sequence d^n and t is a spanning tree of g, we have

$$\mathbf{P}\left\{(G(\mathrm{d}^n), T(\mathrm{d}^n)) = (g, t)\right\} \propto \frac{\sum_{\gamma \in e(g-t)} 2^{\mathbf{1}_{[\gamma \text{ is a loop}]}} \cdot m_{g-t}(\gamma)}{\prod_{i=1}^n 2^{m_{g-t}(ii)} \cdot \prod_{e \in e(g)} m_{g-t}(e)!}, \tag{34}$$

and for any edge γ of $g - t$,

$$\mathbf{P}\left\{\Gamma(\mathrm{d}^n) = \gamma \mid (G(\mathrm{d}^n), T(\mathrm{d}^n)) = (g, t)\right\} \propto 2^{\mathbf{1}_{[\gamma \text{ is a loop}]}} \cdot m_{g-t}(\gamma).$$

If the graph with edge set $e(g) - (e(t) \cup \gamma)$ is simple, then $g - t$ has at most one loop, and no edge with multiplicity more than two, so

$$\sup_{e \in e(g-t)} \mathbf{P}\left\{\Gamma(\mathrm{d}^n) = e \mid (G(\mathrm{d}^n), T(\mathrm{d}^n)) = (g, t)\right\} \leq \frac{2}{|e(g-t)|}.$$

In particular, writing

$$\mathcal{B}(g, t) = \mathcal{L}(g) \cup \mathcal{N}(g, t) \cup \bigcup_{((ui_i, vj_1), (ui_2, vj_2)) \in \mathcal{M}(g, t)} \{(ui_i, vj_1), (ui_2, vj_2)\},$$

and recalling that the half-edges comprising $\Gamma(\mathrm{d}^n)$ are $\Gamma^-(\mathrm{d}^n)$ and $\Gamma^+(\mathrm{d}^n)$, it follows that

$$\begin{aligned}
&\mathbf{P}\left\{ (\Gamma^-(\mathrm{d}^n), \Gamma^+(\mathrm{d}^n)) \in \mathcal{B}(G_-(\mathrm{d}^n), T(\mathrm{d}^n)) \;\middle|\; (G(\mathrm{d}^n), T(\mathrm{d}^n))\right\} \\
&\leq \frac{2|\mathcal{B}(G_-(\mathrm{d}^n), T(\mathrm{d}^n))|}{|e(G_-(\mathrm{d}^n))|} \mathbf{1}_{[G'_n \text{ simple}]} + \mathbf{1}_{[G'_n \text{ not simple}]} \\
&\leq \frac{4\big(L(G_-(\mathrm{d}^n)) + N(G_-(\mathrm{d}^n), T(\mathrm{d}^n)) + M(G_-(\mathrm{d}^n), T(\mathrm{d}^n))\big)}{|e(G_-(\mathrm{d}^n))|} \mathbf{1}_{[G'_n \text{ simple}]} \\
&\qquad\qquad + \mathbf{1}_{[G'_n \text{ not simple}]} \\
&= \frac{4(L_n + M_n + N_n)}{m - (n-1)} + \mathbf{1}_{[G'_n \text{ not simple}]};
\end{aligned}$$

the last inequality is not tight unless $M(G_-(\mathrm{d}^n), T(\mathrm{d}^n)) = 0$. Now,

$$L_n + M_n + N_n \leq 3\left(L'_n + M'_n + N'_n + \mathbf{1}_{[(\Gamma^-(\mathrm{d}^n), \Gamma^+(\mathrm{d}^n)) \in \mathcal{B}(G_-(\mathrm{d}^n), T(\mathrm{d}^n))]}\right); \tag{35}$$

this inequality is never tight but it suffices for our purposes. Using this bound, and taking expectations in the previous conditional probability bound, we obtain that

$$\begin{aligned}
&\mathbf{P}\left\{(\Gamma^-(\mathrm{d}^n), \Gamma^+(\mathrm{d}^n)) \in \mathcal{B}(G_-(\mathrm{d}^n), T(\mathrm{d}^n))\right\} \\
&\leq \frac{12}{m-(n-1)} \mathbf{E}\left(L'_n + M'_n + N'_n + 1\right) + \mathbf{P}\left\{G'_n \text{ not simple}\right\} \\
&= \frac{12}{m-(n-1)} \mathbf{E}\left(L'_n + M'_n + N'_n + 1\right) + \mathbf{P}\left\{L'_n + M'_n + N'_n > 0\right\}.
\end{aligned}$$

Since $\mathbf{E}(L'_n + M'_n + N'_n) \to 0$, if $m - (n-1) \to \infty$ it follows that

$$\mathbf{P}\left\{(\Gamma^-(\mathrm{d}^n), \Gamma^+(\mathrm{d}^n)) \in \mathcal{B}(G_-(\mathrm{d}^n), T(\mathrm{d}^n))\right\} \xrightarrow{\text{prob}} 0$$

in which case (35) and the fact that $L'_n + M'_n + N'_n \xrightarrow{\text{prob}} 0$ together imply that $L_n + M_n + N_n \xrightarrow{\text{prob}} 0$ as required.

Finally, if $m - n = O(1)$, the fact that $L_n + M_n + N_n \xrightarrow{\text{prob}} 0$ can be seen as follows. Let h be a uniformly random half-edge from $\bigcup_{1 \leq i \leq n}\{i1, \ldots, i(d^n(i) - 1)\}$. Then with high probability, $d^n(v(h)) = O(1)$. Since $|\mathcal{U}| = |\bigcup_{1 \leq i \leq n}\{i1, \ldots, i(d^n(i)-1)\} \setminus \mathcal{S}^n| = 2(m-n) = O(1)$, it follows that with high probability $d^n(v(h)) = O(1)$ for all edges of $\bigcup_{1 \leq i \leq n}\{i1, \ldots, i(d^n(i) - 1)\} \setminus \mathcal{H}$ and that $v(h) \neq v(h')$ for all distinct $h, h' \in \mathcal{U}$. This already implies that $L_n + M_n = 0$ with probability $1 - o(1)$. Finally, by considering Pitman's additive coalescent it is not hard to see that for any vertex v with $d^n(v) = O(1)$, the probability that v is the root of $T^n(v)$ or is adjacent to the root is $o(1)$, and that for any two vertices v, w with $d^n(v) = O(1)$ and $d^n(w) = O(1)$, the probability that v and w are adjacent is $o(1)$. (Verifying the assertions of the last sentence in detail is left to the reader.) This immediately implies that $N_n = 0$ with probability $1 - o(1)$ and so completes the proof. Q.E.D.

§ Acknowledgements

Both authors thank Serte Donderwinkel and an anonymous referee for a very careful reading of the paper and for several useful comments.

References

[ADH] Romain Abraham, Jean-François Delmas and Patrick Hoscheit, A note on the Gromov-Hausdorff-Prokhorov distance between (locally) compact metric measure spaces, *Electron. J. Probab.*, 18: no. 14, 21, 2013, doi: 10.1214/EJP.v18-2116, URL `https://arxiv.org/abs/1202.5464`.

[BLR] Nathanaël Berestycki, Benoît Laslier and Gourab Ray, Critical exponents on Fortuin-Kasteleyn weighted planar maps, *Comm. Math. Phys.*, 355(2):427–462, 2017. ISSN 0010-3616, doi: 10.1007/s00220-017-2933-7.

[B1] Olivier Bernardi, Bijective counting of tree-rooted maps and shuffles of parenthesis systems, *Electron. J. Combin.*, 14(1):Research Paper 9, 36, 2007, URL `http://www.combinatorics.org/Volume_14/Abstracts/v14i1r9.html`.

[B2] Olivier Bernardi, Tutte polynomial, subgraphs, orientations and sandpile model: new connections via embeddings, *Electron. J. Combin.*, 15(1):Research Paper 109, 53, 2008, URL `http://www.combinatorics.org/Volume_15/Abstracts/v15i1r109.html`.

[BMC] Mireille Bousquet-Mélou and Julien Courtiel, Spanning forests in regular planar maps, *J. Combin. Theory Ser. A*, 135:1–59, 2015, ISSN 0097-3165, doi: 10.1016/j.jcta.2015.04.002, URL `https://arxiv.org/abs/1306.4536`.

[BM] Nicolas Broutin and Jean-François Marckert, Asymptotics of trees with a prescribed degree sequence and applications, *Random Structures Algorithms*, 44(3):290–316, 2014, ISSN 1042-9832, doi: 10.1002/rsa.20463, URL `https://arxiv.org/abs/1110.5203`.

[GS] Jason Miller Ewain Gwynne and Scott Sheffield, The Tutte embedding of the mated-CRT map converges to Liouville quantum gravity, arXiv:1705.11161 [math.PR], May 2017, URL `https://arxiv.org/abs/1705.11161`.

[GKW] Catherine Greenhill, Matthew Kwan and David Wind, On the number of spanning trees in random regular graphs, *Electron. J. Combin.*, 21(1):Paper 1.45, 26, 2014, doi: 10.37236/3752, URL `https://www.combinatorics.org/ojs/index.php/eljc/article/view/v21i1p45`.

[GIKM] Catherine Greenhill, Mikhail Isaev, Matthew Kwan and Brendan D. McKay, The average number of spanning trees in sparse graphs with given degrees, *European J. Combin.*, 63:6–25, 2017, ISSN 0195-6698, doi: 10.1016/j.ejc.2017.02.003, URL `https://arxiv.org/abs/1606.01586`.

[GKMW] Ewain Gwynne, Adrien Kassel, Jason Miller and David B. Wilson, Active spanning trees with bending energy on planar maps and SLE-decorated Liouville quantum gravity for $\kappa > 8$, *Comm. Math. Phys.*, 358(3):1065–1115, 2018, ISSN 0010-3616, doi: 10.1007/s00220-018-3104-1, URL `https://arxiv.org/abs/1603.09722`.

[GHS] Ewain Gwynne, Nina Holden and Xin Sun, A mating-of-trees approach for graph distances in random planar maps, *Probab. Theory Related Fields*, 177(3-4):1043–1102, 2020, ISSN 0178-8051, doi: 10.1007/s00440-020-00969-8, URL `https://arxiv.org/abs/1711.00723`.

[HS] Nina Holden and Xin Sun, SLE as a mating of trees in Euclidean geometry, *Comm. Math. Phys.*, 364(1):171–201, 2018, ISSN 0010-3616, doi: 10.1007/s00220-018-3149-1, URL `https://arxiv.org/abs/1610.05272`.

[LSW] Yiting Li, Xin Sun and Samuel S. Watson, Schnyder woods, SLE(16), and Liouville quantum gravity, arXiv:1705.03573, May 2017, URL `https://arxiv.org/abs/1705.03573`.

[M] Brendan D. McKay, Spanning trees in random regular graphs, In: *Proceedings of the Third Caribbean Conference on Combinatorics and Computing* (*Bridgetown*, 1981), pages 139–143, Univ. West Indies, Cave Hill Campus, Barbados, 1981.

[MS] Jason Miller and Scott Sheffield, Liouville quantum gravity spheres as matings of finite-diameter trees, *Ann. Inst. Henri Poincaré Probab. Stat.*, 55(3):1712–1750, 2019, ISSN 0246-0203, doi: 10.1214/18-aihp932, URL `https://arxiv.org/abs/1506.03804`.

[M2] J. W. Moon, *Counting labelled trees*, volume 1969 of *From lectures delivered to the Twelfth Biennial Seminar of the Canadian Mathematical Congress* (*Vancouver*), Canadian Mathematical Congress, Montreal, Que., 1970.

[M3] R. C. Mullin, On the enumeration of tree-rooted maps, *Canadian J. Math.*, 19:174–183, 1967, ISSN 0008-414X, doi: 10.4153/CJM-1967-010-x, URL `https://doi.org/10.4153/CJM-1967-010-x`.

[P] Jim Pitman, Coalescent random forests, *J. Combin. Theory Ser. A*, 85(2):165–193, 1999, ISSN 0097-3165, doi: 10.1006/jcta.1998.2919, URL `https://www.stat.berkeley.edu/~pitman/457.pdf`.

[vdH] Remco van der Hofstad, *Random graphs and complex networks. Vol.* 1, Cambridge Series in Statistical and Probabilistic Mathematics, [43], Cambridge University Press, Cambridge, 2017, ISBN 978-1-107-17287-6, doi: 10.1017/9781316779422,
URL `https://www.win.tue.nl/~rhofstad/NotesRGCN.pdf`.

[WL] T. Walsh and A. B. Lehman, Counting rooted maps by genus. II, *J. Combin. Theory Ser. B*, 13:122–141, 1972, ISSN 0095-8956,
doi: 10.1016/0095-8956(72)90049-4,
URL `https://www.sciencedirect.com/science/article/pii/0095895672900494`.

Department of Mathematics and Statistics,
McGill University,
Montréal, Canada
E-mail address: louigi.addario@mcgill.ca
E-mail address: jordan.barrett@mail.mcgill.ca

Advanced Studies in Pure Mathematics 87, 2021
Stochastic Analysis, Random Fields and Integrable Probability — Fukuoka 2019
pp. 59–88

A survey on determinantal point processes

Guillaume Baverez, Alexander I. Bufetov and Yanqi Qiu

Abstract.

We present a first introduction to determinantal point processes, focussing on multiplicative functionals, Palm theory, and key applications such as random matrices, Young diagrams and random holomorphic functions.

§1. Definitions of determinantal point processes

Let E be a locally compact σ-compact metric complete seperable space, endowed with a positive σ-finite Radon measure μ. By Hocking and Young [20, Theorem 2-61], we may assume that the metric on E is such that any bounded set is relatively compact.

By definition, a (locally finite) configuration $X = \{x_i\}$ on E is a locally finite subset of E, possibly with multiplicities. A configuration is called simple if all points in it have multiplicity one. We denote by $\mathrm{Conf}(E)$ the set of all configurations on E. The mapping

$$X \mapsto \sum_i \delta_{x_i}$$

embeds $\mathrm{Conf}(E)$ into the space of Radon measures on E. Under the vague topology, the space $\mathrm{Conf}(E)$ is a Polish space, see, e.g., Daley and Vere-Jones [15, Theorem 9.1. IV]. For any Borel subset $B \subset E$, let $\#_B$ be the mapping that associates every configuration X on E to the cardinality of the set $X \cap B$. Then the Borel sigma-algebra on $\mathrm{Conf}(E)$ is the smallest sigma-algebra $\mathcal{F}_E$ that makes all the mappings $X \mapsto \#_B(X)$ measurable, with B ranges over all Borel subsets of E.

Received October 31, 2020.
Revised December 8, 2020.
2010 *Mathematics Subject Classification.* Primary 60G55; Secondary 37D40, 32A36.
Key words and phrases. Determinantal point processes, multiplicative functionals, Palm measures, completeness, minimality.

By definition, a Borel probability measure on $\mathrm{Conf}(E)$ is called a *point process* on E. For further background on point processes, see, e.g., Daley and Vere-Jones [14], Kallenberg [23]. A point process $\mathbb{P}$ on E is called *simple* if

$$\mathbb{P}(\{X \in \mathrm{Conf}(E) \mid X \text{ is simple}\}) = 1.$$

In what follows, we shall only deal with the simple point processes.

How can one define a Borel probability measure $\mathbb{P}$ (with associated expectation $\mathbb{E}$) on $\mathrm{Conf}(E)$? We give two ways of defining Borel probability measures on $\mathrm{Conf}(E)$ which are enough for our purpose of defining determinantal point processes on E.

i) Definition via additive functionals

For determining a Borel probability measure on $\mathrm{Conf}(E)$, one way would be to define the additive l-point correlation functions ρ_l via:

$$\mathbb{E}\left[\sum_{\substack{x_1,\ldots,x_l\in X\\ x_i\neq x_j}} \phi(x_1,\ldots,x_l)\right] = \int_{E^l} \phi(y_1,\ldots,y_l)\rho_l(y_1,\ldots,y_l)d\mu(y_1)\cdots d\mu(y_l)$$

for all bounded compactly supported functions $\phi : E^l \to \mathbb{C}$. However it is not obvious that these correlation functions exist and they define the measure uniquely. We refer to [24, 25, 26] for a systematic treatment of the correlation functions of general point processes.

Definition 1.1. We say that a point process $\mathbb{P}$ on E is determinantal if there exists a two-variable function $K : E \times E \to \mathbb{C}$ such that for all $l \geq 1$, we have

$$\rho_l(y_1,\ldots,y_l) = \det\Big[K(y_i,y_j)\Big]_{1\leqslant i,j\leqslant l}. \tag{1.1}$$

In this case, K determines the measure $\mathbb{P}$ uniquely and is called the *correlation kernel* or simply the *kernel*, and the measure $\mathbb{P}$ is called the determinantal point process (abbr. DPP) induced by K and will be denoted by $\mathbb{P}_K$.

Conversely, to what extent does a determinantal point process determines the kernel? This is essentially an open problem.

ii) Definition via multiplicative functionals

We refer to [38, Section 3] for the theory of Fredholm determinants. Given a trace class operator T on $L^2(E,\mu)$, we will denote its Fredholm determinant by

$$\det(1+T).$$

A bounded integral operator K on $L^2(E,\mu)$ is called *locally trace class* if for any bounded subset $B \subset E$, the compression $\mathbb{1}_B K \mathbb{1}_B$ is of trace class.

Let $g : E \to \mathbb{C}$ be bounded such that $g - 1$ has compact support. Then we define for each configuration $X \in \mathrm{Conf}(E)$, the multiplicative functional by

$$\Psi_g(X) := \prod_{x \in X} g(x).$$

Definition 1.2. A point process $\mathbb{P}$ on E is called determinantal if there exists a locally trace class operator K on $L^2(E,\mu)$ such that if $g : E \to \mathbb{C}$ is a bounded function and $g - 1$ has compact support, then

$$\mathbb{E}[\Psi_g] = \det\big(1 + (g-1)K\mathbb{1}_B\big), \tag{1.2}$$

where B is any bounded subset of E containing the support of $g - 1$.

The formulations (1.1) and (1.2) give two equivalent definitions of a DPP.

For which kernels do such processes exist? In all generality, this question remains open, but we have the following partial answer.

Theorem 1.3 (Macchi[31]). *Let K be a locally trace class operator on $L^2(E,\mu)$. If K is self-adjoint and a strict contraction, then there exists a (unique) DPP induced by K.*

Soshnikov [39] later extended the result to all non-negative constractions (see also the simultaneous work of Shirai-Takahashi [36]).

Theorem 1.4 (Soshnikov [39], Shirai-Takahashi [36]). *Let K be a locally trace class self-adjoint operator on $L^2(E,\mu)$. Then there exists a DPP induced by K if and only if K satisfies the operator order inequalities*

$$0 \le K \le 1.$$

In other words, if and only if K is a non-negative contraction.

Example 1. The *sine process* $\mathbb{P}_{\mathcal{S}}$ is the DPP induced by the orthogonal projection from $L^2(\mathbb{R})$ onto the Paley-Wiener space $\mathscr{PW}$, the space of functions in $L^2(\mathbb{R})$ with Fourier transform supported in $[-\pi,\pi]$. Its kernel has the following explicit expression:

$$\mathcal{S}(x,y) = \frac{\sin \pi(x-y)}{\pi(x-y)}. \tag{1.3}$$

Ghosh [17] proved that $\mathbb{P}_{\mathcal{S}}$-almost any configuration $X \in \mathrm{Conf}(\mathbb{R})$ is a uniqueness set for the Paley-Wiener space $\mathscr{PW}$: if $f \in \mathscr{PW}$ vanishes at every point $x \in X$, then $f = 0$ identically.

In [6] it is proved that $\mathbb{P}_{\mathcal{S}}$-almost every configuration *with one particle removed* is still a uniqueness set for the Paley-Wiener space $\mathscr{PW}$; while $\mathbb{P}_{\mathcal{S}}$-almost every configuration with two particles removed is a zero set for a non-zero Paley-Wiener function.

§2. Limit theorems

In this section, we are concerned with limit theorems for the couting function $\#_{[0,T]}$ and other additive functionals.

Theorem 2.1 (Costin-Lebowitz, Soshnikov). *For any DPP induced by a projection, we have*

$$\frac{\#_{[0,T]} - T}{\sqrt{\mathrm{Var}\#_{[0,T]}}} \overset{law}{\to} \mathcal{N}(0,1).$$

Furthermore, for any additive statistics of the form $S_f(X) = \sum_{x \in X} f(x)$ such that $\mathrm{Var}S_{f(\varepsilon\cdot)} \to \infty$ as $\varepsilon \to 0$, we have

$$\frac{S_{f(\varepsilon\cdot)} - \mathbb{E}[S_{f(\varepsilon\cdot)}]}{\sqrt{\mathrm{Var}S_{f(\varepsilon\cdot)}}} \overset{law}{\to} \mathcal{N}(0,1). \tag{2.1}$$

For the sine process, Johannsson and Soshnikov extended this result to the case where $\mathrm{Var}S_{f(\varepsilon\cdot)}$ has a finite limit as $\varepsilon \to 0$. This remarkable result means that cancellations are so precise that no renormalisation is needed in (2.1).

With this central limit theorem in hand, one may wonder about functional limit theorems: does there exist an analogue of Donsker's invariance principle? We do not expect to fall in the universality class of Brownian motion due to the strong dependence between the increments. This is reflected in the fact that the variance grows much slower than linearly (logarithmically for the counting function).

A functional limit theorem was established in the case of the sine process, for which it is known that

$$\mathrm{Var}(\#_{[0,T]}) = \frac{1}{\pi^2}\log T + o(\log T).$$

For each $N \in \mathbb{N}$, consider the random process on $[0,1]$ defined by

$$\xi_t^N := \frac{\#_{[0,tN]} - tN}{\mathrm{Var}(\#_{[0,tN]})}.$$

We are interested in the integral process $t \mapsto \int_0^t \xi_s^N ds$. First, we have

$$\frac{1}{t}\int_0^t \xi_s^N ds \overset{law}{\to} \mathcal{N}\left(0, \frac{1}{2}\right)$$

as $N \to \infty$, so that the "average" behaviour of $\int_0^t \xi_s^N ds$ is a line with (random) slope. To study the deviations from this behaviour, we fix $\tau \in (0,1)$ and introduce the random variable $\eta^N := \frac{1}{\tau}\int_0^\tau \xi_s^N ds$, which also converges in distribution to a $\mathcal{N}(0, \frac{1}{2})$. We then define z_t^N by

$$\int_0^t \xi_s^N ds = \eta^N t + \frac{z_t^N}{\sqrt{\pi^{-2}\log N}}.$$

The process z_t^N encodes the fluctuations we are interested in.

Theorem 2.2. [8, Theorem 1.1] *The pair (η^N, z^N) converges in distribution to (η, z), where $\eta \sim \mathcal{N}(0, \frac{1}{2})$ and z is an independent centred Gaussian process on $[0,1]$ with covariance kernel*

$$\mathbb{E}[z_t z_s] = \frac{1}{2\pi}\left(\int_0^t\int_0^s - \frac{s}{\tau}\int_0^t\int_0^\tau - \frac{t}{\tau}\int_0^s\int_0^\tau + \frac{st}{\tau^2}\right) H(u,v)dudv,$$

where

$$H(u,v) = \mathcal{G}(u,v) - \mathcal{G}(u,0) - \mathcal{G}(0,v)$$

with

$$\mathcal{G}(t,s) = \frac{1}{\pi}\log\frac{1}{|t-s|}, \qquad t,s \in \mathbb{R}. \tag{2.2}$$

One interesting feature of Theorem 2.2 is the appearance of the correlation function of the trace of the Gaussian free field (GFF) in (2.2). Recall that the GFF in the upper-half plane $\mathbb{H}$ is the centred Gaussian process based on the homogeneous Sobolev space $\dot{H}^1(\mathbb{H})$. Its covariance kernel is the resolvent of the Laplacian with Neumann boundary conditions on $\partial\mathbb{H} = \mathbb{R}$:

$$\mathcal{G}(z,z') = \frac{1}{2\pi}\log\frac{1}{|z-z'|} + \frac{1}{2\pi}\log\frac{1}{|z-\bar{z}'|}, \qquad z, z' \in \mathbb{H}.$$

The GFF has a trace on $\mathbb{R}$, with covariance kernel given by (2.2), the limit of the above equation as $z, z' \to \mathbb{R}$. The trace of the GFF is also characterised as the Gaussian process based on the homogeneous space $\dot{H}^{1/2}(\mathbb{R})$, the trace space of $\dot{H}^1(\mathbb{H})$ on $\mathbb{R}$. In particular, (2.2) is the resolvent of the Dirichlet-to-Neumann operator.

The GFF is a natural Gaussian process in two dimensions and describes the scaling limit of many models of statistical mechanics and random matrix theory. For instance, the eigenvalue counting function of large random matrices from the Gaussian unitary ensemble (GUE) is known to be governed by the GFF. Moreover, the asymptotic behaviour of the spectrum of GUE matrices is described by the sine process. In this respect, it is natural to find a connection between the GFF and the counting function of the sine process in Theorem 2.2.

§3. Ghosh-Peres rigidity

Consider the discrete sine process, i.e. the DPP on the integer lattice $\mathbb{Z}$ (equipped with the counting measure) with kernel

$$\mathcal{S}_\alpha(x,y) = \frac{\sin \alpha(x-y)}{\pi(x-y)}, \qquad x, y \in \mathbb{Z}, \alpha \in (0,\pi).$$

Note that in this case, the space $\mathrm{Conf}(\mathbb{Z})$ can be identified in a natural way with the set $\{0,1\}^{\mathbb{Z}}$ and thus the probability measure $\mathbb{P}_{\mathcal{S}_\alpha}$ can be identified with a Borel probability measure on $\{0,1\}^{\mathbb{Z}}$.

The probability measure $\mathbb{P}_{\mathcal{S}_\alpha}$ satisfies the following property: there exists a Borel function $\zeta : \{0,1\}^{\mathbb{Z}\setminus\{0\}} \to \{0,1\}$ such that

$$\omega_0 = \zeta(\omega_n, n \neq 0), \qquad \mathbb{P}_{\mathcal{S}_\alpha}\text{-a.s.}$$

This result follows from a result of Kolmogorov (1941) on stationary processes which we recall in the following. Consider a sequence of vectors $\xi = (\xi_n)_{n\in\mathbb{Z}}$ living in some Hilbert space with inner-product $\langle \cdot, \cdot \rangle$, such that

$$\langle \xi_k, \xi_l \rangle = \int_0^{2\pi} e^{i\theta(k-l)} \rho_\xi(d\theta), \qquad k, l \in \mathbb{Z},$$

where ρ_ξ is called the spectral measure of ξ. Write the Lebesgue decomposition of the measure ρ_ξ as

$$\rho_\xi(d\theta) = \rho_\xi^{ac}(\theta)d\theta + \rho_\xi^{s}(d\theta).$$

Kolmogorov (see also [32, Lemma 5.2.3] and [30, Remark 5.17]) showed that

$$\xi_0 \in \overline{\mathrm{Span}((\xi_n)_{n\neq 0})} \Leftrightarrow \int_0^{2\pi} \frac{d\theta}{\rho_\xi^{ac}(\theta)} = \infty.$$

To do this, he considered additive statistics of the form

$$S_n = \sum_{k\in\mathbb{Z}} \alpha_k^{(n)} \xi_k$$

such that $\alpha_0^{(n)} = 1$ and $\|S_n\| \to 0$.

Definition 3.1. A statioinary sequence $\xi = (\xi_n)_{n\in\mathbb{Z}}$ in a Hilbert space is said to be minimal if

$$\xi_0 \in \overline{\mathrm{Span}((\xi_n)_{n\neq 0})}.$$

Applying Kolmogorov's criterion, we obtain

Proposition 3.2 ([7]). *Let $X = (X_n)_{n\in\mathbb{Z}}$ be a stationary stochastic process. If*

$$\sup_{N\geq 1}\left(N\sum_{|n|\geq N}\left|\mathrm{Cov}(X_0, X_n)\right|\right) < \infty \quad \textit{and} \quad \sum_{n\in\mathbb{Z}}\mathrm{Cov}(X_0, X_n) = 0.$$

Then X is minimal.

Definition 3.3. A stationary point process $\mathbb{P}$ on $\mathbb{R}$ is called **linearly rigid**, if for any $\lambda > 0$, the stationary stochastic process $(\#_{[n\lambda, n\lambda+\lambda)})_{n\in\mathbb{Z}}$ defined on $(\mathrm{Conf}(\mathbb{R}), \mathbb{P})$ is minimal.

Using Proposition 3.2, we show the following

Theorem 3.4 ([7]). *Consider the stationary DPP $\mathbb{P}_{K_B}$ on $\mathbb{R}$ induced by the kernel $K_B(x,y) = \widehat{\mathbb{1}_B}(x-y)$ and assume that*

$$\sup_{R>0}\left(R\int_{|\xi|\,\geqslant\, R}|\widehat{\mathbb{1}_B}(\xi)|^2 d\xi\right) < \infty.$$

Then $\mathbb{P}_{K_B}$ is linearly rigid.

The linear rigidity for stationary point processes on $\mathbb{R}$ is a stronger version of the Ghosh-Peres rigidity for point processes on a general space.

Definition 3.5 (Ghosh [17], Ghosh-Peres [18]). A point process $\mathbb{P}$ on E is called **rigid**, if for any bounded Borel set $B \subset E$, there exists a measurable function $F_B : \mathrm{Conf}(E) \setminus \mathbb{R}$ such that

$$\#(X \cap B) = F_B(X \setminus B), \quad \text{for } \mathbb{P}\text{-almost every } X \in \mathrm{Conf}(E).$$

We shall discuss the Ghosh-Peres rigidity for various determinantal point processes in the following sections.

§4. Palm theory and quasi-symmetry

4.1. Palm theory

Let $\mathbb{P}$ be a probability measure on $\mathrm{Conf}(E)$ and $\mathbb{P}^p$ be the conditional measure (called the reduced Palm measure of $\mathbb{P}$ conditioned at the point p) on the event that there is a particle at site p. This is well-defined in the discrete and it is possible to make sense of this conditional measure in the continuum through limit transitions. Another way of definining $\mathbb{P}^p$ is given through the disintegration formula. More precisely, assume that $\mathbb{P}$ is a simple point process on E such that its first intensity function $\rho_{1,\mathbb{P}}$ exists. Then the family of the reduced Palm measures $\mathbb{P}^p$ (defined for $\rho_{1,\mathbb{P}}(x)d\mu$-almost every p) can be obtained by: for any Borel function $F : E \times \mathrm{Conf}(E) \to \mathbb{R}^+$, we have

$$\int_{\mathrm{Conf}(E)} \left[\sum_{x \in X} F(x, X \setminus \{x\}) \right] \mathbb{P}(\mathrm{d}X)$$
$$= \int_E \rho_{1,\mathbb{P}}(x)\mu(\mathrm{d}x) \int_{\mathrm{Conf}(E)} F(x, X)\mathbb{P}^x(\mathrm{d}X).$$

More generally, assume that the n-th correlation functions $\rho_{n,\mathbb{P}}$ exists. Then the family $\mathbb{P}^{p_1,\dots,p_n}$ of the reduced Palm measures $\mathbb{P}^{p_1,\dots,p_n}$ can be defined through the following disintegration formula: for any Borel function $F : E^n \times \mathrm{Conf}(E) \to \mathbb{R}^+$, we have

$$\int_{\mathrm{Conf}(E)} \left[\sum_{x \in X^n}^{\#} F(x, X \setminus \{x_1, \dots, x_n\}) \right] \mathbb{P}(\mathrm{d}X)$$
$$= \int_{E^n} \rho_{n,\mathbb{P}}(x_1, \dots, x_n)\mu^{\otimes n}(\mathrm{d}x_1 \cdots \mathrm{d}x_n) \int_{\mathrm{Conf}(E)} F(x, X)\mathbb{P}^{x_1,\dots,x_n}(\mathrm{d}X),$$

where $\sum^{\#}$ is the summation over all *ordered n-tuples* $(x_1, \dots, x_n)$ with distinct coordinates $x_1, \dots, x_n \in X$.

By [37, Lemma 6.4], the correlation functions $\rho_n^{x_1,\dots,x_m}$ of the Palm measures $\mathbb{P}^{x_1,\dots,x_m}$ and the correlation functions ρ_n of the original measure $\mathbb{P}$ satisfy the following relation

$$\rho_m(x_1, \dots, x_m)\rho_n^{x_1,\dots,x_m}(y_1, \dots, y_n) = \rho_{m+n}(x_1, \dots, x_m, y_1, \dots, y_n).$$

Theorem 4.1 (Shirai-Takahashi [37]). *Let $\mathbb{P}_K$ be a DPP induced by a kernel K. Then we have*

$$\mathbb{P}_K^p = \mathbb{P}_{K^p},$$

where

$$K^p(x, y) = K(x, y) - \frac{K(x,p)K(p,y)}{K(p,p)}.$$

In particular, if K is the projection onto L, then K^p is the projection onto the subspace

$$L(p) = \{\varphi \in L, \, \varphi(p) = 0\}. \tag{4.1}$$

4.2. Quasi-symmetry

Equivalence of Palm measures of a point process of the same orders is closely related to the quasi-symmetry of this point process. Note first that any Borel automorphism $T : E \to E$ induces a natural map, denoted again by T, on $\mathrm{Conf}(E)$ by sending each configuration $X = \{x_i\} \in \mathrm{Conf}(E)$ to the configuration $T(X) := \{T(x_i)\}$.

Proposition 4.2 ([5]). *Let $\mathbb{P}$ be a point process on E. Let $T : E \to E$ be a Borel automorphism admitting a bounded subset $B \subset E$ such that $T(x) = x$ for all $x \in E \setminus B$. Assume that for any $l \in \mathbb{N}$,*

- *the correlation measures ρ_l and $\rho_l \circ T$ are equivalent;*
- *for any two collections $\{q_1, \ldots, q_l\}$ and $\{T(q_1), \ldots, T(q_l)\}$ of distinct points of E, the Palm measures $\mathbb{P}^{q_1,\ldots,q_l}$ and $\mathbb{P}^{T(q_1),\ldots,T(q_l)}$ are equivalent.*

Then the measures $\mathbb{P}$ and $\mathbb{P} \circ T$ on $\mathrm{Conf}(E)$ are equivalent, and for $\mathbb{P}$-almost every configuration $X \in \mathrm{Conf}(E)$ such that $X \cap B = \{q_1, \ldots, q_l\}$ we have

$$\frac{d\mathbb{P} \circ T}{d\mathbb{P}}(X) = \frac{d\mathbb{P}^{T(q_1),\ldots,T(q_l)}}{d\mathbb{P}^{q_1,\ldots,q_l}}(X \setminus \{q_1, \ldots, q_l\}) \times \frac{d\rho_l \circ T}{d\rho_l}(q_1, \ldots, q_l).$$

4.2.1. *Determinantal point processes with integrable kernels.* In this subsection, we focus on the case when E is the real line $\mathbb{R}$ or a suitable subset of real line.

For concreteness, consider the discrete sine processes $\mathbb{P}_{\mathcal{S}_\alpha}$ induced by the kernel

$$\mathcal{S}_\alpha(m, n) = \frac{\sin(\pi\alpha(m-n))}{\pi(m-n)}, \qquad m, n \in \mathbb{Z}.$$

Then for all bounded $B \subset \mathbb{Z}$, $\#_B$ is determined by the values of the process outside of B. The action of the infinite symmetric group S_∞ does not preserve $\mathbb{P}_{\mathcal{S}}$, but we have for all transposition σ_{pq} exchanging the points p and q, we have:

$$\frac{d\mathbb{P}_{\mathcal{S}_\alpha} \circ \sigma_{pq}}{d\mathbb{P}_{\mathcal{S}_\alpha}} = \prod_{x \in X} \left(\frac{x-q}{x-p}\right)^2, \tag{4.2}$$

where this product is to be understood in terms of principal value. Notice that this Radon-Nikodym derivative does not depend on α.

A similar formula holds for all kernels of integrable type of the form

$$\Pi(x,y) = \frac{A(x)B(y) - A(y)B(x)}{x-y}.$$

We now pass to the precise statements in the continuous setting (the discrete setting is similar). Let μ be a σ-finite Borel measure on $\mathbb{R}$. Let $L \subset L^2(\mathbb{R},\mu)$ be a closed subspace such that the corresponding operator Π of the orthogonal projection onto L is locally of trace class and admits a kernel, denoted again by $\Pi(\cdot,\cdot)$. We shall make the following assumptions.

Assumption 1.

- There exists a set $U \subset \mathbb{R}$ with $\mu(\mathbb{R} \setminus U) = 0$ such that for any $q \in U$, the function $v_q(x) = \Pi(x,q)$ lies in $L^2(\mathbb{R},\mu)$ and for any $f \in L^2(\mathbb{R},\mu)$,

$$(\Pi f)(q) = \langle f, v_q \rangle_{L^2(\mathbb{R},\mu)}.$$

- The diagonal values $\Pi(q,q)$ are defined for all $q \in U$ and

$$\Pi(q,q) = \langle v_q, v_q \rangle_{L^2(\mathbb{R},\mu)},$$

  and, for any bunded Borel subset $B \subset \mathbb{R}$, we have

$$\operatorname{tr}(\mathbb{1}_B \Pi \mathbb{1}_B) = \int_B \Pi(q,q)\,d\mu(q).$$

- For any $q \in U$ and any $\varphi \in L$ satisfying $\varphi(q) = 0$, we have

$$\frac{\varphi(x)}{x-q} \in L^2(\mathbb{R},\mu)$$

Assumption 2. If $p \in U$ and $\varphi \in L$ are such that $\varphi(q) = 0$, then there exists $\varphi \in L$ such that

$$\varphi = (x-p)\psi.$$

Assumption 3. The kernel Π satisfies

$$\int_{\mathbb{R}} \frac{\Pi(x,x)}{1+x^2}\,d\mu(x) < \infty.$$

Theorem 4.3 ([5]). *Let μ be a continuous measure on $\mathbb{R}$, that is, μ has no atoms. Let Π be a kernel inducing a locally trace-class operator of orthogonal projection onto a closed subspace $L \subset L^2(\mathbb{R}, \mu)$ and satisfying Assumptions* 1, 2, 3. *Then for any $l \in \mathbb{N}$ and two l-tuples of distinct points*

$$(p_1, \ldots, p_l), (q_1, \ldots, q_l) \in U^l,$$

the Palm measures $\mathbb{P}_{\Pi}^{p_1,\ldots,p_l}$ and $\mathbb{P}_{\Pi}^{q_1,\ldots,q_l}$ are equivalent and we have

$$\frac{d\mathbb{P}_{\Pi}^{p_1,\ldots,p_l}}{d\mathbb{P}_{\Pi}^{q_1,\ldots,q_l}}(X) = \overline{\Psi}(p_1, \ldots, p_l; q_1, \ldots, q_l)(X),$$

where $\overline{\Psi}(p_1, \ldots, p_l; q_1, \ldots, q_l)(X)$ is the regularized multiplicative functional given by

(4.3)
$$\overline{\Psi}(p_1, \ldots, p_l; q_1, \ldots, q_l)(X) := \lim_{R\to\infty, \varepsilon\to 0} \overline{\Psi}_{R,\varepsilon}(p_1, \ldots, p_l; q_1, \ldots, q_l)(X),$$

with the convergence holds in $L^1(\mathrm{Conf}(\mathbb{R}), \mathbb{P}_{\Pi}^{q_1,\ldots,q_l})$ and

$$\overline{\Psi}_{R,\varepsilon}(p_1, \ldots, p_l; q_1, \ldots, q_l)(X) := C(R, \varepsilon) \prod_{\substack{x\in X, |x|\le R, \\ \min|x-q_i|\ge\varepsilon}} \prod_{i=1}^{l} \left(\frac{x-p_i}{x-q_i}\right)^2.$$

Here $C(R, \varepsilon)$ is the normalization constant such that

$$\int_{\mathrm{Conf}(R)} \overline{\Psi}_{R,\varepsilon}(p_1, \ldots, p_l; q_1, \ldots, q_l) d\mathbb{P}_{\Pi}^{q_1,\ldots,q_l} = 1.$$

It is difficult in general to say that two DPPs are equivalent as measures. However the case is easier when the Radon-Nikodym derivative is a multiplicative functional. If Π is the projection onto L and $g : E \to \mathbb{R}_+$ is a bounded function such that $g - 1$ has a compact support, then

$$\frac{\Psi_g \mathbb{P}_\Pi}{\int \Psi_g d\mathbb{P}_\Pi} = \mathbb{P}_{\Pi^g}, \tag{4.4}$$

where Π^g is the projection onto $\sqrt{g}L$. See [2] for the details.

The key ingredient in the proof of Theorem 4.3 is the *weak division property* of the subspace $L \subset L^2(\mathbb{R}, \mu)$ which we informally explain here. Suppose Π is of the form

$$\Pi(x, y) = \frac{A(x)B(y) - A(y)B(x)}{x - y}. \tag{4.5}$$

Let $\varphi \in L$. Then for any site p, we have

$$\varphi(p) = 0 \Rightarrow \frac{\varphi(x)}{x-p} \in L. \tag{4.6}$$

In fact, [12] proved that (4.6) implies (4.5) and moreover that L is a space of holomorphic functions. Recall the definition of the subspace $L(p)$ defined in (4.1). The relation (4.6) implies in particular

$$L(p) = \left(\frac{x-p}{x-q}\right) L(q).$$

More generally, we have

$$L(p_1, \dots, p_l) = \prod_{i=1}^{l} \left(\frac{x-p_i}{x-q_i}\right) \cdot L(q_1, \dots, q_l). \tag{4.7}$$

Thus, intuitively, Theorem 4.3 follows from the relations (4.4) and (4.7). Of course, in this case, the definition of the usual multiplicative functional does not work and it requires substantial efforts to prove that the regularized multiplicative functional can be indeed defined as in (4.3).

This idea of obtaining the relation between Palm measures from the relation of the corresponding subspaces is also adopted in the study of quasi-invariance results for determinantal point processes corresponding to Hilbert spaces of holomorphic functions on the plane and on the disc, see Section 6 below.

4.3. Conditional measures

Let K be a bounded self-adjoint locally trace class positive contractive operator on $L^2(E, \mu)$. Assume that $B \subset E$ is a bounded subset such that the operator $1 - \chi_B K$ is invertible. Then

$$\mathbb{P}_K(X \cap B = \emptyset) > 0.$$

Let $\mathbb{P}_K^B$ be the conditional measure on the event that there is no particle in B, i.e.

$$\mathbb{P}_K^B := \mathbb{P}_K(\cdot | X \cap B = \emptyset).$$

Since

$$\mathbb{1}_{\{X \cap B = \emptyset\}} = \Psi_{\mathbb{1}_{E \setminus B}}(X),$$

we have

$$\mathbb{P}_K^B = \frac{\mathbb{1}_{\{X \cap B = \emptyset\}} \mathbb{P}_K}{\mathbb{P}_K(X \cap B = \emptyset)} = \frac{\Psi_{\mathbb{1}_{E \setminus B}} \mathbb{P}_K}{\int \Psi_{\mathbb{1}_{E \setminus B}} d\mathbb{P}_K}.$$

Using [2] or [4, Proposition 2.1] or [5, Proposition 2.3], the measure $\mathbb{P}_K^B$ is again determinantal and governed by kernel

$$K^{E\setminus B} = \mathbb{1}_{E\setminus B} K (1 - \mathbb{1}_B K)^{-1} \mathbb{1}_{E\setminus B}.$$

Although it is not possible to write an explicit formula for this kernel $K^{E\setminus B}(x, y)$, we have a very useful interpretation of $K^{E\setminus B}$ as an operator: when K is an orthogonal projection onto a subspace L, then $K^{E\setminus B}$ is the orthogonal projection onto the closure of the subspace

$$\mathbb{1}_{E\setminus B} L := \{\mathbb{1}_{E\setminus B}\varphi \mid \varphi \in L\} \subset L^2(E, \mu).$$

For $B \subset E$ bounded, the conditional measure

$$\mathbb{P}_K^{[Y;B]} := \mathbb{P}_K(\cdot | X_{|B} = Y \cap B)$$

conditioned on the event that $X_{|B} = Y \cap B$ is also determinantal for $\mathbb{P}_K$-almost every configuration Y. To see this, first iterate the Palm construction on the number of particles in B, then condition on the event that there is no more particles in B.

The previous result holds even for the conditioning on an unbounded subset.

Theorem 4.4 ([11]). *Let K be a bounded self-adjoint locally trace class positive contractive operator on $L^2(E, \mu)$. Then for any Borel subset $W \subset E$ (not necessarily bounded), there exists a version of conditional measures such that $\mathbb{P}_K^{[X;W]}$ is determinantal for $\mathbb{P}_K$-almost every X.*

Remark 4.5. *The determinantal property is not closed under weak limits. The link between the kernel and the measure seems delicate, e.g., do kernels converge if the measures converge?*

In fact, in [11], a special family of conditional kernels $K^{[X;B]}$ are explicitly constructed (via limiting procedure and martingale theory) such that

$$\mathbb{P}_K^{[X;W]} = \mathbb{P}_{K^{[X;W]}} \quad \text{for } \mathbb{P}_K\text{-almost every } X.$$

In particular, if K is an orthogonal projection onto a subspace $L \subset L^2(E, \mu)$, the conditional kernels $K^{[X;W]}$ are particularly simple to describe. More precisely, if $W = B \subset E$ is bounded, then we may set $K^{[X;B]}$ to be the orthogonal projection onto the space

$$\mathbb{1}_{E\setminus B} L(X \cap B) = \{\mathbb{1}_{E\setminus B} h : h \in L(X \cap B)\}$$

where

$$L(X \cap B) = \{h \in L : h_{|X\cap B} = 0\},$$

where the evaluation of a function h at a point is understood as

$$h(x) = \langle h, K(\cdot, x)\rangle.$$

More generally, for a Borel subset $W \subset E$, let $B_1 \subset \cdots \subset B_n \subset \cdots \subset W$ be an increasing exhausting sequence of bounded Borel subsets of W. Then for $\mathbb{P}_K$-almost every $X \in \mathrm{Conf}(E)$, we can take the conditional kernel $K^{[X,W]}$ to be

$$K^{[X,W]} = \lim_{n\to\infty} \mathbb{1}_{E\setminus W} K^{[X,B_n]} \mathbb{1}_{E\setminus W},$$

where the convergence takes place in the space of locally trace class operators on $L^2(E,\mu)$.

If K is an orthogonal projection onto a subspace $L \subset L^2(E,\mu)$, Lyons and Peres conjectured that

$$\overline{\mathrm{Span}}^{L^2(E,\mu)}\{K(\cdot,x) : x \in X\} = L \quad \text{for } \mathbb{P}_K\text{-almost every } X. \tag{4.8}$$

An equivalent statement is that $\mathbb{P}_K$-almost every X satisfies the uniqueness property: for all $f \in L$,

$$f_{|X} \equiv 0 \Rightarrow f \equiv 0. \tag{4.9}$$

Relation (4.8) was proved by Lyons in the discrete case [28], and by Ghosh in the rigid case [17]. The general case was treated in [11].

Let us outline the proof of the relation (4.8) in the general case. Suppose by contradiction that there exists a subset $\Omega_0 \subset \mathrm{Conf}(E)$ with $\mathbb{P}_K(\Omega_0) > 0$ such that for any $X \in \Omega_0$, the relation (4.8) is violated. Then we can find $X \in \Omega_0$ such that all the following are satisfied:

(1) There exists a non-zero element f in

$$L(X) = \{h \in L : h_{|X} = 0\}.$$

(2) We can find a bounded subset $B \subset E \setminus X$ such that $\mathbb{1}_B f$ is a non-zero element of $L^2(B,\mu)$.

(3) Using our particularly chosen conditional kernels $K^{[X,E\setminus B]}$, we have

$$\mathbb{P}_K(\cdot|X, E\setminus B) = \mathbb{P}_{K^{[X,E\setminus B]}}.$$

(ii) By the explicit description of the kernel $K^{[X,E\setminus B]}$, we can show that the assumption $f \in L(X)$ implies that the function $\mathbb{1}_B f$ is a fixed point of our specific operator $K^{[X,E\setminus B]}$, that is,

$$K^{[X,E\setminus B]}(\mathbb{1}_B f) = \mathbb{1}_B f.$$

(iii) Since $\mathbb{1}_B f$ is a non-zero element in $L^2(B,\mu)$, we have

$$\mathbb{P}_K(\#_B = 0 | X, E\setminus B) = \mathbb{P}_{K^{[X,\,E\setminus B]}}(\#_B = 0) = \det(1 - K^{[X,\,E\setminus B]}) = 0.$$

(iv) Since X has no particles in B, the Fubini theorem implies that

$$\mathbb{P}_K(\#_B = 0 | X, E\setminus B) > 0.$$

This contradiction settles the Lyons-Peres completeness conjecture.

§5. Young diagrams

5.1. Preliminaries

A *partition* λ of a non-negative integer n is a sequence $\lambda_1 \geqslant \lambda_2 \cdots$ of non-negative integers such that $\sum_i \lambda_i = n$ (note that the sum has only finitely many non-zero terms). A *Young diagram* with shape λ is an array such that the ith row has λ_i cells (see Figure 1). The collection of Young diagrams with n cells is denoted $\mathbb{Y}_n$, and it is indexed by the partitions of n. It is known that $\#\mathbb{Y}_n \sim \exp(\frac{2\pi}{\sqrt{6}}\sqrt{n})$ as $n \to \infty$. One can obtain a Young diagram with $n+1$ cells from a Young diagram with n cells by adding a cell in a way that respects the definition: this yields the Young graph of Figure 2, showing all the ways to obtain a given diagram the empty one.

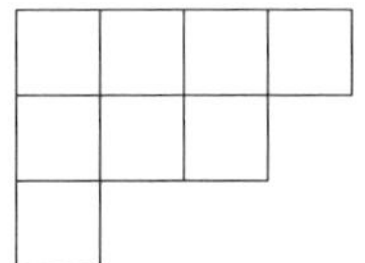

Fig. 1. The Young diagram with 8 cells of shape $(4,3,1)$.

The *Plancherel measure* $\mathbb{P}\ell^{(n)}$ on $\mathbb{Y}_n$ is the measure defined by

$$\mathbb{P}\ell^{(n)}(\lambda) = \frac{\dim^2 \lambda}{n!}$$

for all partitions λ of n, where $\dim \lambda$ is the number of paths leading from $\emptyset$ to λ in the Young graph (equivalently, the number of Young tableaux of shape λ), see Figure 2. This number has a representation theoretical interpretation: it is the dimension of the irreducible representation of the symmetric group indexed by λ. Standard results from representation theory ensure that $\sum_\lambda \dim^2 \lambda = \#\mathfrak{S}_n = n!$, i.e. $\mathbb{P}\ell^{(n)}$ is a probablity measure.

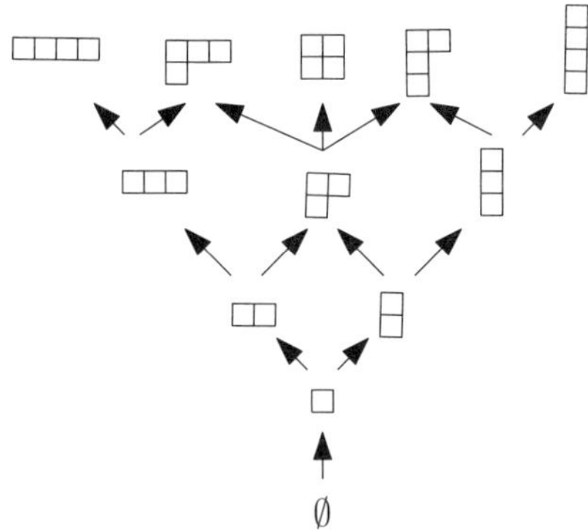

Fig. 2. The Young graph up to level 4.

One can understand a Young diagram as a piecewise linear function on $\mathbb{R}$ by rotating it as in Figure 3: this representation is the so-called "Russian convention". When rescaled properly, samples from the Plancherel measure converge to a deterministic limit as $n \to \infty$. This is the so-called limit shape theorem of Vershik-Kerov [40] and Logan-Shepp [27], and the limit shape involves the anti-derivative of arcsin. The limit shape theorem is understood as a law of large number and can be seen as an analogue of Wigner's semi-circle law.

Vershik and Kerov formulated the following conjecture, proved in [3].

Theorem 5.1. *There exists $H > 0$ such that for any $\varepsilon > 0$,*

$$\mathbb{P}\ell^{(n)}\left(\left\{\lambda, \left|-\frac{\log \mathbb{P}\ell^{(n)}(\lambda)}{\sqrt{n}} - H\right| < \varepsilon\right\}\right) \underset{n\to\infty}{\to} 1.$$

This result is reminiscent to a large deviation principle, and H is interpreted as an entropy. This theorem means that $\mathbb{P}\ell^{(n)}$ concentrates on small sets on which it is equidistributed. Otherwise stated, there exists $\widetilde{\mathbb{Y}}_n \subset \mathbb{Y}_n$ such that $\#\widetilde{\mathbb{Y}}_n = e^{\sqrt{n}(H+o(1))}$ as $n \to \infty$ and

$$\forall \lambda \in \widetilde{\mathbb{Y}}_n, \, \mathbb{P}\ell^{(n)}(\lambda) = e^{-\sqrt{n}(H+o(1))}.$$

5.2. Poissonisation of the Plancherel measure

In order to connect Plancherel measures with DPPs, we would like to encode Young diagrams as subsets $X \subset \mathbb{Z} + \frac{1}{2}$. This will allow us to understand Young diagrams as elements of $\mathrm{Conf}(\mathbb{R})$. We use the Russian representation and we put a particle (resp. a hole) when the path encoding λ goes down (resp. up), as in Figure 3. Conversely, any configuration of holes and particles where the number of holes in $\mathbb{Z}_{\geqslant 0}+\frac{1}{2}$

and particles in $\mathbb{Z}_{\leqslant 0} - \frac{1}{2}$ are equal encodes a Young diagram. In other words, the natural action of the infinite symmetric group generates all diagrams from the empty one.

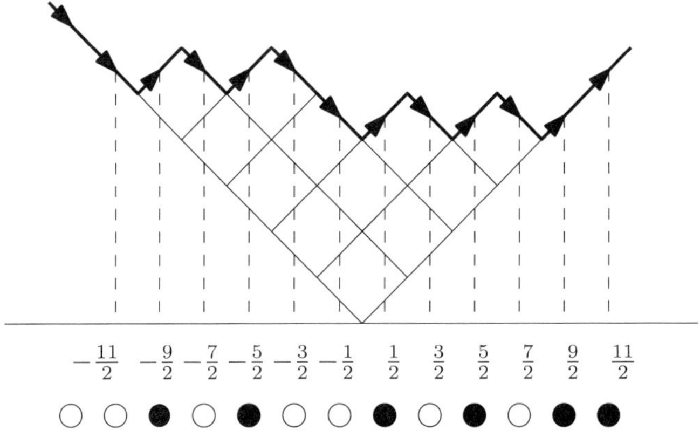

Fig. 3. The Young diagram $(5, 4, 2, 1)$ in the Russian representation and its configuration of holes and particles.

The *poissonised Plancherel measure* is defined, for each $\eta > 0$, by

$$\mathrm{Pois}_\eta := \exp(-\eta) \sum_{n=0}^{\infty} \frac{\eta^n}{n!} \mathbb{P}\ell^{(n)}.$$

Note that Pois_η is determined by the value it takes on sets of the form $\{k_1, \ldots, k_l \in X\} = \{\omega, \omega_{k_1} = \cdots = \omega_{k_l} = 1\}$. It was shown that the correlation functions of Pois_η have a determinantal form, given explicitely by [22, 1]

$$\mathrm{Pois}_\eta\big(\{k_1, \ldots, k_l \in X\}\big) = \det \mathcal{J}_\eta\big((k_r, k_s)_{1 \leqslant r,s \leqslant l}\big),$$

where

$$\mathcal{J}_\eta(x, y) = \sqrt{\eta}\, \frac{J_{x+1}(2\sqrt{\eta}) J_y(2\sqrt{\eta}) - J_x(2\sqrt{\eta}) J_{y+1}(2\sqrt{\eta})}{x - y}$$

is the kernel of an orthogonal projection and J is the Bessel function.

Borodin-Okounkov-Olshanski proved

$$\mathbb{P}\ell^{(n)}\big(\alpha\sqrt{n} + k_1, \ldots, \alpha\sqrt{n} + k_l \in X\big) \underset{n\to\infty}{\to} \det\big(\mathcal{S}_\alpha(k_i, k_j)\big),$$

where $\mathcal{S}_\alpha(k, l) = \frac{\sin a(k-l)}{\pi(k-l)}$ is the discrete sine kernel from Section 3, with $\cos\alpha = \frac{a}{2}$.

§6. DPPs associated with Hilbert spaces of holomorphic functions

In this section, we discuss various determinantal point processes on a domain of complex Euclidean space $\mathbb{C}^d$ induced by kernels of certain reproducing kernel Hilbert spaces on that domain.

6.1. Random holomorphic functions in the unit disc

Consider the series

$$g_{\mathbb{D}}(z) := \sum_{n=0}^{\infty} a_n z^n, \tag{6.1}$$

where a_n are i.i.d. $\mathcal{N}_{\mathbb{C}}(0,1)$ random variables. This series defines a random holomorphic function in the unit disc $\mathbb{D}$ (viewed as the Lobachevsky plane). It was shown by Peres and Virág [33] that the zeros set $Z(g_{\mathbb{D}})$ of the random holomorphic function g has a determinantal structure: the distribution of $Z(g_{\mathbb{D}})$ is given by the determinantal point process on $\mathbb{D}$ (equipped with the normalized Lebesgue measure $dz = \frac{1}{\pi}dxdy$) induced by the Bergman kernel

$$K_{\mathbb{D}}(z, w) = \frac{1}{(1 - z\bar{w})^2}.$$

Recall that the kernel $K_{\mathbb{D}}$ is the reproducin kernel of the Bergman space defined by

$$\mathbb{A}^2(\mathbb{D}) = \left\{ f : \mathbb{D} \to \mathbb{C} \,\middle|\, f \text{ is holomorphic and } \int_{\mathbb{D}} |f(z)|^2 dz < \infty \right\}.$$

More generally, for any $p > 0$, the L^p-version Bergman space is defined by

$$\mathbb{A}^p(\mathbb{D}) = \left\{ f : \mathbb{D} \to \mathbb{C} \,\middle|\, f \text{ is holomorphic and } \int_{\mathbb{D}} |f(z)|^p dz < \infty \right\}.$$

Lemma 6.1. *If $0 < p < 2$, then almost surely, the holomorphic function $g_{\mathbb{D}}$ defined in (6.1) belongs to the class $\mathbb{A}^p(\mathbb{D})$. On the other hand, almost surely, $g_{\mathbb{D}}$ does not belong to the class $\mathbb{A}^2(\mathbb{D})$.*

Proof. First suppose that $0 < p < 2$, then by using the unitary invariance of the joint Gaussian distribution $(a_n)_{n=0}^{\infty}$, we have

$$\mathbb{E}\left[\int_{\mathbb{D}} |g_{\mathbb{D}}(z)|^p dz\right] = \int_{\mathbb{D}} \mathbb{E}(|g_{\mathbb{D}}(z)|^p) dz = \int_{\mathbb{D}} \mathbb{E}\left[\left|\left(\sum_{n=0}^{\infty} |z^n|^2\right)^{1/2} a_0\right|^p\right] dz$$
$$= \mathbb{E}(|a_0|^p) \int_{\mathbb{D}} \frac{dz}{(1-|z|^2)^{p/2}} = \frac{2\mathbb{E}(|a_0|^p)}{2-p} < \infty.$$

It follows that, almost surely,

$$\int_{\mathbb{D}} |g_{\mathbb{D}}(z)|^p dz < \infty.$$

That means, almost surely, $g_{\mathbb{D}}$ belongs to the class $\mathbb{A}^p(\mathbb{D})$.

On the other hand, if $p = 2$, then

$$\int_{\mathbb{D}} |g_{\mathbb{D}}(z)|^2 dz = \sum_{n=0}^{\infty} \frac{|a_n|^2}{n+1}.$$

Thus by Kolmogorov's Three Series Theorem, it is easy to see that

$$\int_{\mathbb{D}} |g_{\mathbb{D}}(z)|^2 dz = \infty \quad a.s.$$

In other words, almost surely, we have $g_{\mathbb{D}} \notin \mathbb{A}^2(\mathbb{D})$. Q.E.D.

A subset $X \subset \mathbb{D}$ is called an $\mathbb{A}^2(\mathbb{D})$-*uniqueness set* if a function $f \in \mathbb{A}^2(\mathbb{D})$ satisfying $f_{|X} \equiv 0$ must be the zero function, where $f_{|X}$ means the restriction of the function f on the subset $X \subset \mathbb{D}$. A closely related notion is the $\mathbb{A}^2(\mathbb{D})$-*zero sets*: a subset $X \subset \mathbb{D}$ is called an $\mathbb{A}^2(\mathbb{D})$-zero set if there exists a non-identically Bergman function $f \in \mathbb{A}^2(\mathbb{D})$ such that X coincides with the set $Z(f)$ of zeros of f. Then a subset $X \subset \mathbb{D}$ is an $\mathbb{A}^2(\mathbb{D})$-uniqueness set if and only if it is not an $\mathbb{A}^2(\mathbb{D})$-zero set.

In general, it is a quite difficult problem to determine whether a subset $X \subset \mathbb{D}$ is an $\mathbb{A}^2(\mathbb{D})$-uniqueness set or not. The reader is referred to [19, Chapter 4] for more details on this topic. Here we recall a notion of density for subsets of $\mathbb{D}$. For any finite subset F of the unit circle $\mathbb{T}$, define the corresponding Stolz star domain $\mathfrak{s}_F$ as

$$\mathfrak{s}_F := \bigcup_{z \in F} \operatorname{conv}\Big(\{z\} \cup \{w \in \mathbb{D} : |w| \leq 1/\sqrt{2}\}\Big),$$

where $\mathrm{conv}(A)$ denotes the Euclidean convex hull in the plane of a subset $A \subset \mathbb{C}$. We also associate the following quantity $\widehat{k}(F)$ to any finite subset $F \subset \mathbb{T}$:

$$\widehat{k}(F) := 1 - \sum_k \frac{|I_k|}{2\pi} \log \frac{|I_k|}{2\pi},$$

where $\{I_k\}_k$ be the complementary arcs of the subset F in the unit circle $\mathbb{T}$. For a countable subset $X \subset \mathbb{D}$ without accumulation points in the interior of the disc, following [19, Chapter 4, Definition 4.9], we define the upper-density of X by

$$D^+(X) := \limsup_{\widehat{k}(F)\to\infty} \frac{1}{\widehat{k}(F)} \sum_{x \in \mathfrak{S}_F \cap X} \frac{1-|x|^2}{2}.$$

By [19, Theorem 4.31 and Corollary 4.38], the upper-density of a subset $X \subset \mathbb{D}$ is closely related to the uniqueness properties:

(6.2)

$$X \text{ is an } \mathbb{A}^{2+\varepsilon}(\mathbb{D})\text{-zero set for some } \varepsilon > 0 \text{ if and only if } D^+(X) < 1/2;$$
$$X \text{ is an } \mathbb{A}^{2-\varepsilon}(\mathbb{D})\text{-zero set for all } \varepsilon > 0 \text{ if and only if } D^+(X) \leq 1/2.$$

However, in the critical situation when the equality $D^+(X) = 1/2$ holds, one can not determine whether X is an $\mathbb{A}^2(\mathbb{D})$-uniqueness set or not.

The following result has been conjectured by Lyons and Peres (see [28] and [29]) and has been confirmed in [11].

Theorem 6.2. *Almost surely, $Z(g_{\mathbb{D}})$ is an $\mathbb{A}^2(\mathbb{D})$-uniqueness set.*

As an immediate consequence of Lemma 6.1 and Theorem 6.2, we have

Corollary 6.3. *Almost surely, the upper-density of $Z(g_{\mathbb{D}})$ is given by*

$$D^+(Z(g_{\mathbb{D}})) = \frac{1}{2}. \tag{6.3}$$

Proof. Lemma 6.1 and Theorem 6.2 together imply that almost surely, the set $Z(g_{\mathbb{D}})$ is an $\mathbb{A}^p(\mathbb{D})$-zero set for all $p < 2$ and is not an $\mathbb{A}^2(\mathbb{D})$-zero set. Consequently, by the relation (6.2) between the uniqueness properties and the upper-density, we obtain the desired almost sure equality (6.3). Q.E.D.

Another stronger property than the uniqueness property is the sampling property. Recall that a discrete subset $X \subset \mathbb{D}$ is called an $\mathbb{A}^2(\mathbb{D})$-*sampling set* if there exist $C_1, C_2 > 0$ such that for any $f \in \mathbb{A}^2(\mathbb{D})$ we have

$$C_1\|f\|^2_{\mathbb{A}^2(\mathbb{D})} \leq \sum_{z\in X} |f(z)|^2(1-|z|^2)^2 \leq C_2\|f\|^2_{\mathbb{A}^2(\mathbb{D})}.$$

By definition, an $\mathbb{A}^2(\mathbb{D})$-sampling set is also an $\mathbb{A}^2(\mathbb{D})$-uniqueness set.

Proposition 6.4. *Almost surely, the subset $Z(g_{\mathbb{D}})$ is not an $\mathbb{A}^2(\mathbb{D})$-sampling set.*

To see that the set $Z(g_{\mathbb{D}})$ is not sampling, we will use Seip [35, Theorem 7.1], which says that any $\mathbb{A}^2(\mathbb{D})$-sampling set is a finite union of sets uniformly separated with respect to the Lobachevskian distance.

Proposition 6.5. *Almost surely, $Z(g_{\mathbb{D}})$ can not be expressed as a finite union of uniformly separated sets.*

Proposition 6.5 is a consequence of the following ergodicity result. The action on $\mathbb{D}$ of the group $\mathrm{Aut}(\mathbb{D})$ of the isometries of the Lobachevsky plane can be lifted naturally to the action on $\mathrm{Conf}(\mathbb{D})$.

Lemma 6.6. *If $\gamma \in \mathrm{Aut}(\mathbb{D})$ is either hyperbolic or parabolic, then the dynamical system $(\mathrm{Conf}(\mathbb{D}), \mathbb{P}_{K_{\mathbb{D}}}, \gamma)$ is strongly mixing.*

6.2. Palm measures of DPPs with Bergman kernels on $\mathbb{D}$

Recall that two measures are called *equivalent* if they are mutually absolutely continuous.

The Palm measures of the determinantal point process $\mathbb{P}_{K_{\mathbb{D}}}$ have very nice properties: the Palm measures of $\mathbb{P}_{K_{\mathbb{D}}}$ of arbitrary orders are all equivalent and the Radon-Nikodym derivatives between them are given in regularized multiplicative functionals.

For emphasizing which particular properties of $K_{\mathbb{D}}$ are involved, we state the results in this subsection in the more general situation of the weighted Bergman kernels on $\mathbb{D}$. Consider a weight function $\omega : \mathbb{D} \to \mathbb{R}^+$, that is, ω is a non-negative function on $\mathbb{D}$ such that

$$0 < \int_{\mathbb{D}} \omega(z)dz < \infty.$$

Define $\mathbb{A}^2(\mathbb{D}, \omega)$ as the set of all holomorphic functions $f : \mathbb{D} \to \mathbb{C}$ such that

$$\|f\|^2_\omega := \int_{\mathbb{D}} |f(z)|^2\omega(z)dz < \infty.$$

The weight ω will be called *Bergman-admissible* if for any compact subset $S \subset \mathbb{D}$,

$$\sup_{z\in S} \sup_{f\in\mathbb{A}^2(\mathbb{D},\omega)\setminus\{0\}} \frac{|f(z)|}{\|f\|_\omega} < \infty.$$

Let ω be a Bergman-admissible weight on $\mathbb{D}$. Then $A^2(\mathbb{D},\omega)$ is closed in $L^2(\mathbb{D},\omega dz)$ and is called the *weighted Bergman space* associated with the weight ω. It is a reproducing kernel Hilbert space whose reproducing kernel will be denoted by $K_\omega(\cdot,\cdot)$.

The reproducing kernel K_ω is clearly locally trace class and it represents an orthogonal projection on $L^2(\mathbb{D},\omega dz)$. Therefore, K_ω induces a DPP. Let $\mathfrak{p}\in\mathbb{D}^\ell$ be an ℓ-tuple of distinct points in $\mathbb{D}$ and denote by $\mathbb{P}^{\mathfrak{p}}_{K_\omega}$ the reduced Palm measures of $\mathbb{P}_{K_\omega}$ at $\mathfrak{p}$.

For an ℓ-tuple $\mathfrak{p}=(p_1,\dots,p_\ell)$ of distinct points in $\mathbb{D}$, set

$$b_{\mathfrak{p}}(z)=\prod_{j=1}^{\ell}\frac{z-p_j}{1-\bar{p}_j z}.$$

Theorem 6.7 ([10]). *Let ω be a Bergman-admissible weight on $\mathbb{D}$ such that*

$$\int_{\mathbb{D}}(1-|z|)^2K_\omega(z,z)\omega(z)dz<\infty. \tag{6.4}$$

Let $\mathfrak{p}\in\mathbb{D}^\ell$ and $\mathfrak{q}\in\mathbb{D}^k$ be two tuples of distinct points in $\mathbb{D}$. Then

1) *For $\mathbb{P}^{\mathfrak{q}}_{K_\omega}$-almost every configuration $X\in\mathrm{Conf}(\mathbb{D})$, the following limit exists:*

$$S_{\mathfrak{p},\mathfrak{q}}(X):=\lim_{r\to1^-}\left(\sum_{z\in X:|z|\le r}\log\left(\frac{|b_{\mathfrak{p}}(z)|}{|b_{\mathfrak{q}}(z)|}\right)-m_{\mathfrak{p},\mathfrak{q},r}\right),$$

where

$$m_{\mathfrak{p},\mathfrak{q},r}:=\int_{\mathrm{Conf}(\mathbb{D})}\sum_{z\in\mathcal{Z}:|z|\le r}\log\left(\frac{|b_{\mathfrak{p}}(z)|}{|b_{\mathfrak{q}}(z)|}\right)d\mathbb{P}^{\mathfrak{q}}_{K_\omega}(\mathcal{Z}).$$

Moreover, we have

$$\mathbb{E}_{\mathbb{P}^{\mathfrak{q}}_{K_\omega}}(e^{2S_{\mathfrak{p},\mathfrak{q}}})=\int_{\mathrm{Conf}(\mathbb{D})}e^{2S_{\mathfrak{p},\mathfrak{q}}(X)}d\mathbb{P}^{\mathfrak{q}}_{K_\omega}(X)<\infty.$$

2) *We have the following explicit formula for the Radon-Nikodym derivative:*

$$\frac{d\mathbb{P}^{\mathfrak{p}}_{K_\omega}}{d\mathbb{P}^{\mathfrak{q}}_{K_\omega}}(X)=\frac{e^{2S_{\mathfrak{p},\mathfrak{q}}(X)}}{\mathbb{E}_{\mathbb{P}^{\mathfrak{q}}_{K_\omega}}(e^{2S_{\mathfrak{p},\mathfrak{q}}})}$$

for $\mathbb{P}^{\mathfrak{q}}_{K_\omega}$-almost every configuration $X\in\mathrm{Conf}(\mathbb{D})$.

Corollary 6.8. *Let ω be a Bergman-admissible weight on $\mathbb{D}$ such that*

$$\int_{\mathbb{D}} (1-|z|)^2 K_\omega(z,z)\omega(z)dz < \infty.$$

Then the Palm measures of all orders of $\mathbb{P}_{K_\omega}$ are equivalent.

Remark. *For the weight $\omega \equiv 1$, the result that all Palm measures of $\mathbb{P}_{K_{\mathbb{D}}}$ are equivalent is due to Holroyd and Soo* [21].

Remark. *It is a natural question whether the condition* (6.4) *can be removed from Theorem* 6.7.

6.3. Palm measures of DPPs with Bergman kernels on $\mathbb{C}$

In the situation of DPPs associated to weighted Bergman kernels on the whole plane $\mathbb{C}$, the relation between Palm measures is quite different from that on the unit disc.

Let $\psi : \mathbb{C} \to \mathbb{R}$ be a C^2-smooth function and equip the complex plane $\mathbb{C}$ with the measure $e^{-2\psi(z)}dz$. Assume that there exist positive constants $m, M > 0$ so that

$$m \leq \Delta\psi \leq M, \tag{6.5}$$

where Δ is the Euclidean Laplacian differential operator. Denote by $\mathscr{F}_\psi$ the generalized Fock space with respect to the weight $e^{-2\psi(z)}$:

$$\mathscr{F}_\psi := \left\{ f : \mathbb{C} \to \mathbb{C} \,\middle|\, f \text{ is holomorphic and } \int_{\mathbb{C}} |f(z)|^2 e^{-2\psi(z)} dz < \infty \right\}.$$

Then $\mathscr{F}_\psi$ is a closed subspace of $L^2(\mathbb{C}, e^{-2\psi(z)}dz)$ and is a reproducing kernel Hilbert space. Thus the reproducing kernel K_ψ of $\mathscr{F}_\psi$ induces a DPP $\mathbb{P}_{K_\psi}$ on $\mathbb{C}$. The condition (6.5) implies in particular the following useful Christ's pointwise estimate (cf. [13], [16] and [34, Theorem 3.2]) for the reproducing kernel K_ψ: there are contants $\delta, C > 0$ such that

$$|K_\psi(z,w)|^2 e^{-2\psi(z)-2\psi(w)} \leq Ce^{-\delta|z-w|} \quad \text{for all } z, w \in \mathbb{C}.$$

For any tuple $\mathfrak{p} \in \mathbb{C}^\ell$ of *distinct* points in $\mathbb{C}$, we denote by $\mathbb{P}^{\mathfrak{p}}_{K_\psi}$ the reduced Palm measures of $\mathbb{P}_{K_\psi}$ conditioned at $\mathfrak{p}$.

Theorem 6.9 ([10]). *Let $\psi : \mathbb{C} \to \mathbb{R}$ be a C^2-smooth function satisfying* (6.5) *and let $\mathfrak{p}, \mathfrak{q} \in \mathbb{C}^\ell$ be any two tuples of distinct points in $\mathbb{C}$. Then*

1) *For $\mathbb{P}^{\mathfrak{q}}_{K_\psi}$-almost every configuration $X \in \mathrm{Conf}(\mathbb{C})$, the following limit exists:*

$$\Sigma_{\mathfrak{p},\mathfrak{q}}(X) := \lim_{R\to\infty}\left\{ \sum_{z\in X:|z|\le R} \log\left|\frac{(z-p_1)\cdots(z-p_\ell)}{(z-q_1)\cdots(z-q_\ell)}\right| - \int_{\mathrm{Conf}(\mathbb{C})} \sum_{z\in \mathcal{Z}:|z|\le R} \log\left|\frac{(z-p_1)\cdots(z-p_\ell)}{(z-q_1)\cdots(z-q_\ell)}\right| d\mathbb{P}^{\mathfrak{q}}_{K_\psi}(\mathcal{Z})\right\}.$$

Moreover, we have

$$\mathbb{E}_{\mathbb{P}^{\mathfrak{q}}_{K_\psi}}(e^{2\Sigma_{\mathfrak{p},\mathfrak{q}}}) = \int_{\mathrm{Conf}(\mathbb{C})} e^{2\Sigma_{\mathfrak{p},\mathfrak{q}}(\mathcal{Z})} d\mathbb{P}^{\mathfrak{q}}_{K_\psi}(\mathcal{Z}) < \infty.$$

2) *The Palm measures $\mathbb{P}^{\mathfrak{p}}_{K_\psi}$ and $\mathbb{P}^{\mathfrak{q}}_{K_\psi}$ are equivalent. Moreover, the Radon-Nikodym derivative between them is given by*

$$\frac{d\mathbb{P}^{\mathfrak{p}}_{K_\psi}}{d\mathbb{P}^{\mathfrak{q}}_{K_\psi}}(X) = \frac{e^{2\Sigma_{\mathfrak{p},\mathfrak{q}}(X)}}{\mathbb{E}_{\mathbb{P}^{\mathfrak{q}}_{K_\psi}}(e^{2\Sigma_{\mathfrak{p},\mathfrak{q}}})}$$

for $\mathbb{P}^{\mathfrak{q}}_{K_\psi}$-almost every $X \in \mathrm{Conf}(\mathbb{C})$.

6.4. Palm measures of DPPs with Bergman kernels on domains of $\mathbb{C}^d$

Let U be a non-empty connected open subset of the d-dimensional complex space $\mathbb{C}^d$. Consider a weight ω on U, that is, ω is a Borel function $\omega : U \to (0,\infty)$ such that

$$0 < \int_U \omega(z) dm_{2d}(z) < \infty,$$

where dm_{2d} stands for the Lebesgue measure on $\mathbb{C}^d \simeq \mathbb{R}^{2d}$. Assume moreover that that for any relatively compact subset $B \subset U$ satisfying $m_{2d}(B) > 0$ we have

$$\operatorname*{essinf}_{z\in B} \omega(z) > 0. \tag{6.6}$$

Define the weighted Bergman space $A^2(U,\omega)$ as the set of all holomorphic functions $f : U \to \mathbb{C}$ such that

$$\int_U |f(z)|^2 \omega(z) dm_{2d}(z) < \infty.$$

Then $A^2(U,\omega)$ is closed in $L^2(U,\omega dV)$ and admits a reproducing kernel, denoted by K_ω. The kernel K_ω induces a determinantal point process $\mathbb{P}_{K_\omega}$ on U.

Definition 6.10. We say that U has the Liouville property if any bounded holomorphic functions on U is a constant function.

Theorem 6.11 ([9]). *Assume that $U \subset \mathbb{C}^d$ is a domain without Liouville property and ω is a weight on U satisfying* (6.6). *Then the determinantal measure $\mathbb{P}_{K_\omega}$ is equivalent to its arbitrary reduced Palm measure, of any order.*

Let us outline the proof of Theorem 6.11. For proving Theorem 6.11, we need to prove, for any given tuple $\mathfrak{p} \in U^\ell$ of distinct points, the following two relations:

- $\mathbb{P}^{\mathfrak{p}}_{K_\omega} \ll \mathbb{P}_{K_\omega}$;
- $\mathbb{P}_{K_\omega} \ll \mathbb{P}^{\mathfrak{p}}_{K_\omega}$.

Step 1: $\mathbb{P}^{\mathfrak{p}}_{K_\omega} \ll \mathbb{P}_{K_\omega}$.

The proof of the relation $\mathbb{P}^{\mathfrak{p}}_{K_\omega} \ll \mathbb{P}_{K_\omega}$ relies on the proof that $\mathbb{P}_{K_\omega}$ is *deletion tolerant*. The notion of deletion tolerance is given in the following

Definition 6.12 (Deletion tolerance [21]). A point process $\mathbb{P}$ on E is called **deletion tolerant**, if for any relatively compact Borel subset $B \subset E$ and $\mathbb{P}$-almost every configuration $X \in \mathrm{Conf}(E)$, we have

$$\mathbb{P}(\#_B = 0 | X, B^c) > 0.$$

The notion of deletion tolerance comes to play a role in the proof of the relation $\mathbb{P}^{\mathfrak{p}}_{K_\omega} \ll \mathbb{P}_{K_\omega}$ via the following general result on deletion tolerant DPPs with self-adjoint kernel.

Proposition 6.13 ([9]). *Let $\mathbb{P}_K$ be a DPP on a Polish space E induced by a self-adjoint kernel K. If $\mathbb{P}_K$ is deletion tolerant, then any reduced Palm measure of $\mathbb{P}_K$ is absolutely continuous with respect to $\mathbb{P}_K$.*

The proof of Proposition 6.13 relies on [29, Theorem 3.8] for the monotone coupling between DPP and an equivalent formulation of deletion tolerance in [21, Theorem 1.1].

Proposition 6.14 ([9]). *Assume that $U \subset \mathbb{C}^d$ is a domain without Liouville property and ω is a weight on U satisfying* (6.6). *Then $\mathbb{P}_{K_\omega}$ is deletion tolerant.*

The proof of Proposition 6.14 relies on the module structure of the weighted Bergman space $A^2(U,\omega)$ over the algebra $H^\infty(U)$ of all bounded holomorphic functions on U and is outlined as follows. We need to show that for any relatively compact $B \subset U$ with positive Lebesgue measure, $\mathbb{P}_{K_\omega}(\#_B = 0|X, B^c) > 0$ for $\mathbb{P}_{K_\omega}$-almost every X. Since for $\mathbb{P}_{K_\omega}$-almost every X, the conditional measure $\mathbb{P}_{K_\omega}(\cdot|X, B^c)$ is determinantal and is induced by an explicitly described *trace class* positive contraction $K_\omega^{[X,B^c]}$, that is,

$$\mathbb{P}_{K_\omega}(\cdot|X, B^c) = \mathbb{P}_{K_\omega^{[X,B^c]}}.$$

Therefore,

$$\mathbb{P}_{K_\omega}(\#_B = 0|X, B^c) = \det(1 - \chi_B K_\omega^{[X,B^c]} \chi_B) = \det(1 - K_\omega^{[X,B^c]})$$

and so we need to show that for $\mathbb{P}_{K_\omega}$-almost every configuration $X \in \mathrm{Conf}(U)$, the trace class positive contractive operator $K_\omega^{[X,B^c]}$ is strictly contractive. That is, we only need to show that 1 is not an eigenvalue of $K_\omega^{[X,B^c]}$.

Lemma 6.15 ([9]). *For $\mathbb{P}_{K_\omega}$-almost every configuration $X \in \mathrm{Conf}(U)$, the space*

$$V_1(K_\omega^{[X,B^c]}) = \{h \in L^2(B) : K_\omega^{[X,B^c]}(h) = h\}$$

is an $H^\infty(U)$-module.

The assumption that U is without Liouville property implies that $H^\infty(U)$ contains a non-constant function, then the space $V_1(K_\omega^{[X,B^c]})$ is either zero or has infinite dimension, the latter impossible by compactness of $K_\omega^{[X,B^c]}$.

Step 2: $\mathbb{P}_{K_\omega} \ll \mathbb{P}^{\mathfrak{p}}_{K_\omega}$.

The proof of the relation $\mathbb{P}_{K_\omega} \ll \mathbb{P}^{\mathfrak{p}}_{K_\omega}$ relies on the proof that $\mathbb{P}_{K_\omega}$ is *insertion tolerant.* In fact, we need both the notion of number insertion tolerance and insertion tolerance, whose definitions are given in the following

Definition 6.16. A point process $\mathbb{P}$ on E is called **number insertion tolerant** if for any relatively compact Borel subset $B \subset E$ with $\mu(B) > 0$, we have

$$\mathbb{P}(\#_B > 0|X, B^c) > 0 \quad \text{for } \mathbb{P}\text{-almost every } X.$$

The stronger notion of insertion tolerance is defined for point processes on subsets of Euclidean spaces. Recall that given two point processes $\mathbb{P}_1$ and $\mathbb{P}_2$ on U, their convolution $\mathbb{P}_1 * \mathbb{P}_2$ is defined by

$$\mathbb{P}_1 * \mathbb{P}_2 = \mathrm{Law}(\mathscr{X}_1 \cup \mathscr{X}_2),$$

with $\mathscr{X}_1, \mathscr{X}_2$ random configurations independently sampled with $\mathbb{P}_1, \mathbb{P}_2$ respectively.

Definition 6.17 (Holroyd and Soo [21]). A point process $\mathbb{P}$ on U is called insertion tolerant if for any relatively compact $S \subset U$ with positive Lebesgue measure,

$$\mathbb{P} * \delta_{\mathrm{Unif}(S)} \ll \mathbb{P},$$

where $\delta_{\mathrm{Unif}(S)}$ denotes the point process of one single random particle with uniform distribution $\mathrm{Unif}(S)$ on S.

Proposition 6.18 ([9]). *The determinantal point process $\mathbb{P}_{K_\omega}$ and the determinantal point process $\mathbb{P}^{\mathfrak{p}}_{K_\omega}$ are insertion tolerant.*

The proof of Proposition 6.18 uses both the module structure of our underlying Hilbert space and the *real analyticity* of the well chosen conditional kernels $(K^{\mathfrak{p}}_{\omega})^{[X,B^c]}(z,w)$.

For completing the proof of the relation $\mathbb{P}_{K_\omega} \ll \mathbb{P}^{\mathfrak{p}}_{K_\omega}$, we still need an extra result on the monotone coupling for $\mathbb{P}_{K_\omega}$ and $\mathbb{P}^{\mathfrak{p}}_{K_\omega}$, that is, the difference random configuration is *conditionally diffusive*: let $\mathscr{X}$ and $\mathscr{Y}$ be any two coupled random configurations on U defined on a common probability space $(\Omega, \mathcal{B}, \mathbf{P})$, such that

$$\mathrm{Law}(\mathscr{X}) = \mathbb{P}_{K_\omega}, \ \mathrm{Law}(\mathscr{Y}) = \mathbb{P}^{\mathfrak{p}}_{K_\omega} \text{ and } \mathscr{Y} \subset \mathscr{X} \quad \mathbf{P}\text{-almost surely.}$$

Then for $\mathbb{P}^{\mathfrak{p}}_{K_\omega}$-almost every Y,

$$\mathrm{Law}(\mathscr{X} \setminus \mathscr{Y} | \mathscr{Y} = Y) \ll \delta_{\emptyset} + \sum_{n=1}^{\infty} (\pi^{(n)})_*((dV|_U)^{\otimes n}),$$

where $(dV|_U)^{\otimes n}$ is the Lebesgue measure on U^n and $\pi^{(n)} : U^n \to \mathrm{Conf}(U)$ is the map sending each $(x_1, \dots, x_n)$ to the configuration $\{x_1, \dots, x_n\}$.

Acknowledgements.

GB is supported by the EPSRC grant EP/L016516/1 for the University of Cambridge CDT. AB's research is supported by the European Research Council (ERC) under the European Union Horizon 2020 research and innovation programme, grant 647133 (ICHAOS), by the Agence Nationale de Recherche, project ANR-18-CE40-0035, and by the Russian Foundation for Basic Research, grant 18-31-20031. The research of YQ is supported by grants NSFC Y7116335K1, NSFC 11801547 and NSFC 11688101 of National Natural Science Foundation of China.

References

[1] Alexei Borodin, Andrei Okounkov and Grigori Olshanski, Asymptotics of Plancherel measures for symmetric groups, *J. Amer. Math. Soc.*, 13(3): 481–515, 2000.

[2] Alexander I. Bufetov, On multiplicative functionals of determinantal processes, *Uspekhi Mat. Nauk*, 67(1(403)):177–178, 2012.

[3] Alexander I. Bufetov, On the Vershik-Kerov conjecture concerning the Shannon-McMillan-Breiman theorem for the Plancherel family of measures on the space of Young diagrams, *Geom. Funct. Anal.*, 22(4):938–975, 2012.

[4] Alexander I. Bufetov, Infinite determinantal measures, *Electron. Res. Announc. Math. Sci.*, 20:12–30, 2013.

[5] Alexander I. Bufetov, Quasi-symmetries of determinantal point processes, *Ann. Probab.*, 46(2):956–1003, 2018.

[6] Alexander I. Bufetov, The sine-process has excess one, *arXiv e-prints*, arXiv:1912.13454, December 2019.

[7] Alexander I. Bufetov, Yoann Dabrowski and Yanqi Qiu, Linear rigidity of stationary stochastic processes, *Ergodic Theory Dynam. Systems*, 38(7): 2493–2507, 2018.

[8] Alexander I. Bufetov and Andrey V. Dymov, A functional limit theorem for the sine-process, *Int. Math. Res. Not. IMRN*, (1):249–319, 2019.

[9] Alexander I. Bufetov, Shilei Fan and Yanqi Qiu, Equivalence of Palm measures for determinantal point processes governed by Bergman kernels, *Probab. Theory Related Fields*, 172(1-2):31–69, 2018.

[10] Alexander I. Bufetov and Yanqi Qiu, Determinantal point processes associated with Hilbert spaces of holomorphic functions, *Comm. Math. Phys.*, 351(1):1–44, 2017.

[11] Alexander I. Bufetov, Yanqi Qiu and Alexander Shamov, Kernels of conditional determinantal measures and the proof of the Lyons-Peres Conjecture, *arXiv e-prints*, to appear in *J. Eur. Math. Soc.*, arXiv:1612.06751, December 2016.

[12] Alexander I. Bufetov and Roman V. Romanov, Division subspaces and integrable kernels, *Bull. Lond. Math. Soc.*, 51(2):267–277, 2019.

[13] Michael Christ, On the $\overline{\partial}$ equation in weighted L^2 norms in $\mathbf{C}^1$, *J. Geom. Anal.*, 1(3):193–230, 1991.

[14] D. J. Daley and D. Vere-Jones, *An introduction to the theory of point processes. Vol.* I, Probability and its Applications (New York), Springer-Verlag, New York, second edition, 2003, Elementary theory and methods.

[15] D. J. Daley and D. Vere-Jones, *An introduction to the theory of point processes. Vol.* II, Probability and its Applications (New York), Springer, New York, second edition, 2008, General theory and structure.

[16] Henrik Delin, Pointwise estimates for the weighted Bergman projection kernel in $\mathbf{C}^n$, using a weighted L^2 estimate for the $\overline{\partial}$ equation, *Ann. Inst. Fourier (Grenoble)*, 48(4):967–997, 1998.

[17] Subhroshekhar Ghosh, Determinantal processes and completeness of random exponentials: the critical case, *Probab. Theory Related Fields*, 163(3-4):643–665, 2015.

[18] Subhroshekhar Ghosh and Yuval Peres, Rigidity and tolerance in point processes: Gaussian zeros and Ginibre eigenvalues, *Duke Math. J.*, 166(10): 1789–1858, 2017.

[19] Haakan Hedenmalm, Boris Korenblum and Kehe Zhu, *Theory of Bergman spaces*, volume 199 of *Graduate Texts in Mathematics*, Springer-Verlag, New York, 2000.

[20] John G. Hocking and Gail S. Young, *Topology*, Addison-Wesley Publishing Co., Inc., Reading, Mass.-London, 1961.

[21] Alexander E. Holroyd and Terry Soo, Insertion and deletion tolerance of point processes, *Electron. J. Probab.*, 18:no. 74, 24, 2013.

[22] Kurt Johansson, Discrete orthogonal polynomial ensembles and the Plancherel measure, *Ann. of Math.* (2), 153(1):259–296, 2001.

[23] Olav Kallenberg, *Random measures*, Akademie-Verlag, Berlin; Academic Press, Inc., London, fourth edition, 1986.

[24] A. Lenard, Correlation functions and the uniqueness of the state in classical statistical mechanics, *Comm. Math. Phys.*, 30:35–44, 1973.

[25] A. Lenard, States of classical statistical mechanical systems of infinitely many particles. I, *Arch. Rational Mech. Anal.*, 59(3):219–239, 1975.

[26] A. Lenard, States of classical statistical mechanical systems of infinitely many particles. II. Characterization of correlation measures, *Arch. Rational Mech. Anal.*, 59(3):241–256, 1975.

[27] B. F. Logan and L. A. Shepp, A variational problem for random Young tableaux, *Advances in Math.*, 26(2):206–222, 1977.

[28] Russell Lyons, Determinantal probability measures, *Publ. Math. Inst. Hautes Études Sci.*, (98):167–212, 2003.

[29] Russell Lyons, Determinantal probability: basic properties and conjectures, In: *Proceedings of the International Congress of Mathematicians—Seoul* 2014. *Vol.* IV, pages 137–161, Kyung Moon Sa, Seoul, 2014.

[30] Russell Lyons and Jeffrey E. Steif, Stationary determinantal processes: phase multiplicity, Bernoullicity, entropy, and domination, *Duke Math. J.*, 120(3):515–575, 2003.

[31] Odile Macchi, The coincidence approach to stochastic point processes, *Advances in Appl. Probability*, 7:83–122, 1975.

[32] Nikolai K. Nikolski, *Operators, functions, and systems: an easy reading. Vol.* 2, volume 93 of *Mathematical Surveys and Monographs*, American Mathematical Society, Providence, RI, 2002. Model operators and systems, Translated from the French by Andreas Hartmann and revised by the author.

[33] Yuval Peres and Bálint Virág, Zeros of the i.i.d. Gaussian power series: a conformally invariant determinantal process, *Acta Math.*, 194(1):1–35, 2005.

[34] Alexander P. Schuster and Dror Varolin, Toeplitz operators and Carleson measures on generalized Bargmann-Fock spaces, *Integral Equations Operator Theory*, 72(3):363–392, 2012.
[35] Kristian Seip, On Korenblum's density condition for the zero sequences of $A^{-\alpha}$, *J. Anal. Math.*, 67:307–322, 1995.
[36] Tomoyuki Shirai and Yoichiro Takahashi, Fermion process and Fredholm determinant, In: *Proceedings of the Second ISAAC Congress, Vol.* 1 (*Fukuoka*, 1999), volume 7 of *Int. Soc. Anal. Appl. Comput.*, pages 15–23, Kluwer Acad. Publ., Dordrecht, 2000.
[37] Tomoyuki Shirai and Yoichiro Takahashi, Random point fields associated with certain Fredholm determinants. I: Fermion, Poisson and boson point processes, *J. Funct. Anal.*, 205(2):414–463, 2003.
[38] Barry Simon, *Trace ideals and their applications*, volume 35 of *London Math. Soc. Lecture Note Ser.*, Cambridge University Press, Cambridge-New York, 1979.
[39] A. Soshnikov, Determinantal random point fields, *Uspekhi Mat. Nauk*, 55(5(335)):107–160, 2000.
[40] A. M. Veršik and S. V. Kerov, Asymptotic behavior of the Plancherel measure of the symmetric group and the limit form of Young tableaux, *Dokl. Akad. Nauk SSSR*, 233(6):1024–1027, 1977.

Guillaume BAVEREZ:
Centre for Mathematical Sciences,
Wilberforce Road, Cambridge CB3 0WA, United Kingdom
E-mail address: gb539@cam.ac.uk

Alexander I. BUFETOV:
Aix-Marseille Université,
Centrale Marseille, CNRS, Institut de Mathématiques de Marseille, UMR7373, 39 Rue F. Joliot Curie 13453, Marseille, France;
Steklov Mathematical Institute of RAS, Moscow, Russia
E-mail address: bufetov@mi.ras.ru, alexander.bufetov@univ-amu.fr

Yanqi QIU:
Institute of Mathematics and Hua Loo-Keng Key Laboratory of Mathematics, AMSS, Chinese Academy of Sciences,
Beijing 100190, China
E-mail address: yanqi.qiu@amss.ac.cn; yanqi.qiu@hotmail.com

Advanced Studies in Pure Mathematics 87, 2021
Stochastic Analysis, Random Fields and Integrable Probability — Fukuoka 2019
pp. 89–136

Invariance of polymer partition functions under the geometric RSK correspondence

Ivan Corwin

Abstract.

We prove that the values of discrete directed polymer partition functions involving multiple non-intersecting paths remain invariant under replacing the background weights by their images under the geometric RSK correspondence. This result is inspired by a recent and remarkable identity proved by Dauvergne, Orthmann and Virág which is recovered as the zero-temperature, semi-discrete limit of our main result.

§1. Introduction

The *Robinson-Schensted-Knuth* (*RSK*) *correspondence* [dBR38, Sch61, Knu70] plays an central role in combinatorics and symmetric function theory [Sta99, Sag01]. In the past twenty years, starting with work of [Joh00, BDJ99], it has also taken on a key role in the study of certain models in integrable probability such as $(1+1)$-dimensional *last passage percolation* (*LPP*). Greene's theorem [Gre74] provides the link between the RSK correspondence and LPP. The push-forward under the RSK correspondence of an array of independent and identically distributed geometric random variables produces a measure on pairs of semi-standard Young tableaux which can be described in terms of Schur symmetric polynomials.

This connection between geometric weight LPP and the *Schur process/measure* [Oko01, OR03] combined with the *determinantal* structure behind these measures serves as a starting point for computing asymptotics. Greene's theorem implies that the last passage times from the origin to various points along down-right (i.e., *space-like*) paths can be

Received February 21, 2020.
Revised October 16, 2020.
2010 *Mathematics Subject Classification.* 05E99.
Key words and phrases. Directed polymers, RSK correspondence, Kardar-Parisi-Zhang.

read-off from directly from the output of the RSK correspondence. In the geometric weight setting, the asymptotic joint distribution of these last passage times is described by the *Airy$_2$ process* [PS02, Joh03]. The Airy$_2$ process is the top curve of the Airy line ensemble [CH14] which records the limit of the entire Schur process. The lower curves of the Airy line ensemble relate to limits of last passage times for multipaths.

These asymptotics fit into the study of the *Kardar-Parisi-Zhang (KPZ) universality class* [Cor12, Qua12, QS15]. Based on the aforementioned LPP results, the Airy$_2$ process has been understood to be the universal process which governs the fixed starting point and spatially varying ending point distribution of the KPZ universality class. The one-parameter Airy$_2$ process was conjectured in [CQR15] to be a marginal of similarly conjectural two-parameter process termed the *Airy sheet* which should describe the limiting joint law of LPP (and other KPZ class model) fluctuations under spatially varying both the starting and ending points. Given the Airy sheet, one can construct the full KPZ fixed point (i.e., the universal space-time limit of models in the KPZ universality class) – see also [MQR17].

[DOV18, Theorem 1.3] constructs the *Airy sheet* from the *Airy line ensemble*. Instead of studying geometric weight LPP, [DOV18] focuses on a semi-discrete limit known as Brownian LPP. The starting point for the work of [DOV18] is a remarkable generalization of Greene's theorem in the semi-discrete setting which shows how the collection of last passage times from spatially varying starting and endpoint points is encoded simply from the (semi-discrete) RSK correspondence output. Their result, [DOV18, Proposition 4.1], is Corollary 3.5 in this text.

Our present paper addresses the question of whether [DOV18, Proposition 4.1] generalize to the *discrete* RSK correspondence and to the *geometric* RSK correspondence? Theorem 2.4 answers both of these questions in the affirmative by providing a generalization of [DOV18, Proposition 4.1] for the discrete geometric RSK correspondence.

The *geometric RSK (gRSK) correspondence* [Kir01, NY04, O'C12, COSZ14, OSZ14] (also sometimes called the *tropical RSK correspondence*) is the image of the usual RSK correspondence under replacement of the $(\max, +)$ semi-ring by the $(+, \times)$ semi-ring. It is important to note that the word *geometric* has been used in two ways so far in this introduction – in reference to geometric random variables in LPP and in reference to this geometric lifting of the RSK correspondence. We will generally use *geometric* in the second sense in this paper.

Greene's theorem generalizes to the gRSK correspondence and provides a relationship to directed polymer partition functions (instead of

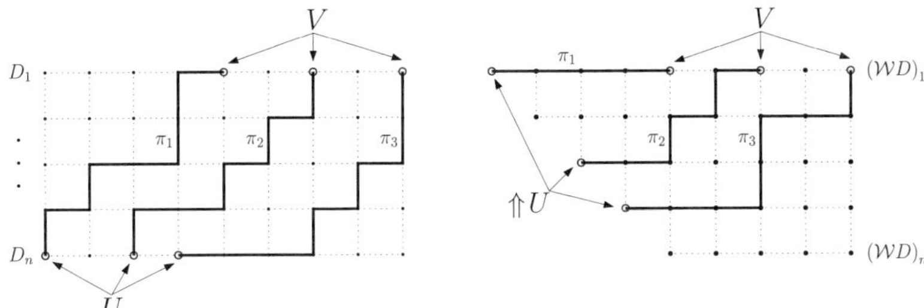

Fig. 1. An illustration of many of the terms used to formulate our main result, Theorem 2.4 when $n = 5$ and $k = 3$. The mapping from U to $\Uparrow U$ as defined in Theorem 2.4. The circles correspond to the U and V on the left or $\Uparrow U$ and V on the right, and π_1, π_2 and π_3 represent possible non-intersecting paths which go from U to V on the left, or $\Uparrow U$ to V on the right. Also illustrated is a portion (left-justified) of the domain of D on the left and the domain of $\mathcal{W}D$ on the right. To the ith row, we associate the function D_i on the left and $(\mathcal{W}D)_i$ on the right.

LPP as for usual RSK). The special geometric random variables which, under RSK, produced the Schur process, have an analog for gRSK too – the image of inverse-gamma random variables under the gRSK correspondence maps to the *Whittaker process/measure* [O'C12, COSZ14, OSZ14].

Our main result is a generalization of [DOV18, Proposition 4.1] to the discrete geometric RSK correspondence. We briefly explain this here, leaving precise definitions to the main text.

Consider any $n \in \mathbb{Z}_{\geq 2}$, $k \in \mathbb{Z}_{\geq 1}$, $U = \big((u_1, n), \dots, (u_k, n)\big)$ and $V = \big((v_1, 1), \dots, (v_k, 1)\big)$ with $u_1 < \cdots < u_k \in \mathbb{Z}_{\geq 1}$ and $v_1 < \cdots < v_k \in \mathbb{Z}_{\geq 1}$. Define $U \to V$ to be the set of multipaths $\pi = (\pi_1, \dots, \pi_k)$ where each π_i is a directed lattice path starting at (u_i, n) and ending at $(v_i, 1)$, and the π_i are pairwise non-intersecting. Assume that $U \to V$ is non-empty. The left-side of Figure 1 illustrates a choice of U, V and $\pi \in U \to V$ when $n = 5$ and $k = 3$.

Consider any functions $D_1, \dots, D_n : \mathbb{Z}_{\geq 1} \to (0, \infty)$ with the convention that $D_i(0) \equiv 1$, and write $D = (D_1, \dots, D_n)$. Treating a multipath π as the union of all points along the various π_i, we define the partition function from U to V with respect to D to be

$$D[U \to V] = \sum_{\pi \in U \to V} \prod_{(x,m) \in \pi} d_{x,m} \qquad \text{where} \quad d_{x,m} = \frac{D_m(x)}{D_m(x-1)}.$$

Write $\mathcal{W}D = \big((\mathcal{W}D)_1, \ldots, (\mathcal{W}D)_n\big)$ where, for $1 \le i \le n$ we define $(\mathcal{W}D)_i : \mathbb{Z}_{\ge i} \to (0, \infty)$ so that for all $N \in \mathbb{Z}_{\ge 1}$ and $1 \le \ell \le n \wedge N$,

$$\prod_{r=1}^{\ell} (\mathcal{W}D)_r(N) = D\Big[\big([\![1,\ell]\!], n\big) \to \big([\![N-\ell+1, N]\!], 1\big)\Big],$$

where $\big([\![a,b]\!], m\big) := \big((a,m), \ldots, (b,m)\big)$. (Note that in Section 2.1 we define $\mathcal{W}$ in a different manner via discrete geometric Pitman transforms and it is not until Corollary 2.14 that we relate $\mathcal{W}$ to the above right-hand side.)

Finally, define $\Uparrow U$ as in Figure 1 by lifting up the points in U until they are in the domain of definition for the $\mathcal{W}D$. We may now state our main theorem.

Theorem 1.1 (Theorem 2.4). *$D[U \to V] = (\mathcal{W}D)[\Uparrow U \to V]$.*

Section 2.2 shows that $D \mapsto \mathcal{W}D$ is half of the gRSK correspondence. **Thus, up to the action of the $\Uparrow$ operator on U, the partition functions for multipaths are invariant under the gRSK correspondence**. Remark 2.5 describes the new complexities encountered in formulating and proving this theorem, most of which come from the discrete setup.

After initially posting this paper we learned of three alternative proofs of our main result, Theorem 2.4. The first is essentially already present in the work of Noumi and Yamada [NY04]. We summarize this in Section 2.2.5. The second is due to Konstantin Matveev and relies on the Desnanot-Jacobi identity for matrix minors. The proof was proposed after the author gave a talk on this subject in the NYC Integrable Probability Seminar. With Matveev's permission, we record this in Section 2.2.4. The third will appear in forthcoming work of Duncan Dauvergne [Dau20] (which also contains some extensions of the results of [BGW19]) and relies on other manipulations involving matrix minors.

None of these three proofs were apparent (or have particularly clear analogs) in the zero-temperature semi-discrete setting as in [DOV18]. By lifting to the geometric and discrete setting, we gain new tools, namely we can relate partition functions to matrix minors, and then make use of various manipulations based on that. In the zero-temperature limit (which relates to the usual RSK correspondence) one loses tools like Lindström-Gessel-Viennot, and in the semi-discrete limit, one loses the matrices entirely. That is not to say that these other proofs cannot be implemented in the various degenerations, and figuring out how that works seems like a valuable problem to consider.

Outline. Section 2 focuses on discrete polymers. The pertinent definitions and Theorem 2.4 are given in Section 2.1, along with the proof of the theorem. Section 2.2 relates the operator $\mathcal{W}$ and the discrete geometric analogs of the Pitman transform to the gRSK correspondence and row insertion. Section 2.3 contains the zero-temperature limit of Theorem 2.4. This relates to the usual RSK correspondence and last passage percolation. Section 3 contains analogous results for the semi-discrete version of the gRSK correspondence. In that setting, the proof Theorem 3.4 (which is the semi-discrete analog of Theorem 2.4) is considerably simpler. A reader may benefit from browsing that proof first before the proof of Theorem 2.4. Section 4 closes with three brief remarks/questions. The first relates to braid relations for the discrete (geometric) Pitman transforms, the second considers a KPZ-equation analog of the Airy sheet construction, and the third notes some other polymer invariances recently proved and queries whether our main theorem has a lifting to the level of stochastic vertex models.

Notation. For an integer $m \in \mathbb{Z}$, we write $\mathbb{Z}_{\geq m}$ to represent all integers $n \geq m$. For two integers $m \leq n$, define $[\![m, n]\!] := \{m, \dots, n\}$ to be the set of all integers between m and n, inclusive of the endpoints. For two real numbers x and y, we use shorthand $x \wedge y = \min(x, y)$ and $x \vee y = \max(x, y)$. Other notation will be introduced where it is used.

Acknowledgements. I. Corwin wishes to thank Duncan Dauvergne for graciously informing him about the work in progress [Dau20] and pointing him to reread certain parts of the work of Noumi and Yamada [NY04]; Konstantin Matveev for providing an alternative third proof of Theorem 2.4 and granting him permission to produce that proof herein; and Alan Hammond, Neil O'Connell, Leonid Petrov, Mark Rychnovsky and Xuan Wu for valuable comments on a first version of this article. I. Corwin was partially supported by the NSF grants DMS:1811143 and DMS:1664650, and the Packard Fellowship for Science and Engineering.

§2. Discrete polymers

2.1. Partition function invariance

Before stating our main result, Theorem 2.4, we first introduce non-intersecting (multi)paths, partition functions and the Pitman transform.

Definition 2.1 (Non-intersecting (multi)paths). Fix $n \in \mathbb{Z}_{\geq 2}$. We will consider the graph with vertices $\Lambda_n := \mathbb{Z}_{\geq 1} \times [\![1, n]\!]$ and edges between nearest-neighbors. We will call the second coordinate in Λ_n the *level* and will plot the points $(x, \ell) \in \Lambda_n$ as in Figure 2 so that level $\ell = 1$ is on the top and level $\ell = n$ is on the bottom. In some of our discussion in what follows, we will consider subsets of Λ_n where vertices (and also the edges incident to them) have been removed from the bottom right corner of Λ_n in such a way that the removed vertices form a Young diagram under the French convention. The path and partition function definitions that follow readily generalize to that setting.

A *path* in Λ_n is a finite ordered sequence of nearest neighbor, up-right connected points $\pi = \big((a_1, b_1), \ldots, (a_L, b_L)\big)$ in Λ_n. Precisely, for any $L \in \mathbb{Z}_{\geq 1}$ we assume that $(a_i, b_i) \in \Lambda_n$ for $i \in [\![1, L]\!]$ and further we assume that for all $i \in [\![1, L-1]\!]$ either

- $a_i - a_{i+1} = 1$ and $b_i = b_{i+1}$ or
- $a_i = a_{i+1}$ and $b_{i+1} - b_i = 1$.

The *starting point* of such a path is (a_1, b_1) and the *ending point* is (a_L, b_L). We will generally use π to denote a path and we will not label the ordered vertices traversed along the path (thus the (a_i, b_i) notation will not be used below and was just introduced to clarify the definition of a path). We will often deal with multiple paths $\pi_1, \ldots, \pi_k$ in which case we will write $\pi = (\pi_1, \ldots, \pi_k)$ and call π a *multipath.* Two paths π_1 and π_2 are called *non-intersecting* if, as sets, $\pi_1 \cap \pi_2 = \emptyset$. In words, π_1 and π_2 do not touch at any vertices in Λ_n (including the starting and ending points). A multipath $\pi = (\pi_1, \ldots, \pi_k)$ is non-intersecting if for each $1 \leq i \neq j \leq k$, π_i and π_j are non-intersecting. We will assume that all multipaths are non-intersecting. Figure 2 depicts two non-intersecting paths (hence a multipath): π_1 from (u_1, n) to $(v_1, 1)$ and π_2 from (u_2, n) to $(v_2, 1)$.

For any $k \in \mathbb{Z}_{\geq 1}$, and pair of k starting and ending points

$$U = \big((u_i, \ell_i)\big)_{i \in [\![1,k]\!]} \qquad \text{and} \qquad V = \big((v_i, m_i)\big)_{i \in [\![1,k]\!]}$$

we define the set of (non-intersecting) multipaths π from U to V

$$U \to V := \Big\{\pi : \forall i \in [\![1, k]\!], \pi_i \text{ starts at } (u_i, \ell_i) \text{ and ends at } (v_i, m_i)\Big\}.$$

If the set $U \to V$ is non-empty, then we say that the pair (U, V) constitute an *endpoint pair.*

Our next definition builds upon non-intersecting paths and defines a partition function. The definition of the weight in (2.1) may seem a bit odd. However, if we define $d_{x,m} = \frac{D_m(x)}{D_m(x-1)}$, then the partition functions defined below match those generally studied in the context of directed polymers with environment given by the $d_{x,m}$. We will return to this in Section 2.2.

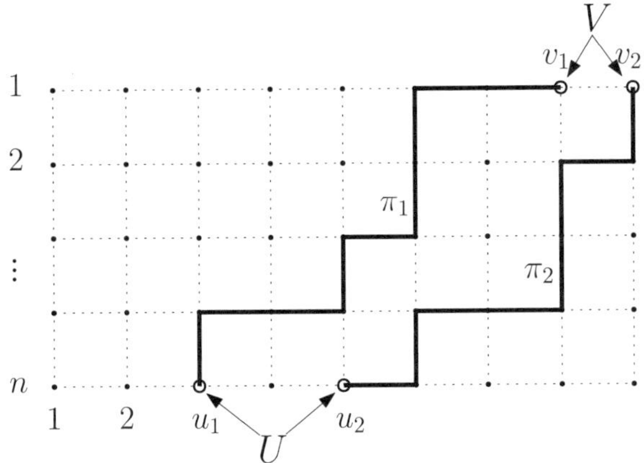

Fig. 2. A subset of the lattice Λ_n with two non-intersecting paths π_1 and π_2 depicted. The starting points $U = \big((3,n),(5,n)\big)$ and ending points $V = \big((8,1),(9,1)\big)$ (depicted by circles) constitute an endpoint pair (U,V) since the set $U \to V$ of non-intersecting paths from U to V is non-empty.

Definition 2.2 (Partition functions)**.** Fix any n weakly increasing integers $r_1 \le \cdots \le r_n$ and then fix any n functions $D_1,\ldots,D_n$ where $D_i : \mathbb{Z}_{\ge r_i} \to (0,\infty)$, for $i \in [\![1,n]\!]$. We will adopt the convention that $D_i(r_i - 1) = 1$ for all $i \in [\![1,n]\!]$. Write $D = (D_1,\ldots,D_n)$ to denote the n-tuple of such functions. We say that a point (x,m) is *in the domain of* D if $m \in [\![1,n]\!]$ and $x \ge r_m$. For any set U of points in the domain of D, the *weight* of U with respect to D is

$$D\big[U\big] := \prod_{(x,m)\in U} \frac{D_m(x)}{D_m(x-1)}. \tag{2.1}$$

For a single path π composed entirely of points in the domain of D, the weight of π with respect to D is given as above by treating π as a set of points. Explicitly,

$$D\big[\pi\big] := \prod_{(x,m)\in \pi} \frac{D_m(x)}{D_m(x-1)}. \tag{2.2}$$

For a multipath $\pi = (\pi_1,\ldots,\pi_k)$ composed of points in the domain of D,

$$D\big[\pi\big] := D\big[\pi_1\big]\cdots D\big[\pi_k\big].$$

This definition is consistent with (2.1) if we treat π as the union of the points in each π_i. For any endpoint pair (U,V) with both U and V in the domain of D, the *partition function* from U to V with respect to D is

$$D[U \to V] := \sum_{\pi \in U \to V} D[\pi].$$

It will also be useful to have half-open versions of the partition function:

(2.3)
$$D(U \to V] := \frac{D[U \to V]}{D[U]} \qquad \text{and} \qquad D[U \to V) := \frac{D[U \to V]}{D[V]}.$$

Our final set of definitions deal with the Pitman transform and its tensorization.

Definition 2.3 (Discrete geometric Pitman transform). For any $r \in \mathbb{Z}$ and any functions $f, g : \mathbb{Z}_{\geq r} \to (0, \infty)$ define functions $(g \odot f) : \mathbb{Z}_{\geq r} \to (0, \infty)$ and $(f \otimes g) : \mathbb{Z}_{\geq r+1} \to (0, \infty)$ by

$$(g \odot f)(x) := f(x) \cdot \sum_{m=r}^{x} \frac{g(m)}{f(m-1)} \qquad \text{for } x \geq r$$

$$(f \otimes g)(x) := g(x) \cdot \left(\sum_{m=r}^{x} \frac{g(m)}{f(m-1)} \right)^{-1} \qquad \text{for } x \geq r+1$$

where we have adopted the convention that $f(r-1) = 1$ in the denominator when $m = r$. Define the operator $\mathcal{T}$ which acts on the pair (f, g) as

$$\mathcal{T}(f, g) := \Big((g \odot f), (f \otimes g) \Big). \tag{2.4}$$

$\mathcal{T}$ acts on the tensor-product of two functions from $\mathbb{Z}_{\geq r} \to (0, \infty)$ and returns the tensor product of two functions, the first from $\mathbb{Z}_{\geq r} \to (0, \infty)$ and the second from $\mathbb{Z}_{\geq r+1} \to (0, \infty)$.

For any $m \in [\![1, n-1]\!]$ and any n weakly increasing integers $r_1 \leq \cdots \leq r_n$ with $r_m = r_{m+1} = r$ for some r, define the operator $\mathcal{T}_{r,m}$ which acts on the n-fold tensor product $D = (D_1, \ldots, D_n)$ of functions $D_i : \mathbb{Z}_{\geq r_i} \to (0, \infty)$, $i \in [\![1, n]\!]$, as

(2.5)
$$\mathcal{T}_{r,m} D := \Big(D_1, \ldots, D_{m-1}, D_{m+1} \odot D_m, D_m \otimes D_{m+1}, D_{m+2}, \ldots, D_n \Big).$$

$\mathcal{T}_{r,m}$ acts as $\mathcal{T}$ in the m and $m+1$ slot, and this action is well-defined since by assumption both $D_m, D_{m+1} : \mathbb{Z}_{\geq r} \to (0, \infty)$.

Now consider n functions $D_1, \ldots, D_n : \mathbb{Z}_{\geq 1} \to (0, \infty)$ as in Definition 2.2 and write $D = (D_1, \ldots, D_n)$. Using the operators $\mathcal{T}_{r,m}$ from (2.5), define the operators $\mathcal{S}_r$ for $r \in [\![1, n-1]\!]$ and the operator $\mathcal{W}$ which act on D via

(2.6)
$$\mathcal{S}_r D := \mathcal{T}_{r,r}\mathcal{T}_{r,r+1} \cdots \mathcal{T}_{r,n-2}\mathcal{T}_{r,n-1}D, \quad \mathcal{W}D := \mathcal{S}_{n-1}\mathcal{S}_{n-2} \cdots \mathcal{S}_2\mathcal{S}_1 D.$$

$\mathcal{S}_r$ acts on D by first applying (in that order) $\mathcal{T}_{r,n-1}$ through $\mathcal{T}_{r,r}$. Likewise, $\mathcal{W}$ acts on D by first applying (in that order) $\mathcal{S}_1$ through $\mathcal{S}_{n-1}$. Notice that the sequence of applications of the $\mathcal{T}$ operators in $\mathcal{W}$ is well-defined. Indeed, in terms of the values of $(r_1, \ldots, r_n)$ which set the domain of D, initially, as assumed, D has $r_i = i \wedge 1$. After applying $\mathcal{T}_{1,n-1}$ to D, the value of r_n becomes 2 (all others stay the same). Continuing in this manner, we see that for $r \in [\![1, n-1]\!]$, $\mathcal{S}_r \cdots \mathcal{S}_1 D$ has domain specified by $r_i = i \wedge (r+1)$ for $i \in [\![1, n]\!]$. In other words, when we apply the $\mathcal{T}_{r,m}$ operator, the domain changes from having $r_m = r_{m+1} = r$ to having $r_m = r$ and $r_{m+1} = r + 1$. Figure 1 shows how the original domain for D transforms into the domain for $\mathcal{W}D$. There is a graphical depiction of this sequence of applications of $\mathcal{T}$ which is discussed and used in Section 2.2. Also therein we explain how this operator $\mathcal{W}$ is related to the geometric RSK correspondence.

We are now ready to state the main result of this paper. There is a semi-discrete version of this which can be found as Theorem 3.4 in Section 3.1. Also, there are zero-temperature versions of this result which are stated as Corollary 2.3 in Section 2.3 in the discrete case, and as Corollary 3.3 in Section 3.3 in the semi-discrete case. In fact, all of these results can be derived as a limit of our main theorem below.

Theorem 2.4. *Let* $U = \big((u_i, n)\big)_{i \in [\![1,k]\!]}$ *and* $V = \big((v_i, 1)\big)_{i \in [\![1,k]\!]}$ *be any endpoint pair and define* $\Uparrow U := \big((u_i, n \wedge u_i)\big)_{i \in [\![1,k]\!]}$ *(see Figure 1 for an illustration). Then, for any* n *functions* $D_1, \ldots, D_n : \mathbb{Z}_{\geq 1} \to (0, \infty)$, *writing* $D = (D_1, \ldots, D_n)$ *we have*

$$D\big[U \to V\big] = \big(\mathcal{W}D\big)\big[\Uparrow U \to V\big] \tag{2.7}$$

where the operator $\mathcal{W}$ *is defined in* (2.6).

Remark 2.5. We follow the same three step program as used to prove [DOV18, Proposition 4.1] (stated here as Corollary 3.3 in Section 3.3), though encounter some new complexities due to the discreteness of our present setting. In Step 1 we consider $n = 2$ and $k = 1$ where $U = (u, 2)$ and $V = (v, 1)$ (note that for simpler notation in the $k = 1$ case we drop the extra parenthesis since really $U = \big((u, 2)\big)$ and likewise

$V = ((v,1))$). The proof of this special case reduces to a summation identity which we readily checked by induction. Step 2 extends the result to $n = 2$ and $k \geq 1$ by a decomposition of $D[U \to V]$ into a product over terms to which the result of Step 1 can be applied. Step 3 extends the result to $n \geq 2$ and $k \geq 1$ via a decomposition of $D[U \to V]$ into a sum of products of terms to which the result from Step 2 can be applied. Besides needing to prove a new summation identity in Step 1 (which is rather simple), the most significant new complexity in this proof arises from the fact that the operators $\mathcal{T}_{r,m}$ shift the domain of the functions upon which they act. Keeping track of this effect and constructing a suitable decomposition which works in these deformed domains constitutes the main challenge of the proof (which mainly arises in Step 3).

Besides proving the above theorem, it was not immediately obvious how to correctly formulate the discrete geometric generalization of [DOV18, Proposition 4.1]. Two things provided some guidance in this regard. The first was a set of online notes by Y. Pei available at `https://toywiki.xyz/` in which the discrete non-geometric Pitman transform was discussed. The second was the work of [COSZ14] on the geometric RSK correspondence. Indeed, in Section 2.2 we explain how the discrete geometric Pitman transform and $\mathcal{W}$ operator are related to the geometric RSK correspondence.

Definition 2.6. Theorem 2.4 introduces the notation $\Uparrow U$. In the proof we will need a refinement of it that we present here. For any integers r and m, define the operator $\Uparrow_{r,m}$ which acts on sets U of points $U = \big((u_i,\ell_i)\big)_{i\in[\![1,k]\!]}$ (for k arbitrary) as follows: if $(r, m+1) \in U$, then $\Uparrow_{r,m} U$ is composed of all points in U except that $(r, m+1)$ is replaced by (r,m) (if $(r,m) \in U$ as well, then $\Uparrow_{r,m} U$ is U with the point $(r, m+1)$ removed); if $(r, m+1) \notin U$, then $\Uparrow_{r,m} U = U$. In a similar spirit to (2.6), we define

$$\Uparrow_r := \Uparrow_{r,r} \Uparrow_{r,r+1} \cdots \Uparrow_{r,n-2} \Uparrow_{r,n-1}, \qquad \Uparrow := \Uparrow_{n-1} \Uparrow_{n-2} \cdots \Uparrow_2 \Uparrow_1. \tag{2.8}$$

The $\Uparrow U$ defined in (2.8) matches $\Uparrow U$ defined in the statement of Theorem 2.4.

Proof of Theorem 2.4. We prove this in three steps.

Step 1 ($n = 2$, $k = 1$ case of (2.7)). In this case, $D = (D_1, D_2)$, $U = (u,2)$ and $V = (v,1)$ (actually, $U = \big((u,2)\big)$ and $V = \big((v,1)\big)$, but we will drop the outer parenthesis in this case). Having in mind our use of this step later in Step 3 (where we deal with $n \geq 2$), we may consider a slightly more general situation where, for some $r \geq 1$, $D_1, D_2 : \mathbb{Z}_{\geq r}$.

In that case, what we seek to prove that for any $r \le u \le v$,

$$D\big[(u,2) \to (v,1)\big] = \big(\mathcal{T}_{r,1}D\big)\big[\big(u, 2 \wedge (u-r+1)\big) \to (v,1)\big]. \tag{2.9}$$

Note that $\big(u, 2 \wedge (u-r+1)\big) = \Uparrow_{r,1} U$. It suffices to prove (2.9) with $r=1$ since everything can be trivially shifted into larger r. So, for the rest of this step, let us prove (2.9) under the assumption that $r=1$. In that case $\big(u, 2 \wedge (u-r+1)\big) = \Uparrow_{1,1} U = \Uparrow U$.

The left-hand side of (2.9) can be written explicitly as

$$D\big[(u,2) \to (v,1)\big] = \frac{D_1(v)}{D_2(u-1)} \sum_{m=u}^{v} \frac{D_2(m)}{D_1(m-1)} \tag{2.10}$$

where we have used the earlier assumed convention that $D_i(0) = 1$ for $i = 1$ and 2.

Clearly there are two cases to consider, when $u = 1$ and when $u > 1$. When $u = 1$, the identity we seek to prove reads (writing $\mathcal{T}$ in place of $\mathcal{T}_{1,1}$)

$$D\big[(1,2) \to (v,1)\big] = \big(\mathcal{T}D\big)\big[(1,1) \to (v,1)\big]. \tag{2.11}$$

Since there is only one path from $(1,1)$ to $(v,1)$, the right-hand side of (2.11) equals the weight of the path from $(1,1)$ to $(v,1)$ with respect to $\mathcal{T}D$. Observing the telescoping of the product in (2.2) and using the assumed convention that $(\mathcal{T}D)_1(0) = 1$, we find that the right-hand side of (2.11) equals $(\mathcal{T}D)_1(v)$. It follows from Definition 2.3 that $(\mathcal{T}D)_1(v) = (D_2 \odot D_1)(v)$ which equals the expression in (2.10), thus proving the case $u = 1$.

The case $x > 1$ is just a bit harder. Now $\Uparrow U = U = (u,2)$ so, using (2.10), we seek to prove

$$\frac{D_1(v)}{D_2(u-1)} \sum_{m=u}^{v} \frac{D_2(m)}{D_1(m-1)} = \frac{\big(\mathcal{T}D\big)_1(v)}{\big(\mathcal{T}D\big)_2(u-1)} \sum_{m=u}^{v} \frac{\big(\mathcal{T}D\big)_2(m)}{\big(\mathcal{T}D\big)_1(m-1)}. \tag{2.12}$$

Keeping in mind our convention that $D_i(0) = 1$, we introduce the notation

$$G_m := \frac{D_2(m)}{D_1(m-1)} \quad \text{and} \quad G_{a,b} := \sum_{m=a}^{b} G_m \quad \text{for integers } 1 \le a < b.$$

We can rewrite $(\mathcal{T}D)_1$ and $(\mathcal{T}D)_2$ via this notation as:

$$\big(\mathcal{T}D\big)_1(v) = D_1(v) G_{1,v} \quad \text{and} \quad \big(\mathcal{T}D\big)_2(u-1) = D_2(u-1)\big(G_{1,u-1}\big)^{-1}.$$

Using this and canceling the common factor $\frac{D_1(v)}{D_2(u-1)}$ from both sides, (2.12) reduces to

$$G_{u,v} = G_{1,v}G_{1,u-1}\sum_{m=u}^{v}\frac{G_m}{G_{1,m}G_{1,m-1}},$$

or equivalently (after moving the $G_{1,v}G_{1,u-1}$ terms to the left-hand side)

$$\frac{G_{u,v}}{G_{1,v}G_{1,u-1}} = \sum_{m=u}^{v}\frac{G_m}{G_{1,m}G_{1,m-1}}. \tag{2.13}$$

(2.13) is an exact summation identity which follows by induction in v. The base case $v = u$ is easily checked. Assuming the identity up to v, to show (2.13) for $v + 1$ it suffices to verify that

$$\frac{G_{u,v+1}}{G_{1,v+1}G_{1,u-1}} = \frac{G_{u,v}}{G_{1,v}G_{1,u-1}} + \frac{G_{v+1}}{G_{1,v+1}G_{1,v}}.$$

Decomposing all terms into $G_{1,u-1}$, $G_{u,v}$ and G_{v+1} (i.e., $G_{u,v+1} = G_{u,v} + G_{v+1}$, $G_{1,v+1} = G_{1,u-1} + G_{u,v} + G_{v+1}$, and $G_{1,v} = G_{1,u-1} + G_{u,v}$) and cross-multiplying immediately yields the eqaulity. This completes the proof of (2.13) and hence also the proof of (2.12) and Step 1.

Step 2 ($n = 2$, $k \geq 1$ case of (2.7)). Now $D = (D_1, D_2)$, $U = \big((u_i, 2)\big)_{i\in[\![1,k]\!]}$ and $V = \big((v_i, 1)\big)_{i\in[\![1,k]\!]}$. As in Step 1, we will consider $D_1, D_2 : \mathbb{Z}_{\geq r} \to (0,\infty)$ for some $r \geq 1$, and assume that all $u_i, v_i \geq r$. In this step we seek to prove that

$$D\big[U \to V\big] = \big(\mathcal{T}_{r,1}D\big)\big[\Uparrow_{r,1}U \to V\big]. \tag{2.14}$$

Recall that $\Uparrow_{r,1}$ is given in Definition 2.6. In particular, $\Uparrow_{r,1}U = \big((u_i, 2\wedge (u_i - r + 1))\big)_{i\in[\![1,k]\!]}$. As in Step 1, it suffices to prove (2.14) for $r = 1$, in which case $\Uparrow_{1,1}U = \Uparrow U$. For the rest of the proof we assume $r = 1$. In that case (2.14) involves $\mathcal{T}_{1,1}$ which we abbreviate as $\mathcal{T}$.

The proof of (2.14) is based on a decomposition of $D[U \to V]$ and $(\mathcal{T}D)[\Uparrow U \to V]$ into factors which can be matched by applying the $k = 1$ result from Step 1. We start by partitioning the integers $[\![u_1, v_k]\!]$ into disjoint intervals that we call type 0, type 1 or type 2. With the convention that $v_0 = -\infty$ and $u_{k+1} = +\infty$, for each $i \in [\![1,k]\!]$, let $a_i = u_i \vee (v_{i-1} + 1)$ and $b_i = (u_{i+1} - 1) \wedge v_i$. If $a_i < b_i$, then we say that the interval $[\![a_i, b_i]\!]$ is of type 1, and we write type_1 to be the set of all such type 1 intervals. Likewise, for each $i \in [\![1,k]\!]$, any integer c such that $y_{i-1} < c < x_i$ will be called type 0. All integers which are not type 0 and are not in intervals of type 1 will be called type 2 and we

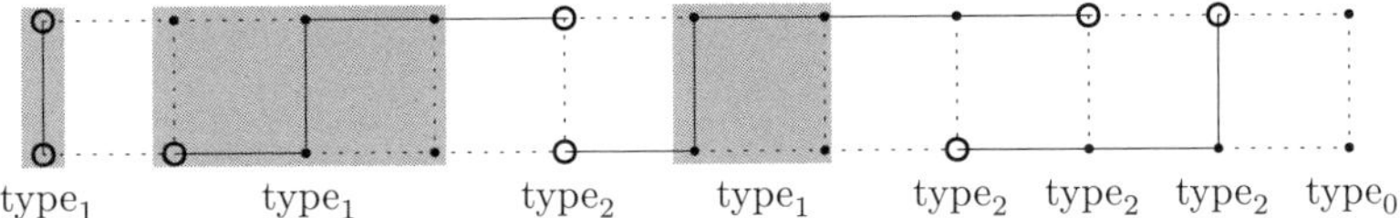

Fig. 3. U is represented by the open discs on the bottom level and V on the top level Intervals of type 1 are shaded in grey; integers of types 0 and 2 are also labeled, but not shaded. Notice that the first type 1 interval is actually composed of just a single integer, while the second one contains three consecutive integers and the third one contains two consecutive integers. Each type 1 interval represents a place where multipaths $\pi \in U \to V$ may vary in terms of which points are included in π.

write type_2 for the set of all such type 2 integers. Figure 3 provides an illustration for this partitioning.

Using the decomposition into types, we can write

(2.15)
$$D\big[U \to V\big] = \prod_{[\![a,b]\!] \in \mathrm{type}_1} D\big[(a,2) \to (b,1)\big] \prod_{c \in \mathrm{type}_2} D\big[(c,2) \to (c,1)\big].$$

To verify this note that each type 1 term can be expanded as a sum of single-path weights over paths from the bottom left to top right of the interval in question (type 1 intervals which are composed of a single integer correspond to a single path which goes straight up from the bottom layer to the top layer). All type 2 terms contribute a product of weights $D\big[(c,1)\big]D\big[(c,2)\big]$, and all type 0 terms contribute a multiplicative factor of 1. This expansion shows that the right-hand side of (2.15) can be written as a sum over certain subsets of vertices of products of the weights of the vertices. By the construction of the three types, it is easy to see that these subsets are precisely in bijective correspondence with the set of multipaths in $U \to V$, hence proving (2.15). See Figure 3 for an illustration.

Given the decomposition in (2.15), we may now apply the result of Step 1 to each term. In the product over type_1 terms, note that the result of Step 1 implies that

$$D\big[(a,2) \to (b,1)\big] = \big(\mathcal{T}D\big)\big[(a, 2 \wedge a) \to (b,1)\big].$$

Thus, we can replace each term in that product by the corresponding term involving $\mathcal{T}D$. For the terms in the type_2 product, we also have that

$$D\big[(c,2) \to (c,1)\big] = \big(\mathcal{T}D\big)\big[(c, 2 \wedge c) \to (c,1)\big].$$

This can be seen quite directly, or as a simple application of the result of Step 1 when $a=b$. Putting this all together we find that the right-hand side of (2.15) equals

$$\prod_{[\![a,b]\!]\in\mathtt{type}_1}(\mathcal{T}D)\big[(a,2\wedge a)\to(b,1)\big]\prod_{c\in\mathtt{type}_2}(\mathcal{T}D)\big[(c,2\wedge c)\to(c,1)\big]$$
$$=(\mathcal{T}D)\big[\Uparrow U\to V\big],$$

The equality above is a similar decomposition as in (2.15), the main difference being that U is replaced by $\Uparrow U$. This completes the proof of (2.14) (with $r=1$) and hence Step 2.

Step 3 ($n\geq 2$, $k\geq 1$ case of (2.7)). In this case, $D=(D_1,\ldots,D_n)$, $U=\big((u_i,n)\big)_{i\in[\![1,k]\!]}$ and $V=\big((v_i,1)\big)_{i\in[\![1,k]\!]}$. The key to this proof is a decomposition of $D[U\to V]$ into a sum of products to which we can apply the $n=2$ result proved earlier in Step 2. Since this decomposition is a bit involved, we will first explicitly work out the $n=3$ proof in that hope that it may make it easier to understand the general n proof.

Before embarking on the $n=3$ case it will be useful to rewrite the identity (2.14) proved in Step 2 in terms of the notation that we will use below. Consider any $m\in[\![1,n-1]\!]$ and any sequence of weakly increasing integers $r_1\leq\cdots\leq r_n$ such that $r_m=r_{m+1}=r$ for some r. Further, consider any n functions $\widetilde{D}=(\widetilde{D}_1,\ldots,\widetilde{D}_n)$ where $\widetilde{D}_i:\mathbb{Z}_{\geq r_i}\to(0,\infty)$ for $i\in[\![1,n]\!]$, and any endpoint pair (W,Z) where $W=\big((w_i,m+1)\big)_{i\in[\![1,k]\!]}$, $Z=\big((z_i,m)\big)_{i\in[\![1,k]\!]}$ and all $w_i,z_i\geq r$. Then, (2.14) of Step 2 implies that

$$\widetilde{D}\big[W\to Z\big]=\big(\mathcal{T}_{r,m}\widetilde{D}\big)\big[\Uparrow_{r,m}W\to Z\big]. \tag{2.16}$$

We return now to proving (2.7) with $n=3$. We will refer to Figure 4 in describing certain steps the proof. Recall that for $n=3$, $\mathcal{W}=\mathcal{S}_2\mathcal{S}_1=\mathcal{T}_{2,2}\mathcal{T}_{1,1}\mathcal{T}_{1,2}$. Thus, in order to prove the equality in (2.7), we will prove three intermediate equalities, relating an expression with D to one with $\mathcal{T}_{1,2}D$, then relating that to an expression with $\mathcal{T}_{1,1}\mathcal{T}_{1,2}D$ and finally relating that to an expression with $\mathcal{T}_{2,2}\mathcal{T}_{1,1}\mathcal{T}_{1,2}D$. Collecting these equalities together will prove (2.7).

We start with the following decomposition (recall half-open partition functions from (2.3))

$$D\big[U\to V\big]=\sum_Z D\big[U\to W^+\big)D\big[W^+\to W\big)\times D\big[W\to Z\big]D\big(Z\to Z^-\big]D\big(Z^-\to V\big], \tag{2.17}$$

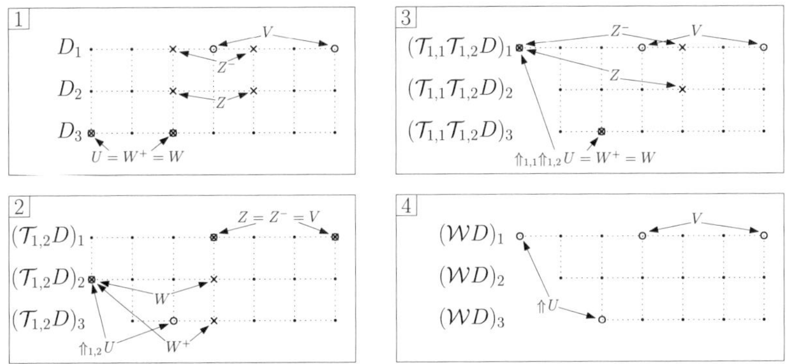

Fig. 4. Sets of points used in the proof of (2.7) when $n = 3$ and $k = 2$.

with notation that we now define. The summation in (2.17) is over $Z = \big((z_i, 2)\big)_{i\in[\![1,k]\!]}$. Given Z, we define $Z^- := \big((z_i, 1)\big)_{i\in[\![1,k]\!]}$ (i.e., points in Z^- match those in Z except they have a level one less). The W and W^+ are simply set equal to U. (It may seem unnecessary to include W and W^+ here, and it is. However, in the general n case they will be important, so we keep them to match with the notation eventually used there.) The summation in (2.17) is restricted to only those Z such that (U, W^+), (W^+, W), (W, Z), (Z, Z^-) and (Z^-, V) are all endpoint pairs. Note that due to the assumption that $W = W^+ = U$ and the relationship between Z and Z^-, these five conditions reduce to two: that (W, Z) and (Z^-, V) are endpoint pairs. Furthermore, the summand in the right-hand side of (2.17) reduces to $D[U \to Z]D(Z \to Z^-]D(Z^- \to V]$. Display box 1 of Figure 4 illustrates the choices of W, W^+, Z, Z^- in this decomposition.

To see why the decomposition in (2.17) holds, recall that $D[U \to V]$ is equal to a sum of weights of multipaths π from U to V. We can partition that sum based on the locations Z of the exit points for the multipath π going from level 2 to level 1 (hence Z^- has the meaning as the entry point locations into level 1). The sum over weights of multipaths π which have a given value of Z factorizes precisely as in (2.17), thus proving the decomposition. Notice that the use of the half-open and half-closed path sums defined in (2.3) is important here to ensure that the correct weight factors arise in our decomposition.

The decomposition in (2.17) has isolated the dependence on D_2 and D_3 in the right-hand side summand. Indeed, since $D[U \to W^+) = D[W^+ \to W) = 1$, they do not depend on D_2 or D_3, and clearly $D(Z \to Z^-] = D[Z^-]$ and $D(Z^- \to V]$ only depend on D_1. This leaves $D[W \to Z]$ as the only term which involves D_2 and D_3. We now apply (2.16) with $r = 1$, $m = 2$, and $\widetilde{D} = D$. This implies that $D[W \to Z] = (\mathcal{T}_{1,2}D)[\Uparrow_{1,2}W \to Z]$. Since $(\mathcal{T}_{1,2}D)_1 = D_1$, we may replace D by $D' = (\mathcal{T}_{1,2}D)$ in the other terms in the summand of (2.17) without changing their values (they are all either constant or only depend on D_1). Moreover, since $D[U \to W^+) = D[W^+ \to W) = 1$, we may replace these terms by $D[\Uparrow_{1,2}U \to \Uparrow_{1,2}W^+) = D[\Uparrow_{1,2}W^+ \to \Uparrow_{1,2}W) = 1$. This implies that, denoting $D' = \mathcal{T}_{1,2}D$, $U' = \Uparrow_{1,2}U$, $W'^+ = \Uparrow_{1,2}W^+$ and $W' = \Uparrow_{1,2}W$,

$$D\big[U \to V\big] = \sum_Z D'\big[U' \to W'^+\big)D'\big[W'^+ \to W'\big) \times D'\big[W' \to Z\big]D'\big(Z \to Z^-\big]D'\big(Z^- \to V\big],$$

where we now assume that $U' = W'^+ = W'$ and where the sum over Z is such that (W', Z) and (Z^-, V) are endpoint pairs (note that the condition that (W', Z) is an endpoint pair imposes the same restriction on Z as the condition that (W, Z) is an endpoint pair). This right-hand side is a decomposition of $D'[U' \to V]$ in the same spirit as (2.17), hence we have shown that (going back to $\mathcal{T}_{1,2}D$ in place of D' and $\Uparrow_{1,2}U$ in place of U')

$$D\big[U \to V\big] = \big(\mathcal{T}_{1,2}D\big)\big[\Uparrow_{1,2}U \to V\big]. \tag{2.18}$$

This is the first of our three intermediate equalities.

In order to relate the right-hand side of (2.18) to an expression involving $\mathcal{T}_{1,1}\mathcal{T}_{1,2}D$, we employ another decomposition. As above, let us denote $D' = \mathcal{T}_{1,2}D$ and $U' = \Uparrow_{1,2}U$. Then,

$$D'\big[U' \to V\big] = \sum_W D'\big[U' \to W^+\big)D'\big[W^+ \to W\big) \times D'\big[W \to Z\big]D'\big(Z \to Z^-\big]D'\big(Z^- \to V\big],$$

with notation that we now define. The summation in (2.18) is over $W = \big((w_i, 2)\big)_{i \in [\![1,k]\!]}$. Given W, we define $W^+ = \big((w_i, (w_i{+}1) \wedge 3)\big)_{i \in [\![1,k]\!]}$ (i.e., as long as the first coordinate is ≥ 2, every point in W corresponds to a point in W^+ with one larger level; if $(1, 2) \in W$ then $(1, 2) \in W^+$ as well). The Z and Z^- are simply set equal to V. The summation in (2.18) is restricted to only those W such that $('U, W^+)$, (W^+, W), (W, Z), (Z, Z^-) and (Z^-, V) are all endpoint pairs. Note that due to the assumption that $Z = Z^- = V$ and the relationship between W and $W+$, these five conditions reduce to two: that $('U, W^+)$ and (W, Z) are endpoint pairs. Furthermore, the summand in the right-hand side of

(2.18) reduces to $D'[U' \to W^+)D'[W^+ \to W)D'[W \to Z]$. Display box 2 of Figure 4 illustrates the choices of W, W^+, Z, Z^- in this decomposition. The same reasoning as for the first decomposition (2.17) applies in justifying its validity.

Now, notice that the decomposition in (2.18) has isolated the dependence on D'_1 and D'_2 in the right-hand side summand. Indeed, since $D'(Z \to Z^-] = D'(Z^- \to V] = 1$, they do not depend on D'_1 or D'_2, and clearly $D'[U' \to W^+)$ and $D'[W^+ \to W)$ only depend on D'_3 (remember the notation defined in (2.3) to see this). This leaves $D'[W \to Z]$ as the only term which involves D'_1 and D'_2. We now apply (2.16) with $r = 1$, $m = 1$, and $\widetilde{D} = D'$. This implies that $D'[W \to Z] = (\mathcal{T}_{1,1}D')[\Uparrow_{1,1}W \to Z]$. Since $(\mathcal{T}_{1,1}D')_3 = D'_3$, we may replace D' by $D'' = (\mathcal{T}_{1,1}D')$ in the other terms in the summand of (2.18) without changing their values (they are all either constant or only depend on D'_3). Now observe that

$$D''\big[U' \to W^+\big) = D''\big[\Uparrow_{1,1}U' \to \Uparrow_{1,1}W^+\big),$$
$$D''\big[W^+ \to W\big) = D''\big[\Uparrow_{1,1}W^+ \to \Uparrow_{1,1}W\big),$$

and that the condition that (U', W^+) and (W, Z) are endpoint pairs is equivalent to the condition that $(\Uparrow_{1,1}U, \Uparrow_{1,1}W^+)$ and $(\Uparrow_{1,1}W, Z)$ are endpoint pairs. Thus, writing $U'' = \Uparrow_{1,1}U'$, $W'^+ = \Uparrow_{1,1}W^+$ and $W' = \Uparrow_{1,1}W$ we have shown that

$$D'\big[U' \to V\big] = \sum_{W'} D''\big[U'' \to W'^+\big)D''\big[W'^+ \to W'\big) \times D''\big[W' \to Z\big]D''\big(Z \to Z^-\big]D''\big(Z^- \to V\big],$$

where $Z = Z^- = V$ and where the sum is restricted to W' such that (U'', W'^+) and (W', Z) are endpoint pairs. This, however, is a decomposition for $D''[U'' \to V]$, and hence we have shown that (going back to $\mathcal{T}_{1,1}\mathcal{T}_{1,2}D$ in place of D'' and $\Uparrow_{1,1}\Uparrow_{1,2}U$ in place of U'')

$$(\mathcal{T}_{1,2}D)\big[\Uparrow_{1,2}U \to V\big] = (\mathcal{T}_{1,1}\mathcal{T}_{1,2}D)\big[\Uparrow_{1,1}\Uparrow_{1,2}U \to V\big]. \tag{2.19}$$

This is the second of our three intermediate equalities.

The final step is to use a decomposition similar to that of (2.17) to relate the right-hand side of (2.19) above to an expression involving $\mathcal{T}_{2,2}\mathcal{T}_{1,1}\mathcal{T}_{1,2}D$. As above, let us denote $D'' = \mathcal{T}_{1,1}\mathcal{T}_{1,2}D$ and $U'' = \Uparrow_{1,1}\Uparrow_{1,2}U$. Then

$$D''\big[U'' \to V\big] = \sum_{Z} D''\big[U'' \to W^+\big)D''\big[W^+ \to W\big) \times D''\big[W \to Z\big]D''\big(Z \to Z^-\big]D''\big(Z^- \to V\big], \tag{2.20}$$

with notation that we now define. The summation is over sets of k points Z which are either on level 2 or the point $(1,1)$. For each point on level 2 in Z, there is a point on level 1 in Z^- with the same first coordinate; and if $(1,1) \in Z$ then $(1,1) \in Z^-$ as well. The W and W^+ are simply set to be U''. The summation in (2.20) is over Z such that (U, W^+), (W^+, W), (W, Z), (Z, Z^-) and (Z^-, V) are all endpoint pairs. Note that due to the assumption that $U'' = W^+ = W$ and the relationship between Z and Z^-, these five conditions reduce to two: that (W, Z) and (Z^-, V) are endpoint pairs. Furthermore, the summand in the right-hand side of (2.17) reduces to $D''[U'' \to Z]D''(Z \to Z^-]D''(Z^- \to V]$. Display box 3 of Figure 4 illustrates the choices of W, W^+, Z, Z^- in this decomposition. The same reasoning as for the first decomposition (2.17) applies in justifying its validity.

The decomposition in (2.20) has isolated the dependence on D''_2 and D''_3 in the right-hand side summand. Indeed, since $D''[U'' \to W^+) = D''[W^+ \to W) = 1$, they do not depend on D''_2 or D''_3, and clearly $D''(Z \to Z^-]$ and $D''(Z^- \to V]$ only depend on D''_1. This leaves $D''[W \to Z]$ as the only term which involves D''_2 and D''_3. We claim that by applying (2.16) with $r = 2$, $m = 2$, and $\widetilde{D} = D''$, we can show that

$$D''\big[W \to Z\big] = \big(\mathcal{T}_{2,2}D''\big)\big[\Uparrow_{2,2}W \to Z\big]. \tag{2.21}$$

This equality requires a bit of explanation. If $(1,1) \notin W$ (i.e., if $(1,1) \notin U$) then W may only include points on level 3 and Z may only include points on level 2. In this case, we may directly apply (2.16) to conclude the equality (2.21). On the other hand, if $(1,1) \in W$, it must also be in Z. Thus, if we write $\hat{W}$ and $\hat{Z}$ for W and Z with the point $(1,1)$ removed, then $D''[W \to Z] = D''\big[(1,1)\big]D''[\hat{W} \to \hat{Z}]$. We may apply (2.16) (with $r = 2$, $m = 2$, and $\widetilde{D} = D''$) to show that $D''[\hat{W} \to \hat{Z}] = (\mathcal{T}_{2,2}D'')[\Uparrow_{2,2}\hat{W} \to \hat{Z}]$. Finally, since $(\mathcal{T}_{2,2}D'')_1 = D''_1$, we have that $D''\big[(1,1)\big] = (\mathcal{T}_{2,2}D'')\big[(1,1)\big]$, which proves (2.21).

Using the fact that $(\mathcal{T}_{2,2}D'')_1 = D''_1$ again, we may replace D'' by $D''' = \mathcal{T}_{2,2}D''$ in all of the other terms in the summand of (2.20) without changing their values (they are all either constant or only depend on D''_1). Moreover, since $D''[U'' \to W^+) = D''[W^+ \to W) = 1$, we may replace these terms by $D'''[\Uparrow_{2,2}U'' \to \Uparrow_{2,2}W^+) = D''[\Uparrow_{2,2}W^+ \to \Uparrow_{2,2}W) = 1$. This implies that, denoting $D''' = \mathcal{T}_{2,2}D''$, $U''' = \Uparrow_{2,2}U''$, $W'^+ = \Uparrow_{2,2}W^+$ and $W' = \Uparrow_{2,2}W$,

$$\begin{aligned} D\big[U \to V\big] = \sum_{Z} & D'''\big[U''' \to W'^+\big)D'''\big[W'^+ \to W'\big) \\ & \times D'''\big[W' \to Z\big]D'''\big(Z \to Z^-\big]D'''\big(Z^- \to V\big], \end{aligned}$$

where we now assume that $U''' = W'^{+} = W'$ and where the sum over Z is such that (W', Z) and (Z^{-}, V) are endpoint pairs (note that the condition that (W', Z) is an endpoint pair imposes the same restriction on Z as the condition that (W, Z) is an endpoint pair). This right-hand side is a decomposition of $D'''[U''' \to V]$ in the same spirit as (2.20), hence we have shown (going back to $\mathcal{T}_{2,2}\mathcal{T}_{1,1}\mathcal{T}_{1,2}D$ in place of D''' and $\Uparrow_{2,2}\Uparrow_{1,1}\Uparrow_{1,2}U$ in place of U''')

$$(2.22)\qquad \big(\mathcal{T}_{1,1}\mathcal{T}_{1,2}D\big)\big[\Uparrow_{1,1}\Uparrow_{1,2}U \to V\big] = \big(\mathcal{T}_{2,2}\mathcal{T}_{1,1}\mathcal{T}_{1,2}D\big)\big[\Uparrow_{2,2}\Uparrow_{1,1}\Uparrow_{1,2}U \to V\big].$$

Combining together (2.18), (2.19) and (2.22) proves (2.7) for $n = 3$ and general k.

Now we turn to address the general n case. This proceeds by a sequence of $\binom{n}{2}$ replacements through which we transition from D to $\mathcal{W}D$. The following lemma is the key to this scheme. Figure 5 may be helpful in understanding the hypotheses in the statement of the lemma.

Lemma 2.7. *Fix any $m \in [\![1, n-1]\!]$, any n weakly increasing integers $r_1 \leq \cdots \leq r_n$ with (for some r) $r = r_m = r_{m+1} < r_{m+2}$ (if $m = n-1$ then by convention set $r_{n+1} = +\infty$), and any n functions $D = (D_1, \ldots, D_n)$ such that $D_i : \mathbb{Z}_{\geq r_i} \to (0,\infty)$, $i \in [\![1,n]\!]$. Fix any $k \in \mathbb{Z}_{\geq 1}$ and any k-tuple of points U such that every point in U is either of the form (u, n) for some $u \geq r_n$, or of the form (r_i, i) for some $i \in [\![1,n]\!]$ such that $r_i < r_{i+1}$. Finally, fix another k-tuple of points $V = \big((v_i, 1)\big)_{i\in[\![1,k]\!]}$ such that (U, V) form an endpoint pair. Then*

$$(2.23)\qquad D\big[U \to V\big] = \big(\mathcal{T}_{r,m}D\big)\big[\Uparrow_{r,m}U \to V\big].$$

Let us assume this lemma for the moment and conclude the proof of (2.7). The left-hand side of (2.7) fits into the setup of Lemma 2.7 with $r_1 = \cdots = r_n = 1$ and hence $U = \big((u_i, n)\big)_{i\in[\![1,k]\!]}$. Now (as prescribed by the definition of $\mathcal{W}$) sequentially apply the lemma with $r = 1$ and $m = n-1$ through 1, and then $r = 2$ and $m = n-1$ through 2, and so forth until finally $r = n-1$ and $m = n-1$. Each application amounts to applying the relevant $\mathcal{T}_{r,m}$ to the D functions (or rather, to the image of the D functions under previous application of $\mathcal{T}$ operators) and the relevant $\Uparrow_{r,m}$ operator to the U points (or rather, to the image of the U points under previous application of $\Uparrow$ operators). It is clear that the output of each step in this sequence is suitable for applying the next step of the lemma. Recalling the definition of $\mathcal{W}$ and $\Uparrow$ in terms of the $\mathcal{T}_{r,m}$ and $\Uparrow_{r,m}$, we find that this procedure shows that

$$D\big[U \to V\big] = \big(\mathcal{W}D\big)\big[\Uparrow U \to V\big],$$

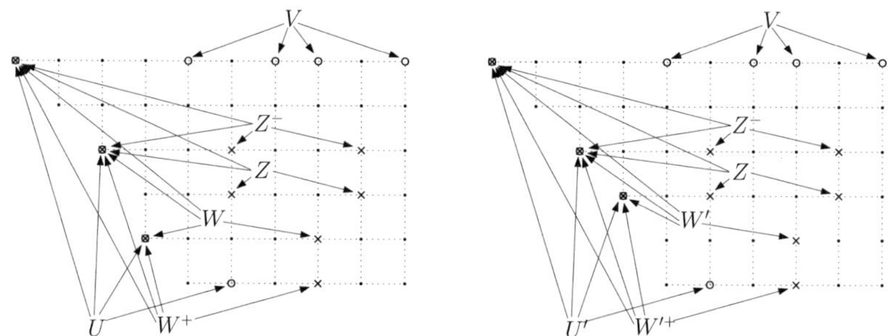

Fig. 5. An illustration of the decomposition (2.24) in Lemma 2.7. On the left, we have $r_1 = 1$, $r_2 = 2$, $r_3 = 3$, $r_4 = 4$, $r_5 = 4$, $r_6 = 5$ so the domain of D satisfies the hypotheses of Lemma 2.7 with $m = 4$ and $r = 4$. On the right, all of the r_i remain the same except $r_5 = 5$ now. On the left, the starting points U satisfy the hypotheses of Lemma 2.7. Notice that graphically, the hypothesis on U corresponds to having every point at the "bottom" of the domain of D (no other points in the domain of D below the points of U). Also shown in the figure are the choices of W^+, W, Z, and Z^- for the decomposition (2.24) on the left and (2.26) on the right.

precisely as claimed in (2.7). Thus, to complete this step we are left to prove Lemma 2.7.

Proof of Lemma 2.7. We appeal to the following decomposition which simultaneously captures all of the decompositions used in the $n = 3$ proof given earlier. Assuming the hypotheses on the r_i, U and V from the statement of Lemma 2.7, we have that

$$D\big[U \to V\big] = \sum_{W,Z} D\big[U \to W^+\big) D\big[W^+ \to W\big) \times D\big[W \to Z\big] D\big(Z \to Z^-\big] D\big(Z^- \to V\big], \tag{2.24}$$

where all sets U, W^+, W, Z, Z^-, V contain k points and additionally satisfy

- W contains all points in U of level $\leq m+1$, and for each point in U of level $> m+1$, there is a unique point in W of level $m+1$;
- W^+ contains all points in U of level $\leq m+1$, and for each point in U of level $> m+1$, there is a unique point in W^+ of level $m+2$;

- $W \setminus U$ and $W^+ \setminus U$ are in the following correspondence: for each $(w, m+1) \in W \setminus U$ there is a point $(w, m+2) \in W^+ \setminus U$;
- If $m = 1$, then $Z = Z^- = V$.
- If $m \geq 2$, Z contains all points in U of level $\leq m-1$, and for each point in U of level $\geq m+1$, there is a unique point in Z of level m;
- If $m \geq 2$, Z^+ contains all points in U of level $\leq m-1$, and for each point in U of level $\geq m+1$, there is a unique point in Z^+ of level $m-1$;
- $Z \setminus V$ and $Z^- \setminus U$ are in the following correspondence: for each $(z, m) \in Z \setminus U$ there is a point $(z, m-1) \in Z^- \setminus U$;
- The sum over W and Z is restricted so as to satisfy the above conditions as well as the condition that (U, W^+), (W^+, W), (W, Z), (Z, Z^-), and (Z^-, V) are all endpoint pairs.

Note that the hypotheses on the r_i and U from the statement of Lemma 2.7 imply that there are no points in U of level m (since we have assumed $r_m = r_{m+1}$ and all points of the form $(r_i, i) \in U$ have $r_i < r_{i+1}$). This implies that the sets above all have exactly k points. (2.24) follows by expanding each term on the right-hand side and then verifying that the resulting summation puts each term in bijection with a multipath with the corresponding weight.

The key fact that is evident from the decomposition (2.24) is that the only term on the right-hand side inside the summation over W and Z which depends on the functions D_m and D_{m+1} is $D[W \to Z]$. We can use this fact along with (2.16) (with $\widetilde{D} = D$ and r and m as specified in the statement of Lemma 2.7) to show that

$$D\big[W \to Z\big] = \big(\mathcal{T}_{r,m} D\big)\big[\Uparrow_{r,m} W \to Z\big]. \tag{2.25}$$

Now, there are two cases to consider: when $(r, m+1) \in U$ or when $(r, m+1) \notin U$.

When $(r, m+1) \in U$, we also have $(r, m+1) \in W$ and $(r, m+1) \in W^+$. If we define $U' = \Uparrow_{r,m} U$, $W' = \Uparrow_{r,m} W$, $W'^+ = \Uparrow_{r,m} W^+$ and $D' = \mathcal{T}_{r,m} D$, then it follows from (2.25) and (2.24) that

$$\begin{aligned} D\big[U \to V\big] = \sum_{W', Z} & D'\big[U' \to W'^+\big) D'\big[W'^+ \to W'\big) \\ & \times D'\big[W' \to Z\big] D'\big(Z \to Z^-\big] D'\big(Z^- \to V\big]. \end{aligned} \tag{2.26}$$

We have used the fact that the only term in (2.24) which depends on the functions D_m and D_{m+1} is $D[W \to Z]$, and hence replacing D by D' does not effect the other terms. We have also used the fact that since

$(r, m+1)$ is common to U, W and W^+, replacing it by (r, m) does not change the value of the first two terms in the summand in (2.24).

When $(r, m+1) \notin U$, we also have $(r, m+1) \notin W$ and $(r, m+1) \notin W^+$. In that case, $W' = \Uparrow_{r,m} W = W$ and likewise $U' = \Uparrow_{r,m} U = U$ and $W'^+ = \Uparrow_{r,m} W^+ = W^+$. Thus, setting $D' = \mathcal{T}_{r,m} D$, by the same reasoning as above we find that the decomposition (2.26) holds.

It remains to observe that the right-hand side of (2.26) is a decomposition of $D'[U' \to V]$. The only difference relative to the decomposition in (2.24) is that U is replaced by U' and that W' and W'^+ now contain points from U' of level $\le m$. Recalling that $D' = \mathcal{T}_{r,m} D$ and $U' = \Uparrow_{r,m} U$ we see that we have proved (2.23) and hence the lemma. ■

This completes the proof of the third step, and hence the proof of Theorem 2.4. ■

2.2. Relation to geometric RSK correspondence

The operator $\mathcal{W}$ defined via the (discrete geometric) Pitman transform in (2.6) is closely related to the geometric RSK correspondence. We recall the geometric RSK correspondence in Section 2.2.1 and state the relationship with $\mathcal{W}$ in Section 2.2.2. Finally, in Section 2.2.3 we consider a special choice for the functions D (related to inverse-gamma random variables) and briefly recall how, via [COSZ14], this gives rise to a Markovian structure for $(\mathcal{W}D)(t)$.

2.2.1. *Recalling the geometric RSK.* We recall the geometric RSK correspondence (gRSK) as defined in [COSZ14, Definitions 2.1 and 2.2] (see also [Kir01, NY04]).

Definition 2.8. Fix any $N \in \mathbb{Z}_{\ge 1}$. Let $1 \le \ell \le N$. Consider two *words* $\xi = (\xi_\ell, \dots, \xi_N)$ and $b = (b_\ell, \dots, b_N)$ with strictly positive real entries. *Geometric row insertion* of the word b into the word ξ transforms the pair (ξ, b) into a new pair (ξ', b') where $\xi' = (\xi'_\ell, \dots, \xi'_N)$ and $b' = (b'_{\ell+1}, \dots, b'_N)$ (i.e., b' has one fewer entry and the index starts at $\ell+1$ instead of ℓ). The transformation is notated and defined as follows:

$$\xi \overset{b}{\underset{b'}{\longrightarrow}} \xi' \quad \text{where} \quad \begin{cases} \xi'_\ell = b_\ell \xi_\ell, & \\ \xi'_k = b_k(\xi'_{k-1} + \xi_k), & \ell+1 \le k \le N, \\ b'_k = b_k \dfrac{\xi_k \xi'_{k-1}}{\xi_{k-1} \xi'_k}, & \ell+1 \le k \le N. \end{cases} \tag{2.27}$$

If $\ell = N$, the output word b' is empty. In addition to $\xi \in (0, \infty)^{N-\ell+1}$ we admit the case $\xi = (1, 0, \dots, 0)$. This will correspond to row insertion into an initially empty word. For $k \geq 1$ we denote such a word as $e_1^{(k)} = (1, 0, \dots, 0)$, where k is the total number of coordinates for the vector. The notation and definition are extended so that

$$(2.28) \qquad e_1^{(N-\ell+1)} \overset{b}{\underset{}{\longrightarrow}} \xi' \quad \text{where} \quad \xi'_k = \prod_{i=\ell}^{k} b_i, \quad \ell \leq k \leq N.$$

This is consistent with (2.27) except that output b' is not defined and hence not displayed in the diagram above.

For N fixed, consider a semi-infinite array of strictly positive real numbers $m = \big(m_{j,n} : j \in [\![1, N]\!], n \geq 1\big)$. For integers $1 \leq a \leq b \leq N$ and $1 \leq c \leq d$, we write $m_{[\![a,b]\!],[\![c,d]\!]} := \big(m_{j,n} : j \in [\![a, b]\!], n \in [\![c, d]\!]\big)$ for the corresponding subarray. In particular we will generally fix N in which case we denote the nth row of m by $m_{[\![1,N]\!],[\![n]\!]} = (m_{1,n}, \dots, m_{N,n})$ and the first n rows by $m_{[\![1,N]\!],[\![1,n]\!]}$. See Figure 6 for an illustration of this notation.

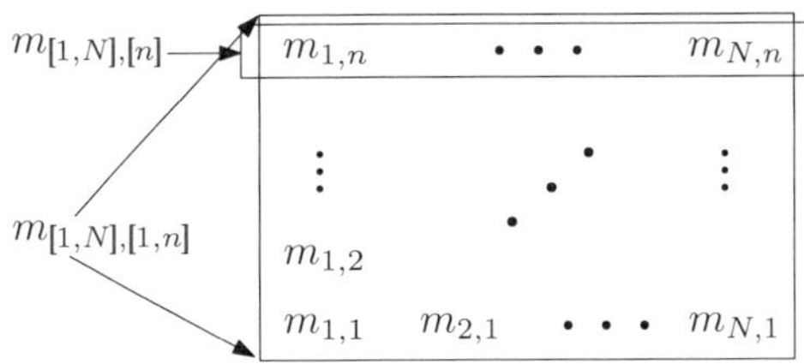

Fig. 6. The first n rows and N columns of the array m.

The *geometric RSK* correspondence is a bi-rational map between $m_{[\![1,N]\!],[\![1,n]\!]}$ and another N by n matrix of strictly positive real numbers. As we will explain in what follows, under the gRSK, $m_{[\![1,N]\!],[\![1,n]\!]}$ maps to a pair of *geometric* Young Tableaux (P, Q) of the same *shape*. Let us start by defining the P-geometric Young tableaux which is a map from $m_{[\![1,N]\!],[\![1,n]\!]}$ to an array of the form

$$(2.29) \qquad z(n) = \big(z_{k,\ell}(n) : 1 \leq \ell \leq k \leq N \text{ and } \ell \in [\![1, N \wedge n]\!]\big).$$

When $n \geq N$, $z(n)$ is a triangular array, and when $n \leq N$ it is a trapezoidal array. For $\ell \in [\![1, N \wedge n]\!]$, denote by $z_\ell(n) = \big(z_{\ell,\ell}(n), \dots, z_{N,\ell}(n)\big)$ the ℓ-th diagonal and denote by $\mathrm{sh}(z(n)) = \big(z_{N,\ell}(n)\big)_{\ell \in [\![1,n \wedge N]\!]}$ the bottom row of $z(n)$, which we call the *shape* of $z(n)$. See Figure 7 for an illustration of this notation.

We now describe how to construct the P tableaux $z(n)$ from the matrix $m_{[\![1,N]\!],[\![1,n]\!]}$ through insertion of the rows of $m_{[\![1,N]\!],[\![1,n]\!]}$ into an

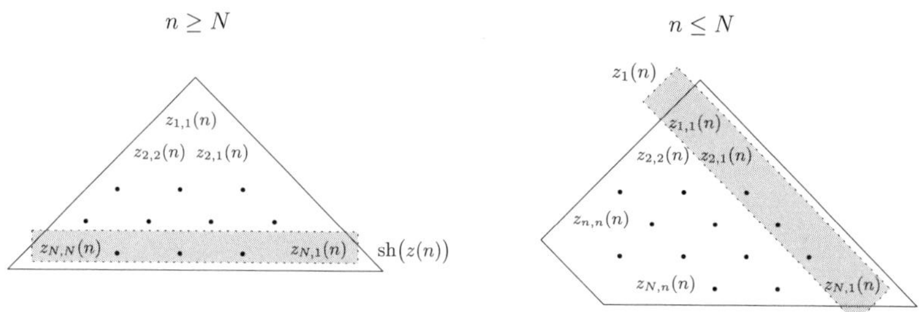

Fig. 7. The array $z(n)$. When $n \geq N$ it is triangular and when $n \leq N$ it is trapezoidal. Also depicted on the left is the shape $\mathrm{sh}(z(n))$ (i.e., bottom row), and on the right is the first diagonal $z_1(n)$.

empty tableaux. For each $n \in \mathbb{Z}_{\geq 1}$, let $a_1(n) = m_{[\![1,N]\!],[\![n]\!]}$. Then the mapping is given graphically on the left side of Figure 8 (the right side contains an example when $n = 4$ and $N = 3$). It is also possible to define this mapping through operators, though it is no more informative (and perhaps less so) than the graphical definition which we stick with.

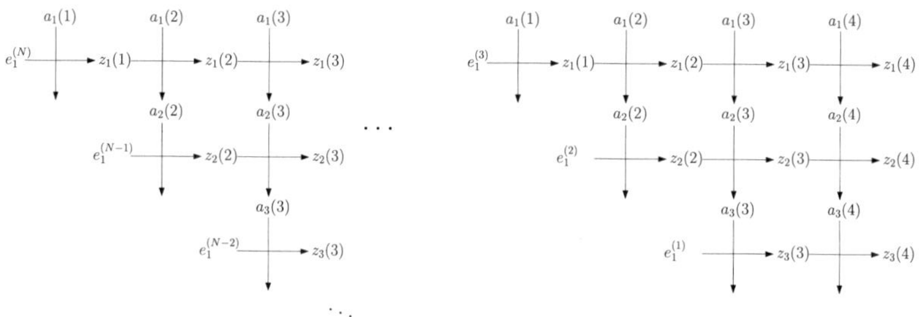

Fig. 8. On the left: The matrix $m_{[\![1,N]\!],[\![1,n]\!]}$ is inserted into an empty tableaux. The inputs on the top are specified by setting $a_1(n) = m_{[\![1,N]\!],[\![n]\!]}$. Everything else is computed sequentially (from top-left down and to the right) via the geometric row insertion. The output matrix $z(n)$ is read off from its diagonals $z_1(n), \ldots, z_{n\wedge N}(n)$ on the right-side of the figure. On the right: the specific case when $n = 4$ and $N = 3$.

For later uses, let us call the above defined mapping P. The Q tableaux is defined in terms of the sequence of shapes of $z(1), \ldots, z(n)$. Specifically,

$$P(m_{[\![1,N]\!],[\![1,n]\!]}) := z(n) \qquad Q(m_{[\![1,N]\!],[\![1,n]\!]}) := \big(\mathrm{sh}(z(t))\big)_{t\in[\![1,n]\!]}.$$

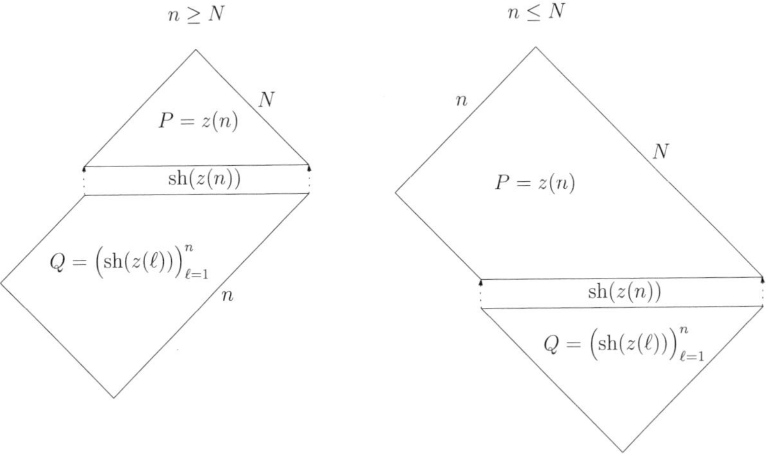

Fig. 9. The P and Q tableaux agree in that the shape (bottom row) of the P tableaux matches the top row of the Q tableaux (where the shapes are listed from that of $z(n)$ down to $z(1)$). On the right this is shown when $n \geq N$ and on the right is the case $n \leq N$.

Figure 9 shows how the P and Q tableaux may be joined together to form another n by N matrix of strictly positive real numbers. Notice that the definition of the Q tableaux implies that it is consistent as n varies so that for $n' > n$, the first n elements in the sequence of shapes that defines $Q(m_{[\![1,N]\!],[\![1,n']\!]})$ equals $Q(m_{[\![1,N]\!],[\![1,n]\!]})$.

Generalizing Greene's theorem for the usual RSK correspondence, [NY04] showed that the geometric RSK also has a partition function interpretation. We will follow the exposition of [COSZ14, Section 2.2] and define a mapping from the weight matrix $m_{[\![1,N]\!],[\![1,n]\!]}$ to an array $\tilde{z}(n)$ of the same dimensions as $z(n)$ defined in (2.29). The result quoted below in Proposition 2.10 shows that $z(n)$ and $\tilde{z}(n)$ are equal, and thus (in light of the way $\tilde{z}$ is defined) provides a partition function interpretation for $z(n)$.

Definition 2.9. For $1 \leq \ell \leq k \leq N$ let $\Pi^{\ell}_{n,k}$ denote the set of ℓ-tuples $\pi = (\pi_1, \ldots, \pi_\ell)$ of non-intersecting lattice paths in $\mathbb{Z}^2$ with the property that for $1 \leq r \leq \ell$, π_r is a lattice path from $(1, r)$ to $(n, k+r-\ell)$. The term lattice path means that between nearest-neighbor lattice points in $\mathbb{Z}$, the path either takes a unit step up or right; the term non-intersecting means that the paths do not touch, even at lattice points. For any ℓ-tuple of non-intersecting lattice paths $\pi = (\pi_1, \ldots, \pi_\ell)$, define their *weights* to be

$$wt(\pi) = \prod_{r=1}^{\ell} \prod_{(i,j)\in\pi_r} m_{i,j}.$$

For $1 \le \ell \le k \le N$ let

$$\tau_{k,\ell}(n) = \sum_{\pi\in\Pi^{\ell}_{n,k}} wt(\pi).$$

For $0 \le n < \ell < k \le N$ the set $\Pi^{\ell}_{n,k}$ is empty and hence we set that $\tau_{k,\ell}(n) = 0$ in that case. When $\ell = k$, there exists only a single $\pi \in \Pi^{\ell}_{n,k}$. In fact, whenever $0 \le n < \ell \le k \le N$, the set $\Pi^{\ell}_{n,k}$ is only non-empty in the case that $k = \ell$. In otherwords,

$$\tau_{k,\ell}(n) = \delta_{k,\ell}\tau_{k,n}(n) \quad \text{whenever} \quad 0 \le n < \ell \le k \le N,$$

where $\delta_{k,\ell}$ is the Kronecker delta function. Finally, for $\ell = 0$ we adopt the convention that $\tau_{k,0} = 1$ for $1 \le k \le N$.

Having defined the multi-path partition function $\tau_{k,\ell}(n)$, we can now define the array $\tilde{z}(n)$ by the relation that for all indices k and ℓ,

$$\tilde{z}_{k,1}(n) \cdots \tilde{z}_{k,\ell}(n) = \tau_{k,\ell}(n).$$

Proposition 2.10 (Proposition 2.5 of [COSZ14]). *The mapping* $m_{[\![1,N]\!],[\![1,n]\!]} \mapsto z(n)$ *from Definitions* 2.8 *and* $m_{[\![1,N]\!],[\![1,n]\!]} \mapsto \tilde{z}(n)$ *from Definition* 2.9 *are the same, i.e.,* $z(n) = \tilde{z}(n)$.

Proof. In [COSZ14], the first of these mappings is denoted (with m replaced therein by d) by $\emptyset \leftarrow m^{[1]} \leftarrow m^{[2]} \leftarrow \cdots \leftarrow m^{[n]}$ and the second is denoted by $P_{n,N}(m^{[1,n]})$. Proposition 2.5 of [COSZ14] then provides the equality, following methods used in [NY04]. ∎

2.2.2. *Rewriting the geometric RSK via the geometric Pitman transform.* We now explain how the geometric row insertion and hence geometric RSK can be rewritten in terms of the Pitman transform introduced earlier in Definition 2.3. For the usual RSK correspondence, this is explained, for instance, in notes of Pei available at `https://toywiki.xyz/`. In lifting this to the geometric setting, there are some subtleties which arise.

Let us recall the Pitman transform from Definition 2.3. For any $r \in \mathbb{Z}$ and any functions $f, g : \mathbb{Z}_{\ge r} \to (0, \infty)$ define functions $(g \odot f)$:

$\mathbb{Z}_{\geq r} \to (0,\infty)$ and $(f \otimes g) : \mathbb{Z}_{\geq r+1} \to (0,\infty)$ by

$$(g \odot f)(x) := f(x) \cdot \sum_{m=r}^{x} \frac{g(m)}{f(m-1)} \qquad \text{for } x \geq r$$

$$(f \otimes g)(x) := g(x) \cdot \left(\sum_{m=r}^{x} \frac{g(m)}{f(m-1)} \right)^{-1} \qquad \text{for } x \geq r+1$$

where we have adopted the convention that $f(r-1) = 1$ in the denominator when $m = r$. The operator $\mathcal{T}$ defined in (2.4) can be encoded graphically as

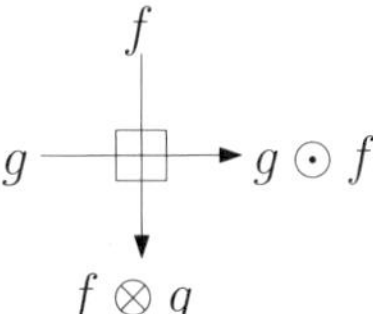

noting that the output on the right-side is still a function from $\mathbb{Z}_{\geq r} \to (0,\infty)$ whereas the bottom output is a function from $\mathbb{Z}_{\geq r+1} \to (0,\infty)$.

The next lemma rewrites the geometric row insertion in terms of the Pitman transform.

Lemma 2.11. *Let $\ell \in [\![1, N]\!]$ and consider two words $\xi = (\xi_\ell, \ldots, \xi_N)$ and $b = (b_\ell, \ldots, b_N)$ with strictly positive real entries. Compute ξ' and b' (where $\xi' = (\xi'_\ell, \ldots, \xi'_N)$ and $b' = (b'_{\ell+1}, \ldots, b'_N)$) via the geometric row insertion of b into ξ as depicted in the left-hand side in Figure 10. Now define functions $\xi, B : \mathbb{Z}_{\geq \ell} \to (0,\infty)$ such that $\xi(k) = \xi_k$ for all $k \in [\![\ell, N]\!]$ and such that for all $k \in [\![\ell, N]\!]$ we have $B(k) = \prod_{j=\ell}^{k} b_j$. The values of the functions ξ and B for arguments larger than N do not matter and can be set to 1 for concreteness. Now, compute ξ' and B' via the Pitman transform as depicted in the right-hand side in Figure 10. Then $\xi'(k) = \xi'_k$ for all $k \in [\![\ell, N]\!]$ and $B'(k) = \prod_{j=\ell+1}^{k} b'_j$ for all $k \in [\![\ell+1, N]\!]$. In other words, the geometric row insertion and Pitman transform produce the same result.*

Proof. Owing to (2.27) we can compute explicitly ξ'_k for all $k \in [\![\ell, N]\!]$:

$$\xi'_k = \sum_{j=\ell}^{k} b_j \cdots b_k \xi_j = B(k) \sum_{j=\ell}^{k} \frac{\xi(j)}{B(j-1)} = (\xi \odot B)(k) = \xi'(k). \tag{2.30}$$

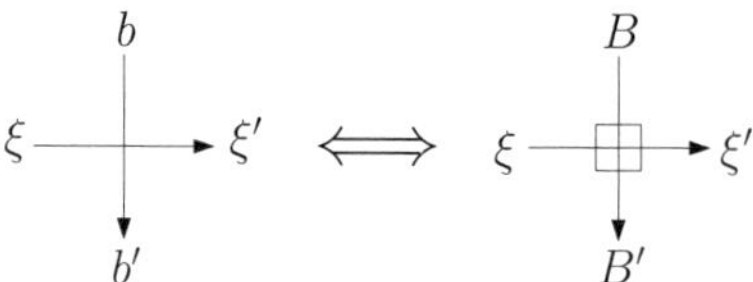

Fig. 10. The equivalence between the geometric row insertion and the Pitman transform as shown in Lemma 2.11.

The first equality is easily shown by induction in k, the second is verified by substituting the definitions of the functions B and ξ, the third equality is by definition of $\odot$ and the final equality is the definition of the function ξ'.

Again appealing to (2.27), we know that $b'_j = b_j \frac{\xi_j \xi'_{j-1}}{\xi_{j-1}\xi'_j}$ for all $j \in [\![\ell+1, N]\!]$. Taking the product of this equality over $j \in [\![\ell+1, k]\!]$, we see that

$$\prod_{j=\ell+1}^{k} b'_j = \frac{B(k)}{B(\ell)} \frac{\xi(k)\xi'(\ell)}{\xi(\ell)\xi'(k)} = \frac{B(k)\xi(k)}{\xi'(k)} = \big(B \otimes \xi\big)(k) = B'(k).$$

The first equality comes from the explicit formula for b'_j recalled above and the definition of the functions B, ξ and ξ', the second equality comes cancelations by substituting the equality $\xi'(\ell) = B(\ell)\xi(\ell)$, the third equality uses the formula for $\xi'(k)$ from (2.30) (after the second equality sign) and the definition of $\otimes$ and the final equality is the definition of the function B'. ■

From Lemma 2.11, we rewrite the entire gRSK correspondence via the Pitman transform.

Lemma 2.12. *Fix any $n, N \in \mathbb{Z}_{\geq 1}$ and vectors $d_1, \ldots, d_n \in (0,\infty)^N$ so that $d_i = (d_{i,1}, \ldots, d_{i,N})$. Compute vectors $\widetilde{D}_1, \ldots, \widetilde{D}_n$ via the gRSK correspondence as depicted in the left-hand side of Figure 11. Define functions $D_1, \ldots, D_n : \mathbb{Z}_{\geq 1} \to (0,\infty)$ such that for each $i \in [\![1,n]\!]$ and for each $k \in [\![1,N]\!]$ we have $D_i(k) = \prod_{j=1}^{k} d_{i,j}$. For $k > N$ the value of the D_i does not matter and can be set to 1 for concreteness. Compute n functions $\widetilde{D}_1, \ldots, \widetilde{D}_n$ where $\widetilde{D}_i : \mathbb{Z}_{\geq i} \to (0,\infty)$ for $i \in [\![1,n]\!]$ via the Pitman transform as depicted in the right-hand side of Figure 11. The, the outcome of these two calculations match in the sense that $\widetilde{D}_{i,j} = \widetilde{D}_i(j)$ as long as $1 \leq i \leq j \leq N$ and $i \in [\![1, n \wedge N]\!]$.*

Proof. The follows immediately from Lemma 2.11 along with the insertion rule (2.28) used on the boundary in the gRSK correspondence. ■

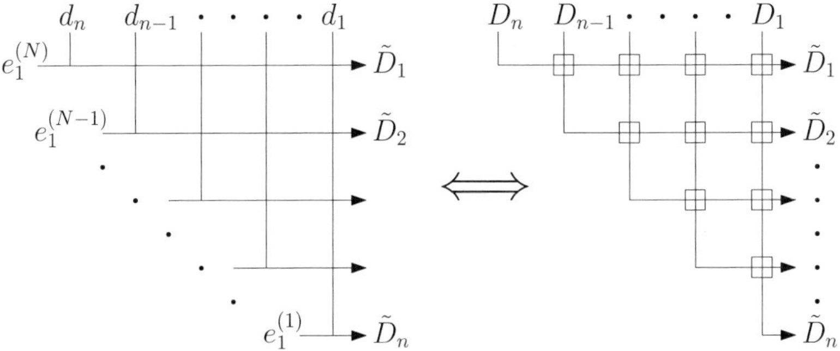

Fig. 11. The equivalence between the gRSK correspondence (on the left) and Pitman transform (on the right) as shown in Lemma 2.12. We have dropped all internal labels. The outgoing lines (coming out to the right and below each vertex) carries the output from each vertex to the next vertices. In the depiction of the gRSK correspondence, we have dropped the arrows coming out below the vertices in which the e_1 vectors are inserted. This is justified since that output is empty anyway.

The Pitman transform depicted on the right-hand side of Figure (11) is a graphical implementation of the $\mathcal{W}$ operator from (2.6). Thus, in light of Lemma 2.12, we have the following.

Corollary 2.13. *Let $\widetilde{D} = (\widetilde{D}_1, \ldots, \widetilde{D}_n)$ denote the output on the right-hand side of* (11) *with given input functions $D_1, \ldots, D_n$. Then*

$$\widetilde{D} = \mathcal{W}D.$$

Combining Corollary 2.13 and Proposition 2.10 yields a partition function formula for $\mathcal{W}D$.

Corollary 2.14. *Fix any $n, N \in \mathbb{Z}_{\geq 1}$ and any functions $D_1, \ldots, D_n\colon \mathbb{Z}_{\geq 1} \to (0, \infty)$, writing $D = (D_1, \ldots, D_n)$. Then, for $\ell \in [\![1, n \wedge N]\!]$,*

$$\prod_{r=1}^{\ell} (\mathcal{W}D)_r(N) = D\Big[\big([\![1, \ell]\!], n\big) \to \big([\![N-\ell+1, N]\!], 1\big)\Big],$$

where we have used the shorthand that for $a \leq b \in \mathbb{Z}$, $\big([\![a, b]\!], m\big) := \big((a, m), \ldots, (b, m)\big)$ (see Figure 12*). Letting $d_{i,j} := \frac{D_i(j)}{D_i(j-1)}$ (with $D_i(0) \equiv 1$) and defining the N by n array $\widetilde{d}$ via $\widetilde{d}_{i,j} = d_{i,n-j+1}$,*

$$\big(\mathcal{W}D\big)(N) = sh\big(P(\widetilde{d}_{[\![1,N]\!],[\![1,n]\!]})\big).$$

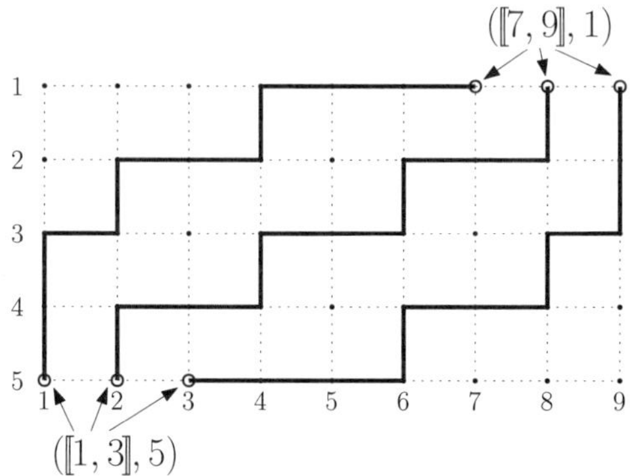

Fig. 12. Corollary 2.14 provides a partition function representation for $\mathcal{W}D$. Illustrated here is the endpoint pair (and a multipath between them) associated to $n = 5$, $N = 9$ and $\ell = 3$.

Consequently, letting $\widetilde{d}^{\top}$ be the transpose of $\widetilde{d}$ so $\widetilde{d}^{\top}_{i,j} = \widetilde{d}_{j,i}$, we have that for any N

$$\Big((\mathcal{W}D)(t)\Big)_{t\in[\![1,N]\!]} = Q\big(\widetilde{d}^{\top}_{[\![1,n]\!],[\![1,N]\!]}\big).$$

2.2.3. *Inverse-gamma weights.* A random variable X has *inverse-gamma distribution with parameter $\theta > 0$* if it is supposed on the positive reals where it has density relative to Lebesgue measure given by $\frac{1}{\Gamma(\theta)}x^{-\theta-1}e^{-\frac{1}{x}}dx$.

If the array m from Definition 2.8 is filled with iid inverse-gamma distributed random variables, [COSZ14, Theorem 3.9] shows that the shape $\mathrm{sh}(z(n))$ of $m_{[\![1,N]\!],[\![1,n]\!]}$ under the gRSK correspondence evolves as a Markov process in n with an explicit transition kernel (between n and $n+1$). It follows directly from the factorized nature of the transition kernel that $\big(\mathrm{sh}(z(n))\big)_{n\in\mathbb{Z}_{\geq 1}}$ enjoys the structure of a (discrete) Gibbsian line ensemble – see [Wu19, JO19, BCD] for details. In Section 4.2 we mention how a special limit of this Gibbsian structure is important in the Airy sheet construction of [DOV18] from the Airy line ensemble.

2.2.4. *Proof of Theorem* 2.4 *via Desnanot-Jacobi.* This proof was communicated to us by Konstantin Matveev. It has the flavor of methods used to parameterize totally positive matrices (see, e.g. [FZ00] and references therein), though the matrices presently are totally non-negative, for which the parameterization is more involved.

Recall that we seek to prove that $D[U \to V] = (\mathcal{W}D)[\Uparrow U \to V]$. We will first relate this equality, by means of the Lindström-Gessel-Viennot lemma, to an equality of matrices M and $\widetilde{M}$ whose entries are the $k = 1$

versions of this identity (when $|U| = |V| = 1$). The idea to proving the equality of those matrices is to match certain minor determinants and then use the Desnanot-Jacobi identity to inductively match all of the others.

Before commencing with the proof, let us recall the Desnanot-Jacobi identity. For a matrix M, we write $M_{U,V}$ for the minor of M comprised of rows U and columns V. For instance, $M_{[\![i,i+k-1]\!],[\![j,j+k-1]\!]}$ is the minor composed of rows i through $i+k-1$, and columns j through $j+k-1$. Writing $|M|$ for the determinant M, the Desnanot-Jacobi identity states that

$$\begin{aligned} (2.31) \qquad & \left|M_{[\![i,i+k-1]\!],[\![j,j+k-1]\!]}\right| \cdot \left|M_{[\![i-1,i+k-2]\!],[\![j-1,j+k-2]\!]}\right| \\ &= \left|M_{[\![i-1,i+k-1]\!],[\![j-1,j+k-1]\!]}\right| \cdot \left|M_{[\![i,i+k-2]\!],[\![j,j+k-2]\!]}\right| \\ &\quad + \left|M_{[\![i-1,i+k-2]\!],[\![j,j+k-1]\!]}\right| \cdot \left|M_{[\![i,i+k-1]\!],[\![j-1,j+k-2]\!]}\right|. \end{aligned}$$

To ease notations, let us define $\widetilde{D} = \mathcal{W}D$ and from D and $\widetilde{D}$ we define vertex weights $d_{x,m} = D_m(x)/D_m(x-1)$ and $\widetilde{d}_{x,m} = \widetilde{D}_m(x)/\widetilde{D}_m(x-1)$. Note that while $d_{x,m}$ is defined for all $x \in \mathbb{Z}_{\geq 1}$ and $m \in [\![1,n]\!]$, $\widetilde{d}_{x,m}$ is only defined when we additionally assume that $x \geq m$. Let us also define matrices M and $\widetilde{M}$ by (see 13 for an illustration)

$$M_{u,v} := D\big[(u,n) \to (v,1)\big] \qquad \widetilde{M}_{u,v} := \widetilde{D}\big[(u,n \wedge u) \to (v,1)\big].$$

Our first observation is that by applying the Lindström-Gessel-Viennot (LGV) lemma, it follows that

$$D\big[U \to V\big] = \big|M_{U,V}\big|, \qquad \text{and} \qquad \widetilde{D}\big[\Uparrow U \to V\big] = \big|\widetilde{M}_{U,V}\big|.$$

We have slightly abused notation above. On the left-hand sides of the above equations, we take $U = \big((u_1,n),\ldots,(u_k,n)\big)$ and $V = \big((v_1,1),\ldots,(v_k,1)\big)$ with $u_1 < \cdots < u_k$ and $v_1 < \cdots < v_k$; on the right-hand sides we take $U = (u_1,\ldots,u_k)$ and $V = (v_1,\ldots,v_k)$, where $u,v \in \mathbb{Z}_{\geq 1}$. Note that the LGV lemma is typically stated for edge-weighted graphs, not vertex-weighted. However, in our present case we can transfer our vertex weights to edge weights in the following way. To every edge entering a vertex from the left or below, associate the edge with the weight of the vertex. For vertices on the bottom or left-diagonal of the diagrams in Figure 13, attach an extra vertical edge below them and transfer the vertex weight to that edge. Then, the edge-weighted partition function from u (where the starting vertex for u is now shifted down by one to be below the newly added edge) to v is equal to the original vertex-weighted partition function. The same holds for the partition functions

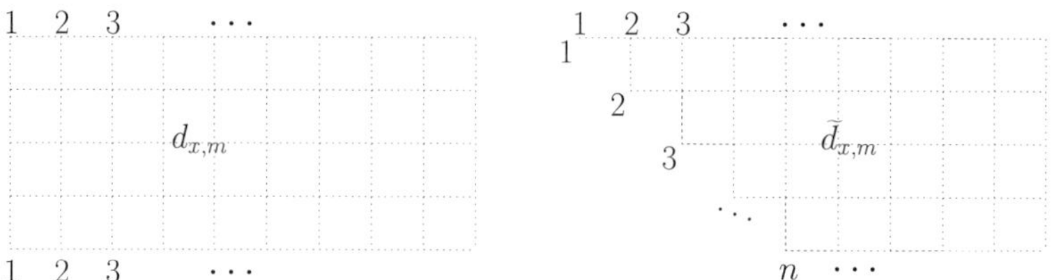

Fig. 13. On the left we depict the matrix M and on the right the matrix $\widetilde{M}$. The rows are indexed by the numbers on the bottom (and left-diagonal in the case of $\widetilde{M}$) and the columns are indexed by the numbers on the top of the diagrams. The matrix elements with row u and column v are equal to the sum over all directed paths from u to v of the product of the vertex weights over the paths. On the left, the vertex weights are given by the $d_{x,m}$ and on the right by the $\tilde{d}_{x,m}$.

for ensembles of non-intersecting multipaths from U to V, hence we can apply LGV.

In light of the above observation and application of the LGV lemma, it suffices to prove equality of the matrix elements of M and $\widetilde{M}$. To prove this we will appeal to certain a priori equalities of minor determinants of M and $\widetilde{M}$, along with the Desnanot-Jacobi identity. Let us start by recording the a priori obvious equalities:

(1) For all $j,k\in\mathbb{Z}_{\geq 1}$, $\left|M_{[\![1,k]\!],[\![j,j+k-1]\!]}\right| = \left|\widetilde{M}_{[\![1,k]\!],[\![j,j+k-1]\!]}\right|$. This is from the definition of $\widetilde{D}$.

(2) $\left|M_{[\![i,i+k-1]\!],[\![j,j+k-1]\!]}\right|$ and $\left|\widetilde{M}_{[\![i,i+k-1]\!],[\![j,j+k-1]\!]}\right|$ are strictly positive if and only if $k\leq n$ and $i\leq j$, or $k>n$ and $i=j$. In the other cases, the minor determinants are zero. This follows because these determinants present partition functions which involve positive weights. In the case when $k>n$, the only non-zero partition function is when $i=j$.

(3) $\left|M_{[\![i,i+k-1]\!],[\![i,i+k-1]\!]}\right| = \left|\widetilde{M}_{[\![i,i+k-1]\!],[\![i,i+k-1]\!]}\right|$. This follows from the calculation depicted in Figure 14 and is a consequence of the upper-triangularity of the matrices in question.

(4) $\left|M_{[\![i,i+k-1]\!],[\![j,j+k-1]\!]}\right| = \left|\widetilde{M}_{[\![i,i+k-1]\!],[\![j,j+k-1]\!]}\right| = 0$ if $i<j$ and $k>n$. This is because it is not possible to connect off-set starting and ending points by more than n disjoint paths.

We will now inductively show that for all $i,j,k\in\mathbb{Z}_{\geq 1}$

$$\left|M_{[\![i,i+k-1]\!],[\![j,j+k-1]\!]}\right| = \left|\widetilde{M}_{[\![i,i+k-1]\!],[\![j,j+k-1]\!]}\right|. \tag{2.32}$$

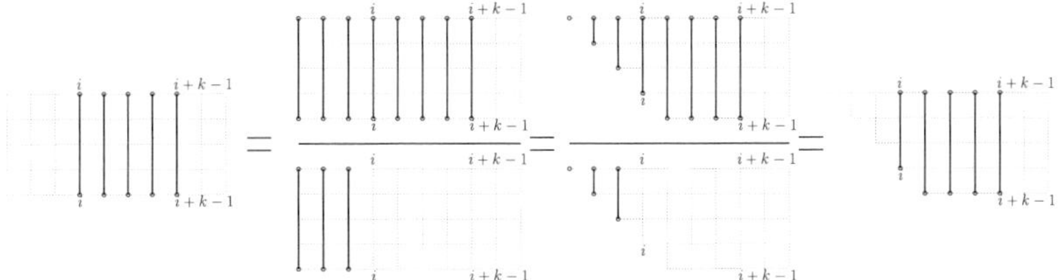

Fig. 14. On the left we represent (in light of the LGV lemma) the partition function equal to $|M_{[\![i,i+k-1]\!],[\![i,i+k-1]\!]}|$. There is only one set of paths (as depicted in the figure) which contributes and the weights used are those of $d_{x,m}$. The first equality rewrites the contribution of that path as a ratio of two partition functions (still using the $d_{x,m}$ weights) which start at the left edge of the diagram. Both of these also only have single paths which contribute. The second equality follows by observing that both the numerator and denominator are equal to the corresponding numerator and denominator for the diagram with the corner removed and with weights $\widetilde{d}_{x,m}$. This equality was observe as the first a prior equality and is from the definition of $\widetilde{D}$. The final equality is again reliant on the fact that only one path contributes to the partition functions.

We prove this by induction. As a base case, from the a priori equalities listed earlier we know that this is true for $i = j$ and all k, as well as for $i = 1$ and all $j \geq i$ and k. The induction is on $(i, j + k - 1)$ in lexicographic order and relies on the Desnanot-Jacobi identity (2.31). As long as $|M_{[\![i-1,i+k-2]\!],[\![j-1,j+k-2]\!]}| > 0$, we may divide both sides of (2.31) by it, yielding

$$(2.33)\quad \begin{aligned} &\left|M_{[\![i,i+k-1]\!],[\![j,j+k-1]\!]}\right| \\ &= \left|M_{[\![i-1,i+k-2]\!],[\![j-1,j+k-2]\!]}\right|^{-1} \\ &\quad\times \Big(\left|M_{[\![i-1,i+k-1]\!],[\![j-1,j+k-1]\!]}\right| \cdot \left|M_{[\![i,i+k-2]\!],[\![j,j+k-2]\!]}\right| \\ &\qquad + \left|M_{[\![i-1,i+k-2]\!],[\![j,j+k-1]\!]}\right| \cdot \left|M_{[\![i,i+k-1]\!],[\![j-1,j+k-2]\!]}\right|\Big) \end{aligned}$$

which can be graphically depicted in Figure 15.

On the left-hand side of (2.33) we are dealing with paths which start left-justified at i and ending right-justified at $j+k-1$. On the right-hand side of (2.33) we either have the starting point i is diminished to $i-1$, or that it remains i but the ending point is diminished to $j+k-2$. Note that the number of paths may change, but we are more interested in

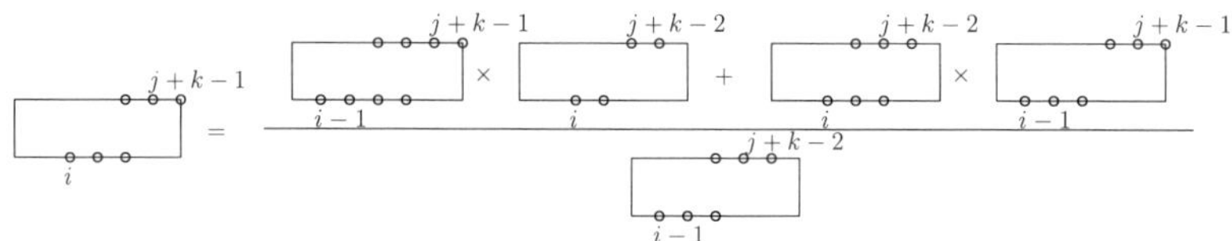

Fig. 15. A graphical depiction of the identity (2.33) which is essentially the Desnanot-Jacobi identity (2.31) up to dividing by one of the terms on the left. The diagrams represent minor determinants of the matrix M. If the bottom label is i and the top label is $j+k-1$ and the number of circles $\circ$ is j, then the diagram represents the determinant $\left|M_{[\![i,i+k-1]\!],[\![j,j+k-1]\!]}\right|$ (as on the left-hand side of the equation). Notice that the number of circles and the starting and ending points differ in each term on the right-hand side, but in manner lexicographically dominated by $(i, j+k-1)$.

keeping track of the left starting point and right ending point since that is what we induct upon. Inductively from the based cases, we establish the positivity of the denominator and that we may replace the M by $\widetilde{M}$ on the right-hand side of (2.33). Indeed, the base case is sufficient for this induction since repeated application of (2.33) eventually yields expressions of the form $\left|M_{[\![\tilde{i},\tilde{i}+\tilde{k}-1]\!],[\![\tilde{j},\tilde{j}+\tilde{k}-1]\!]}\right|$ with either $\tilde{i}=1$ or $\tilde{i}=\tilde{j}$.

To summarize, via the above described induction, we show the equality desired in (2.32), precisely as needed to prove the theorem.

2.2.5. *Proof of Theorem* 2.4 *via the Noumi-Yamada matrix encoding of gRSK.* This proof essentially comes from combining the results of Theorem 1.7 and Section 2.3 of [NY04]. As indicated in the introduction, we appreciate Duncan Dauvergne pointing us towards this. For completeness, we describe the proof in the notation of our present paper.

From a function $E:\mathbb{Z}_{\geq r}\to(0,\infty)$ (with the convention that $E(r-1)\equiv 1$) define $e_x = E(x)/E(x-1)$ for $x\in\mathbb{Z}_{\geq r}$. From this we define a matrix $H_r(E):\mathbb{Z}^2_{\geq 1}\to[0,\infty)$ whose i,j entry is equal to the partition function from i on the bottom to j on the top of the diagram depicted in Figure 16. More concretely, for $i=j\in[\![1,r-1]\!]$ the entries are 1 and for $i,j\in\mathbb{Z}_{\geq r}$ with $i\leq j$, the entry is the partial product $e_i\cdots e_j$; all other entries are zero.

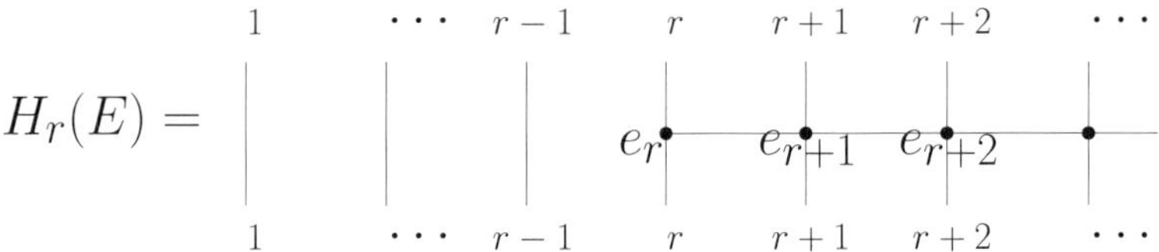

Fig. 16. The matrix entry (i, j) of $H_r(E)$ corresponds to the partition function from i on the bottom to j on the top. The weights of the edges are all 1 and the weights of the vertices are as labeled.

The starting point for this proof is the observation made in Section 2.2 of [NY04] that geometric row insertion can be related to matrix factorization via these H_r matrices. For our purposes, this observation boils down to the following fact: For any $r \in \mathbb{Z}_{\geq 1}$ and $G, F : \mathbb{Z}_{\geq r} \to (0, \infty)$, the geometric Pitman transforms $(g \odot f) : \mathbb{Z}_{\geq r} \to (0, \infty)$ and $(f \otimes g) : \mathbb{Z}_{\geq r+1} \to (0, \infty)$ defined in Definition 2.3 satisfy the matrix identity

$$H_r(G)H_r(F) = H_{r+1}(F \otimes G)H_r(G \odot F). \tag{2.34}$$

This identity is readily checked from definitions, and in fact is essentially the same as the result from Step 1 of the proof of Theorem 2.4 given in Section 2.1.

As shown in Corollary 2.13, the gRSK correspondence can be written in terms of the Pitman transforms (see also the right-hand side of Figure 11). Given this and (2.34) it follows immediately that D and $\widetilde{D} := \mathcal{W}D$ satisfy the matrix identity

$$H_1(D_n) \cdots H_1(D_1) = H_n(\widetilde{D}_n) \cdots H_1(\widetilde{D}_1). \tag{2.35}$$

Multiplying the H matrices is equivalent to appending the diagrams and computing matrix entries as partition functions. From this, it follows that the (u, v) entry corresponding to the left-hand side of (2.35) are precisely $D\big[(u, n) \to (v, 1)\big]$ and the (u, v) entry corresponding to the right-hand side of (2.35) are precisely $\widetilde{D}\big[(u, n \wedge u) \to (v, 1)\big]$. The general U and V result follows (as in the Matveev proof) by an application of the Lindström-Gessel-Viennot lemma.

2.3. Zero-temperature limit

All of the results in Sections 2.1 and 2.2 admit zero-temperature limits. We focus here only on the limit of Theorem 2.4. For an *inverse-temperature* $\beta \in (0, \infty)$ and $f_1, \ldots, f_n : \mathbb{Z}_{\geq 1} \to \mathbb{R}$ (with the convention that $f_i(0) \equiv 0$) define $D_1^\beta, \ldots, D_n^\beta : \mathbb{Z}_{\geq 1} \to (0,\infty)$ via $D_i(x) := e^{\beta f_i(x)}$. For an endpoint pair (U,V), define (note that the β^{-1} superscript below is not an exponent, but rather a label for the variant of $f[U \to V]$ scaled as below)

$$f\big[U \to V\big]^{\beta^{-1}} := \beta^{-1} \log\Big(D^\beta \big[U \to V\big]\Big).$$

For any set of points in the domain of D, observe that

$$f\big[U\big]^{\beta^{-1}} := \beta^{-1} \log\Big(D^\beta \big[U\big]\Big) = \sum_{(x,m)\in U} \big(f_m(x) - f_m(x-1)\big).$$

Thus, by Laplace's method we can extract a *zero-temperature limit*

$$\begin{aligned} &\lim_{\beta\to\infty} f\big[U \to V\big]^{\beta^{-1}} \\ &= \max_{\pi \in U \to V} \Bigg(\sum_{(x,m)\in\pi} \big(f_m(x) - f_m(x-1)\big)\Bigg) =: f\big[U \to V\big]^0. \end{aligned}$$

We can also define zero-temperature limits of the operator $\mathcal{T}$ and $\mathcal{W}$ from Definition 2.3. In particular, for any two functions $f, g : \mathbb{Z}_{\geq r} \to \mathbb{R}$, we define the zero-temperature Pitman transform via (we put a superscript 0 to denote that this is a zero-temperature version)

$$\begin{aligned} \big(g \odot f\big)^0(x) &:= f(x) + \max_{m\in[\![r,x]\!]} \big(g(m) - f(m-1)\big) \qquad \text{for } x \geq r \\ \big(f \otimes g\big)^0(x) &:= g(x) - \max_{m\in[\![r,x]\!]} \big(g(m) - f(m-1)\big) \qquad \text{for } x \geq r+1 \end{aligned}$$

(where by convention $f(r-1) = 0$ when $m = r$). Then, the operator $\mathcal{T}^0$ is defined as

$$\mathcal{T}^0(f,g) := \Big(\big(g \odot f\big)^0, \big(f \otimes g\big)^0\Big).$$

As in Definition 2.3 we define $\mathcal{T}^0_{r,m}$ and then define

$$\mathcal{S}^0_r f := \mathcal{T}^0_{r,r}\mathcal{T}^0_{r,r+1}\cdots\mathcal{T}^0_{r,n-2}\mathcal{T}^0_{r,n-1} f, \qquad \mathcal{W}^0 f := \mathcal{S}^0_{n-1}\mathcal{S}^0_{n-2}\cdots\mathcal{S}^0_2\mathcal{S}^0_1 f.$$

Recalling the notation $\Uparrow\! U$ from the statement of Theorem 2.4, we may now state the immediate zero-temperature limit of that result.

Corollary 2.15. *Let $U = \big((u_i, n)\big)_{i\in[\![1,k]\!]}$ and $V = \big((v_i, 1)\big)_{i\in[\![1,k]\!]}$ be any endpoint pair. Then, for any n functions $f_1, \ldots, f_n : \mathbb{Z}_{\geq 1} \to \mathbb{R}$, writing $f = (f_1, \ldots, f_n)$ we have*

$$f\big[U \to V\big]^0 = \big(\mathcal{W}^0 f\big)\big[\Uparrow\! U \to V\big]^0.$$

The zero-temperature limit of the gRSK correspondence and its relation to the Pitman transform is essentially contained in the earlier mentioned online notes of Pei available at `https://toywiki.xyz/`. There are analogs of the results mentioned in Section 2.2.3 when the inverse-gamma random variables are replaced by geometric or exponential random variables – see the discussion after [COSZ14, Proposition 4.1].

§3. Semi-discrete polymers

The results contained in this section could be derived as limits of the discrete results from Section 2. However, as the proofs are considerably simpler in this semi-discrete (semi-continuous) setting, we will provide complete and direct proofs. In comparing the results and proofs below to those in Section 2 a reader will also notice that we will now be working with logarithmic variables (i.e., the logarithms of the limits of variables from the discrete setting). We do this so as to provide a direct comparison with the zero temperature work of [DOV18], as well as to directly connect our work to that of [O'C12].

Before stating our semi-discrete result, Theorem 3.4, we first introduce the semi-discrete version of non-intersecting (multi)paths, partition functions and the Pitman transform.

3.1. Partition function invariance

Definition 3.1 (Non-intersecting (multi)paths). Fix $n \in \mathbb{Z}_{\geq 2}$. We will consider paths in the semi-discrete space $\mathbb{R} \times [\![1, n]\!]$. For any $u < v \in \mathbb{R}$ and $m \leq \ell \in [\![1, n]\!]$, we call a function $\pi : [u, v] \to [\![m, \ell]\!]$ a *path* with *starting points* (u, m) and *ending points* (v, ℓ) if π is nonincreasing, cadlag on (u, v) and satisfies $\pi(u) = \ell$ and $\pi(v) = m$. We will often be interest in multiple paths $\pi_1, \ldots, \pi_k$ with respective starting and endpoint points

$$U = \{(u_i, \ell_i)\}_{i \in [\![1,k]\!]} \qquad \text{and} \qquad V = \{(v_i, \ell_i)\}_{i \in [\![1,k]\!]},$$

in which case we will write $\pi = (\pi_1, \ldots, \pi_k)$ and call π a *multipath* from U to V. Two paths π_1 and π_2 are called *non-intersecting* if for all $t \in (u_1, v_2) \cap (u_2, v_2)$, $\pi_1(t) < \pi_2(t)$. This condition enforces that paths are disjoint in the interior of their common domain of definition. A multipath π is non-intersecting if for each $1 \leq i \neq j \leq k$, π_i and π_j are non-intersecting. We will assume that all multipaths are non-intersecting. We denote the set of (non-intersecting) multipaths π from U to V by $U \to V$. If the set $U \to V$ is non-empty, then we say that the pair (U, V) constitute an *endpoint pair*.

Definition 3.2 (Partition functions). Fix n continuous functions $f_1, \ldots, f_n : \mathbb{R}_{\geq 0} \to \mathbb{R}$ centered so that $f_i(0) = 0$ and write $f = (f_1, \ldots, f_n)$. To a single path π from (u, ℓ) to (v, m) we associated a *energy* to π with respect to f given by

$$\int df \circ \pi := \int_u^v f'_{\pi(t)}(t)dt.$$

Equivalently, to any path π we can associate *jump times* $u = t_\ell < \cdots < t_m < t_{m-1} = v$ so that for all $j \in [\![m, \ell]\!]$, t_j is the first time that π is at level j. Then

$$\int df \circ \pi = \sum_{j=m}^{\ell} f_j(t_{j-1}) - f_j(t_j).$$

To a (multi)path $\pi = (\pi_1, \ldots, \pi_k)$ we associated an energy

$$\int df \circ \pi := \sum_{i=1}^{k} \int df \circ \pi_i.$$

For any endpoint pair (U, V) we associate the *free energy*

$$f\big[U \to V\big] := \log\left(\int_{\pi \in U \to V} e^{\int df \circ \pi}\right) \tag{3.1}$$

whose exponential is called the *partition function.* In (3.1), the integral over the set of $\pi \in U \to V$ should be understood as Lebesgue integral over the simplex of all possible jump times which define multipaths connecting U to V.

Definition 3.3 (Semi-discrete geometric Pitman transform). Define $n-1$ operators $\mathcal{T}_1, \ldots, \mathcal{T}_n$ which act on n-tuples of functions $f_1, \ldots, f_n$, which we write, as $f(t) = \big(f_1(t), \ldots, f_n(t)\big)$ as

$$\big(\mathcal{T}_i f\big)(t) := f(t) + \left(\log \int_0^t e^{f_{i+1}(s) - f_i(s)} ds\right)\big(e_i - e_{i+1}\big)$$

where $e_1, \ldots, e_n$ are basis vectors. Using the $\mathcal{T}$ operators, we define operators $\mathcal{S}_1, \ldots, \mathcal{S}_{n-1}$ and the operator $\mathcal{W}$ which act on functions f as

$$\mathcal{S}_r f := \mathcal{T}_r \mathcal{T}_{r+1} \cdots \mathcal{T}_{n-1} f, \qquad \text{and} \qquad \mathcal{W} f := \mathcal{S}_{n-1} \mathcal{S}_{n-2} \cdots \mathcal{S}_1 f. \tag{3.2}$$

Theorem 3.4. *For any endpoint pair (U, V) and collection of n functions $f = (f_1, \ldots, f_n)$ as in Definition 3.2,*

$$f\big[U \to V\big] = \big(\mathcal{W} f\big)\big[U \to V\big]. \tag{3.3}$$

In contrast to discrete case Theorem 2.4, we do not need to shift the starting points U here.

Proof. The proof of Theorem 3.4 follows the same three step program as we used in proving Theorem 2.4 (and as used to prove [DOV18, Proposition 4.1]). This proof is much closer to that of [DOV18, Proposition 4.1]. It is only in the first step that there is any real deviation.

Step 1 ($n = 2$ and $k = 1$ case of (3.9)). In this case $f = (f_1, f_2)$, $U = (u, 2)$ and $V = (v, 1)$. We seek to prove (3.9), which now reads (writing $\mathcal{T}$ in place of $\mathcal{T}_1$)

$$f\big[(u,2) \to (v,1)\big] = (\mathcal{T}f)\big[(u,2) \to (v,1)\big].$$

From the definition of $\mathcal{T}$,

$$(\mathcal{T}f)_1(t) = f_1(t) + \log \int_0^t e^{f_2(s)-f_1(s)}ds,$$
$$(\mathcal{T}f)_2(t) = f_2(t) - \log \int_0^t e^{f_2(s)-f_1(s)}ds.$$

Using this we may rewrite (3.9) as the identity

$$s(u,v) - s(0,v) - s(0,u) = \log\left(\int_u^v e^{f_2(t)-f_1(t)-2s(0,t)}dt\right), \tag{3.4}$$

where we have defined

$$s(x,y) := \log \int_x^y e^{f_2(s)-f_1(s)}ds.$$

Denoting $g(s) = f_2(s) - f_1(s)$, we can rewrite the integral on the right-hand side of (3.4) as

$$\int_u^v e^{f_2(t)-f_1(t)-2s(0,t)}dt = \int_u^v \frac{e^{g(t)}dt}{\left(\int_0^t e^{g(s)}ds\right)^2} = \frac{-1}{\int_0^t e^{g(s)}ds}\bigg|_u^v = \frac{1}{\int_0^u e^{g(s)}ds} - \frac{1}{\int_0^v e^{g(s)}ds}. \tag{3.5}$$

The first equality is simply a rewriting via g, the second and third equalities evaluate the integration. It remains to match the result of the above calculation with the exponentiated left-hand side of (3.4) which is evaluated as

$$\exp\Big(s(u,v) - s(0,v) - s(0,u)\Big) = \frac{\int_u^v e^{g(s)}ds}{\int_0^v e^{g(s)}ds \int_0^u e^{g(s)}ds}.$$

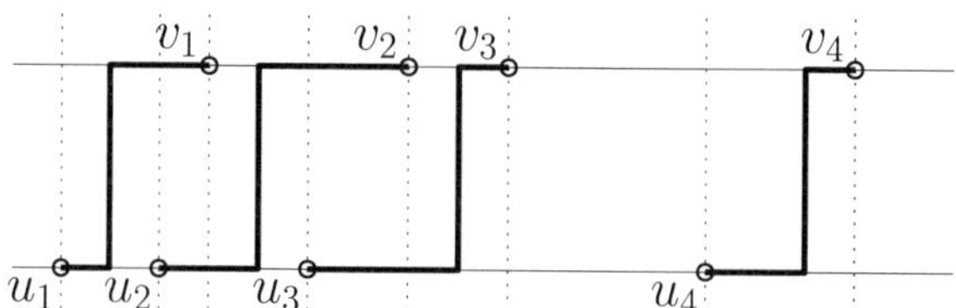

Fig. 17. The interval $(v_3, x_4]$ is type 0; $(u_1, u_2]$, $(v_1, u_3]$, $(v_2, v_3]$ and $(u_4, v_4]$ are type 1; and $(u_2, v_1]$ and $(u_3, v_2]$ are type 2. A multipath $\pi \in U \to V$ is shown and the number of paths matches the type of each interval.

Since $u < v$, the right-hand side matches the right-hand side of (3.5) completing Step 1.

Step 2 ($n = 2$ and $k \geq 1$ case of (3.9)). Now $f = (f_1, f_2)$, $U = \{(u_i, 2)\}_{i \in [\![1,k]\!]}$ and $V = \{(v_i, 1)\}_{i \in [\![1,k]\!]}$. This step is shown along the same lines as [DOV18, Lemma 4.3]. In order that U and V constitute an endpoint pair, it is necessary that the u's and v's are ordered so that connecting them with pairs of non-intersecting paths is possible. The implies that $v_i \leq u_{i+2}$ for $i \in [\![1, k-2]\!]$. Define $(z_1, \ldots, z_{2k})$ to be the ordering of the union of $\{u_1, \ldots, u_k\}$ and $\{v_1, \ldots, v_k\}$. We say that the interval $(z_i, z_{i+1}]$ is of "type 2" if $z_i = u_{j+1}$ and $z_{i+1} = v_j$ for some $j \in [\![1, k-1]\!]$. We say that the interval $(z_i, z_{i+1}]$ is of "type 1" if either of the following holds (we temporarily employ the notational convention that $v_0 = 0$ and $u_{k+1} = \infty$): $z_i = u_j$, $z_{i+1} = u_{j+1}$ and $v_{j-1} < u_j < u_{j+1} < v_j$ for some $j \in [\![1, k-1]\!]$; or $z_i = u_j$, $z_{i+1} = v_j$ and $v_{j-1} < u_j < v_j < u_{j+1}$ for some $j \in [\![1, k]\!]$. Finally, we say that the interval $(z_i, z_{i+1}]$ is of "type 0" if $z_i = v_j$ and $z_{i+1} = u_{j+1}$ for some $j \in [\![1, k-1]\!]$. For $r \in \{0, 1, 2\}$, if the interval $(z_i, z_{i+1}]$ is of type r this means that for all paths $\pi \in U \to V$, π necessarily has exactly r paths traversing that interval. See Figure 17 for an example.

The value of the above partitioning of $(u_1, \ldots, v_n]$ into the z intervals is that on the boundary of each interval, every $\pi \in U \to V$ must take the same values. Thus, the integral which defines $f[U \to Y]$ factorizes into a product of integrals over each of the z intervals. Consequently, the produces the following decomposition:

$$
\begin{aligned}
&f[U \to Y] \\
&= \sum_{i=1}^{2k-1} f\Big[\big((z_i, 2), (z_i, 1)\big) \to \big((z_{i+1}, 2), (z_{i+1}, 1)\big)\Big] \mathbf{1}\big\{(z_i, z_{i+1}] \text{ is type } 2\big\} \\
&\quad + \sum_{i=1}^{2k-1} f\big[(z_i, 2) \to (z_{i+1}, 1)\big] \mathbf{1}\big\{(z_i, z_{i+1}] \text{ is type } 1\big\}.
\end{aligned}
$$

Type 0 intervals do not contribute to this sum. By Step 1, we may replace $f\big[(z_i, 2) \to (z_{i+1}, 1)\big]$ above by $(\mathcal{W}f)\big[(z_i, 2) \to (z_{i+1}, 1)\big]$ without changing the value. As follows directly from definitions, for the type 2 terms we may replace f by $\mathcal{W}f$ without changing the value. From these two replacements we conclude that $f[U \to Y] = (\mathcal{W}f)[U \to Y]$, completing Step 2.

Step 3 ($n \geq 2$ and $k \geq 1$ case of (3.9)). Now $f = (f_1, \dots, f_n)$, $U = \big\{(u_i, 2)\big\}_{i\in[\![1,k]\!]}$ and $V = \big\{(v_i, 1)\big\}_{i\in[\![1,k]\!]}$. We claim that for all $m \in [\![1, n-1]\!]$,

$$f\big[U \to V\big] = (\mathcal{T}_m f)\big[U \to V\big]. \tag{3.6}$$

Observe that since $\mathcal{W}$ is written as a composition of $\mathcal{T}_m$ operators, applying (3.6) repeatedly for various values of m yields the desired result (3.9). Note that it is apparent from (3.6) that the order in which we apply the $\mathcal{T}_m$ does not matter. This is different than in the discrete case, and is related to the braid relations discussed in Section 4.1.

To prove (3.6), we utilize the following decomposition:

$$f\big[U \to V\big] = \log \int_{W,Z} e^{f[U\to W^+]+f[W\to Z]+f[Z^-\to V]} dW dZ. \tag{3.7}$$

In the above equation, we have taken $W = (w_i, m+1)_{i\in[\![1,k]\!]}$, $W^+ = (w_i, m+2)_{i\in[\![1,k]\!]}$, $Z = (z_i, m)_{i\in[\![1,k]\!]}$, and $Z^- = (z_i, m-1)_{i\in[\![1,k]\!]}$, the integration is over all W and Z for which (U, W^+), (W, Z), and (Z^-, V) are all endpoint pairs, and the dW and dZ are Lebesgue measure on the k-tuples $(w_1, \dots, w_k)$ and $(z_1, \dots, z_k)$. The decomposition is valid as written as long as $m \in [\![2, n-2]\!]$. To deal with the boundary cases $m = 1$ and $m = n-1$ we need to modify the decomposition slightly. For $m = 1$, we set $Z = V$, and drop the integration in dZ and the term $[Z^- \to V]$ from (3.7). For $m = n-1$, we set $W = U$, and drop the integration in dW and the term $[U \to W^+]$ from (3.7). This decomposition is illustrated in Figure 18.

Now observe that inside the exponential in the integrand on the right-hand side of (3.7), the term $f[U \to W^+]$ depends only on $f_n, \dots, f_{m+2}$, the term $f[W \to Z]$ depends only on f_{m+1} and f_m and the term $f[Z^- \to V]$ depends only on $f_{m-1}, \dots, f_1$. Since $\mathcal{T}_m$ acts as the identity on all functions in f except f_m and f_{m+1}, it follows immediately that $f[U \to W^+] = (\mathcal{T}_m f)[U \to W^+]$ and $f[Z^- \to V] = (\mathcal{T}_m f)[Z^- \to V]$. Meanwhile, using the result of Step 2 it follows that $f[W \to Z] = (\mathcal{T}_m f)[W \to Z]$. Putting this together we see that we can replace f by $(\mathcal{T}_m f)$ in the right-hand side of (3.7) without changing the value. Thus, using that decomposition backwards, now with $(\mathcal{T}_m f)$ in place of f, we

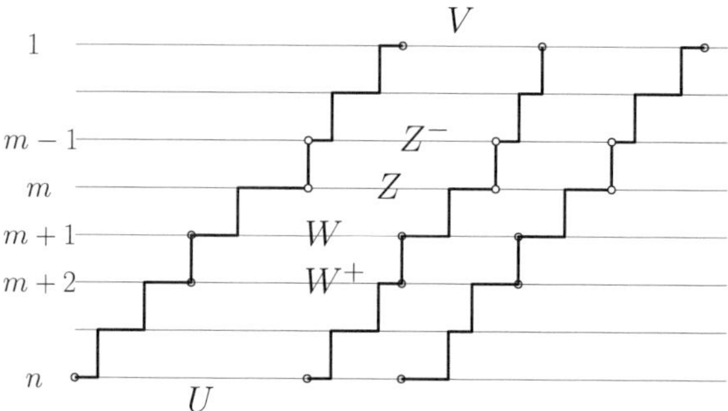

Fig. 18. The decomposition (3.7). The circles on level 1 correspond to points in V, on level $m-1$ to Z^-, on level m to Z, on level $n+1$ to W and on level $m+2$ to W^+. Notice that the set Z^- sits immediately above Z and the set W^+ sits immediately below W. The boundary cases, when $m=1$ or $m=n+1$ are not shown here.

conclude (3.6). This completes Step 3 and thus also completes the proof of the theorem. ■

3.2. Relation to geometric RSK correspondence

As explained at the bottom of page 445 of [O'C12] (see also [BBO05]), the operator $\mathcal{W}$ defined in (3.8) is related to a semi-discrete geometric RSK correspondence. In particular, $\mathcal{W}$ admits the following path formula: For all $k \in [\![1,n]\!]$ and $t \in [0,\infty)$

$$\sum_{i=1}^{k} (\mathcal{W}f)_i(t) = f\big[(0,n)^k \to (t,1)^k\big].$$

[O'C12, Theorem 3.1 and Corollary 4.1] shows that f_i are taken to be independent Brownian motions, then $(\mathcal{W}f)(t)$ evolves in t as diffusion with generator given by the Doob h-transform of the quantum Toda Hamiltonian with h given by the class-one Whittaker function. Based on this result, [CH16, Proposition 3.4] showed that $(\mathcal{W}f)(t)$ enjoys the H-Brownian Gibbs property for an exponential interaction Hamiltonian $H(x)=e^x$.

3.3. Zero-temperature limit

For $f=(f_1,\dots,f_n)$ fixed define (note that the β^{-1} superscript below is not an exponent, but rather a label for the variant of $f[U\to V]$ scaled as below)

$$f[U \to V]^{\beta^{-1}} := \beta^{-1}(\beta f)[U \to V].$$

Then, observe that by Laplace's method we can extract a *zero-temperature limit*

$$\lim_{\beta \to \infty} f[U \to V]^{\beta^{-1}} = \sup_{\pi \in U \to V} \int df \circ \pi =: f[U \to V]^0.$$

We can also define zero-temperature limits of the operators $\mathcal{T}_m$ and $\mathcal{W}$ from Definition 3.3. Define $n-1$ operators $\mathcal{T}_1^0, \dots, \mathcal{T}_n^0$ which act on n-tuples of functions $f_1, \dots, f_n$, which we write, as $f(t) = \big(f_1(t), \dots, f_n(t)\big)$ as

$$\big(\mathcal{T}_i^0 f\big)(t) := f(t) + \sup_{s \in [0,t]} \big(f_{i+1}(s) - f_i(s)\big)\big(e_i - e_{i+1}\big)$$

where $e_1, \dots, e_n$ are basis vectors. Using the $\mathcal{T}^0$ operators, we define operators $\mathcal{S}_1^0, \dots, \mathcal{S}_{n-1}^0$ and the operator $\mathcal{W}^0$ which act on functions f as

$$\mathcal{S}_r^0 f := \mathcal{T}_r^0 \mathcal{T}_{r+1}^0 \cdots \mathcal{T}_{n-1}^0 f, \quad \text{and} \quad \mathcal{W}^0 f := \mathcal{S}_{n-1}^0 \mathcal{S}_{n-2}^0 \cdots \mathcal{S}_1^0 f. \tag{3.8}$$

The following is an immediate corollary of Theorem 3.4 under the same zero-temperature limit. It was proved earlier as [DOV18, Proposition 4.1].

Corollary 3.5. *For any endpoint pair (U, V) and collection of n functions $f = (f_1, \dots, f_n)$ as in Definition* 3.2,

$$f^0[U \to V] = (\mathcal{W}^0 f)[U \to V]^0. \tag{3.9}$$

§4. Some remarks

4.1. Braid relations

The semi-discrete Pitman transform operators $\mathcal{T}_1, \dots, \mathcal{T}_{n-1}$ in Definition 3.3 satisfy braid relations [O'C12, BBO05] $\mathcal{T}_i \mathcal{T}_{i+1} \mathcal{T}_i = \mathcal{T}_{i+1} \mathcal{T}_i \mathcal{T}_{i+1}$. This does not seem to hold for the discrete Pitman transform operators from Definition 2.3. For instance, when $n = 3$, the operator $\mathcal{W} = \mathcal{T}_{2,2} \mathcal{T}_{2,1} \mathcal{T}_{1,2}$ but it is easy to see that it does not equal $\mathcal{T}_{1,2} \mathcal{T}_{2,1} \mathcal{T}_{2,2}$. In fact, the latter composition of operators is not even well-defined due to the changing of the domains under the application of the $\mathcal{T}$ operators. The question here is whether there is any sense in which the braid relations generalize to the discrete setting.

4.2. Airy / KPZ sheet limits

[DOV18, Proposition 4.1] (Corollary 3.5 herein) serves as the starting point in [DOV18] for the construction of the Airy sheet. When $k = 1$ and the endpoint pair (U, V) has $U = (u, n)$ and $V = (v, 1)$, the result says that the point-to-point last passage time (i.e., $f^0[U \to V]$) for all choices of $u \le v\mathbb{Z}_{\ge 1}$ can be recovered from the Pitman transform $\mathcal{W}^0 f$ through another last passage time calculation. [DOV18] calls $\mathcal{W}^0 f$ the *melon* of f. When the f_i are Brownian motions, the melon $\mathcal{W}^0 f$ is an n-particle Dyson Brownian motion (DBM). This fact was shown in [BJ02, OY02], generalizing results of [Bar01, GTW01] which matched the one-time marginal of $(\mathcal{W}^0 f)_1$ to the top eigenvalue of an $n \times n$ GUE random matrix. It can also be seen as a limit of the results mentioned briefly, at the discrete geometric level, in Section 2.2.3. The Brownian choice of the f_i is known as a *solvable* or *integrable* model due to its special structural properties such as just mentioned.

Much of [DOV18] is devoted to showing that under KPZ $n \to \infty$ scaling, the point-to-point last passage time fluctuation has a limit as a two parameter process defined by varying the starting and ending points (the limits of the u and v). [DOV18, Theorem 1.3] claims that the limit, termed the *Airy sheet* from [CQR15], can be defined directly (not just as a limit) through the *Airy line ensemble* [PS02, Joh03, CH14] which, itself, arises as the $n \to \infty$ edge limit of the DBM $\mathcal{W}^0 f$.

[DOV18] relies on fine control over particle locations and gaps in DBM as $n \to \infty$ as well as its convergence to the Airy line ensemble. This control, some of which is in the companion paper [DV18], comes from two main tools – the *Brownian Gibbs property* [CH14] and the *determinantal* structure [AvM05] of the DBM and Airy line ensemble.

The question here is whether a similar analysis can be performed based on Theorem 2.4. It seems most like that in the discrete zero-temperature setting. As described at the end of Section 2.3, when the weights in last passage percolation are geometric (i.e., the solvable choice of weights), Dyson Brownian motion is replaced by another Markov process which has a discrete version of the Brownian Gibbs property and which is determinantal.

In the positive-temperature cases under the solvable choices of inverse-gamma distributed weights (in the discrete case, as in Section 2.2.3) or under Brownian f_i (in the semi-discrete case, as in Section 3.2), the $\mathcal{W}D$ and $\mathcal{W}f$ functions still enjoy relatively simple Gibbs properties (see Sections 2.2.3 and 3.2 for references to precise statements). However, $\mathcal{W}D$ and $\mathcal{W}f$ are not known to be determinantal point processes. Instead, they relate to *Whittaker measures* [O'C12, COSZ14, BC14]. While there has been much success (for example [BC14, BCF14, BCR13, KQ18]) in extracting asymptotics for the top-labeled particle of Whittaker measures, there are essentially no results providing control over the lower particles or particle gaps. This poses an immediate impediment to adapting the approach used in [DV18, DOV18].

It is known that under special *weak noise* scaling, the discrete [AKQ14] and semi-discrete [Nic16] positive temperature directed polymer models have limits to the KPZ stochastic PDE. The Gibbsian line ensembles which arise from the solvable models likewise have scaling limits under the special weak noise scaling to the KPZ line ensemble [CH16] (see also [OW16, CN17, Wu19]). Thus, it is natural to speculate that the KPZ sheet (that we define in a moment) can be recovered directly from the KPZ line ensemble and that this relationship is facilitated through taking a limit of Theorems 2.4 and 3.4. The KPZ sheet is easy to define as a function of space-time white noise ξ. For $x \in \mathbb{R}$ let $h(t;x,y) = \log Z(t;x,y)$ be the Hopf-Cole solution to the KPZ equation, where Z solves the multiplicative stochastic heat equation

$$\partial_t Z = \frac{1}{2}\partial_y^2 Z + \xi Z, \qquad h(0;x,y) = \delta_{x=y}.$$

The KPZ sheet (for a fixed time t) is the two-parameter function $(x,y) \mapsto h(t;x,y)$.

4.3. Other invariances

Recently there have been two papers [BGW19, Pet19] which have studied other types of invariances of polymer models and stochastic vertex models. At face-value, those results do not seem directly related to those studied here, though it is enticing to search for a way to unify them. For instance, stochastic vertex models [CP16, BP18] are known to generalize the solvable polymer models. It would be very interesting to find a lifting of our results into that setting. Indeed, there has been plenty of study in [Pei17, BP16, PM17, BM18] of generalizations of the gRSK correspondence up the hierarchy of Macdonald processes to the q-Whittaker or Hall-Littlewood level (the gRSK relates to the Whittaker level) which relate to stochastic vertex models. However, at that level the correspondence is no-longer bijective and instead involves some randomization. This presents an immediate impediment to even formulating an analog of our main results and deserves further consideration.

References

[AKQ14] T. Alberts, K. Khanin and J. Quastel, The intermediate disorder regime for directed polymers in dimension 1 + 1, *Ann. Probab.*, 42(3):1212–1256, 2014.

[AvM05] M. Adler and P. van Moerbeke, PDEs for the joint distributions of the Dyson, Airy and Sine processes, *Ann. Probab.*, 33(4):1326–1361, 2005.

[Bar01] Y. Baryshnikov, GUEs and queues, *Probab. Theory Related Fields*, 119:256–274, 2001.

[BBO05] P. Biane, P. Bougerol and N. O'Connell, Littelmann paths and Brownian paths, *Duke Math. J.*, 130(1):127–167, 10 2005.

[BC14] A. Borodin and I. Corwin, Macdonald processes, *Probab. Theory Related Fields*, 158(1):225–400, 2014.

[BCD] G. Barraquand, I. Corwin and E. Dimitrov, Spatial tightness at the edge of Gibbsian line ensembles, *arXiv*:2101.03045, 2021.

[BCF14] A. Borodin, I. Corwin and P. Ferrari, Free energy fluctuations for directed polymers in random media in 1 + 1 dimension, *Comm. Pure Appl. Math.*, 67(7):1129–1214, 2014.

[BCR13] A. Borodin, I. Corwin and D. Remenik, Log-gamma polymer free energy fluctuations via a Fredholm determinant identity, *Comm. Math. Phys.*, 324:215–232, 2013.

[BDJ99] J. Baik, P. Deift and K. Johansson, On the distribution of the length of the longest increasing subsequence of random permutations, *J. Amer. Math. Soc.*, 12:1119–1178, 1999.

[BGW19] A. Borodin, V. Gorin and M. Wheeler, Shift-invariance for vertex models and polymers, *arXiv*:1912.02957, 2019.

[BJ02] P. Bougerol and T. Jeulin, Paths in Weyl chambers and random matrices, *Probab. Theory Related Fields*, 124(4):517–543, 2002.

[BM18] A. Bufetov and K. Matveev, Hall-Littlewood RSK field, *Selecta Math.*, 24:4839–4884, 2018.

[BP16] A. Borodin and L. Petrov, Nearest neighbor Markov dynamics on Macdonald processes, *Adv. Math.*, 300:71–155, 2016.

[BP18] A. Borodin and L. Petrov, Higher spin six vertex model and symmetric rational functions, *Selecta Math.*, 24(2):751–874, Apr 2018.

[CH14] I. Corwin and A. Hammond, Brownian Gibbs property for Airy line ensembles, *Invent. Math.*, 195(2), 2014.

[CH16] I. Corwin and A. Hammond, KPZ line ensemble, *Probab. Theory Related Fields*, 166(1-2):67–185, 2016.

[CN17] I. Corwin and M. Nica, Intermediate disorder directed polymers and the multi-layer extension of the stochastic heat equation, *Electron. J. Probab.*, 22:49, 2017.

[Cor12] I. Corwin, The Kardar-Parisi-Zhang equation and universality class, *Random Matrices Theory Appl.*, 1(1):1130001, 76, 2012.

[COSZ14] I. Corwin, N. O'Connell, T. Seppäläinen and N. Zygouras, Tropical combinatorics and Whittaker functions, *Duke Math. J.*, 163(3): 513–563, 2014.

[CP16] I. Corwin and L. Petrov, Stochastic higher spin vertex models on the line, *Commun. Math. Phys.*, 343(2):651–700, 2016.

[CQR15] I. Corwin, J. Quastel and D. Remenik, Renormalization fixed point of the KPZ universality class, *J. Stat. Phys.*, 160(4):815–834, 2015.

[Dau20] D. Dauvergne, Hidden invariance of last passage percolation and directed polymers, *arXiv*:2002.09459, 2020.

[dBR38] G. de B. Robinson, On the representations of the symmetric group, *Amer. J. Math.*, 60(3):745–760, 1938.

[DOV18] D. Dauvergne, J. Ortmann and B. Virág, The directed landscape, *arXiv*:1812.00309, 2018.

[DV18] D. Dauvergne and B. Virág, Basic properties of the Airy line ensemble, *arXiv*:1812.00311, 2018.

[FZ00] S. Fomin and A. Zelevinsky, Total positivity: Tests and parametrizations, *Mathematical Intelligencer.*, 22:22–33, 2000.

[Gre74] C. Greene, An extension of Schensted's theorem., *Adv. Math.*, 14:254–265, 1974.

[GTW01] J. Gravner, C. Tracy and H. Widom, Limit theorems for height fluctuations in a class of discrete space and time growth models, *J. Stat. Phys.*, 102(5):1085–1132, 2001.

[JO19] S. Johnston and N. O'Connell, Scaling limits for non-intersecting polymers and Whittaker measures, *J. Stat. Phys.*, 179:354–407, 2019.

[Joh00] K. Johansson, Shape fluctuations and random matrices, *Comm. Math. Phys.*, 209(2):437–476, 2000.

[Joh03] K. Johansson, Discrete polynuclear growth and determinantal processes, *Comm. Math. Phys.*, 242:277–329, 2003.

[Kir01] A. N. Kirillov, Introduction to tropical combinatorics, In: *Physics and Combinatorics* 2000, *Proceedings of the Nagoya* 2000 *International Workshop*, pages 85–150, Nagoya, Japan, 2001, World Scientific.

[Knu70] D. Knuth, Permutations, matrices, and generalized young tableaux, *Pacific J. Math.*, 34:709–727, 1970.

[KQ18] A. Krishnan and J. Quastel, Tracy-Widom fluctuations for perturbations of the log-gamma polymer in intermediate disorder, *Ann. Appl. Probab.*, 28(6):3736–3764, 2018.

[MQR17] K. Matetski, J. Quastel and D. Remenik, The KPZ fixed point, *arXiv*:1701.00018, 2017.

[Nic16] M. Nica, Intermediate disorder limits for multi-layer semi-discrete directed polymers, *arXiv*:1609.00298, 2016.

[NY04] M. Noumi and Y. Yamada, Tropical Robinson-Schensted-Knuth correspondence and birational Weyl group actions, In: *Representation Theory of Algebraic Groups and Quantum Groups*, pages 371–442, Mathematical Society of Japan, 2004.

[O'C12] N. O'Connell, Directed polymers and the quantum Toda lattice, *Ann. Probab.*, 40(2):437–458, 2012.

[Oko01] A. Okounkov, Infinite wedge and random partitions *Selecta Math.*, 7(1):57, Apr 2001.

[OR03] A. Okounkov and N. Reshetikhin, Correlation function of Schur process with application to local geometry of a random 3-dimensional Young diagram, *J. Amer. Math. Soc.*, 16:581–603, 2003.

[OSZ14] N. O'Connell, T. Seppäläinen and N. Zygouras, Geometric RSK correspondence, Whittaker functions and symmetrized random polymers, *Invent. Math.*, 197:361–416, 2014.

[OW16] N. O'Connell and J. Warren, A multi-layer extension of the stochastic heat equation, *Comm. Math. Phys.*, 341(1):1–33, 2016.

[OY02] N. O'Connell and M. Yor, A representation for non-colliding random walks, *Electron. Commun. Probab.*, 7:1–12, 2002.

[Pei17] Y. Pei, A q-Robinson-Schensted-Knuth Algorithm and a q-Polymer, *Electron. J. Combin.*, 24(4), 2017.

[Pet19] L. Petrov, Parameter permutation symmetry in particle systems and random polymers, *arXiv*:1912.06067, 2019.

[PM17] L. Petrov and K. Matveev, q-randomized Robinson-Schensted-Knuth correspondences and random polymers, *Ann. Inst. H. Poinc. D*, 4: 1–123, 2017.

[PS02] M. Prähofer and H. Spohn, Scale invariance of the PNG droplet and the Airy process, *J. Stat. Phys.*, 108(5):1071–1106, 2002.

[QS15] J. Quastel and H. Spohn, The one-dimensional KPZ equation and its universality class, *J. Stat. Phys.*, 160(4):965–984, 2015.

[Qua12] J. Quastel, Introduction to KPZ, In: *Current developments in mathematics*, 2011, pages 125–194, Int. Press, Somerville, MA, 2012.

[Sag01] B. Sagan, *The symmetric group*, Springer-Verlag, Berlin, 2001.

[Sch61] C. Schensted, Longest increasing and decreasing subsequences, *Cand. J. Math.*, 13:179–191, 1961.

[Sta99] R. Stanley, *Enumerative Combinatorics, Volume* 2, Cambridge University Press, 1999.

[Wu19] X. Wu, Tightness of discrete Gibbsian line ensembles with exponential interaction Hamiltonians, *arXiv*:1909.00946, 2019.

Columbia University,
Department of Mathematics,
2990 Broadway, New York, NY 10027, USA
E-mail address: corwin@math.columbia.edu

Advanced Studies in Pure Mathematics 87, 2021
Stochastic Analysis, Random Fields and Integrable Probability — Fukuoka 2019
pp. 137–154

Complex networks: structure and functionality

Frank den Hollander

Abstract.

The goal of the present paper is to describe three topics in the area of complex networks that constituted a mini-course delivered at the 12th Mathematical Society of Japan – Seasonal Institute, held at Kyushu University in Fukuoka, Japan, 31/07–09/08, 2019. The topics are: (I) Spectra of adjacency matrices, (II) Equivalence of ensembles, (III) Exploration and mixing times. Topics (I) and (II) are part of the ongoing attempt to understand *structure* of networks, topic (III) to elucidate *functionality* of networks.

§1. Outline

In Section 2, which is based on joint work with Arijit Chakrabarty, Rajat Hazra and Matteo Sfragara [7], we consider inhomogeneous *Erdős-Rényi random graphs* on n vertices. We study the *empirical spectral distribution* of the *adjacency matrix* A_n in the limit as $n \to \infty$ in a regime that interpolates between sparse and dense. In particular, we show that the empirical spectral distribution of A_n when properly scaled converges to a deterministic limit weakly in probability. For the special case where the connectivity probability between two vertices has the *product property*, we give an explicit characterisation of the limit distribution. The result is applied to statistical inference of *sociability patterns in social networks*.

In Section 3, which is based on joint work with Diego Garlaschelli, Michel Mandjes, Joey de Mol, Andrea Roccaverde, Tiziano Squartini and Nicos Starreveld [9], [10], [11], [12], [13], we consider random graphs subject to topological constraints. We compare two probability distributions on the set of simple graphs on n vertices induced by a given

Received December 30, 2019.
Revised March 22, 2020.
2010 *Mathematics Subject Classification.* 60C05, 60K35.
Key words and phrases. Complex networks, random graphs, spectrum adjacency matrix, ensemble equivalence, exploration, mixing time.

constraint: (1) The *microcanonical ensemble*, where the constraint is *hard*, i.e., has to be satisfied for every realisation of the graph; (2) The *canonical ensemble*, where the constraint is *soft*, i.e., has to be satisfied on average. We say that *breaking of ensemble equivalence* occurs in the limit as $n \to \infty$ when the *relative entropy* of the two ensembles per vertex (in the sparse regime), respectively, per edge (in the dense regime) is strictly positive. We present two examples of constraints where breaking of ensemble equivalence occurs, namely, when the constraint is on the degree sequence and when the constraint is on the total number of edges and triangles. The result is applied to model selection for real-world networks.

In Section 4, which is based on joint work with Luca Avena, Hakan Guldas and Remco van der Hofstad [1], [2], we consider the mixing time of random walks on random graphs. Many real-world networks, such as WWW, are dynamic in nature. It is therefore natural to study random walks on *dynamic random graphs.* We consider random walk on a random graph with prescribed degrees. We investigate what happens when at each unit of time a fraction α_n of the edges is randomly rewired, where n is the number of vertices. We identify *three regimes* for the mixing time in the limit as $n \to \infty$, depending on the choice of α_n. These regimes exhibit surprising behaviour. The results are relevant for Google PageRank.

§2. Spectra of adjacency matrices

Spectra of random matrices have been analysed for almost a century. In recent years, many interesting results have been derived for spectra of random matrices associated with networks. The question that will be addressed in this section is: *What can be said about the spectrum of the adjacency matrix of a large inhomogeneous Erdős-Rényi random graph?*

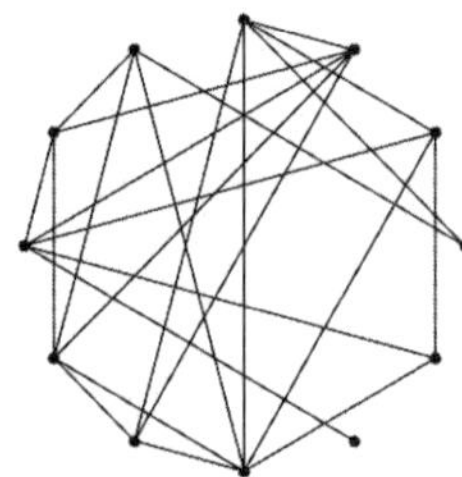

Fig. 1. Erdős-Rényi random graph.

2.1. Setting

Let $(\varepsilon_N)_{N\in\mathbb{N}}$ be a sequence of positive numbers such that

$$\lim_{N\to\infty} \varepsilon_N = 0, \qquad \lim_{N\to\infty} N\varepsilon_N = \infty.$$

Let $f\colon [0,1]\times[0,1] \to [0,\infty)$ be a continuous function such that $f(x,y) = f(y,x)$ for all $x,y \in [0,1]$. Fix $N \in \mathbb{N}$, and consider the inhomogeneous Erdős-Rényi random graph ER_N on N vertices where an edge is placed between the pair of vertices $\{i,j\}$ with probability

$$\varepsilon_N\, f\left(\frac{i}{N}, \frac{j}{N}\right), \quad 1 \le i,j \le N,$$

independently for different edges. Write $\mathbb{P}$ for the law of ER_N.

Let A_N be the adjacency matrix of ER_N. Write

$$\lambda_i(A_N), \qquad 1 \le i \le N,$$

for the real eigenvalues of A_N. The *empirical spectral distribution* of A_N is defined as

$$\mathrm{ESD}(A_N) = \frac{1}{N}\sum_{i=1}^{N} \delta_{\lambda_i(A_N)},$$

which is a random probability distribution on $\mathbb{R}$.

2.2. Scaling

Theorem 1. *There exists a compactly supported symmetric probability measure μ on $\mathbb{R}$ such that, weakly in $\mathbb{P}$-probability,*

$$\lim_{N\to\infty} \mathrm{ESD}\left(A_N \big/ \sqrt{N\varepsilon_N}\right) = \mu.$$

Furthermore, if

$$\min_{x,y\in[0,1]} f(x,y) > 0,$$

then μ is absolutely continuous with respect to Lebesgue measure. The density of μ can be characterised implicitly via an integral equation for its Stieltjes transform.

(A weaker version of the above theorem also appeared in [18].)

It is possible to identify μ when

$$f(x,y) = r(x)r(y), \qquad x,y \in [0,1],$$

for some continuous function $r\colon [0,1] \to [0,\infty)$.

Theorem 2. *If f is of product form, then*

$$\mu = \mu_r \boxtimes \mu_s,$$

where

$$\begin{aligned}\mu_r &= \mathrm{LAW}[r(U)], \quad U = \mathrm{UNIF}[0,1],\\ \mu_s &= \textit{standard Wigner semicircle law},\end{aligned}$$

and $\boxtimes$ denotes free multiplicative convolution.

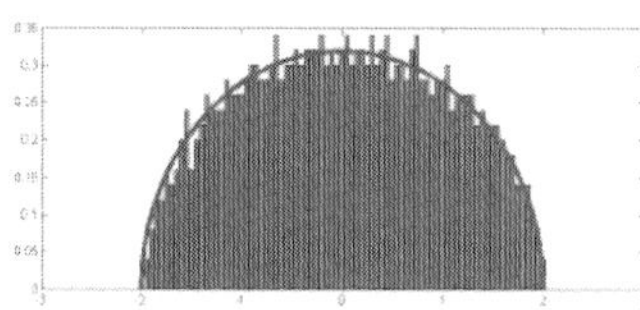

Fig. 2. Wigner semicircle law.

In free probability, the Wigner semicircle law takes over the role of the normal law in classical probability. The so-called free cumulants replace the classical cumulants, in the sense that partitions are replaced by non-crossing partitions. Just as the cumulants of degree ≥ 2 are all zero if and only if the distribution is normal, the free cumulants of degree ≥ 2 are all zero if and only if the distribution is the Wigner semicircle law.

Theorem 3. *Theorems* 1–2 *can be generalized to the situation where the function f is random, depends on N and converges to a deterministic limit as $N \to \infty$.*

Key ingredients of the proof are: centering, Gaussianisation, perturbation, decoupling, and combinatorics from free probability.

2.3. Application 1: social networks

Consider a community of N individuals, represented by the vertices in ER_N. Data is available about which individuals are acquainted. Based on this data, the sociability pattern of the community has to be inferred statistically.

Let ρ denote a probability measure on $[0,\infty)$ with bounded support. Let $(R_i)_{1\leq i\leq N}$ be i.i.d. random variables drawn from ρ. Think of R_i as the sociability index of individual i. Pick N so large that

$$0 \leq \varepsilon_N R_i R_j \leq 1 \qquad \forall 1 \leq i,j \leq N.$$

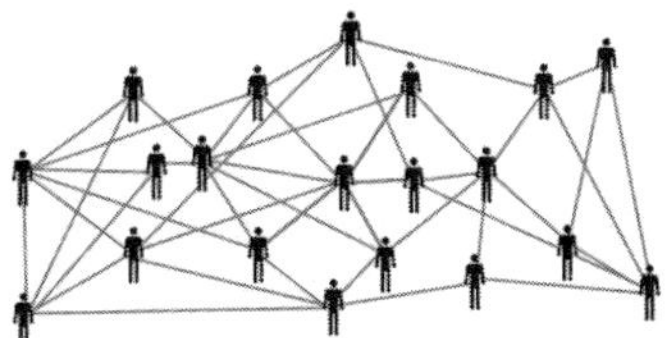

Fig. 3. A social network.

Suppose that i, j are acquainted with probability $\varepsilon_N R_i R_j$, which is represented by an edge in ER_N between vertices i, j. The data that is available is the adjacency matrix A_N. The statistical inference problem is to estimate ρ from A_N. To standardise ρ, we assume that

$$\int_0^\infty x\rho(dx) = 1.$$

Since, weakly $\mathbb{P}$-a.s.,

$$\lim_{N\to\infty} \frac{1}{N}\sum_{i=1}^{N} \delta_{R_i} = \rho,$$

Theorem 3 gives that, weakly in $\mathbb{P}$-probability,

$$\lim_{N\to\infty} \mathrm{ESD}\left(A_N \big/ \sqrt{N\varepsilon_N}\right) = \rho \boxtimes \mu_s.$$

In practice, ε_N is unknown, which can be worked around by arguing that, weakly in $\mathbb{P}$-probability,

$$\lim_{N\to\infty} \mathrm{ESD}\left(\sqrt{\frac{N}{\mathrm{Tr}(A_N^2)}}\, A_N\right) = \rho \boxtimes \mu_s.$$

The procedure is that $\rho \boxtimes \mu_s$ can be *statistically estimated* from A_N. Subsequently, ρ can be estimated because the moments of $\rho \boxtimes \mu_s$ are functions of the moments of ρ and μ_s. Indeed, since the moments of μ_s are known, the moments of ρ can be recursively computed from the moments of $\rho \boxtimes \mu_s$. Since ρ is compactly supported, it can in turn be computed via its moments.

2.4. Application 2: configuration model

Let $\mathcal{S}_N$ be the set of simple graphs on N vertices. We fix the degrees of all the vertices, namely, vertex i has degree d_i^*, where

$$\vec{d}_N^{\,*} = \{d_i^*\}_{1\le i\le N}$$

is a sequence of positive integers of which we only require that it is graphical, i.e., there is at least one simple graph matching these degrees. The Gibbs canonical ensemble P_N is the unique probability distribution on $\mathcal{S}_N$ with the following two properties:

(I) The average degree of vertex i, defined by

$$\sum_{G\in\mathcal{S}_N} d_i(G)P_N(G),$$

equals d_i^* for all $i \leq i \leq N$.

(II) The entropy of P_N, defined by

$$-\sum_{G\in\mathcal{S}_N} P_N(G)\log P_N(G),$$

is maximal.

P_N models a random graph of which we have *no prior information* other than the average degrees.

Property (II) forces P_N to take the form [14]

$$P_N(G) = \frac{1}{Z_N(\vec{\theta}_N^*)}\exp\left[-\sum_{i=1}^N \theta_i^* d_i(G)\right], \qquad G \in \mathcal{S}_N,$$

where $\vec{\theta}_N^* = \{\theta_i^*\}_{1\leq i\leq N}$ is the unique sequence of Lagrange multipliers such that property (I) is satisfied. Reparametrisation yields

$$P_N(G) = \prod_{1\leq i<j\leq N} (p_{ij}^*)^{A_N[G](i,j)}\,(1-p_{ij}^*)^{1-A_N[G](i,j)},\ G \in \mathcal{S}_N,$$

where $A_N[G]$ is the adjacency matrix of G, and

$$p_{ij}^* = \frac{x_i^* x_j^*}{1+x_i^* x_j^*}, \quad x_i^* = e^{-\theta_i^*}, \qquad 1 \leq i \neq j \leq N.$$

Property (I) requires that

$$d_i^* = \sum_{\substack{1\leq j\leq N\\ j\neq i}} p_{ij}^*, \qquad 1 \leq i \leq N,$$

which constitutes a set of N equations for N unknowns.

Abbreviate

$$m_N = \max_{1\leq i\leq N} d_i^*.$$

We focus on the regime

$$\lim_{N\to\infty} m_N = \infty, \qquad \lim_{N\to\infty} m_N/\sqrt{N} = 0.$$

It turns out that in this regime

$$p^*_{ij} = [1+o(1)]\,\frac{d^*_i d^*_j}{\sigma_N}, \qquad N\to\infty,$$

with

$$\sigma_N = \sum_{1\le i\le N} d^*_i.$$

Pick

$$\varepsilon_N = m_N^2/\sigma_N.$$

Then

$$\lim_{N\to\infty} \varepsilon_N = 0, \qquad \lim_{N\to\infty} N\varepsilon_N = \infty,$$

and

$$p^*_{ij} = [1+o(1)]\,\varepsilon_N\,(d^*_i/m_N)(d^*_j/m_N).$$

Under the assumption that

$$\lim_{N\to\infty} \frac{1}{N}\sum_{i=1}^{N} \delta_{d^*_i/m_N} = \rho$$

for some probability measure ρ, Theorem 3 gives that, weakly in $\mathbb{P}$-probability,

$$\lim_{N\to\infty} \mathrm{ESD}\left(A_N\Big/\sqrt{N\varepsilon_N}\right) = \rho \boxtimes \mu_s.$$

This identifies the scaling of the ESD for the network that is modeled by the soft configuration model as a function of the imposed average degrees.

Challenges for the future are:

- ▷ What can we say in the sparse regime and in the dense regime?
- ▷ How can we deal with more general classes of random graphs?

§3. Equivalence of ensembles

3.1. Statistical physics

Systems consisting of a very large number of interacting particles can be described by *statistical ensembles*, i.e., probability distributions on spaces of configurations. Two important examples are:

I. *micro-canonical* ensemble.
II. *canonical* ensemble.

The former fixes the energy of the system, the latter fixes the average energy of the system, with temperature as the control parameter. The two ensembles capture physically different microscopic situations. For both the entropy is maximal subject to the constraint. The canonical ensemble is easier to compute with than the micro-canonical ensemble, because the constraint is soft rather than hard.

In textbooks of statistical physics the two ensembles are assumed (!) to be *thermodynamically equivalent*, i.e., to have the same macroscopic behaviour. Here the idea is that for large systems the energy is typically close to its average value. This assumption is certainly reasonable for systems with interactions that are short-ranged. But, counterexamples have been found for systems with interactions that are long-ranged.

3.2. Complex networks

We will be interested in large random graphs, i.e., the two ensembles live on the set $\mathcal{S}_N$ of all simple graphs with N vertices where $N \to \infty$.

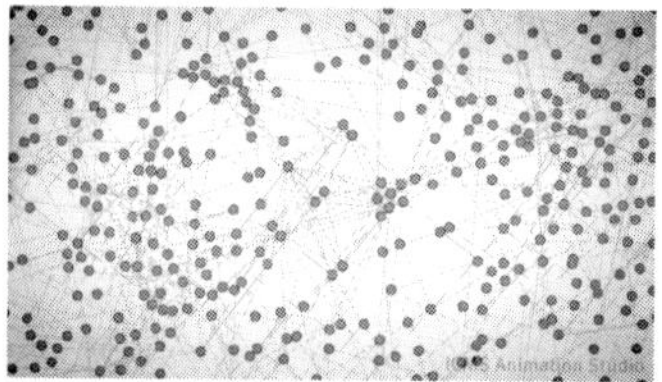

Fig. 4. A realisation of a large random graph.

Given are a vector-valued function $\vec{C}$ on $\mathcal{S}_N$, and a specific vector $\vec{C}^*$ called the constraint.

I. The *micro-canonical ensemble* is defined by

$$P_N^{\mathrm{mic}}(G) = \begin{cases} 1/\Omega_{\vec{C}^*} & \text{if } \vec{C}(G) = \vec{C}^*, \\ 0 & \text{else,} \end{cases}$$

where $\Omega_{\vec{C}^*} = \big|\{G \in \mathcal{S}_N \colon \vec{C}(G) = \vec{C}^*\}\big|$.

II. The *canonical ensemble* is defined by

$$P_N^{\mathrm{can}}(G) = \frac{1}{\mathcal{N}(\vec{\theta}^*)}\, \mathrm{e}^{-\vec{\theta}^* \cdot \vec{C}(G)},$$

where $\mathcal{N}(\vec{\theta}^*)$ is the normalising constant and $\vec{\theta}^*$ is to be chosen such that $\sum_{G\in\mathcal{S}_N} \vec{C}(G)P_N^{\text{can}}(G) = \vec{C}^*$.

Interpretation:

- P_N^{mic} models a random graph of which no information is available other than the *constraint*.
- P_N^{can} models a random graph of which no information is available other than the *average constraint*.

Which of the two ensembles should be used to model a real-world network depends on the *a priori knowledge* that is available about the network.

3.3. Ensemble equivalence

P_N^{mic} and P_N^{can} are said to be equivalent when their *relative entropy per vertex* defined by

$$s_N\left(P_N^{\text{mic}} \mid P_N^{\text{can}}\right) = \frac{1}{N} \sum_{G\in\mathcal{S}_N} P_N^{\text{mic}}(G) \log\left(\frac{P_N^{\text{mic}}(G)}{P_N^{\text{can}}(G)}\right)$$

tends to zero as $N \to \infty$. Because in both ensembles all $G \in \mathcal{S}_N$ such that $\vec{C}(G) = \vec{C}^*$ have the same probability, we get the simpler formula

$$s_N\left(P_N^{\text{mic}} \mid P_N^{\text{can}}\right) = \frac{1}{N} \log\left(\frac{P_N^{\text{mic}}(G^*)}{P_N^{\text{can}}(G^*)}\right)$$

for any G^* such that $\vec{C}(G^*) = \vec{C}^*$. This greatly *simplifies* the computation, since we need not carry out the sum over $\mathcal{S}_N$ and only need to compute with a single graph G^*.

As shown in [17], relative entropy is the sharpest tool to detect breaking of ensemble equivalence. In the remainder we illustrate breaking of ensemble equivalence via a number of examples.

3.4. Constraint of the degree sequence

In the configuration model, each vertex gets a prescribed number of half-edges, which are paired off randomly to form edges.

Consider a graph $G = (V, E)$ with vertex set $V = \{1, \dots, N\}$ and edge set E such that all the vertices have prescribed degrees. In other words, consider the constraint

$$\vec{C}^* = \vec{d}_N^* = (d_1^*, \dots, d_N^*) \in \mathbb{N}_0^N.$$

Suppose that the degrees are moderate, corresponding to what is called the sparse regime:

$$\max_{1\le i\le N} d_i^* = o(\sqrt{N}), \qquad N \to \infty.$$

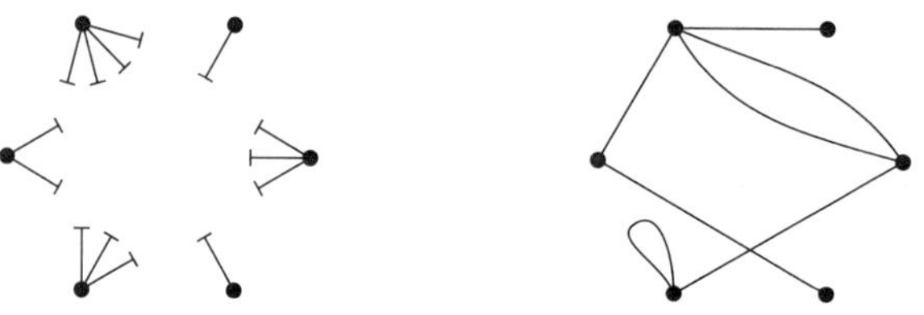

Fig. 5. Example with $N = 6$ and $\vec{d}_N = (1, 3, 1, 3, 2, 4)$.

Let

$$f_N = \frac{1}{N} \sum_{i=1}^{N} \delta_{d_i^*} = \text{empirical degree distribution.}$$

Define

$$g(k) = \log\left(\frac{k!}{k^k e^{-k}}\right), \qquad k \in \mathbb{N}_0.$$

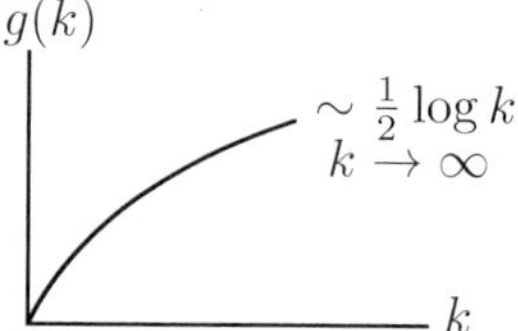

Fig. 6. Picture of $k \mapsto g(k)$.

Theorem 4. *Suppose that*

$$\lim_{N\to\infty} \|f_N - f\|_{\ell^1(g)} = 0$$

for some limiting degree distribution f. Then

$$s_\infty = \lim_{N\to\infty} s_N\left(P_N^{\mathrm{mic}} \mid P_N^{\mathrm{can}}\right) = \|f\|_{\ell^1(g)}.$$

The interpretation is that each vertex with degree k contributes an amount $g(k)$ to the relative entropy. Note that there is breaking of ensemble equivalence for all $f \neq \delta_0$, i.e., breaking is the rule rather than the exception. The proof is based on graph counting (micro-canonical) and percolation theory (canonical).

It turns out that $g(k)$ is the relative entropy of Dirac(k) with respect to Poisson(k). What this says is that, in the limit as $N \to \infty$,

- Micro-canonical ensemble: vertices have a *fixed* degree.

- Canonical ensemble: vertices have a *random* degree.

We illustrate the above with two examples.

Example 1. $f_N = \delta_k$ with $k = o(\sqrt{N})$. For k-regular graphs:

$$s_\infty = g(k) > 0.$$

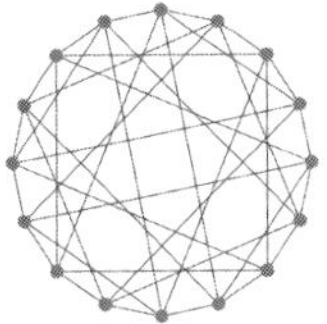

Fig. 7. The 5-regular graph.

Example 2. $f_N(k) = C_N\, k^{-\tau}$, $1 \leq k \leq k_{\text{cutoff}}(N)$, with $k_{\text{cutoff}}(N) = o(\sqrt{N})$ and $\tau \in (1, \infty)$ a tail exponent. For scale-free graphs:

$$s_\infty \approx \frac{1}{2(\tau - 1)} + \frac{1}{2}\log(2\pi) > 0.$$

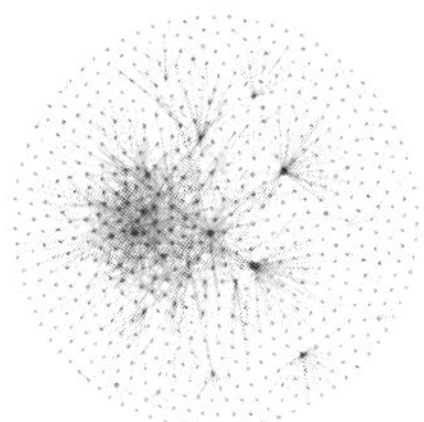

Fig. 8. A graph with hubs.

3.5. Constraint of the total number of edges and triangles

Interesting behaviour shows up when we pick

$$\begin{aligned} \vec{C}^* &= (\text{number of edges}, \text{number of triangles}) \\ &= \left(T_1^* \binom{N}{2}, T_2^* \binom{N}{3}\right), \qquad T_1^*, T_2^* \in [0, 1]. \end{aligned}$$

This corresponds to the so-called dense regime, in which the number of edges per vertex is of order N. The quantity of interest is now

$$s_\infty = \lim_{N\to\infty} \frac{1}{N^2} \log\left(\frac{P_N^{\text{mic}}(G^*)}{P_N^{\text{can}}(G^*)}\right),$$

where we scale by N^2 instead of N.

Theorem 5. *See the figure below.*

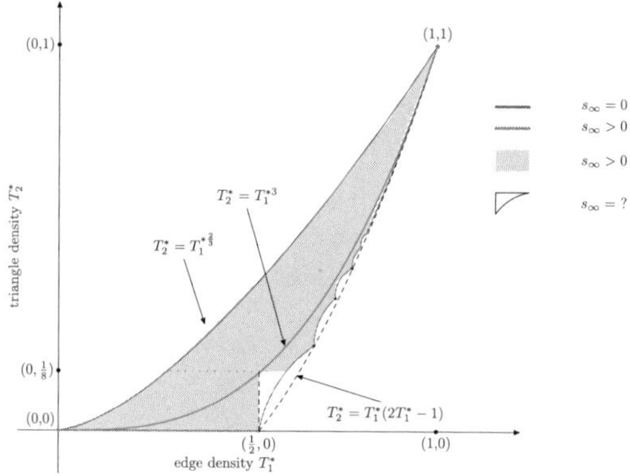

Fig. 9. Between the blue curves the edge-triangle densities are admissible [16]. Breaking of ensemble equivalence occurs everywhere except on the red curves. In the white region between the red curve and the lower blue curve it is not known what happens.

Breaking of ensemble equivalence occurs when (T_1^*, T_2^*) is *frustrated.* The proof is based on the theory of graphons, which are continuum limits of adjacency matrices of graphs [5], [6]. We derive a variational formula for s_∞ with the help of the large deviation principle for graphons associated with the Erdős-Rényi random graph [8].

What happens close the line $T_2^* = T_1^{*3}$? It turns out that anomalous behaviour shows up:

Theorem 6. *For $T_1^* \in (0,1)$,*

$$\lim_{\epsilon\downarrow 0} \epsilon^{-1}\, s_\infty(T_1^*, T_1^{*3} + \epsilon) = C^+ \in (0,\infty),$$
$$\lim_{\epsilon\downarrow 0} \epsilon^{-2/3}\, s_\infty(T_1^*, T_1^{*3} - \epsilon) = C^- \in (0,\infty),$$

where C^+, C^- are computable functions of T_1^* that are, however, not so easy to identify.

In summary, we have obtained a complete classification of breaking of ensemble equivalence in random graphs with constraints on the degree sequence, respectively, the total number of edges and triangles. Breaking occurs when the number of constraints is extensive or when the constraints are frustrated.

Challenges for the future are:

- ▷ Can we estimate the relative entropy away from the Erdős-Rényi curve?
- ▷ What happens when other constraints than edge-triangles are considered?

§4. Exploration and mixing times

4.1. Searching on networks

Search algorithms on networks are important tools for the organisation of large data sets. A key example is Google PageRank, which assigns a weight to each element of a hyperlinked set of documents, such as the World Wide Web, with the purpose of measuring its relative importance within the set.

The weights are assigned via exploration and are obtained recursively. A hyperlink counts as a vote of support: a page that is linked to by many pages with a high rank receives a high rank itself.

Fig. 10. An example of Google PageRank weights in a small network. The size of each face is proportional to the total size of the other faces pointing to it. (Picture taken from Wikipedia.)

4.2. Searching on complex networks

- ▷ Networks are modelled as *graphs*, consisting of a set of vertices and a set of edges connecting pairs of vertices.

- ▷ Complex networks are modelled as *random graphs*, where the vertices and the edges are chosen according to some probability distribution.
- ▷ Search algorithms are modelled as *random walks*, moving along the network by randomly picking an edge incident to the vertex currently visited and jumping to the vertex at the other end.

The question we ask is: *How long does it take the random walk to explore the random graph properly*? The answer to this question is important because it tells us how long the search algorithm must run.

The *mixing time* of a random walk is the time it needs to approach its stationary distribution. For random walks on *static* random graphs, the mixing time has been the subject of intensive study. However, since many networks are dynamic in nature, it is natural to study random walks on *dynamic* random graphs. This line of research is very recent in the mathematics literature.

As we saw in Section 3, the configuration model is a random graph with a prescribed degree sequence. It is popular because of its mathematical tractability and its flexibility in modelling real-world networks. In what follows we consider a discrete-time dynamic version of the configuration model, where at each unit of time a certain fraction of the edges is *rewired*.

Static version. Let $\mathcal{G}(\vec{d}_N)$ denote the set of all graphs on N vertices with a prescribed degree sequence

$$\vec{d}_N = (d_i)_{i=1}^N, \quad \sum_{i=1}^N d_i = \text{even}.$$

We draw a random graph uniformly from the set $\mathcal{G}(\vec{d}_N)$. The outcome may have self-loops and multiple edges. The *stationary distribution* of the random walk equals

$$\pi(i) = \frac{d_i}{\sum_{j=1}^N d_j}, \qquad 1 \leq i \leq N,$$

and does *not* depend on the outcome of the graph, provided this is connected. One way to generate the random graph is by randomly pairing half-edges (recall Section 3).

For random walk on the static configuration model, the mixing time is known to be

$$[1 + o(1)]\, c \log N, \quad N \to \infty,$$

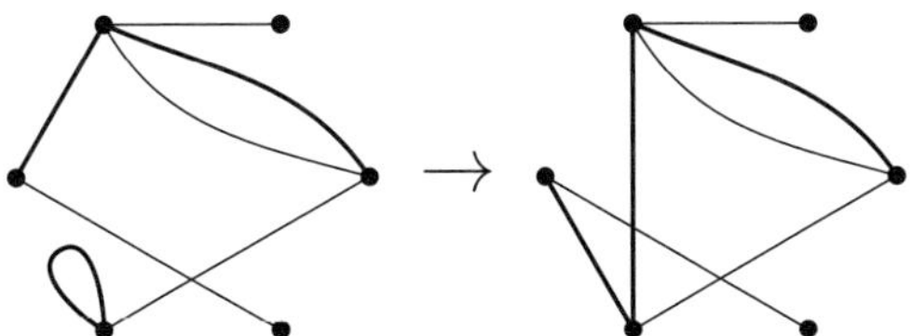

Fig. 11. A transition for the dynamic configuration model. Bold edges on the left are the ones chosen to be rewired. Bold edges on the right are the newly formed edges.

with

$$\frac{1}{c} = \sum_{m \in \mathbb{N}} p^*(m) \log m, \; p^*(m) = \frac{1}{Z^*}(m+1)p(m), \; Z^* = \sum_{m \in \mathbb{N}} (m+1)p(m),$$

where p is the limiting *empirical degree distribution* given $p = \lim_{N \to \infty} N^{-1} \sum_{i=1}^{N} \delta_{d_i}$, subject to certain *regularity assumptions* on the degrees [15], [3], [4].

Dynamic version. For fixed N, draw a starting graph η and a starting vertex i, and proceed as follows. At each time $t \in \mathbb{N}$:

(1) Draw edges randomly with probability $\alpha_N \in (0, 1)$.
(2) Rewire these edges by breaking them into half-edges and pairing these half-edges again randomly.
(3) After the rewiring, let the random walk make a step to a randomly chosen neighbouring vertex.

We make the following assumptions:

- The degrees must be *moderate*, i.e., not too large.
- The random walk is *non-backtracking*, i.e., immediate jumps back along edges are not allowed.
- $\lim_{N \to \infty} \alpha_N = 0$, i.e., the dynamics is *slow*.

4.3. Mixing time

Let $\mathbb{P}_{\eta,i}$ denote probability with respect to the joint process of random graph and random walk with starting graph η and starting vertex i. Let X_t denote the location of the random walk at time $t \in \mathbb{N}$, and write

$$\mathcal{D}_{\eta,i}(t) = \frac{1}{2} \sum_{j=1}^{N} \left| \mathbb{P}_{\eta,i}(X_t = j) - \pi(j) \right|$$

to denote total variation distance between the distribution of X_t and the stationary distribution π. It turns out that there is are *three regimes*:

(1) $\lim_{N\to\infty} \alpha_N (\log N)^2 = \infty$: supercritical regime.
(2) $\lim_{N\to\infty} \alpha_N (\log N)^2 = \beta \in (0,\infty)$: critical regime.
(3) $\lim_{N\to\infty} \alpha_N (\log N)^2 = 0$: subcritical regime.

Theorem 7. *With high probability, i.e., for a set of (η, i) with probability tending to 1 as $N \to \infty$, the following hold.*

(1) *Supercritical regime*:

$$\mathcal{D}_{\eta,i}(s/\sqrt{\alpha_N}) = e^{-s^2/2} + o(1), \quad s \in [0,\infty).$$

(2) *Critical regime*:

$$\mathcal{D}_{\eta,i}(s \log N) = \begin{cases} e^{-\beta s^2/2} + o(1), & s \in [0,c), \\ o(1), & s \in [c,\infty). \end{cases}$$

(3) *Subcritical regime*:

$$\mathcal{D}_{\eta,i}(s \log N) = \begin{cases} 1 - o(1), & s \in [0,c), \\ o(1), & s \in [c,\infty). \end{cases}$$

Here, c is the constant in the static version.

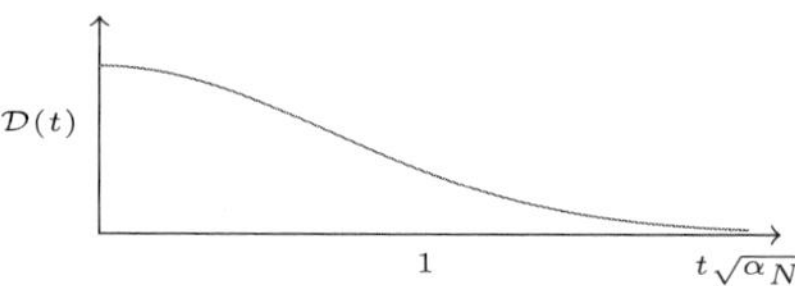

Fig. 12. Scaled mixing profile in the supercritical regime.

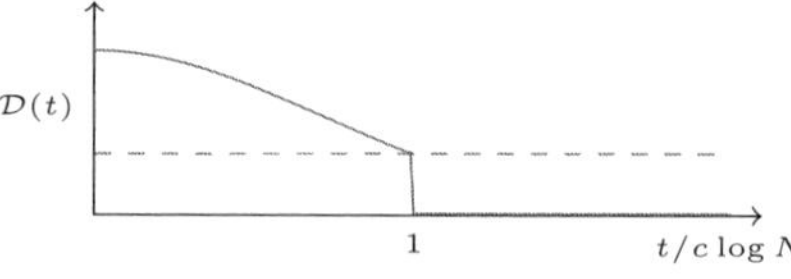

Fig. 13. Scaled mixing profile in the critical regime.

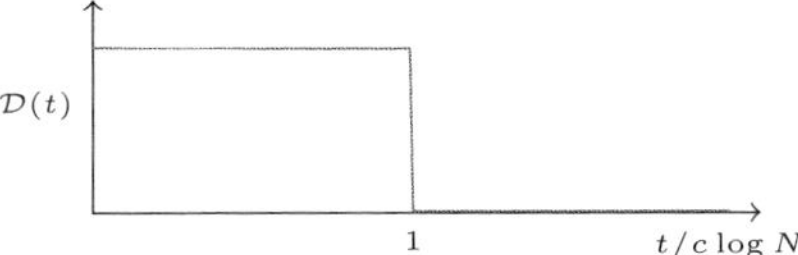

Fig. 14. Scaled mixing profile in the subcritical regime.

In the supercritical regime the mixing time is of order

$$1/\sqrt{\alpha_N} \ll \log N,$$

and does not depend on the degree sequence. In the critical regime and the subcritical regime the mixing time is of order $\log N$ and depends on the degree sequence.

The proof is based on a *stopping time* argument: the first time the random walk moves along an edge that has been relocated is close to a strong uniform time.

Challenges for the future are:

- ▷ What effect do hubs have on the mixing time?
- ▷ What happens when only edges touched by the random walk can be rewired?

References

[1] L. Avena, H. Guldas, R. van der Hofstad and F. den Holllander, Mixing times for random walks on dynamic configuration models, Ann. Appl. Probab., 28 (2018), 1977–2002.

[2] L. Avena, H. Guldas, R. van der Hofstad and F. den Holllander, Random walks on dynamic configuration models: a trichotomy, Stoch. Proc. Appl., 129 (2019), 3360–3375.

[3] A. Ben-Hamou and J. Salez, Cutoff for nonbacktracking random walks on sparse random graphs, Ann. Probab., 45 (2017), 1752–1770.

[4] N. Berestycki, E. Lubetzky, Y. Peres and A. Sly, Random walks on the random graph, Ann. Probab., 46 (2018), 456–490.

[5] C. Borgs, J. T. Chayes, L. Lovász, V. T. Sós and K. Vesztergombi, Convergent graph sequences I: Subgraph frequencies, metric properties, and testing, Adv. Math., 219 (2008), 1801–1851.

[6] C. Borgs, J. T. Chayes, L. Lovász, V. T. Sós and K. Vesztergombi, Convergent sequences of dense graphs II: Multiway cuts and statistical physics, Ann. Math., 176 (2012), 151–219.

[7] A. Chakrabarti, R. S. Hazra, F. den Hollander and M. Sfragara, Spectra of adjacency and Laplacian matrices of inhomogeneous Erdős-Rényi random graphs, Random Matrices Theory Appl., 10 (2021), 2150009.

[8] S. Chatterjee and S. R. S. Varadhan, The large deviation principle for the Erdős-Rényi random graph, European J. Combin., 32 (2011), 1000–1017.

[9] D. Garlaschelli, F. den Hollander, J. de Mol and T. Squartini, Breaking of ensemble equivalence in networks, Phys. Rev. Lett., 115 (2015), 268701.

[10] D. Garlaschelli, F. den Hollander and A. Roccaverde, Ensemble nonequivalence in random graphs with modular structure, J. Phys. A: Math. Theor., 50 (2017), 015001.

[11] D. Garlaschelli, F. den Hollander and A. Roccaverde, Covariance structure behind breaking of ensemble equivalence, J. Stat. Phys., 173 (2018), 644–662.

[12] F. den Hollander, M. Mandjes, A. Roccaverde and N. J. Starreveld, Ensemble equivalence for dense graphs, Electron. J. Probab., 23 (2018), Paper no. 12, 1–26.

[13] F. den Hollander, M. Mandjes, A. Roccaverde and N. J. Starreveld, Breaking of ensemble equivalence for perturbed Erdős-Rényi random graphs, submitted to Random Struct. Algor.

[14] E. T. Jaynes, Information theory and statistical mechanics, Phys. Rev., 106 (1957), 620–630.

[15] E. Lubetzky and A. Sly, Cut-off phenomena for random walks on random regular graphs, Duke Math. J., 153 (2010), 475–510.

[16] C. Radin and L. Sadun, Singularities in the entropy of asymptotically large simple graphs, J. Stat. Phys., 158 (2015), 853–865.

[17] H. Touchette, Equivalence and nonequivalence of ensembles: Thermodynamic, macrostate, and measure levels, J. Stat. Phys., 159 (2015), 987–1016.

[18] Y. Zhu, Graphon approach to limiting spectral distributions of Wigner-type matrices, Random Structures Algorithms, 56 (2020), 251–279.

Mathematical Institute,
Leiden University,
P.O. Box 9512, 2300 RA Leiden, The Netherlands
E-mail address: `denholla@math.leidenuniv.nl`

Advanced Studies in Pure Mathematics 87, 2021
Stochastic Analysis, Random Fields and Integrable Probability — Fukuoka 2019
pp. 155–172

Some notes on random walks in two dimensions

Gregory F. Lawler

Abstract.

We consider several problems about random walks, or equivalently, discrete harmonic function on the planar integer lattice. The questions deal with the rate of convergence of quantities to their Brownian counterparts.

§1. Introduction

In this note we consider the rate of convergence of some random walk quantities to their Brownian counterpart. This came up in recent work [3] where random walk in a slit plane was revisited. Let S_j be a simple random walk in the plane lattice $\mathbb{Z}^2 = \mathbb{Z} + i\mathbb{Z}$ and let

$$\tau = \min\{j \geq 0 : S_j \in \mathbb{N} := \{0, 1, 2, \ldots\}\},$$
$$\xi_R = \min\{j \geq 0 : |S_j| \geq R\}.$$

Let

$$v(z) = \lim_{R\to\infty} R^{1/2}\, \mathbb{P}^z\{\xi_R < \tau\}.$$

The existence of the limit has been proved, see, e.g., [1]. Given the existence, we see that that v is positive and discrete harmonic on $\mathbb{Z}^2 \setminus \mathbb{N}$ and vanishes on $\mathbb{N}$. Our focus is the scaling limit and the rate of convergence to the limit.

The corresponding problem for Brownian motion is easier. If B_t is a standard complex Brownian motion and

$$\tau^* = \min\{t \geq 0 : B_t \in [0, \infty)\},$$
$$\xi^*_R = \min\{t \geq 0 : |B_t| = R\},$$

Received January 20, 2020.
Revised May 20, 2020.
2010 *Mathematics Subject Classification.* 60J10.
Key words and phrases. random walk, discrete harmonic functions.
Research supported by NSF grant DMS-1513036.

then using conformal invariance of Brownian motion, it is straightforward to show that as $R \to \infty$,

$$\mathbb{P}^z\{\xi_R^* < \tau_R^*\} \sim \frac{4}{\pi\sqrt{R}} r^{1/2} \sin(\theta_z/2) \qquad \text{where } z = re^{i\theta_z}.$$

This leads to the natural conjecture that $v(z) \sim (4/\pi)\,|z|^{1/2} \sin(\theta_z/2)$. In [3] this was established with an estimate for the error term,

$$v(z) = \frac{4}{\pi}\,|z|^{1/2} \sin(\theta_z/2) \left[1 + O(|z|^{-1})\right], \quad z \to \infty.$$

In this paper we will demonstrate the technique used by extending this result to other wedges. We will also consider a problem about discrete harmonic measure. For this latter problem, again the form of the asymptotics is known, see e.g., [1], and our emphasis is the bound on the error. Finally, we derive a difference estimate for the potential kernel.

To state the estimate for wedges we need some notation. We fix $0 \leq \theta_1 < \theta_2 \leq \theta_1 + 2\pi$, and all quantities and constants will depend on θ_1, θ_2. Let $D = D(\theta_1, \theta_2)$ be the corresponding wedge domain,

$$D = \{re^{i\theta} \in \mathbb{C} : r > 0, \theta_1 < \theta < \theta_2\},$$

with boundary $\partial D = \partial_1 \cup \partial_2$, $\partial_j = \{re^{i\theta_j} : r \geq 0\}$. Let

$$\alpha = \frac{\pi}{\theta_2 - \theta_1}, \qquad \theta_z = \arg(z) - \theta_1,$$

and let

$$g(z) = z^\alpha\, e^{-i\alpha\theta_1} = |z|^\alpha \left[\cos(\alpha\theta_z) + i\,\sin(\alpha\theta_z)\right]$$

which is a conformal transformation of D onto the upper half plane $\mathbb{H}$. Let ϕ be the positive harmonic function on D given by

$$\phi(z) = \operatorname{Im}[g(z)] = |z|^\alpha \sin(\alpha\theta_z). \tag{1}$$

We extend ϕ continuously to $\mathbb{C}$ by setting $\phi \equiv 0$ on $\mathbb{C} \setminus D$.

Let $A = A(D)$ be the unbounded connected (in $\mathbb{Z}^2$) component of $\mathbb{Z}^2 \cap D$; it includes all but a finite number of points in $\mathbb{Z}^2 \cap D$. We write $\partial A = \{z \in \mathbb{Z}^2 : \operatorname{dist}(z, \partial A) = 1\}$ for the outer boundary of A; $\partial_i A = \{z \in A : \operatorname{dist}(z, \partial A) = 1\}$ for the inner boundary; and $\overline{A} = A \cup \partial A$. Note that ϕ is defined on $\overline{A}$: on A, it is given by (1) and $\phi \equiv 0$ on ∂A.[1]

[1]If θ_z is near 2π, it is possible that there exist edges in the lattice that intersect ∂D but whose endpoints are both in A, say connecting z with w. In this case we add two new points z', w' to ∂A so that z is adjacent to w' rather than w and w is adjacent to z' rather than z. We then set $\phi(w') = \phi(z') = 0$.

If $z \in A$, let

$$s_z = s_{z,A} = \frac{\operatorname{dist}(z, \partial A)}{|z|} \asymp \frac{\operatorname{dist}(z, \partial D) + 1}{|z|}.$$

If S_j is a simple random walk, let $\tau_A = \min\{j \geq 0 : S_j \notin A\}$. We write Δ for the continuous Laplacian and

$$\mathcal{L}f(z) = \frac{1}{4} \sum_{|w-z|=1} [f(w) - f(z)]$$

for the discrete Laplacian. We write $G_A(z, w)$ for the Green's function for the random walk killed when it leaves A,

$$G_A(z, w) = \sum_{j=0}^{\infty} \mathbb{P}^z\{S_j = w; j < \tau_A\}.$$

Define

$$\Psi(z) = \Psi_{\phi,A}(z) = \sum_{w \in A} G_A(z, w)\, |\mathcal{L}\phi(w)|.$$

Since A is connected, $\Psi(z)$ is either infinite for all $z \in A$ or it is finite for all $z \in A$. Hence we can write $\Psi < \infty$. If $\Psi < \infty$, we define

$$\tilde{\Psi}(z) = \tilde{\Psi}_{\phi,A}(z) = \sum_{w \in A} G_A(z, w)\, \mathcal{L}\phi(w),$$

and note that for $z \in A$, $\mathcal{L}\tilde{\Psi}(z) = -\mathcal{L}\phi(z)$. The following is a standard corollary of this.

Proposition 1. *Let A be a connected proper subset of $\mathbb{Z}^2$ and ψ a function on $\overline{A}$ such that $\Psi = \Psi_{\psi,A} < \infty$. Let $f(z) = f_{\psi,A}(z) = \psi(z) + \tilde{\Psi}(z)$. Then f is discrete harmonic in A; equals ψ on ∂A; and satisfies*

$$|f(z) - \psi(z)| \leq \Psi(z), \quad z \in \overline{A}.$$

If f is bounded (in particular, if A is finite), $f(z) = \mathbb{E}^z[\psi(S_{\tau_A})]$.

We call D and the corresponding A *nice* if θ_1, θ_2 are integer multiples of $\pi/4$. In the case of nice D we have $A = \mathbb{Z}^2 \cap D$ and with probability one, the first point in ∂A visited by a random walk starting in A is on ∂D. In other words, there is no "overshoot" when leaving the domain.

Theorem 2. *Suppose A is a wedge domain, $\phi, \Psi = \Psi_{A,\phi}, \tilde{\Psi} = \tilde{\Psi}_{A,\phi}$ are defined as above, and $f(z) = \phi(z) + \tilde{\Psi}(z)$. Then*

$$f(z) = \frac{\pi}{4} \lim_{R \to \infty} R^{\alpha}\, \mathbb{P}^z\{\xi_R < \tau_A\}. \tag{2}$$

Moveover, there exists $c < \infty$ such that for all $z \in A$,

$$|f(z) - \phi(z)| \leq c\,[1 + s_z\, e_z]|z|^{\alpha-1}, \tag{3}$$

where $e_z = \log|z|$ if $\alpha = 1/2$ and $e_z = 0$ otherwise. If D is a nice domain, we have the stronger estimate

$$|f(z) - \phi(z)| \leq c\, s_z\, |z|^{\alpha-2+(2-2\alpha)_+}. \tag{4}$$

We remark that there are several nice examples where ϕ is actually discrete harmonic in which case we get $f_\phi = \phi$,

$$\theta_1 = 0, \quad \theta_2 = \pi, \quad \phi(x+iy) = y,$$
$$\theta_1 = 0, \quad \theta_2 = \frac{\pi}{2}, \quad \phi(x+iy) = 2xy,$$
$$\theta_1 = \frac{\pi}{4}, \quad \theta_2 = \frac{3\pi}{4}, \quad \phi(x+iy) = y^2 - x^2.$$

The second result will be an estimate of the harmonic measure of a particular subset of $\mathbb{Z}^2 = \mathbb{Z} + i\mathbb{Z}$. Let $U_n = \{-n, -n+1, \ldots, n-1, n\}$ be the line segment of length $2n$ on the real axis centered at the origin. Let $\mathrm{hm}_n = \mathrm{hm}_{U_n}$ denote harmonic measure from infinity. To be more precise, let S_k denote a usual two-dimensional random walk and let $\tau_n = \min\{k \geq 0 : S_k \in U_n\}$. Then

$$\mathrm{hm}_n(x) := \mathrm{hm}_{U_n}(x) = \lim_{|z|\to\infty} \mathbb{P}^z\{S_{\tau_n} = x\}.$$

Theorem 3. *If $x \in U_n$, then*

$$\mathrm{hm}_n(x) = \frac{1}{\pi\,\sqrt{(n+1)^2 - |x|^2}}\left[1 + O\left(\frac{1}{n+1-|x|}\right)\right].$$

The statement of the theorem is shorthand for the following: there exists $c < \infty$ such that for all positive integers n and all $x \in U_n$

$$\left|\pi\,\sqrt{(n+1)^2 - |x|^2}\,\mathrm{hm}_n(x) - 1\right| \leq \frac{c}{n+1-|x|}.$$

Note that the error gets worse as one approaches the endpoints of U_n. Indeed, at the tip the error is of the same order of magnitude as the leading term. The leading order term is not surprising or new; it is what Brownian motion would predict. The emphasis here is on the error term.

As a final example, we use the techniques to establish an estimate about the potential kernel for simple random walk. Recall that $a(z)$

is the unique function on $\mathbb{Z}^2$ with $a(0) = 0$; $\mathcal{L}a(z) = 0, z \neq 0$; and $a(z) \sim \frac{2}{\pi} \log |z|$, $z \to \infty$. It is known (see, e.g., [2, Theorem 4.4.4]) and we will assume that

$$a(z) = \frac{2}{\pi} \log |z| + k_0 + e(z), \quad k_0 = \frac{\log 8 + 2\gamma}{\pi}$$

where γ is Euler's constant and $|e(z)| \leq c\,|z|^{-2}$. For applications one often wishes to "differentiate" a.

While the following could be derived from the techniques in [2], it is not, and since it is useful we will show how to derive it here. Let $\nabla_w f(z) = f(z + w) - f(z)$, and note that

$$\nabla_w a(z) = \nabla_w \psi(z) + \nabla_w e(z),$$

where $\psi(z) = \log |z|$.

Theorem 4. *There exists $c < \infty$ such that if $|w| \leq |z|/2$,*

$$|\nabla_w e(z)| \leq \frac{c\,|w|}{|z|^3}.$$

One of the purposes of this note is to highlight the relatively easy principle behind the proofs.

- A bounded (continuous) harmonic function in a ball of radius R about the origin may not be discrete harmonic, but the discrete Laplacian at the origin is of order $O(R^{-4})$.
- The contribution from the boundary effect is different depending on whether or not the discrete outer boundary lies on the continuous boundary.

The following simple estimate is the key to estimate the quantity Ψ in Proposition 1.

Proposition 5. *There exists $c < \infty$ such that if $r > 1$ and ϕ is continuous on $\{|z| \leq r\}$ and (continuous) harmonic in the interior, then*

$$|\mathcal{L}\phi(0)| \leq \frac{c}{r^4}\,\|\phi\|_\infty.$$

Proof. We may assume $\|\phi\|_\infty \leq 1$. Using $\Delta\phi(0) = 0$ and Taylor's theorem, we see that

$$|\mathcal{L}\phi(0)| \leq c\left[\max_{|x|\leq 1}\left|\partial_{xxxx}\phi(x)\right| + \max_{|y|\leq 1}\left|\partial_{yyyy}\phi(iy)\right|\right].$$

The right-hand side is $O(r^{-4})$ by differentiating the expression for ϕ in terms of the Poisson kernel. Q.E.D.

§2. Proof of Theorem 2

We fix θ_1, θ_2 and hence A, and allow all quantities and constants to depend on θ_1, θ_2. We write $G(z,w)$ for $G_A(z,w)$ and τ for τ_A. Recall that for $z \in A$,

$$s_z = \frac{\mathrm{dist}(z, \partial A)}{|z|} \asymp |z|^{-1} \vee \sin\left(\frac{\pi\,(\arg(z) - \theta_1)}{\theta_2 - \theta_1}\right).$$

In [1, Section 8] it was shown that the "cone" exponents for random walk are the same as for Brownian motion. In the two dimensional case, these are the wedge exponents. We state the result we need here. This is actually more than we need for our proofs here. We mention that that result below with $\beta \le \alpha$ replaced with $\beta < \alpha$ could be proved for exponents using strong approximation of random walk to Brownian motion and that would be sufficient for our purposes here. Let

$$\xi_r = \min\{j : |S_j| \ge r\}, \quad \hat{\xi}_r = \min\{j : |S_j| \le r\}.$$

Lemma 6. *There exists $c < \infty$ such that for every $z \in A$,*

$$\mathbb{P}^z\{\hat{\xi}_{|z|/2} < \tau\} + \mathbb{P}^z\{\xi_{2|z|} < \tau\} \le c\, s_z,$$
$$\sum_{|w| \le 4|z|} G(z,w) \le c\, s_z\, |z|^2.$$

Proof. For the first inequality, it suffices to show that $\mathbb{P}^z\{\sigma_z < \tau\} \le c\, s_z$ where $\sigma_z = \min\{j : |S_j - z| \le |z|/2\}$. This follows from the "gambler's ruin estimate", see, for example, [2, Section 5.1]. Similarly, we show that $\mathbb{E}[\sigma_z \wedge \tau] \le c\, s_z\, |z|^2$.

For the second expression, we first claim that there exists $c < \infty$ such that for all z, ζ,

$$\sum_{|w| \le 4|z|} G(\zeta, w) \le c\, |z|^2. \tag{5}$$

Indeed, let

$$K = \max_{\zeta} \sum_{|w| \le 4|z|} G(\zeta, w) = \max_{|\zeta| \le 4|z|} \sum_{|w| \le 4|z|} G(\zeta, w).$$

It is easy to see that there exists $\delta > 0$ such that for all z and all $|\zeta| \le 4|z|$, the probability that the random walk starting at z leaves A in the next $|z|^2/\delta$ steps is at least δ. This implies that $K \le (|z|^2/\delta) + (1-\delta)\, K$. from which (5) follows. Given this, the second inequality follows by splitting the visits to $A \cap \{|w| \le 4|z|\}$ into those that occur before time σ_z and those that occur afterwards. Q.E.D.

Lemma 7. *For every $\beta \le \alpha$, there exists $c < \infty$ such that for every z and every $r \le 1/2$,*

$$\mathbb{P}^z\{\xi_{|z|/r} < \tau\} \le c\, s_z r^\beta,$$
$$\mathbb{P}^z\{\hat{\xi}_{r|z|} < \tau\} \le c\, s_z\, r^\beta.$$

Proof. From Lemma 6, we can see it suffices to prove the estimates without the s_z factor on the right-hand side. The first estimate is then proved in [1, Section 8].

For the second inequality, we can use a path reversal argument. Consider the set

$$A^* = A \cup \left[\mathbb{Z}^2 \setminus \{w : 2r|z| \le |w| \le |z|/2\}\right].$$

Let z' be a point with $|z'| \le r|z| + 1$. By symmetry we know that $G_{A^*}(z, z') = G_{A^*}(z', z)$. If $|w| \le r|z|$, then $G_{A^*}(w, z') \ge c$. Hence $G_{A^*}(z, z')$ is bounded below by a constant times the probability that a random walk starting at ζ reaches $\{|w| \le r|z|\}$ without leaving A^*. Conversely, $G_{A^*}(z', z)$ is bounded above by the probability that the walker starting at z' reaches $\{|w| \ge |z|/2\}$ without leaving A^*. The latter probability is bounded above by cr^α and hence so is the former. Q.E.D.

Suppose B_t is a standard complex Brownian motion starting at z and

$$T = \min\{t : B_t \notin D\}, \quad \sigma_R = \min\{t : |B_t| = R\}.$$

If $\Theta_t = \arg(B_t) - \theta_1$ then by the optional sampling theorem,

$$\begin{aligned}\phi(z) = \mathbb{E}^z\left[\phi(B_{T\wedge\sigma_R})\right] &= \mathbb{P}^z\{\sigma_R < T\}\,\mathbb{E}^z\left[\phi(B_{T\wedge\sigma_R}) \mid \sigma_R < T\right] \\ &= R^\alpha\,\mathbb{P}^z\{\sigma_R < T\}\,\mathbb{E}^z\left[\sin(\alpha\Theta_{\sigma_R}) \mid \sigma_R < T\right].\end{aligned}$$

Using conformal invariance, we can see that if $\hat{\Theta}_j = \arg(B_t) - \theta_1$,

$$\lim_{R\to\infty} \mathbb{E}^z\left[\sin(\alpha\hat{\Theta}_{\sigma_R}) \mid \sigma_R < T\right] = \int_0^\pi [\sin x]\,\frac{\sin x}{2}\,dx = \frac{\pi}{4}.$$

Lemma 8. *For every $z \in A$, if $\hat{\Theta}_j = \arg(S_j) - \theta_1$,*

$$\lim_{R\to\infty} \mathbb{E}^z\left[\sin(\alpha\hat{\Theta}_{\xi_R}) \mid \xi_R < \tau_A\right] = \frac{\pi}{4}.$$

Equivalently,

$$\lim_{R\to\infty} R^{-\alpha}\,\mathbb{E}^z\left[\phi(S_{\xi_R}) \mid \xi_R < \tau_A\right] = \frac{\pi}{4}.$$

Sketch of proof. This is proved in the same way as in [3, Lemma 3.9]. There are two major steps. We do not need to worry about the rate of convergence.

- Consider h-processes in $A \cap \{|z| < R\}$ conditioned so that $\xi_R < \tau_A$. Then we claim that if $|z|, |w| < R^{1-\epsilon}$, then we can couple h-processes starting at z, w, respectively so that the variation distance between their exit measures differs by R^{-u}. This is done by showing that there exists $\epsilon > 0$ such that if $|z|, |w| < K$, then the variation distance of the hitting distribution of $\{|\zeta| > 2K\}$ for h-processes starting at the two points is bounded above by $1 - \epsilon$. There are three parts to this.
  - A separation lemma is used to see that random walks starting near the boundary conditioned to reach radius $2K$ before leaving A have a good chance of getting away from the boundary.
  - The hitting distribution of the discrete ciricle of radius $2K$ is compable to K^{-1} for each point that can be reached and is away from the boundary.
  - A Harnack inequality shows that the probability to reach distance R without leaving A is comparable for points on the circle of radius $2K$ that are away from ∂A.

  In other words, we can couple the paths with probability at least ϵ. By doing this along shells with $K = 2^j$ we get the result.
- We then consider the hitting distribution for the h-process starting at $z = R^{1-\epsilon} \exp\{i\frac{\theta_1+\theta_2}{2}\} + O(1)$. For this starting point, we can use the KMT or dyadic coupling [2, Chapter 7] to show that the exit distributions of the random walk and Brownian motion *given that the paths reach the circle of radius R before leaving the wedge* are close. Q.E.D.

We will first show how to derive (2) given (3) and (4). Let $A_R = A \cap \{|z| < R\}$. Applying Proposition 1 we see that

$$\mathbb{E}^z\left[\phi(S(\tau_A \wedge \xi_R))\right] = \phi(z) + \sum_w G_{A_R}(z,w)\,\mathcal{L}\phi(w).$$

Since $\Psi < \infty$, we have

$$\lim_{R\to\infty} \sum_w G_{A_R}(z,w)\,\mathcal{L}\phi(w) = \sum_w G_A(z,w)\,\mathcal{L}\phi(w).$$

Since $\phi \equiv 0$ on ∂A, we have using Lemma 8,

$$\begin{aligned}\mathbb{E}^z\left[\phi(S(\tau_A \wedge \xi_R))\right] &= \mathbb{E}^z\left[\phi(S(\tau_A \wedge \xi_R)); \xi_R < \tau_A\right] \\ &= \mathbb{P}^z\{\xi_R < \tau_A\}\, \mathbb{E}^z\left[\phi(S(\tau_A \wedge \xi_R)); \xi_R < \tau_A\right] \\ &\sim \frac{\pi}{4}\, R^\alpha\, \mathbb{P}^z\{\xi_R < \tau_A\}.\end{aligned}$$

Therefore,

$$\begin{aligned}\lim_{R\to\infty} R^\alpha\, \mathbb{P}^z\{\xi_R < \tau_A\} &= \frac{4}{\pi}\left[\phi(z) + \sum_w G_A(z,w)\, \mathcal{L}\phi(w)\right] \\ &= \frac{4}{\pi}\, f(z).\end{aligned}$$

The remainder of this section will be devoted to proving (3) and (4), and by Proposition 1 it suffices to give the upper bounds on Ψ. We start with a simple estimate.

Proposition 9. *There exists $c < \infty$ such that*

$$|\mathcal{L}\phi(z)| \le c\,|z|^{\alpha-4}, \quad z \in A \setminus \partial_i A, \tag{6}$$

$$|\mathcal{L}\phi(z)| \le c\,|z|^{\alpha-1}, \quad z \in \partial_i A. \tag{7}$$

If D is a nice domain, then

$$|\mathcal{L}\phi(z)| \le c\,|z|^{\alpha-4}, \quad z \in \partial_i A. \tag{8}$$

Proof. There exists $\delta > 0$ (depending only on $\theta_2 - \theta_1$) such that if $z \in A$, then the ball of radius $\delta|z|$ about z intersects at most one of ∂_1 or ∂_2. Let U be the open disk of radius $\delta|z|$ about z. Using Schwarz reflection we can extend ϕ in U to be a (continuous) harmonic function ϕ^* in U (if U does not intersect ∂D, use $\phi^* = \phi$). Also $\|\phi^*\|_\infty \le c\,|z|^\alpha$ in U. Using Proposition 5, we see that $|\mathcal{L}\phi^*(z)| \le c\,|z|^{\alpha-4}$. If $z \in A \setminus \partial_i A$ or D is a nice domain and $z \in \partial_i A$, we have $\mathcal{L}\phi^*(z) = \mathcal{L}\phi(z)$. This gives (6) and (8).

If D is not a nice domain, $\mathcal{L}\phi^*(z) \ne \mathcal{L}\phi(z)$ for $z \in \partial_i A$. In this case we use a weaker estimate. Since ϕ is a nonnegative function,

$$|\mathcal{L}\phi(z)| \le \max_{|w-z|\le 1} \phi(w).$$

Then (7) follows by noting that $\phi(w) \le c|w|^{\alpha-1}$ if $\operatorname{dist}(w, \partial D) \le 2$.

Q.E.D.

For every $k \in \mathbb{Z}$, if D is not a nice domain, we let

$$K_k = K_{k,z,A} = \{w \in A \setminus \partial_i A : 2^{k-1}|z| < |w| \leq 2^k |z|\},$$
$$K'_k = K'_{k,z,A} = \{w \in \partial_i A : 2^{k-1}|z| < |w| \leq 2^k |z|\}.$$

Note that these sets are empty if $2^k\,|z| < 1$. If D is a nice domain, we let

$$K_k = K_{k,z,A} = \{w \in A : 2^{k-1}|z| < |w| \leq 2^k |z|\},$$

and let K'_k be empty.

We claim that for all integers k,

$$\sum_{w \in K_k} G(z,w) \leq c\, s_z\, 2^{-\alpha|k|}\, [2^k\,|z|]^2, \tag{9}$$

If $|k| \leq 1$, this follows from Lemma 6. If $|k| \geq 2$, then the probability to reach K_k before leaving A is bounded above by $c\, s_k\, 2^{-\alpha|k|}$ and the expected number of visits to K_k given at least one visit is bounded by $c[2^k\,|z|]^2$.

Also, we claim

$$\sum_{w \in K'_k} G(z,w) \leq c\, s_z\, 2^{-\alpha|k|}, \qquad \text{if } |k| \geq 2. \tag{10}$$

Indeed, the probability to reach K_k before leaving A is bounded above by $c\, s_k\, 2^{-\alpha|k|}$ and the expected number of visits to K_k given at least one visit is at most 4 (since every time one is at a point in $\partial_i A$, there is a probability at least 1/4 of leaving A in the next step). That last argument also gives the following.

$$\sum_{w \in K'_k} G(z,w) \leq 4, \qquad \text{if } |k| \leq 1. \tag{11}$$

We will be using the estimates from Proposition 9. We first consider $k \geq -1$. If $w \in K_k$,

$$|\mathcal{L}\phi(w)| \leq c\,(2^k |z|)^{\alpha-4},$$

and hence, using (9),

$$\sum_{w \in K_k} |\mathcal{L}\phi(z)|\, G_A(z,w) \leq c\, s_z\, 2^{-2k}\, |z|^{\alpha-2}.$$

Summing over k, we get

$$\sum_{k=-1}^{\infty} \sum_{w \in K_k} |\mathcal{L}\phi(w)|\, G_A(z,w) \leq c\, s_z\, |z|^{\alpha-2}. \tag{12}$$

If $z \in K'_k$, then, $|\mathcal{L}\phi(w)| \leq c\,[|z|2^k]^{\alpha-1}$, and hence, using (10) and (11),

$$\sum_{w \in K'_k} |\mathcal{L}\phi(w)|\, G_A(z,w) \leq c\,|z|^{\alpha-1}\, 2^{-k}.$$

Summing over k, we get

$$\sum_{k=-1}^{\infty} \sum_{w \in K'_k} |\mathcal{L}\phi(w)|\, G_A(z,w) \leq c\,|z|^{\alpha-1}. \tag{13}$$

We now consider $-k$ for $1 < k < \log_2 |z|$. If $w \in K_{-k}$,

$$|\mathcal{L}\phi(w)| \leq c\,(2^{-k}|z|)^{\alpha-4},$$

and hence by (9),

$$\sum_{w \in K_{-k}} |\mathcal{L}\phi(w)|\, G_A(z,w) \leq c\, s_z\, 2^{(2-2\alpha)k}\, |z|^{\alpha-2}.$$

$$\begin{aligned} \sum_{k=2}^{\infty} \sum_{w \in K_{-k}} |\mathcal{L}\phi(w)|\, G_A(z,w) &\leq c\, s_z\, |z|^{\alpha-2} \sum_{k=1}^{\log_2 |z|} 2^{(2-2\alpha)k} \\ &\leq c\, s_z\, |z|^{\alpha-2+(2-2\alpha)_+}. \end{aligned} \tag{14}$$

The second inequality is valid if $\alpha \neq 1$, but if $\alpha = 1$, then ϕ is discrete harmonic and the sum on the left-hand side is zero. If $w \in K'_{-k}$, $|\mathcal{L}\phi(w)| \leq c\,(2^{-k}|z|)^{\alpha-1}$, and hence,

$$\sum_{w \in K'_{-k}} |\mathcal{L}\phi(w)|\, G_A(z,w) \leq c\, s_z\, |z|^{\alpha-1}\, 2^{(1-2\alpha)k}.$$

$$\begin{aligned} \sum_{k=2}^{\infty} \sum_{w \in K_{-k}} |\mathcal{L}\phi(w)|\, G_A(z,w) &\leq c\, s_z\, |z|^{\alpha-1} \sum_{k=2}^{\log_2 |z|} 2^{(1-2\alpha)k} \\ &\leq c\, s_z\, (1+e_z)\, |z|^{\alpha-1}. \end{aligned} \tag{15}$$

§3. Proof of Theorem 3

By symmetry, we may assume $x \geq 0$. If $x = n$, then the conclusion is just $\mathrm{hm}_n(x) \asymp n^{-1/2}$ which is well known so we will assume $0 \leq x < n$.

We start by making some general comments about discrete harmonic measure. If $0 \in U \subset \mathbb{Z}^2$, let a_U denote the unique nonnegative function on $\mathbb{Z}^2$ that vanishes on U; is discrete harmonic on $A := \mathbb{Z}^2 \setminus U$; and satisfies

$$a_U(z) = \frac{2}{\pi} \log |z| + O_U(1), \quad |z| \to \infty.$$

If $U = \{0\}$ this is called the potential kernel. It is known that [2, Propositions 6.4.7 and 6.6.1]

$$a_U(z) = \frac{2}{\pi} \lim_{R \to \infty} (\log R) \, \mathbb{P}^z\{\sigma_R < \tau_A\},$$

and

$$\mathrm{hm}_U(z) = \mathcal{L}\psi_U(z), \quad z \in U.$$

Let $f(z) = z + \sqrt{z^2 - 1}$ which is the conformal transformation of $D := \mathbb{C} \setminus [-1, 1]$ onto $\{|z| > 1\}$ satisfying $f(z) = 2z + O(|z|^{-1})$ as $z \to \infty$.

Let

$$\phi(z) = \log |f(z)| = \mathrm{Re}[\log f(z)].$$

The holomorphic function $\log f$ is well defined locally; ϕ is defined on all of D. Note that ϕ vanishes on $[-1, 1]$ and is harmonic on D. As $z \to \infty$,

$$\phi(z) = \log |z| + \log 2 + O(|z|^{-2}).$$

It can also be given by

$$\phi(z) = \log |z| - \mathbb{E}^z[\log |B_{\bar{\tau}}|]$$

where B_t is a complex Brownian motion and $\bar{\tau} = \min\{t \geq 0 : B_t \in [-1, 1]\}$. Let $\phi_n(z) = \phi(z/n)$ which is the continuous function that vanishes on $[-n, n]$, is harmonic on $\mathbb{C} \setminus [-n, n]$, and satisfies

$$\phi_n(z) = \log |z| + \log 2 - \log n + O_n(|z|^{-2}), \quad z \to \infty.$$

Note that $\phi_n(z) \leq c\,(\delta_z/n)^{1/2}$ where

$$\delta_z = \delta_{z,n} = \mathrm{dist}\{z, \{-n, n\}\} = \min\{|z - n|, |z + n|\}.$$

Let

$$\phi_n^*(z) = \begin{cases} \phi_n(z), & \mathrm{Im}(z) \geq 0 \\ -\phi_n(z), & \mathrm{Im}(z) < 0 \end{cases},$$

and note that if $x \in U_n$, then ϕ_n^* is harmonic in the open disk of radius $n - |x|$ about x.

If $a_n = a_{U_n}$, $A = A_n = \mathbb{Z}^2 \setminus U_n$, $G = G_n = G_{A_n}$, then we claim that

$$a_n(z) = \frac{2}{\pi}\left[\phi_n(z) + \tilde{\Psi}_n(z)\right], \quad \tilde{\Psi}_n(z) = \sum_{w \in A} G(z, w)\, \mathcal{L}\phi_n(z).$$

Indeed, this follows from Proposition 1 and the observation that $a_n - \frac{2}{\pi}\,\phi_n$ is a bounded function.

If $x \in \{0, 1, \ldots, n-1\}$, then $\phi_n(x+i) = \phi_n(x-i), a_n(x+i) = a_n(x-i), \Psi_n(x+i) = \Psi_n(x-i)$ and hence $\mathrm{hm}_n(x) = a_n(x+i)/2$ and

$$\pi\,\mathrm{hm}_n(x) = \frac{\pi}{2}\,a_n(x+i) = \phi_n(x+i) + \tilde{\Psi}_n(x+i).$$

A straightforward computation gives

$$\phi_n(x+i) = \frac{1}{\sqrt{n^2-x^2}}\left[1 + O\left(\frac{1}{n-x}\right)\right].$$

Therefore, to prove the theorem, it suffices to show that

$$\Psi_n(x+i) \le \frac{c}{n^{1/2}\,(n-x)^{3/2}}.$$

This is a particular case of the following proposition that we will prove.

Proposition 10. *There exists $c < \infty$ such that the following is true. If $w = x + iy \in \mathbb{Z}^2 \setminus U_n$, then*

$$\Psi_n(w) \le \frac{c}{n^{1/2}\,\delta_w^{1/2}}. \tag{16}$$

Moreover, if $|x| \le n$,

$$\Psi_n(w) \le \frac{c}{n^{1/2}\,\delta_w^{1/2}}\,\frac{|y|}{\delta_w}. \tag{17}$$

Lemma 11. *There exists $c < \infty$ such that*

$$|\mathcal{L}\phi_n(w)| \le \frac{c}{|w|^4}, \quad |w| \ge 2n,$$
$$|\mathcal{L}\phi_n(w)| \le c\,n^{-1/2}\,\delta_w^{-7/2}, \quad |w| \le 2n, w \notin U_n.$$

Proof. This follows from Lemma 5 by considering various cases. Note that the open disk of radius δ_w about w does not intersect $\{-n, n\}$, and hence the intersection of this disk with $\mathbb{R}$ is either disjoint from or entirely contained in $(-n, n)$.

- If $|w| \ge 2n$, let $\hat{\phi}_n(z) = \phi_n(z) - \phi_n(w)$. Then $\mathcal{L}\phi_n(w) = \mathcal{L}\hat{\phi}_n(w)$ and $\hat{\phi}_n$ is harmonic in the disk of radius $|w|/2$ about w with $\|\hat{\phi}_n\|_\infty \le c$.
- If $|w| \le 2n$ and the disk of radius δ_w about w does not intersect $(-n, n)$, then ϕ_n is harmonic in this disk with $\|\phi_n\|_\infty \le c\,(\delta_w/n)^{1/2}$.

- If $|w| \leq 2n$, and the intersection of the disk of radius δ_w about w with $\mathbb{R}$ is contained in $(-n, n)$, then ϕ_n^* is harmonic in this disk with $\|\phi_n^*\|_\infty \leq c\,(\delta_w/n)^{1/2}$. Also if $w \notin U_n$, $\mathcal{L}\phi_n(w) = \mathcal{L}\phi_n^*(w)$.

Q.E.D.

We will only use the next lemma with $V_n = U_n$ but we state it in more generality.

Lemma 12. *There exists $c < \infty$ such that the following is true. Suppose $n > 0$ and V_n is a connected set of radius at least n including the origin. Let $G = G_{\mathbb{Z}^2 \setminus V_n}$ be the Green's function on the complement of V_n. For every positive integer k, let*

$$C_k = C_{k,n} = \{z : 2^k n \leq |z| < 2^{k+1} n\}.$$

- *If $w \in C_k$, then $a_n(w) \leq ck$,*
- *For every $w \in \mathbb{Z}^2$,*

$$\sum_{z \in C_k} G(w, z) \leq c\,k\,2^{2k}\,n^2.$$

Proof. For the first bullet, see [2, Proposition 6.6.7]. Let

$$Y_{k,n} = \max_w \sum_{z \in C_k} G(w, z).$$

Clearly, the maximum is obtained at some $w \in C_k$. We use the following facts.

- Using the invariance principle, there exists $\epsilon > 0$ such that the probability that the random walk starting in C_k reaches $\{|z| \leq 2^{k-1} n\}$ within $2^{2k} n^2$ steps is greater than ϵ.
- If $|z| \leq 2^{k-1} n\}$, the probability that a random walk starting at z hits V_n before reaching C_k is greater than $c/(k \vee 1)$ (see, e.g., [2, Proposition 6.4.1]).

These imply that

$$Y_{k,n} \leq 2^{2k} n^2 + \left(1 - \frac{c}{k}\right) Y_{k,n},$$

from which the result follows immediately. Q.E.D.

Proof of Proposition 10. Without loss of generality, we will assume $x, y \geq 0$ and hence $\delta_w = |w - n|$.

Let $C_k = C_{k,n}$ be as in Lemma 12. Note that for such $z \in C_k$,

$$\phi_n(z) \asymp k, \quad |\mathcal{L}\phi_n(z)| \leq c\,2^{-4k}\,n^{-4}.$$

Using Lemma 12,

$$\sum_{z\in C_k} G_n(w,z)\,|\mathcal{L}\phi_n(z)| \leq c\,2^{-4k}\,n^{-4} \sum_{z\in C_k} G_n(w,z)$$
$$\leq c\,k\,2^{-2k}\,n^{-2}.$$

Summing over k, we have

$$\sum_{|z|\geq 2n} G_n(w,z)\,|\mathcal{L}\phi_n(z)| \leq \frac{c}{n^2}.$$

Let $J = J_n = \{z : |z| \leq 2n, \delta_z \geq |z|/4\}$. Since ϕ_n is uniformly bounded on J, we have by Lemma 12,

$$\sum_{z\in J} G_n(w,z)\,|\mathcal{L}\phi_n(z)| \leq \frac{c}{n^4} \sum_{z\in J} G_n(w,z) \leq \frac{c}{n^2}.$$

Let

$$K_k = K_{k,n} = \{z \notin U_n : 2^{-k-1}\,n \leq \delta_z < 2^{-k}\,n\}, \quad K = \bigcup_{k=1}^{\infty} K_k.$$

Since $\delta_z \geq 1$ for $z \notin U_n$, we see that K_k is empty if $2^k > n$. Then we have

$$\Psi_n(w) \leq \frac{c}{n^2} + \sum_{k\leq \log_2 n} \Psi_{n,k}(w),$$

where

$$\Psi_{n,k}(w) = \sum_{z\in K_k} G_n(w,z)\,|\mathcal{L}\phi_n(z)|.$$

If $z \in K_k$, Lemma 11 gives $|\mathcal{L}\phi_n(z)| \leq c\,n^{-1/2}\,\delta_z^{-7/2} \leq c\,2^{7k/2}\,n^{-4}$ and using Lemma 12, we have

$$\Psi_{n,k}(w) \leq c\,2^{7k/2}\,n^{-4} \sum_{z\in K_k} G_n(w,z)$$
$$\leq c\,2^{7k/2}\,n^{-4}\left[n^2\,2^{-2k}\right] \leq c\,2^{3k/2}\,n^{-2}.$$

This holds for all w, but we can improve this by letting $\rho_k = \rho_{k,n}$ be the first hitting time of K_k, and noting that

$$\Psi_{n,k}(w) \leq \mathbb{P}^w\{\rho_k < \bar{\tau}\} \sup_{w'} \Psi_{n,k}(w') \leq c\,2^{3k/2}\,n^{-2}\,\mathbb{P}^w\{\rho_k < \bar{\tau}\}.$$

If $\delta_w \geq n/4$, then the Beurling estimate (the $\alpha = 1/2$ of the previous theorem) gives $\mathbb{P}^w\{\rho_k < \bar{\tau}\} \leq c\,2^{-k/2}$. Technically, this estimate applies to a half line rather than U_n, but it is not hard to see this is true for U_n, with perhaps a different constant. This gives $\Psi_{n,k}(w) \leq c\, 2^k\, n^{-2}$. Therefore,

$$\Psi_n(w) \leq \frac{c}{n^2} + \sum_{k \leq \log_2 n} \Psi_{n,k}(w) \leq \frac{c}{n^2}\left[1 + \sum_{k \leq \log_2 n} 2^l\right] \leq \frac{c}{n}.$$

If $w \in K_m$, then the Beurling estimate gives

$$\mathbb{P}^w\{\rho_k < \bar{\tau}\} \leq c\, 2^{-|k-m|/2} \leq c\, 2^{m/2}\, 2^{-k/2}$$

and hence

$$\begin{aligned} \Psi_n(w) &\leq \frac{c}{n^2} + \sum_{k \leq \log_2 n} \Psi_{n,k}(w) \\ &\leq \frac{c}{n^2}\left[1 + 2^{m/2} \sum_{k \leq \log_2 n} 2^l\right] \\ &\leq \frac{c\, 2^{m/2}}{n} \leq \frac{c}{|w-n|^{1/2}\, n^{1/2}}. \end{aligned}$$

Combining these, we get (16).

For the second estimate, assume that $x \leq n$ and $y \leq \delta_w/20 = |w-n|/20$ (if $y \geq \delta_w/20$ this follows from the first estimate). Let $V = \{\zeta : |x - \zeta| < \delta_w/5\}$ and $\eta = \min\{k \geq 0 : S_k \notin V\}$. Then $\Psi_n(w)$ is bounded above by

$$\sum_{|z-x| \leq \delta_w/5} G_V(w,z)\, |\mathcal{L}\phi_n(z)| + \mathbb{P}^w\{\eta < \bar{\tau}\} \max_{|w'-x| \leq 1 + (\delta_z/5)} \Psi_n(w'). \tag{18}$$

Using the harmonicity of ϕ_n^*, we can see that for $|z - x| \leq \delta_w/5$ we have

$$|\mathcal{L}\phi_n(z)| \leq c n^{-1/2}\, \delta_w^{-7/2}.$$

Also, gambler's ruin estimates (see Lemma 6) give

$$\mathbb{P}^w\{\eta < \bar{\tau}\} \leq c(y/\delta_w), \qquad \sum_{|z-x| \leq \delta_w/5} G_V(w,z) \leq c\, y\, \delta_w = c(y/\delta_w)\, \delta_w^2.$$

Combining this with (16) and (18) gives (17). Q.E.D.

§4. Proof of Theorem 4

Throughout this section, we let $A = \mathbb{Z}^2 \setminus \{0\}$, $\psi(z) = \frac{2}{\pi} \log |z|$. By the triangle inequality, it suffices to prove the result for $|w| = 1$ and by symmetry, if suffices to prove the result for $w = i$. We write ∇ for ∇_i. Direct computation shows that

$$\frac{\pi}{2}\, \mathcal{L}\psi(z) = \frac{3}{4|z|^4} - \frac{\mathrm{Re}(z)^4 + \mathrm{Im}(z)^4}{|z|^8} + O(|z|^{-6}),$$
$$|\nabla \mathcal{L}\psi(z)| \leq \frac{c}{|z|^5}.$$

Let $p_A(z, w)$ denote the probability that a random walk starting at z reaches w before reaching 0. Then [2, Proposition 4.4.1], $G_A(z, w) = p_A(z, w)\, G_A(w, w) = 2\, p_A(z, w)\, a(w)$. It is easy to see that $p_A(\infty, w)$ equals $1/2$, and hence $G_A(\infty, w) = a(w)$. Since $a - \psi$ is bounded, Proposition 1 implies

$$a(z) = \psi(z) + \sum_{w \neq 0} G_A(z, w)\, \mathcal{L}\psi(w).$$

In particular,

$$k_0 = \sum_{w \in A} a(w)\, \mathcal{L}\psi(w), \quad e(z) = \sum_{w \in A} \left[G_A(z, w) - a(w)\right] \mathcal{L}\psi(w).$$

Let $U = U_z = z + V_k, U^* = U_z^* = z + V_k^*$ where $k = \lfloor |z|/2 \rfloor$ and

$$V_k = \{x + iy : |x|, |y| < k\}, \quad V_k^* = \{|x| < k, |y| < k - 1\}.$$

For $\zeta \in U$, we can write

$$e(\zeta) = F_1(\zeta) + F_2(\zeta)$$

where

$$F_1(\zeta) = \sum_{w \in U} G_U(z, w)\, \mathcal{L}\psi(w), \quad F_2(\zeta) = \sum_{w' \in \partial U} H_U(\zeta, w')\, e(w').$$

Since F_2 is discrete harmonic on U and $|F_2(\zeta)| \leq c\,|z|^{-2}$, difference estimates for discrete harmonic functions [2, Theorem 6.3.8] give $|\nabla F_2(z)| \leq c\,|z|^{-3}$.

If $\partial U^+ = U \cap \partial U^*$, then we can write

$$F_1(z) = \sum_{w \in U^*} G_{U^*}(z, w)\, \mathcal{L}\psi(w) + \sum_{w' \in U \cap \partial U^*} H_{U^*}(z, w')\, F_1(w').$$

Note that

$$\left|\sum_{w'\in U\cap\partial U^*} H_{U^*}(z,w')\,F_1(w')\right| \le \max_{w'\in\partial_i U} |F_1(w')|$$

$$\le c\,|z|^{-4} \max_{w'\in\partial_i U} \sum_{\zeta\in U} G_U(w',\zeta)$$

$$\le c|z|^{-3}.$$

The last inequality uses the gambler's ruin estimate (Lemma 6). Similarly,

$$F_1(z+i) = \sum_{w\in U^*} G_{U^*+i}(z+i,w+i)\,\mathcal{L}\psi(w+i) + O(|z|^{-3}).$$

Using $G_{U^*+i}(z+i,w+i) = G_{U^*}(z,w)$, we see that $|\nabla F_1(z)|$ is bounded above by

$$O(|z|^{-3}) + \sum_{w\in U^*} G_{U^*}(z,w)\,|\nabla\mathcal{L}\psi(w)| = O(|z|^{-3}).$$

§ Acknowledgment

The author thanks the referee for comments and corrections. Research was supported by NSF grant DMS-1513036.

References

[1] G. Lawler, Properties of Brownian motion and random walk paths, In: *Random Walks*, P. Révész and B. Toth, ed., János Bolyai Society Math. Studies, **9** (1999), 219–258.

[2] G. Lawler and V. Limic, Random Walk: A Modern Introduction, Cambridge Univ. Press, (2010).

[3] C. Beneš, G. Lawler and F. Viklund, Transition probabilities for infinite two-sided loop-erased random walks, Electron. J. Probab., **24**, paper no. 139, (2019).

[4] E. Procaccia, J. Ye and Y. Zhang, Stationary harmonic measures as the scaling limit of truncated harmonic measure, preprint, arXiv:1811.04793.

Department of Mathematics,
University of Chicago,
5734 S. University Ave, Chicago, IL 60637, USA
E-mail address: `lawler@math.uchicago.edu`

Advanced Studies in Pure Mathematics 87, 2021
Stochastic Analysis, Random Fields and Integrable Probability — Fukuoka 2019
pp. 173–198

Compact Brownian surfaces

Grégory Miermont

Abstract.

We study the scaling limit of uniform random quadrangulations of a connected, orientable and compact surface of a fixed topology, when the number of quadrangles tends to infinity, as well as the perimeters of the boundary faces. The limiting random metric spaces are called the (orientable) compact Brownian surfaces. This result generalises the known cases where the surface is either the sphere or the disk. We use an interpretation of a bijection by Chapuy-Marcus-Schaeffer (generalised by Bettinelli) as a way to decompose a quadrangulation along well-chosen geodesics into a finite number of *elementary pieces* whose scaling limits are obtained separately, by using in a crucial way the known results for the sphere and the disk, and their extensions to the non-compact settings of the Brownian plane and half-plane This paper presents the main objects and ideas of the proof, as well as some further properties of the law of the Brownian cylinder. This paper and its pictures are mostly based on joint work with Jérémie Bettinelli.

§1. Introduction

A map is the proper embedding of a graph G into an orientable, connected, compact 2-dimensional surface without boundary, that also satisfies the property that the connected components of the complement of the embedding, which are called the faces of the map, are all homeomorphic to an open disc $\mathbb{D}$. Two maps are always identified if the associated embeddings correspond to each other via a direct homeomorphism between the underlying surfaces. We usually denote maps using bold lowercase letters $\mathbf{m}, \mathbf{q}, \ldots$. With every map (more precisely, with every particular representative of a map) $\mathbf{m}$ we associate its set

Received May 15, 2020.
Revised December 11, 2020.
2010 *Mathematics Subject Classification.* Primary 60D05; Secondary 60F17, 60C05.
Key words and phrases. Random maps, random forests, Brownian surfaces.

of vertices, edges and faces $V(\mathbf{m}), E(\mathbf{m}), F(\mathbf{m})$. Note that every edge $\mathbf{e} \in E(\mathbf{m})$ can be written as $\mathbf{e} = \{e, \bar{e}\}$ where e is an oriented edge and $\bar{e}$ is e with reversed orientation, and we let $e^-, e^+ \in V(\mathbf{m})$ be the origin and target of e. A rooted map if a pair $(\mathbf{m}, e_0)$ where e_0 is an oriented edge of $\mathbf{m}$ (that is usually omitted in the notation).

If e is an oriented edge, the vertex (resp. face) incident to e is by convention e^- (resp. the face that lies to the left of e), and we write $e \in E(f)$. If f is a face of a map $\mathbf{m}$, then the oriented edges incident to f can be arranged in cyclic order (counterclockwise) around f, say as $e_1, e_2, \ldots, e_k, e_{k+1} = e_1$. We call this the facial order around f.

We can view a map $\mathbf{m}$ as a measured metric space by endowing it with the graph distance $d_{\mathbf{m}}$ on $V(\mathbf{m})$ and the counting measure $\mu_b m = \sum_{v \in V(\mathbf{m})} \delta_v$. In this way (which is by no means the only natural one), maps can be seen as discrete metric surfaces of a given topology, and random maps models are therefore natural models of two-dimensional random geometries. Since $\mathbf{m}$ itself is a class of embedded graphs, it is natural to consider the space $(V(\mathbf{m}), d_{\mathbf{m}}, \mu_{\mathbf{m}})$ up to isometries. We will denote by $[X, d, \mu]$ the isometry class of a measured metric space (X, d, μ), and we endow the space of such isometry classes of compact measured spaces (where $\mu(X) < \infty$) with the Gromov-Hausdorff-Prokhorov distance, see for instance [35, Chapter 27][1].

The simplest random map model, which we will consider throughout this paper, is that of bipartite quadrangulations with boundaries. These are bipartite maps (every cycle has an even length) where every face has degree 4, except for a set of marked faces of arbitrary even degrees, called *holes*. We view these holes as boundaries[2] of the map, and call their degrees the *perimeters*. The faces that are not holes are also called *internal faces*. Let $n, g \geq 0$ be fixed integers $\boldsymbol{l} = (l_1, \ldots, l_p)$ a finite sequence of $p \geq 0$ positive integers. We let $\mathbf{Q}_{n,\boldsymbol{l}}^{[g]}$ be the set of rooted bipartite quadrangulations of genus g with n internal faces, and p holes of perimeters $2l_1, \ldots, 2l_p$. We thus informally view an element of $\mathbf{Q}_{n,\boldsymbol{l}}^{[g]}$ as a discrete geometry defined on the (unique up to homeomorphism)

[1]Usually the Gromov-Hausdorff-Prokhorov distance is defined on metric measured spaces where the measure is of total mass one, however it is a minor modification to extend the definition to general finite measures, see the discussion in [35, p.763].

[2]The reader might wonder why we do not define maps with boundaries simply as graphs embedded in surfaces with a boundary. However, this would force the boundaries to form simple curves, and for technical reasons, it is easier to relax this constraint to the price of this slightly convoluted definition.

connected, compact, oriented surface $\mathbf{\Sigma}^{\partial}_{g,p}$ of genus g with p boundaries[3]. Note that by Euler's formula, the number of vertices of any element $\mathbf{q} \in \mathbf{Q}^{[g]}_{n,\boldsymbol{l}}$ is $n + \sum_{i=1}^{p}(l_i - 1) + 2 - 2g$.

We let Q_n be a uniformly chosen random element of $\mathbf{Q}^{[g]}_{n,\boldsymbol{l}}$. Our main result, which is the object of the joint paper [9] in preparation with Jérémie Bettinelli, is the following.

Theorem 1. *Let $\boldsymbol{L} = (L_1, \ldots, L_p)$ be fixed positive real numbers. Assume that $\boldsymbol{l} = \boldsymbol{l}(n)$ satisfies $l_i(n)/\sqrt{2n} \to L_i$ for $1 \leq i \leq p$ as $n \to \infty$. Then the following convergence in law holds for the Gromov-Hausdorff-Prokhorov topology on compact measured metric spaces:*

$$\left[V(Q_n), \left(\frac{9}{8n}\right)^{1/4} d_{Q_n}, \frac{1}{n} \sum_{v \in V(Q_n)} \delta_v\right] \xrightarrow[n\to\infty]{(d)} \mathsf{S}^{[g]}_{\boldsymbol{L}}, \tag{1}$$

where the limit $\mathsf{S}^{[g]}_{\boldsymbol{L}}$ is a random measured metric space that is a.s. homeomorphic to $\mathbf{\Sigma}^{\partial}_{g,p}$.

It is natural to call the limit $\mathsf{S}^{[g]}_{\boldsymbol{L}}$ the *Brownian surface homeomorphic to $\mathbf{\Sigma}^{\partial}_{g,p}$ with area* 1 *and boundary sizes* $\boldsymbol{L}$, since the measure has total mass one. In the case where $(g,p) = (0,0)$, the underlying surface is the sphere and this result is the known convergence of random quadrangulations to the *Brownian map* [29, 31], which in this context is more naturally called the Brownian sphere. When $(g,p) = (0,1)$, the result is also already known [8] and the limit is called the Brownian disk with boundary size L_1. These known cases, as well as non-compact analogues [19, 26, 4] will in fact serve as a basis to prove the general result, as sketched below.

From $\mathsf{S}^{[g]}_{\boldsymbol{L}}$, we can define a version with area $A > 0$ by scaling: writing $\mathsf{S}^{[g]}_{A^{-1/2}\boldsymbol{L}} = [S^{[g]}_{A^{-1/2}\boldsymbol{L}}, d, \mu]$, then $\mathsf{S}^{[g]}_{\boldsymbol{L}}(A)$ has same distribution as $[S^{[g]}_{A^{-1/2}\boldsymbol{L}}, A^{1/4}d, A\mu]$. In this way, $\mathsf{S}^{[g]}_{\boldsymbol{L}}(A)$ is the limit in distribution of the random metric spaces appearing in the left of (1), with same hypotheses except that Q_n is now uniformly chosen among all bipartite quadrangulations with $\lfloor nA \rfloor$ internal quadrangles and p holes of half-perimeters $l_1(n), \ldots, l_p(n)$.

The rest of this paper is organised as follows. Sections 2 and 3 can be seen as a reading guide to [9] giving the main ideas of the proof of Theorem 1, by presenting the decomposition of maps into elementary

[3]Recall that $\mathbf{\Sigma}^{\partial}_{g,p}$ is the connected sum of g tori, from which p disjoint open disks have been removed.

pieces and stating the scaling limit results for these elementary pieces in Section 2, and then sketching the proof for these auxiliary scaling limit results in Section 3. Section 4 contains some further results for the particular case of the Brownian cylinders, that relies on an alternative construction of the space $\mathsf{S}^{[0]}_{(L_1,L_2)}$.

§2. The elementary pieces: quadrilaterals and slices

The proof of Theorem 1 is based on two main ideas.

- Elements of $\mathbf{Q}^{[g]}_{n,l}$ can be viewed as gluings of simpler "elementary pieces" that can be of two types: quadrilaterals and slices, along their boundary parts which turn out to be geodesic segments. This is based on a reinterpretation of the Chapuy-Marcus-Schaeffer bijection [18].
- Elementary pieces satisfy scaling limit results analogous to Theorem 1, which can be deduced from the cases $(g, p) = (0, 0)$ and $(g, p) = (0, 1)$ of Theorem 1 (in fact, it is more convenient to use a non-compact version of these cases studied in [19, 26, 4]).

The fact that the pieces are glued along geodesic segments implies that the gluing and limiting operation commute nicely.

2.1. Quadrilaterals with geodesic boundaries

A discrete quadrilateral with geodesic boundaries is an object of the form $(\mathbf{q}, \rho, \overline{\rho}, x, \overline{x})$, where $\mathbf{q}$ is a plane quadrangulation with one hole and at least one internal face, and $\rho, \overline{\rho}, x, \overline{x}$ are four distinct vertices of $\mathbf{q}$ on the boundary that satisfy the following properties:

- the vertices $\rho, x, \overline{\rho}, \overline{x}$ appear in this clockwise order on the boundary,
- the boundary arcs ρx, $\rho\overline{x}$, $\overline{\rho}x$, $\overline{\rho}\overline{x}$ are all geodesic segments, and moreover, the arcs $\overline{\rho}x$ and $\rho\overline{x}$ are the unique geodesic segments between their respective endpoints in $\mathbf{q}$,
- the boundary arcs ρx and $\overline{\rho}x$ (resp. $\overline{\rho}\overline{x}$ and $\rho\overline{x}$) intersect only at x (resp. $\overline{x}$)
- one has

$$d_{\mathbf{q}}(\overline{\rho}, x) - d_{\mathbf{q}}(\rho, x) = d_{\mathbf{q}}(\overline{\rho}, \overline{x}) - d_{\mathbf{q}}(\rho, \overline{x})$$

  and this common quantity is called the *discrepancy* of $\mathbf{q}$.

The reader might find it surprising that we require the arcs $\overline{\rho}x$ and $\rho\overline{x}$ to be not only geodesic segments, but also the unique ones. However, this is the right combinatorial property that makes quadrilaterals with geodesic boundaries easy to manipulate. This becomes apparent when one glues several such spaces together along their geodesics sides, because this makes the gluing procedure invertible. These gluing operations play an important role in Section 3.3 below.

A quadrilateral $(\mathbf{q}, \rho, \bar{\rho}, x, \bar{x})$ as above is naturally split into a upper half, a lower half and a *ridge* separating the former two, which are informally given by the sets

$$\{z \in V(\mathbf{q}) : d_{\mathbf{q}}(z, x) - d_{\mathbf{q}}(z, \bar{x}) < d_{\mathbf{q}}(\rho, x) - d_{\mathbf{q}}(\rho, \bar{x})\},$$
$$\{z \in V(\mathbf{q}) : d_{\mathbf{q}}(z, x) - d_{\mathbf{q}}(z, \bar{x}) > d_{\mathbf{q}}(\rho, x) - d_{\mathbf{q}}(\rho, \bar{x})\},$$
$$\{z \in V(\mathbf{q}) : d_{\mathbf{q}}(z, x) - d_{\mathbf{q}}(z, \bar{x}) = d_{\mathbf{q}}(\rho, x) - d_{\mathbf{q}}(\rho, \bar{x})\}.$$

However, the exact definition of the ridge (which in general it is only a subset of the latter set) goes as follows. Let

$$\lambda(z) = (d(z, x) - d(\rho, x)) \wedge (d(z, \bar{x}) - d(\rho, \bar{x})), \qquad z \in V(\mathbf{q}).$$

It is a simple exercise to show that $|\lambda(z) - \lambda(z')| = 1$ for every pair of neighbours $z, z' \in V(\mathbf{q})$. Therefore, we can orient canonically every $\mathbf{e} \in E(\mathbf{q})$ in such a way that $\lambda(e^-) = \lambda(e^+) + 1$. Let γ_e be the maximal left-most path starting from e along which λ is strictly decreasing. We let $E(\mathbf{q}, x), E(\mathbf{q}, \bar{x})$ be the set of e such that γ_e ends at x (resp. $\bar{x}$). The ridge $R(\mathbf{q})$ it the set of vertices v such that there exist two edges $e \in E(\mathbf{q}, x), f \in E(\mathbf{q}, \bar{x})$ with $e^- = f^- = v$. With this definition, $R(\mathbf{q})$ can be viewed as a simple path in $\mathbf{q}$ from ρ to $\bar{\rho}$ which is allowed to cross some squares of $\mathbf{q}$ diagonally. The number of quadrangles in the region of $\mathbf{q} \setminus R(\mathbf{q})$ that contains x and $\bar{x}$ are called the *half areas* $a, \bar{a}$ of $\mathbf{q}$ (these can be half-integers due to the fact that some quadrangles might be split in two). The number $\#R(\mathbf{q}) - 1$ is called the *width* of the quadrilateral.

Theorem 2. *Let* $(\mathrm{Quad}_n, \rho, \bar{\rho}, x, \bar{x})$ *be a uniformly chosen random element of the set of discrete quadrilaterals with geodesic boundaries, having half-areas* $a_n, \bar{a}_n$*, width* $\#R(\mathrm{Quad}_n) = w_n$ *and discrepancy* δ_n*, where the sequences* $a_n, \bar{a}_n, w_n, \delta_n$ *satisfy*

$$\frac{a_n}{n} \to A, \qquad \frac{\bar{a}_n}{n} \to \overline{A}, \qquad \frac{w_n}{\sqrt{2n}} \to L, \qquad \left(\frac{9}{8n}\right)^{1/4} \delta_n \to \delta$$

where $A, \overline{A}, L > 0$ *and* $\delta \in \mathbb{R}$*. Then one has the convergence in distribution for the Gromov-Hausdorff-Prokhorov topology*

$$\left(V(\mathrm{Quad}_n), \left(\frac{9}{8n}\right)^{1/4} d_{\mathrm{Quad}_n}, \frac{1}{n} \sum_{v \in V(\mathrm{Quad}_n)} \delta_v \right) \xrightarrow[n \to \infty]{(d)} \mathsf{Quad}_{A, \overline{A}, L, \delta},$$

for some limiting random metric measure space $\mathsf{Quad}_{A,\overline{A},L,\delta}$.

The limiting metric space is homeomorphic to a disc, and as for the discrete quadrilaterals, it comes with four distinguished points $\rho, x, \overline{\rho}, \overline{x}$ appearing in this cyclic order in the boundary, in such a way that the segments $\gamma = [\rho, x], \gamma' = [\overline{\rho}, x], \overline{\gamma}' = [\rho, \overline{x}], \overline{\gamma} = [\overline{\rho}, \overline{x}]$ are geodesic paths with $d(\rho, \overline{x}) - d(\rho, x) = d(\overline{\rho}, \overline{x}) - d(\overline{\rho}, x) = \delta$. We refer the reader to the description of Section 3.3.

In fact, the convergence holds in a stronger sense of the *marked* Gromov-Hausdorff-Prokhorov topology, where the geodesic segments from $\rho, \overline{\rho}$ to $x, \overline{x}$ are distinguished at the origin and end.

It is also useful to consider a *free* version FQuad_L with random areas and discrepancies, obtained by a mixture of the laws $\mathsf{Quad}_{A,\overline{A},L,\delta}$. Precisely, we let $\mathcal{A}, \overline{\mathcal{A}}, \Delta$ be independent random variables, with $\mathcal{A}, \overline{\mathcal{A}}$ both having the law of the first hitting time of $-L$ by standard Brownian motion, and Δ having law $\mathcal{N}(0, L)$, and conditionally given these variables we let FQuad_L have the law of $\mathsf{Quad}_{\mathcal{A},\overline{\mathcal{A}},L,\Delta}$. Those appear naturally in the context of non-compact Brownian surfaces, see Section 3.4.

2.2. Slices with geodesic boundaries

A discrete slice with geodesic boundaries is a tuple $(\mathbf{q}, \rho, \overline{\rho}, x)$ where $\mathbf{q}$ is a plane quadrangulation with one hole and three marked vertices $(\rho, \overline{\rho}, x)$ on the boundary such that

- the vertices $\rho, x, \overline{\rho}$ appear in this clockwise order on the boundary
- the boundary arcs ρx and $\overline{\rho} x$ are geodesic segments, and moreover, the arc $\overline{\rho} x$ is the unique geodesic between $\overline{\rho}$ and x in $\mathbf{q}$,
- the boundary arcs ρx and $\overline{\rho} x$ intersect only at x.

Such objects have appeared in several guises and places in the study of random maps, see in particular [11, 29, 12, 13, 8, 33, 10].

Again, the requirement that the boundary arc $\overline{\rho} x$ be the unique geodesic arc is a convenient combinatorial property that allows to glue several such spaces together, as discussed in [10]. The quantity $d_{\mathbf{q}}(\overline{\rho}, x) - d_{\mathbf{q}}(\rho, x)$ is called the discrepancy of the slice $\mathbf{q}$. Note that the opposite of this quantity is called the *tilt* in [10]. Note also that we do not make any assumption on the boundary arc $\rho\overline{\rho}$, which need not be geodesic. Its length is called the width of the slice.

Theorem 3. *Let* $(\mathrm{Slice}_n, \rho, \overline{\rho}, x)$ *be a uniformly chosen random element of the set of discrete slices with geodesic boundaries with* a_n *internal* *faces, width* $2l_n$ *and discrepancy* δ_n*, where the sequences* a_n, l_n, δ_n *satisfy*

$$\frac{a_n}{n} \to A, \qquad \frac{l_n}{\sqrt{2n}} \to L, \qquad \left(\frac{9}{8n}\right)^{1/4} \delta_n \to \delta,$$

where $A, L > 0$ *and* $\delta \in \mathbb{R}$*. Then one has the convergence in distribution for the Gromov-Hausdorff-Prokhorov topology*

$$\left(V(\mathrm{Slice}_n), \left(\frac{9}{8n}\right)^{1/4} d_{\mathrm{Slice}_n}, \frac{1}{n} \sum_{v \in V(\mathrm{Slice}_n)} \delta_v \right) \xrightarrow[n\to\infty]{(d)} \mathsf{Slice}_{A,L,\delta}.$$

The limiting metric space is a.s. homeomorphic to the disk and has three distinguished points $\rho, x, \overline{\rho}$ on the boundary, so that the segments ρx and $\overline{\rho} x$ are geodesics with $d(\overline{\rho}, x) - d(\rho, x) = \delta$. Here as well, the convergence above holds in the stronger marked Gromov-Hausdorff-Prokhorov topology, where the geodesic segments ρx and $\overline{\rho} x$ are marked a the origin and at the end.

Here again it is useful to consider a free version with random area and discrepancy by letting FSlice_L have the law of $\mathsf{FSlice}_{\mathcal{A},L,\Delta}$, where $\mathcal{A}, \Delta$ are independent, $\mathcal{A}$ having the law of the first hitting time of L by Brownian motion, and Δ having law $\mathcal{N}(0, 3L)$. Note the factor 3 that differs from the case of free quadrilaterals. This difference is explained by the two different normalizations in (5) and (6) below.

2.3. Decomposition in elementary pieces

We finally explain how an element of $\mathbf{Q}_{n,l}^{[g]}$ can be decomposed into elementary pieces. This decomposition is a reformulation of an extension due to Bettinelli [6] of the so-called Chapuy-Marcus-Schaeffer bijection [18]. In fact, we will rather explain how to glue elementary pieces together to create an element of $\mathbf{Q}_{n,l}^{[g]}$ and simply invoke the fact that this operation is bijective.

A *scheme* is a map $\mathbf{s}$ in genus g, with $p+1$ faces denoted by $h_1, \ldots, h_p, f_*$, that satisfies the following conditions

- for every $1 \leq i \leq p$, the boundary of h_i is simple and does not share an edge with $h_j, j \neq i$,
- the degree of every vertex in $\mathbf{s}$ is at least 3,
- the root edge of $\mathbf{s}$ is incident to f_*.

The faces $h_i, 1 \leq i \leq p$ are called the holes of $\mathbf{s}$, and we let $\mathfrak{S}_{g,p}$ be the set of schemes in genus g with p holes.

Note that there is a slight difference between our definition and that of [6], which comes from a variation in the rooting conventions. By the properties of a scheme, any edge $\mathbf{e} \in E(\mathbf{s})$ can be given an orientation e that makes it incident to f_*, which is denoted by $e \in E(f_*)$. We let

$$E_\iota(\mathbf{s}) = \{\mathbf{e} = \{e, \bar{e}\} \in E(\mathbf{s}) : e, \bar{e} \in E(f_*)\}$$

be the set of edges that are incident twice to f_* (that is, whose two orientations are both incident to f_*) and

$$E_b(\mathbf{s}) = \{e \in E(f_*) : \bar{e} \in h_1 \cup \dots \cup h_p\}$$

be those edges whose reversal is incident to one of the holes. We call those edges internal and boundary edges accordingly. We canonically orient these edges in the following way. Let $e_0, e_1, \dots, e_{\deg(f_*)-1}$ be an enumeration of the oriented edges incident to f_*, starting from the root edge of $\mathbf{s}$ and in facial order around f_*, and we continue this sequence cyclically. For $\mathbf{e} \in E(\mathbf{s})$, we let i be the first index such that e_i is an orientation of $\mathbf{e}$, and canonically choose the orientation e_i of $\mathbf{e}$. In this way, the set $E(\mathbf{s}) = E_\iota(\mathbf{s}) \sqcup E_b(\mathbf{s})$ can be viewed as a set of oriented edges, and when we write $e \in E(\mathbf{s})$ we will assume by default that $\mathbf{e} = \{e, \bar{e}\}$ has this specified orientation.

Now with every $e \in E(\mathbf{s})$ we associate en elementary piece as follows:

- if $e \in E_\iota(\mathbf{s})$, we let $(\mathbf{q}_e, \rho_e, \bar{\rho}_e, x_e, \bar{x}_e)$ be a discrete quadrilateral with geodesic boundaries, we let $a_e, a_{\bar{e}}, w_e, \delta_e$ be its half-areas, width and discrepancy, and we let $(\mathbf{q}_{\bar{e}}, \rho_{\bar{e}}, \bar{\rho}_{\bar{e}}, x_{\bar{e}}, \bar{x}_{\bar{e}}) = (\mathbf{q}_e, \bar{\rho}_e, \rho_e, \bar{x}_e, x_e)$,
- if $e \in E_b(\mathbf{s})$ we let $(\mathbf{q}_e, \rho_e, \bar{\rho}_e, x_e)$ be a discrete slice, we let a_e, w_e, δ_e be its area, width and discrepancy,
- the discrepancies must be compatible in the sense that there exists a labeling function $\lambda : V(\mathbf{s}) \to \mathbb{Z}$ such that $\lambda_{e^+} - \lambda_{e^-} = \delta_e$ for every $e \in E(\mathbf{s})$.

The last requirement says that the function $\delta : E(\mathbf{s}) \to \mathbb{Z}$ is a discrete gradient of a function $\lambda : V(\mathbf{s}) \to \mathbb{Z}$: by connectedness of $\mathbf{s}$, such a function is not unique but only defined up to a global additive constant that we can fix by setting $\lambda_{e_0^-} = 0$.

We now proceed to glue together the elementary pieces $\mathbf{q}_e, e \in E(\mathbf{s})$, in the following way. We let γ_e, γ'_e be the geodesic boundary paths respectively from $\rho_e, \bar{\rho}_e$ to x_e in the elementary piece $\mathbf{q}_e$. We assign with every vertex z of γ_e the height $h(z) = d_{\mathbf{q}_e}(\rho_e, z) - \lambda_{e^-}$ and with every vertex z of γ'_e the height $h(z) = d_{\mathbf{q}_e}(\bar{\rho}_e, z) - \lambda_{e^+}$. For every $i \geq 0$ and z belonging to γ'_{e_i}, we let $j \geq i$ be the first index such that γ_{e_j} contains a vertex $\phi(z)$ with $h(\phi(z)) = h(z)$, and we identify the vertices z and $\phi(z)$. Yet otherwise said, the vertices z and z' respectively in γ'_{e_i} and γ_{e_j} are identified if and only if $h(z) = h(z')$ and all heights in $\gamma_{e_k}, \gamma'_{e_k}$

are less than or equal to this common value for k strictly between i and j in cyclic order. Note that $\phi(z)$ is indeed well-defined for every z in γ'_e, and it may belong to γ_e: for instance, if e is such that $h(x_e)$ is maximal, then $\phi(x_e) = x_e$.

Note in particular that in this identification, for every i, the vertices $\bar{\rho}_{e_i}$ and $\rho_{e_{i+1}}$ are identified together, since they have height $-\lambda_{e_i^+} = -\lambda_{e_{i+1}^-}$. Note that if z, z' are two consecutive vertices in γ'_e, with $h(z') = h(z)+1$, then $\phi(z)$ and $\phi(z')$ get identified to two consecutive vertices of some γ_{e_j} with the same heights, and the identification of z, z' with these images has the effect of creating a 2-gon, which we shrink to an edge.

After performing these identifications, we obtain an element of $\mathbf{Q}^{[g]}_{n,\boldsymbol{l}}$ where

$$n = \sum_{e\in E(f_*)} a_e, \qquad l_i = \sum_{e:\bar{e}\in h_i} w_e. \tag{2}$$

By [18, 6] this construction induces a bijective correspondence. This implies that by choosing elementary pieces $(\mathbf{s}, (\mathbf{q}_e, e \in E(\mathbf{s})))$ uniformly at random, subject to the constraints (2), and performing this construction, we obtain a uniformly distributed random element of $\mathbf{Q}^{[g]}_{n,\boldsymbol{l}}$. See Figure.

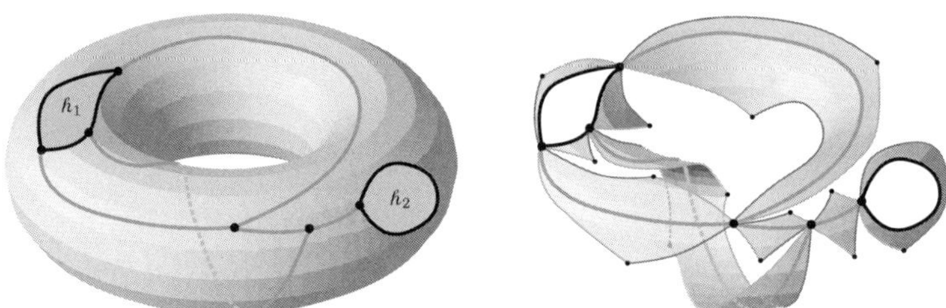

Fig. 1. An illustration of the decomposition into elementary pieces. On the left, a scheme on a two-holed torus. On the right, an assignment of elementary pieces to the edges of the scheme, and the start of the gluing procedure. In order not to overload the picture, the pieces are represented as triangles and quadrilaterals without showing the interior faces.

2.4. Scaling limits of the elementary pieces

Given this and Theorems 2 and 3, proving Theorem 1 amounts to understanding the limiting behavior of the quantities $\mathbf{s}, a_e, w_e, \delta_e$ as $n \to \infty$, when the elementary pieces are chosen uniformly. This was done in [6], again with a slightly different rooting convention which explains the difference in the results.

Let $\mathcal{L}_{\mathbf{s}}$ be the measure on $\mathbb{R}_+^{E(f_*)} \times \mathbb{R}_+^{E(\mathbf{s})} \times \mathbb{R}^{V(\mathbf{s})}$ that is formally written as

$$\prod_{e \in E(f_*)} \mathrm{d}a_e \delta_1\left(\sum_{e \in E(f_*)} a_e\right) \prod_{e \in E(\mathbf{s})} \mathrm{d}w_e \prod_{i=1}^{p} \delta_{L_i}\left(\sum_{\bar{e} \in E(h_i)} w_e\right) \prod_{v \in V(\mathbf{s})} \mathrm{d}\lambda_v \, \delta_0(\lambda_{v_0}),$$

where δ_x is the Dirac measure at x, and which is rigorously defined as follows. Fix an edge $e_{(i)}$ in each h_i. Then $\mathcal{L}_{\mathbf{s}}$ is the image of the measure

$$\prod_{e \in E(f_*), e \neq e_0} \mathrm{d}a_e \prod_{e \in E(\mathbf{s}) \setminus \{\bar{e}_{(1)}, \ldots, \bar{e}_{(p)}\}} \mathrm{d}w_e \prod_{v \in V(\mathbf{s}) \setminus \{e_0^-\}} \mathrm{d}\lambda_v$$
$$\times \mathbf{1}\left\{\sum_{e \in E(f_*), e \neq e_0} a_e \leq 1\right\} \prod_{i=1}^{p} \mathbf{1}\left\{\sum_{\bar{e} \in h_i \setminus \{e_{(i)}\}} w_e \leq L_i\right\}$$

by the mapping that associates with $(a_e, e \in E(f_*), e \neq e_0)$, $(w_e, e \in E(\mathbf{s}) \setminus \{e_{(1)}, \ldots, e_{(p)}\})$ and $(\lambda_v, v \in V(\mathbf{s}) \setminus \{e_0^-\})$ the vector $(a_e, e \in E(f_*)), (w_e, e \in E(\mathbf{s})), (\lambda_v, v \in V(\mathbf{s}))$ where

$$a_{e_0} = 1 - \sum_{e \in E(f_*), e \neq e_0} a_e$$
$$w_{e_{(i)}} = L_i - \sum_{\bar{e} \in E(h_i) \setminus e_{(i)}} w_e$$
$$\lambda_{e_0^-} = 0.$$

We leave to the reader to check that $\mathcal{L}_{\mathbf{s}}$ thus defined does not depend on the choice of $e_0 \in E(f_*), e_{(1)}, \ldots, e_{(p)}$. Finally, we let

$$p_t(x) = \frac{1}{\sqrt{2\pi t}} e^{-x^2/2t}, \qquad q_t(x) = \frac{x}{t} p_t(x),$$

and $\mathfrak{S}_{g,p}^3$ be the set of schemes of genus g with p holes, in which every vertex has degree exactly 3. It is easy to note, from Euler's formula, that for $\mathbf{s} \in \mathfrak{S}_{g,p}^3$, one has

$$|V(\mathbf{s})| = 4g + 2p - 2 \qquad \text{and} \qquad |E(\mathbf{s})| = 6g + 3p - 3.$$

We let $\mu_{g,\boldsymbol{L}}$ be the measure on the set of families of the form $(\mathbf{s},(a_e),(w_e),(\lambda_v))$ where $\mathbf{s}$ is a scheme, and $((a_e),(w_e),(\lambda_v)) \in \mathbb{R}_+^{E(f_*)} \times \mathbb{R}_+^{E(\mathbf{s})} \times \mathbb{R}^{V(\mathbf{s})}$, defined by

$$\mu_{g,\boldsymbol{L}}(\psi) = \frac{1}{\Upsilon_{g,\boldsymbol{L}}} \sum_{\mathbf{s}\in\mathfrak{S}_{g,p}^3} \int d\mathcal{L}_{\mathbf{s}}((a_e),(w_e),(\lambda_v))\psi(\mathbf{s},(a_e),(w_e),(\lambda_v)) \times \prod_{e\in E(f_*)} q_{a_e}(w_e) \prod_{e\in E(\mathbf{s})} p_{\kappa_e^2 w_e}(\lambda_{e^+}-\lambda_{e^-})$$

where κ_e is equal to 1 if $e \in E_\iota(\mathbf{s})$ and to $\sqrt{3}$ if $e \in E_b(\mathbf{s})$, and $\Upsilon_{g,\boldsymbol{L}}$ is the normalising constant that makes $\mu_{g,\boldsymbol{L}}$ a probability measure.

Proposition 4. *Let Q_n be a uniformly chosen random element of $\mathbf{Q}_{n,\boldsymbol{l}}^{[g]}$ and $(S_n,(Q_{e,n}, e \in E(S_n)))$ be the associated scheme and elementary pieces via the above correspondence. We consider that the elementary pieces are marked by their two or four geodesic boundary arcs, according to whether they are discrete slices or quadrilaterals. Then*

$$\left(S_n, \left(\frac{a_{e,n}}{n}\right)_{e\in E(f_*)}, \left(\frac{w_{e,n}}{\sqrt{2n}}\right)_{e\in E(S_n)}, \left(\left(\frac{9}{8n}\right)^{1/4} \lambda_{v,n}\right)_{v\in V(S_n)}\right)$$

converges in distribution to a random vector $(S,(a_e),(w_e),(\lambda_v))$ with law $\mu_{g,\boldsymbol{L}}$. Moreover, this convergence holds jointly with that of the spaces

$$\left(V(Q_{e,n}), \left(\frac{9}{8n}\right)^{1/4} d_{Q_{e,n}}, \frac{1}{n}\sum_{v\in V(Q_{e,n})} \delta_v\right) \xrightarrow[n\to\infty]{(d)} \mathsf{S}_e, \qquad e \in E(S),$$

in the marked Gromov-Hausdorff-Prokhorov topology, where, conditionally given $(S,(a_e),(w_e),(\lambda_v))$, the limiting spaces S_e are elementary pieces with same law as $\mathsf{Quad}_{a_e,a_{\bar{e}},w_e,\delta_e}$ if $e \in E_\iota(S)$ and as $\mathsf{Slice}_{a_e,w_e,\delta_e}$ if $e \in E_b(\mathbf{s})$. Here, we let $\delta_e = \lambda_{e^+} - \lambda_{e^-}$.

Note that there is a slight ambiguity in this statement: in the convergence of the elementary pieces, we let $e \in E(S)$ while we should have written $e \in E(S_n)$ for the left-hand side to make perfect sense. However, since S_n takes its values in the finite set $\mathfrak{S}_{g,p}^3$, the convergence in distribution of S_n means that the probability that S_n equals a given scheme $\mathbf{s}$ converges as $n \to \infty$, and we can work conditionally given this event.

The first part of the statement, about the convergence of the scheme and of the discrete areas, widths and discrepancies of the elementary pieces, is only a slight modification of Proposition 15 in [6]. The second part of the theorem is a direct consequence of the first part, together with Theorems 2 and 3.

2.5. Gluing elementary pieces and proof of Theorem 1

Theorem 1 is now a relatively simple consequence of Proposition 4 and of the discussion of Section 2.3 telling how the elementary pieces should be glued together to obtain an element of $\mathbf{Q}_{n,\boldsymbol{l}}^{[g]}$. While gluing operations for metric spaces can be quite delicate in general, they are much better behaved in the special case where the gluing is performed along geodesic paths, and in particular it commutes with marked Gromov-Hausdorff limits, as stated in the following lemma proved in [9].

Lemma 5. *Let* $(X_n, d_n, \mu_n), (X'_n, d'_n, \mu'_n)$ *be sequences of measured metric spaces marked with a family of geodesic segments* $\Gamma_{1,n}, \ldots, \Gamma_{k,n}$ *and* $\Gamma'_{1,n}, \ldots, \Gamma'_{k',n}$ *with disjoint relative interiors. We assume that these sequences converge in the marked Gromov-Hausdorff-Prokhorov topology to spaces* $(X, d, \mu), (X', d', \mu')$*, marked by sets* $\Gamma_1, \ldots, \Gamma_k$ *and* $(\Gamma'_1, \ldots, \Gamma'_{k'})$*. Then the latter are geodesic segments as well. We further assume that they have disjoint relative interiors. Assume that* $\mathrm{len}(\Gamma_{1,n}) = \mathrm{len}(\Gamma'_{1,n})$*, and consider the metric gluing* Y_n *of* $(X_n, d_n, \mu_n), (X'_n, d'_n, \mu'_n)$ *obtained by identifying the segments* $\Gamma_{1,n}$ *and* $\Gamma'_{1,n}$ *(with a prescribed orientation). We view* Y_n *as being marked by the remaining segments* $\Gamma_{2,n}, \ldots, \Gamma_{k,n}$*,* $\Gamma'_{2,n}, \ldots, \Gamma'_{k',n}$*. Then* Y_n *converges in the marked Gromov-Hausdorff-Prokhorov topology to the metric gluing of* $(X, d, \mu), (X', d', \mu')$ *obtained by* Γ_1 *and* Γ'_1*, and marked with the segments* $\Gamma_2, \ldots, \Gamma_k, \Gamma'_2, \ldots, \Gamma'_{k'}$*.*

A similar statement holds if we consider the gluing of a single space (X_n, d_n, μ_n) with marked geodesic segments $\Gamma_{1,n}, \ldots, \Gamma_{k,n}$ with $k \geq 2$, obtained by identifying the segments $\Gamma_{1,n}$ and $\Gamma_{2,n}$, assuming that they have the same length for every n, and that the segments $\Gamma_{3,n}, \ldots, \Gamma_{k,n}$ remain geodesic after performing this gluing.

§3. Elements of proof of convergence of the elementary pieces

In this section, we briefly explain the idea behind the proof of Theorems 2 and 3. This requires in particular to give a concrete description of the limiting spaces $\mathsf{Quad}_{A,\overline{A},L,\delta}$ and $\mathsf{Slice}_{A,L,\delta}$. To this end, it is useful to describe a general construction of metric spaces based on pairs of functions.

3.1. Pseudometrics and their quotients

For a given set E, a function $d : E \times E \to \mathbb{R}_+ \cup \{\infty\}$ is called a pseudometric if for every $x, y, z \in E$,

- $d(x, y) = d(y, x)$
- $d(x, z) \leq d(x, y) + d(y, z)$
- $d(x, x) = 0.$

We then say that (E, d) is a pseudometric space, and with it we can associate a true metric space (where distances are allowed to be infinite) by considering the equivalence relation $\{d = 0\}$ and endowing the quotient set $E/\{d = 0\}$ with the function induced by d, still denoted by d for convenience.

If (E, d) is a pseudometric space and R is an equivalence relation on E, we can define the quotient pseudometric d/R as being the maximal element of all pseudometrics d' such that $d' \leq d$ and $R \subset \{d' = 0\}$. It is easy to check that it is given by the formula

$$d/R(x,y) = \inf \left\{ \sum_{i=1}^{k} d(x_i, y_i) : \begin{array}{l} k \geq 0,\ x_1, \ldots, x_k, y_1, \ldots, y_k \in E, \\ y_i R x_{i+1}\ \forall i \in \{1, 2, \ldots, k-1\} \end{array} \right\}.$$

In particular, if r is another pseudometric on E, then $\{r = 0\}$ is an equivalence relation on E and it makes sense to define $d/\{r = 0\}$. This is the kind of construction that is used extensively below.

3.2. Pseudometrics from functions

Let I be a closed subinterval of $\mathbb{R}$, or a union of two such intervals. Consider a function $f \in \mathcal{C}(I, \mathbb{R})$ and define, for $s, t \in I$ with $s \leq t$,

$$m_f(s,t) = \inf_{[s,t] \cap I} f. \tag{3}$$

Then define

$$\begin{aligned} d_f(s,t) &= f(s) + f(t) - 2m_f(s \wedge t, s \vee t), \\ \bar{d}_f(s,t) &= d_f(s,t) \mathbf{1}_{\{st \geq 0\}} + \infty \mathbf{1}_{\{st < 0\}}. \end{aligned}$$

These three formulas define pseudometrics on I such that $d_f \leq \bar{d}_f$. The metric pseudometric d_f is classically interpreted [30] as the distance function on an $\mathbb{R}$-tree whose contour process if f, and $\bar{d}_f$ corresponds to the distance function in two $\mathbb{R}$-trees coded by the restrictions of f to the positive and negative axes. If now $f, g \in \mathcal{C}(I, \mathbb{R})$, we let

$$D_{f,g}(s,t) = d_g/\{d_f = 0\}, \quad \overline{D}_{f,g} = \bar{d}_g/\{d_f = 0\},$$

which correspond to the intuitive idea that the tree(s) coded by g is glued (mated) according to the tree coded by f.

3.3. Description of the continuum spaces

We start with the description of continuum quadrilaterals with geodesic boundaries. Fix $A, \overline{A}, L > 0$ and $\delta \in \mathbb{R}$. Let X be the process on

$I = [-\overline{A}, A]$ such that $(X_t, 0 \le t \le A)$ and $(X_{-t}, 0 \le t \le \overline{A})$ are independent first-passage bridges of standard Brownian motion, starting at 0 and hitting $-L$ for the first time at times $A, \overline{A}$ respectively. We let

$$\underline{X}_t = m_X(0 \wedge t, 0 \vee t), \qquad -\overline{A} \le t \le A.$$

By the Bismut decomposition of Brownian excursion, we can alternatively see X as a shifted Brownian excursion, namely as $(e(t+u) - e(t), -t \le u \le \sigma - t)$, where (e, t) is "distributed" under the measure $n(\mathrm{d}e)\mathbf{1}_{[0,\sigma(e)]}(t)\mathrm{d}t$, where n is Itô's excursion measure, and conditioned on $\{e(t) = L, \sigma - t = A, t = \overline{A}\}$.

Conditionally given X, we let Z be a Brownian snake-type process, namely, Z is a centered Gaussian process with covariance kernel specified by

$$\mathbb{E}\big[(Z_s - Z_t)^2 \mid X\big] = d_{X-\underline{X}}(s,t), \qquad s,t \in I. \tag{4}$$

Due to the fact that the trajectories of X are Hölder-continuous, it is easy and well-known that Z admits a continuous version, which is the one we choose to work with. Note that we are going to reuse this definition is several contexts, where the definitions of X and $\underline{X}$ are going to vary. Further, we let $(b(u), 0 \le u \le L)$ be a standard Brownian bridge with length L from 0 to δ, independent of (X, Z). Finally, let

$$W_s = Z_s + b(-\underline{X}_s), \qquad -\overline{A} \le s \le A. \tag{5}$$

Definition 3.1. *The space* $\mathsf{Quad}_{A,\overline{A},L,\delta}$ *is the isometry class of the quotient metric measure space* $([-\overline{A}, A]/\{\overline{D}_{X,W} = 0\}, \overline{D}_{X,W}, \lambda)$ *where* λ *is the image of the Lebesgue measure on* $[-\overline{A}, A]$ *by the canonical projection*

$$\overline{p} : [-\overline{A}, A] \to [-\overline{A}, A]/\{\overline{D}_{X,W} = 0\}.$$

The geodesic boundaries of $\mathsf{Quad}_{A,\overline{A},L,\delta}$ have a relatively simple description. Let $m = \inf_{[0,A]} W$ and $\overline{m} = \inf_{[-\overline{A},0]} W$. Let

$$\begin{aligned}
\Gamma(r) &= \inf\{s \ge 0 : W_s = -r\}, & 0 \le r \le -m \\
\Gamma'(r) &= \sup\{s \ge 0 : W_s = \delta - r\}, & 0 \le r \le \delta - m \\
\overline{\Gamma}(r) &= \inf\{s \le 0 : W_s = \delta - r\}, & 0 \le r \le \delta - \overline{m} \\
\overline{\Gamma}'(r) &= \sup\{s \le 0 : W_s = -r\}, & 0 \le r \le -\overline{m},
\end{aligned}$$

and let $\gamma, \gamma', \overline{\gamma}, \overline{\gamma}'$ be their images by the canonical projection $\overline{p}$. Note that $\gamma, \overline{\gamma}'$ both originate from $\rho = \overline{p}(0)$ and $\gamma', \overline{\gamma}$ both start from $\overline{\rho} = \overline{p}(A) = \overline{p}(-\overline{A})$. Moreover, γ, γ', resp. $\overline{\gamma}, \overline{\gamma}'$, end at the same point $x =$

$\gamma(-m) = \gamma'(\delta - m)$ and $\overline{x} = \overline{\gamma}(-\overline{m}) = \overline{\gamma}'(\delta - \overline{m})$. To see this, one should note that $\Gamma(-m)$ and $\Gamma'(\delta - m)$ are both projected to the point x, which comes from the fact that by definition, W attains its overall minimum m on $[0, A]$ at these two times, which implies easily that $\overline{d}_W(\Gamma(-m), \Gamma'(\delta - m)) = 0$. In fact, it even holds that a.s. $\Gamma(-m) = \Gamma'(\delta - m)$, although we will not need this.

The slices have a very similar, slightly simpler description. Fix again $A, L > 0$ and $\delta \in \mathbb{R}$. Let $(X_t, t \in [0, A])$ be a first-passage bridge of standard Brownian motion starting at 0 and hitting $-L$ for the first time at time A. We also let $\underline{X}_t = m_X(0, t) = \inf_{[0,t]} X$ for $0 \leq t \leq A$. Conditionally given X, let Z be a centered Gaussian process with covariance structure as in (4), let $(b(u), 0 \leq u \leq L)$ be a standard Brownian bridge from 0 to $\delta/\sqrt{3}$ and set

$$W_s = Z_s + \sqrt{3}\, b(-\underline{X}_s), \qquad 0 \leq s \leq A. \tag{6}$$

Note the different normalisation appearing in front of the bridge compared to the preceding situation.

Definition 3.2. *The space* $\mathsf{Slice}_{A,L,\delta}$ *is the isometry class of the quotient metric measure space* $([0, A]/D_{X,W} = 0\}, D_{X,W}, \lambda)$ *where* λ *is the image of the Lebesgue measure on* $[0, A]$ *by the canonical projection*

$$p : [0, A] \to [0, A]/\{D_{X,W} = 0\}.$$

Similarly to the above, the geodesic boundaries of $\mathsf{Slice}_{A,L,\delta}$ are described by letting $m = \inf_{[0,A]} W$,

$$\Gamma(r) = \inf\{s \geq 0 : W_s = -r\}, \qquad 0 \leq r \leq -m$$
$$\Gamma'(r) = \sup\{s \geq 0 : W_s = \delta - r\}, \qquad 0 \leq r \leq \delta - m$$

and letting γ, γ' be their images by the canonical projection p, so that γ, γ' originate from $\rho = p(0)$ and $\overline{\rho} = p(A)$, end at the same point $x = \gamma(-m) = \gamma'(\delta - m)$. The last, non-geodesic part of the boundary of $\mathsf{Slice}_{A,L,\delta}$ is the path from ρ to $\overline{\rho}$ given by the image by p of the path

$$\partial(x) = \inf\{t \geq 0 : X_t = -x\}, \qquad 0 \leq x \leq L.$$

3.4. Connection to non-compact Brownian surfaces

It is useful to view the spaces $\mathsf{Quad}_{A,\overline{A},L,\delta}$ and $\mathsf{Slice}_{A,L,\delta}$, or free versions of these spaces where $A, \overline{A}, \delta$ are randomized appropriately, as subspaces of the non-compact Brownian surfaces considered in [19, 26, 4].

The Brownian plane introduced in [19] and further studied in [20] can be described in the following way. Let $(X_t, t \in \mathbb{R})$ be a two-sided

Brownian motion, so that $(X_t, t \geq 0)$ and $(X_{-t}, t \geq 0)$ are independent standard Brownian motions. Define $\underline{X}_t = m_X(0 \wedge t, 0 \vee t)$ for $t \in \mathbb{R}$. Let W be the snake driven by $X - \underline{X}$, as defined around (4) for $I = \mathbb{R}$.

Definition 3.3. *The Brownian plane* BPlane *is the isometry class of the metric measure space* $(\mathbb{R}/\{D_{X,W} = 0\}, D_{X,W}, \lambda)$ *where* λ *is the image measure of Lebesgue measure on* $\mathbb{R}$ *by the canonical projection* $p : \mathbb{R} \to \mathbb{R}/\{D_{X,W} = 0\}$.

The two equivalent definitions of the Brownian plane in [19, 20] are yet a bit different from the one given here, but the equivalence between these two definitions is an easy exercise in stochastic processes. It is also useful to consider a modified version of the Brownian plane, which can be seen as the plane "cut" along the set γ of $x \in \mathsf{BPlane}$ that can be written as $x = p(s) = p(t)$ for some $s < 0$ and $t > 0$ such that $d_W(s,t) = 0$. This set is a geodesic ray starting from $p(0)$, and this cutting operation simply amounts to changing d_W into $\bar{d}_W$, so that this cut Brownian plane is the isometry class of $(\mathbb{R}/\{\overline{D}_{X,W} = 0\}, \overline{D}_{X,W}, \lambda)$ (where the definition of λ is adapted in an obvious way). This explains the choice of $\overline{D}$-distances below.

For $L \geq 0$, let $T_L = \inf\{t \geq 0 : X_t = -L\}$ and $\overline{T}_L = \sup\{t \leq 0 : X_t = -L\}$. For $0 \leq L < L'$, we consider the restrictions $d_X^{(L,L')}$ and $\bar{d}_W^{(L,L')}$ of the pseudodistances d_X and $\bar{d}_W$ to $[\overline{T}_{L'}, \overline{T}_L] \cup [T_L, T_{L'}]$. It is not difficult to see that these distances are equivalently the metrics associated with the restriction of the functions X, W to $I_{L,L'} = [\overline{T}_{L'}, \overline{T}_L] \cup [T_L, T_{L'}]$ as in Section 3.2, and that the following proposition holds.

Proposition 6. *For every* $0 \leq L < L'$, *the isometry class* $\mathsf{Quad}^{(L,L')}$ *of the metric measure space* $(I_{L,L'}/\{\overline{D}_{X,W}^{(L,L')} = 0\}, \overline{D}_{X,W}^{(L,L')}, \lambda^{(L,L')})$ *has same distribution as* $\mathsf{FQuad}_{L'-L}$.

This statement says that spaces with same distribution as $\mathsf{Quad}_{\mathcal{A},\overline{\mathcal{A}},L\Delta}$ are found within the Brownian plane. We can be more precise than this by noticing that the spaces $\mathsf{Quad}^{(L,L')}$ satisfy a semigroup property for a simple gluing operation. First, we define

$$m_{(L,L')} = \inf_{[T_L, T_{L'}]} (W - W_{T_L}), \qquad \overline{m}_{(L,L')} = \inf_{[\overline{T}_{L'}, \overline{T}_L]} (W - W_{\overline{T}_L})$$

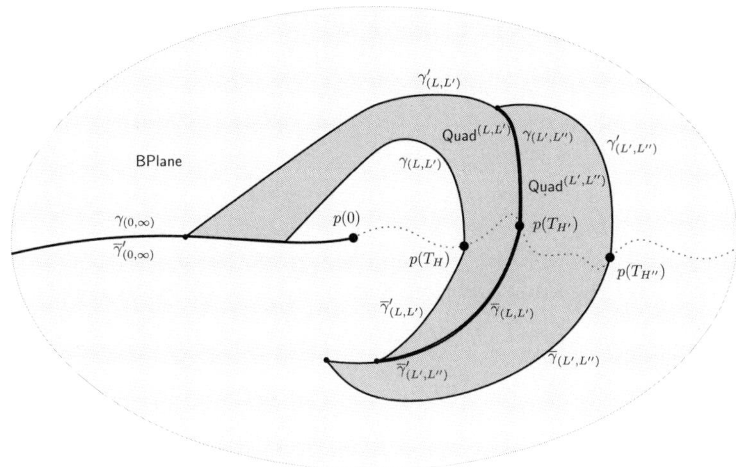

Fig. 2. An illustration of some of the quadrangles with geodesic boundaries sitting in the Brownian plane, as formalized in Proposition 7. The infinite geodesic ray γ is represented on the left, with its two "halves" $\gamma_{(0,\infty)}$ and $\overline{\gamma}'_{(0,\infty)}$ after cutting along it.

and let

$$\begin{aligned}
\Gamma_{(L,L')}(r) &= \inf\{s \geq T_L : W_s - W_{T_L} = -r\}, \quad 0 \leq r \leq W_{T_L} - m_{(L,L')} \\
\Gamma'_{(L,L')}(r) &= \sup\{s \leq T_{L'} : W_s - W_{T_{L'}} = -r\}, \ 0 \leq r \leq W_{T_{L'}} - m_{(L,L')} \\
\overline{\Gamma}_{(L,L')}(r) &= \inf\{s \geq \overline{T}_{L'} : W_s - W_{\overline{T}_{L'}} = -r\}, \ 0 \leq r \leq W_{\overline{T}_{L'}} - \overline{m}_{(L,L')} \\
\overline{\Gamma}'_{(L,L')}(r) &= \sup\{s \leq \overline{T}_L : W_s - W_{\overline{T}_L} = -r\}, \ 0 \leq r \leq W_{\overline{T}_L} - \overline{m}_{(L,L')},
\end{aligned}$$

as well as their projections $\gamma_{(L,L')}, \gamma'_{(L,L')}, \overline{\gamma}_{(L,L')}, \overline{\gamma}'_{(L,L')}$ in $\mathsf{Quad}^{(L,L')}$, which are geodesic segments.

Proposition 7. *For every $0 \leq L \leq L' \leq L''$, the space $\mathsf{Quad}^{(L,L'')}$ is equal to the metric gluing of $\mathsf{Quad}^{(L,L')}$ and $\mathsf{Quad}^{(L',L'')}$ by identifying the initial segments of $\gamma'_{(L,L')}$ and $\gamma_{(L',L'')}$ of length $\mathrm{len}(\gamma'_{(L,L')}) \wedge \mathrm{len}(\gamma_{(L',L'')})$, and the initial segments of $\overline{\gamma}'_{(L,L')}$ and $\overline{\gamma}_{(L',L'')}$ of length $\mathrm{len}(\overline{\gamma}'_{(L,L')}) \wedge \mathrm{len}(\overline{\gamma}_{(L',L'')})$.*

Propositions 6 and 7 together, as well as an application of the Markov property, entails the following.

Theorem 8 (Semigroup property of FQuad). *the gluing of independent random metric spaces of laws* FQuad_L *(with boundary points* $\rho, x, \overline{\rho}, \overline{x}$*) and* $\mathsf{FQuad}_{L'}$ *(with boundary points* $\rho', x', \overline{\rho}', \overline{x}'$*) by identifying the longest possible initial segments of* $\overline{\rho}x$ *with* $\rho' x'$ *and* $\overline{\rho}\overline{x}$ *with* $\rho'\overline{x}'$ *has law* $\mathsf{FQuad}_{L+L'}$.

As before, there is a version of these results for Brownian surfaces with a boundary. We start with the definition of the Brownian half-plane. Let $(X_t, t \in \mathbb{R})$ be a random process such that $(X_t, t \geq 0)$ is a standard Brownian motion and $(X_{-t}, t \geq 0)$ is an independent 3-dimensional Bessel process. Define $\underline{X}_t = \inf_{s \leq t} X_s$, where we stress that the infimum is over the interval $(-\infty, t]$, which makes the definition different from the previous ones.

Let $(b_x, x \in \mathbb{R})$ be an independent two-sided standard Brownian motion. Let Z be a snake process based on $X - \underline{X}$ as in (4) and set

$$W_t = Z_t + \sqrt{3}b(-\underline{X}_t), \qquad t \in \mathbb{R}.$$

Definition 3.4. *The Brownian half-plane* $\mathsf{BHPlane}$ *is the isometry class of the metric measure space* $(\mathbb{R}/\{D_{X,W} = 0\}, D_{X,W}, \lambda)$ *where* λ *is the image measure of Lebesgue measure on* $\mathbb{R}$ *by the canonical projection* $p : \mathbb{R} \to \mathbb{R}/\{D_{X,W} = 0\}$.

For $L \in \mathbb{R}$, let $T_L = \inf\{t \in \mathbb{R} : X_t = -L\}$. For $L < L'$, we consider the restrictions $d_X^{(L,L')}$ and $d_W^{(L,L')}$ of the pseudometrics d_X and d_W to $[T_L, T_{L'}]$. It is not difficult to see that these distances are equivalently the metrics associated with the restriction of the functions X, W to $I_{L,L'} = [T_L, T_{L'}]$ as in Section 3.2, and that the following proposition holds.

Proposition 9. *For every* $L < L'$*, the isometry class* $\mathsf{Slice}^{(L,L')}$ *of the metric measure space* $(I_{L,L'}/\{D_{X,W}^{(L,L')} = 0\}, D_{X,W}^{(L,L')}, \lambda^{(L,L')})$ *has same distribution as* $\mathsf{FSlice}_{L'-L}$.

This statement says that spaces with same distribution as $\mathsf{Slice}_{A,,L\Delta}$ are found within the Brownian half-plane. We can be more precise than this by noticing that the spaces $\mathsf{Slice}^{(L,L')}$ satisfy a semigroup property for a simple gluing operation. First, we define

$$m_{(L,L')} = \inf_{[T_L, T_{L'}]} (W - W_{T_L}),$$

and let

$$\Gamma_{(L,L')}(r) = \inf\{s \geq T_L : W_s - W_{T_L} = -r\}, \qquad 0 \leq r \leq W_{T_L} - m_{(L,L')}$$

$$\Gamma'_{(L,L')}(r) = \sup\{s \leq T_{L'} : W_s - W_{T_{L'}} = -r\}, \quad 0 \leq r \leq W_{T_{L'}} - m_{(L,L')}$$

as well as their projections $\gamma_{(L,L')}, \gamma'_{(L,L')}$ in $\mathsf{Slice}^{(L,L')}$, which are geodesic segments.

Proposition 10. *For every $L \leq L' \leq L''$, the space $\mathsf{Slice}^{(L,L'')}$ is equal to the metric gluing of $\mathsf{Slice}^{(L,L')}$ and $\mathsf{Slice}^{(L',L'')}$ by identifying the initial segments of $\gamma'_{(L,L')}$ and $\gamma_{(L',L'')}$ of length $\mathrm{len}(\gamma'_{(L,L')}) \wedge \mathrm{len}(\gamma_{(L',L'')})$.*

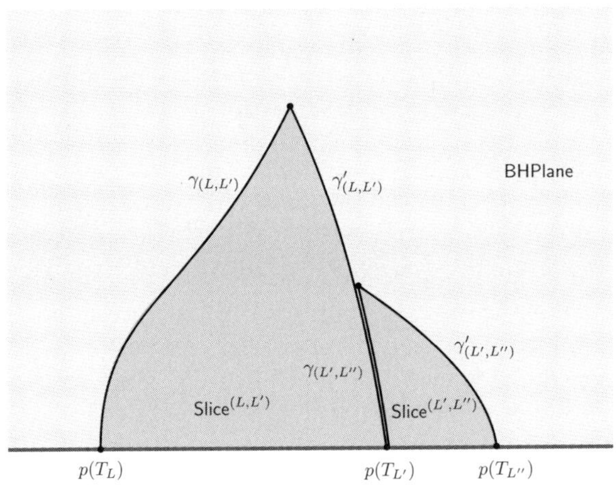

Fig. 3. An illustration from [9] of some of the slices sitting in the Brownian half-plane.

We deduce, similarly to the above, the following.

Theorem 11 (Semigroup property of FSlice). *the gluing of independent random metric spaces of laws FSlice_L (with boundary points $\rho, x, \overline{\rho}$) and $\mathsf{FSlice}_{L'}$ (with boundary points $\rho', x', \overline{\rho}'$) by identifying the longest possible initial segments of $\overline{\rho}x$ with $\rho'x'$ has law $\mathsf{FSlice}_{L+L'}$.*

3.5. Proof of Theorems 2 and 3

We now explain how these descriptions allow to prove the convergence results for discrete elementary pieces. To this end, we use the convergence of two infinite models of random quadrangulations, namely the Uniform Infinite Planar Quadrangulation (UIPQ) and the Uniform Infinite Half-Planar Quadrangulation (UIHPQ) respectively to the Brownian plane and the Brownian half-plane described above. These two random map models can also be described as gluings of discrete quadrilaterals and slices with geodesic boundaries (with a free area) along geodesics in a way that is exactly parallel to the results of Propositions 7 and 10. More precisely, for a given width $l > 0$ we consider a random quadrilateral FQ_l and a random slice FS_l with geodesic boundaries, that are equal to $\mathbf{q}$ (a given quadrilateral or slice with width l) with probability proportional to $(1/12)^{F_{\mathrm{int}}(\mathbf{q})}$, where $F_{\mathrm{int}}(\mathbf{q})$ is the number of (interior) quadrangles of $\mathbf{q}$. Then FQ_l and FS_l can be interpreted as pieces of the UIPQ and the UIHPQ cut along geodesic segments, and after some work this entails the following convergence results:

$$\left[V(\mathrm{FQ}_l), \sqrt{\frac{3}{2l}} d_{\mathrm{FQ}_l}, \frac{2}{l^2} \sum_{v \in V(\mathrm{FQ}_l)} \delta_v\right] \xrightarrow[w \to \infty]{(d)} \mathsf{FQuad}_1$$

$$\left[V(\mathrm{FS}_l), \sqrt{\frac{3}{2l}} d_{\mathrm{FS}_l}, \frac{2}{l^2} \sum_{v \in V(\mathrm{FS}_l)} \delta_v\right] \xrightarrow[l \to \infty]{(d)} \mathsf{FSlice}_1.$$

Theorems 2 and 3 can be seen as conditioned statements of the two convergences above, given the total area (or the areas on both sides of the ridge in the case of quadrilaterals) and discrepancy. The proof of such conditioned statements is obtained by using the semigroup property of FQuad and FSlice explained in the previous section, together with an analogous property of their discrete counterparts, in much the same way as one proves that bridges of random walks converge to Brownian bridges.

§4. An alternative construction of the cylinder

It looks like a very difficult task to use the construction of Brownian surfaces arising in Theorem 1 to obtain any explicit information about their distribution — an example that immediately comes to mind is the 2-point function for instance, that is the law of the distance of two uniformly chosen random points. See Guitter [25] for results on the torus $(g, p) = (1, 0)$. Here we rely on combinatorial results of [5] and [13] (see also [10] for a detailed and pedagogical account of this) allowing us to give an alternative construction of the Brownian cylinders, that are the surfaces $\mathsf{S}_{\boldsymbol{L}}^{[g]}$ with $g = 0$ and $p = 2$. So we will focus on quadrangulations of the cylinder, that are quadrangulations in genus 0 with two holes, and in this section we will assume that these objects are rooted (that is, they carry a distinguished oriented edge) in each hole. This slightly different rooting convention is minor, so we keep the same notation $\mathbf{Q}_{n,(l_1,l_2)}^{[0]}$ to denote such objects.

Fix two integers l_1, l_2 and consider two slices $(\mathbf{q}_1, \rho_1, \bar{\rho}_1, x_1)$ and $(\mathbf{q}_2, \rho_2, \bar{\rho}_2, x_2)$ with a_1 and a_2 quadrangles, with widths $2l_1$ and $2l_2$, and discrepancies $2q$ and $-2q$ respectively. We now identify the (geodesic) boundary arc $\rho_1 x_1$ in $\mathbf{q}_1$ with the initial segment of the (geodesic) boundary arc $\bar{\rho}_1 x_1$ with same length: that is, if $e_1, e_2, \ldots, e_k$ are the consecutive edges of $\rho_1 x_1$, with $k = d_{\mathbf{q}_1}(\rho_1, x_1)$ and $e'_1, \ldots, e'_k, e'_{k+1}, \ldots, e'_{k+2q}$ are the consecutive edges of the arc $\bar{\rho}_1 x_1$, we identify e_i with e'_i for every $1 \leq i \leq k$. This results in a quadrangulation $\mathbf{q}'_1$ with two holes: one of degree $2l_1$, which comes from the boundary arc $\rho_1\bar{\rho}_1$ since its extremities have been identified, and one of degree $2q$ that comes from the arc $e'_{k+1}, \ldots, e'_{k+2q}$, whose extremities are $(e'_k)^-$ and x_1 and are therefore identified. Note that the hole of perimeter $2q$ has a simple boundary, which need not be the case of the hole of perimeter $2l_1$, and, as explained in [10, Chapter 2], the $2q$ is also the minimal length of a cycle separating the two holes of $\mathbf{q}'_1$. Finally, this "minimal" boundary is naturally rooted at the first edge on the boundary arc $\rho_1\bar{\rho}_1$.

Similarly, we identify the boundary arc $\bar{\rho}_2 x_2$ with the initial segment of $\rho_2 x_2$ of same length, resulting in a quadrangulation $\mathbf{q}'_2$ with one hole of perimeter $2l_2$ and one hole of perimeter $2q$ with simple boundary, that is also of minimal length among all cycles separating the two holes of $\mathbf{q}'_2$. Now we glue $\mathbf{q}'_1$ and $\mathbf{q}'_2$ together by identifying their two holes of perimeters $2q$ in such a way that x_1 is glued to the vertex of $\rho_2 x_2$ at distance k from x_2, for some $0 \leq k < 2q$, while preserving the orientation of the two triangles (say that the boundary of both triangles is oriented clockwise). This results in an element of $\mathbf{Q}^{[0]}_{n,(l_1,l_2)}$, where $n = a_1 + a_2$, which is also naturally rooted at the two holes at the first edges of the boundary arcs $\rho_1\bar{\rho}_1$ and $\rho_2\bar{\rho}_2$ of the slices we started from. By the minimality property of the boundaries of length $2q$, this map has the further property that $2q$ is the minimal length of a cycle c separating the two holes f_1, f_2 of $\mathbf{q}$, and a_1 is the minimal number of quadrangles in the part of $\mathbf{q} \setminus c$ containing the first hole, over all cycles c with length $2q$ separating the two holes. We let $m(\mathbf{q}) = 2q$ and $a_1(\mathbf{q}) = a_1$.

A result of Bouttier and Guitter [13] shows that this construction is in fact a one-to-one correspondence.

Proposition 12. *For fixed l_1, l_2, a_1, a_2, q with $a_1 + a_2 = n$, the above construction yields a one-to-one correspondence between, on the one hand triples $(\mathbf{q}_1, \mathbf{q}_2, k)$ where $\mathbf{q}_1$ and $\mathbf{q}_2$ are slices with respectively a_1 and a_2 quadrangles, widths $2l_1$ and $2l_2$ and discrepancies $2q$ and $-2q$, and $0 \leq k < 2q$, and on the other hand cylinder quadrangulations $\mathbf{q} \in \mathbf{Q}^{[0]}_{n,(l_1,l_2)}$ (rooted at the two boundaries), such that $m(\mathbf{q}) = 2q$ and $a_1(\mathbf{q}) = a_1$.*

From this and the known enumeration formulas for slices that can be found for instance in Chapter 2 of [10], the following scaling limit result is derived in [14]. We recall that the probability densities for a stable distribution of exponent 1/2 and parameter L, and a Rayleigh distribution of parameter $L > 0$, are given by

$$q_L(x) = \frac{L}{\sqrt{2\pi x^3}} e^{-L^2/2x} \mathbf{1}_{\{x>0\}}, \qquad r_L(x) = \frac{x}{L} e^{-x^2/2L} \mathbf{1}_{\{x>0\}}.$$

Theorem 13. *Let (L_1, L_2) be two fixed positive real numbers and Q^n be a uniformly chosen random element of $\mathbf{Q}^{[0]}_{n,\boldsymbol{l}}$, where $\boldsymbol{l} = \boldsymbol{l}(n) = (l_1(n), l_2(n))$ satisfies $l_i(n)/\sqrt{2n} \to L_i$ for $i \in \{1,2\}$ as $n \to \infty$. Then we have the convergence in distribution*

$$\left(\left(\frac{9}{8n}\right)^{1/4} m(Q^n), \frac{a_1(Q^n)}{n} \right) \longrightarrow (\mathcal{R}, \mathcal{A}) \tag{7}$$

where $\mathcal{R}$ and $\mathcal{A}$ are independent random variables, $\mathcal{R}$ follows a Rayleigh distribution of parameter $L_{\text{eff}} = 3(L_1^{-1} + L_2^{-1})^{-1}$ and $\mathcal{A}$ has density $q_{L_1}(a) q_{L_2}(1-a)/q_{L_1+L_2}(1)$ for $a \in (0,1)$.

From this statement and a passage to the limit, using the fact that the gluing of the two slices is performed along a geodesic cycle, we obtain that the Brownian cylinder $\mathsf{S}^{[0]}_{(L_1,L_2)}$ can be obtained as follows.

- Let $\mathcal{R}$ be a random variable with Rayleigh distribution of parameter L_{eff}, and work conditionally given $\mathcal{R}$.
- Let $(\mathsf{S}_1, \rho_1, \overline{\rho}_1, x_1)$ and $(\mathsf{S}_2, \rho_2, \overline{\rho}_2, x_2)$ be two free slices with widths L_1 and L_2 and discrepancies $\mathcal{R}$ and $-\mathcal{R}$ respectively, conditioned on the fact that the sum of their areas is 1.
- Glue the boundary arc $\rho_1 x_1$ with the initial segment of $\overline{\rho}_1 x_1$ of same length, and the boundary arc $\overline{\rho}_2 x_2$ with the initial segment of the boundary arc $\rho_2 x_2$ of same length. This has the effect of turning the remaining segments of $\overline{\rho}_1 x_1$ and $\rho_2 x_2$, into cycles C_1 and C_2 of equal length $\mathcal{R}$.
- Let K be a random variable uniform in $[0, \mathcal{R}]$, independent of the previous ones. Glue isometrically the cycles C_1 and C_2 in reversed orientation, in such a way that x_1 is identified with the point of $\rho_2 x_2$ at distance K from x_2.

From the construction, it is a simple exercise to check that the glued cycles induce a cycle in the Brownian cylinder, with a length $\mathcal{R}$ that is minimal among all cycles that separate its two boundaries.

Theorem 14. *Let $L_1, L_2 > 0$ be fixed. Consider the Brownian cylinder $\mathsf{S}^{[0]}_{(L_1,L_2)}$ with boundary sizes L_1, L_2. Then*

(1) *almost surely, there exists a unique non-contractible cycle* γ *with minimal length in* $\mathsf{S}^{[0]}_{(L_1,L_2)}$, *and the length* $\mathrm{len}(\gamma)$ *follows a Rayleigh distribution with parameter* $L_{\mathrm{eff}} = 3L_1L_2/(L_1 + L_2)$

(2) *let* $A_i, i \in \{1,2\}$ *be the area measure of the connected component of* $\mathsf{S}^{[0]}_{(L_1,L_2)} \setminus \mathrm{Im}\,\gamma$ *containing the boundary with length* L_i. *Then* (A_1, A_2) *is independent of* $\mathrm{len}(\gamma)$ *and the law of* A_1 *is* $q_t(L_i)q_{1-t}(L_2)dt\mathbf{1}_{[0,1]}(t)/q_1(L_1 + L_2)$.

The result about uniqueness of the cycle γ comes from the fact that the two geodesic boundaries of the free slices used in the construction are unique, a fact that they inherit from analogous properties of geodesics of the Brownian plane and half-plane. The second result is then a straightforward application of the alternative construction of $\mathsf{S}^{[0]}_{L_1+L_2}$.

What might look surprising in this result is that the law of the shortest non-contractible cycle is independent of the area of the two parts it separates. In fact we can also construct the cylinder of area A by the same method, where this time A_1 follows the probability distribution $q_t(L_i)q_{A-t}(L_2)dt\mathbf{1}_{[0,A]}(t)/q_A(L_1+L_2)$, but $\mathrm{len}(\gamma)$ still follows a Rayleigh distribution with parameter L_{eff}, independent of A. Sending A to 0, we intuitively obtain a planar unicyclic map with two faces of perimeters L_1, L_2. We leave to the reader to check directly in this model that the length of the unique cycle indeed follows a Rayleigh distribution with parameter L_{eff}.

§5. Perspectives

The study of random surfaces obtained by scaling limits of random planar maps is an extremely rich topic, which has developed at a tremendous pace over the past few years. We simply mention some questions in directions related to Brownian surfaces.

One important aspect that we have not touched upon in this article is the question of *universality*, that is, whether large classes of models of random maps converge to the limits considered in this article. Since such universality properties are already known in the cases of the sphere and the disk (see for instance [29, 8, 7, 1, 2, 27, 3]), there is little doubt that the methods presented here can generalize to more general settings, although this requires some specific work.

Second, it would be interesting to understand other statistics for the metric properties of the limiting spaces $\mathsf{S}^{[g]}_{\boldsymbol{L}}$, for instance by extending the results obtained for the Brownian cylinder in the previous section. However, it seems that the recourse to new alternative constructions of known objects to prove such results is a hardly avoidable aspect of the study of random maps, see for instance [15, 20, 22, 23] for similar situations. A further step in this direction is performed in the paper [14] which focuses on the bijective study of *pairs of pants*, that is surfaces of genus zero with three boundaries. Perhaps surprisingly, it is shown in this paper that Rayleigh statistics also arise naturally at the scaling limit when one considers the minimal length of separating cycles in pairs of pants. In this general line of research, let us also mention an intriguing conjecture by Chapuy on the statistics of the Voronoi cells in Brownian surfaces without boundary [16].

It is natural to ask whether similar results to the above can be obtained in non-orientable case. We believe this to be the case, and in this direction, the bijections by Chapuy and Dołega [17] might be the relevant tools.

Finally, it is natural to believe that the limiting spaces $\mathsf{S}_{\boldsymbol{L}}^{[g]}$ can also be constructed as random distance functions on a Riemann surface of a given topology, by using appropriately defined Gaussian free field-like random fields as was done by Miller and Sheffield in the planar case [34, 32, 33]. Section 1.1 in the latter reference contains in particular a discussion on surfaces in higher genera. The papers [21, 24] make precise conjectures for the fields that should be used to define the distance function for the Brownian surfaces (at least without boundary). The question of convergence of the conformal structure induced by the discrete map to the one associated with the continuum ones, generalizing the recent huge program by Holden, Sun and coauthors (see [28] and the references therein) is also a natural and challenging open line of research.

Acknowledgements. Thanks are due to an anonymous referee for a thorough reading and useful suggestions.

References

[1] Céline Abraham, Rescaled bipartite planar maps converge to the Brownian map, *Ann. Inst. Henri Poincaré Probab. Stat.*, 52(2):575–595, 2016.

[2] Louigi Addario-Berry and Marie Albenque, The scaling limit of random simple triangulations and random simple quadrangulations, *Ann. Probab.*, 45(5):2767–2825, 2017.

[3] Marie Albenque, Nina Holden and Xin Sun, Scaling limit of triangulations of polygons, *Electron. J. Probab.*, 25:paper no. 135, 43, 2020.

[4] Erich Baur, Grégory Miermont and Gourab Ray, Classification of scaling limits of uniform quadrangulations with a boundary, *Ann. Probab.*, 47(6):3397–3477, 2019.

[5] Olivier Bernardi and Éric Fusy, Unified bijections for maps with prescribed degrees and girth, *J. Combin. Theory Ser. A*, 119(6):1351–1387, 2012.

[6] Jérémie Bettinelli, Scaling limit of random planar quadrangulations with a boundary, *Ann. Inst. Henri Poincaré Probab. Stat.*, 51(2):432–477, 2015.

[7] Jérémie Bettinelli, Emmanuel Jacob and Grégory Miermont, The scaling limit of uniform random plane maps, *via* the Ambjørn-Budd bijection, *Electron. J. Probab.*, 19:no. 74, 16, 2014.

[8] Jérémie Bettinelli and Grégory Miermont, Compact Brownian surfaces I: Brownian disks, *Probab. Theory Related Fields*, 167(3-4):555–614, 2017.

[9] Jérémie Bettinelli and Grégory Miermont, Compact Brownian surfaces II: Orientable surfaces, in preparation, 2020+.

[10] Jérémie Bouttier, *Cartes planaires et partitions aléatoires*, Université Paris-Sud, 2019, Mémoire d'habilitation à diriger des recherches.

[11] Jérémie Bouttier and Emmanuel Guitter, Statistics in geodesics in large quadrangulations, *J. Phys. A*, 41(14):145001, 30, 2008.

[12] Jérémie Bouttier and Emmanuel Guitter, Planar maps and continued fractions, *Comm. Math. Phys.*, 309(3):623–662, 2012.

[13] Jérémie Bouttier and Emmanuel Guitter, On irreducible maps and slices, *Combin. Probab. Comput.*, 23(6):914–972, 2014.

[14] Jérémie Bouttier, Emmanuel Guitter and Grégory Miermont, Bijective enumeration of planar bipartite maps with three tight boundaries, or how to slice pairs of pants, in preparation, 2020+.

[15] Alessandra Caraceni and Nicolas Curien, Geometry of the uniform infinite half-planar quadrangulation, *Random Structures Algorithms*, 52(3):454–494, 2018.

[16] Guillaume Chapuy, On tessellations of random maps and the t_g-recurrence, *Sém. Lothar. Combin.*, 78B:Art. 79, 12, 2017.

[17] Guillaume Chapuy and Maciej Dołega, A bijection for rooted maps on general surfaces, *J. Combin. Theory Ser. A*, 145:252–307, 2017.

[18] Guillaume Chapuy, Michel Marcus and Gilles Schaeffer, A bijection for rooted maps on orientable surfaces, *SIAM J. Discrete Math.*, 23(3):1587–1611, 2009.

[19] Nicolas Curien and Jean-François Le Gall, The Brownian plane, *J. Theoret. Probab.*, 27(4):1249–1291, 2014.

[20] Nicolas Curien and Jean-François Le Gall, The hull process of the Brownian plane, *Probab. Theory Related Fields*, 166(1-2):187–231, 2016.

[21] François David, Rémi Rhodes and Vincent Vargas, Liouville quantum gravity on complex tori, *J. Math. Phys.*, 57(2):022302, 25, 2016.

[22] Jean-François Le Gall, The Brownian disk viewed from a boundary point, 2020.

[23] Jean-François Le Gall and Armand Riera, Spine representations for non-compact models of random geometry, 2020.

[24] Colin Guillarmou, Rémi Rhodes and Vincent Vargas, Polyakov's formulation of $2d$ bosonic string theory, *Publ. Math. Inst. Hautes Études Sci.*, 130:111–185, 2019.

[25] Emmanuel Guitter, Distance statistics in large toroidal maps, *J. Stat. Mech. Theory Exp.*, 2010(04):P04018, April 2010, Publisher: IOP Publishing.

[26] Ewain Gwynne and Jason Miller, Scaling limit of the uniform infinite half-plane quadrangulation in the Gromov-Hausdorff-Prokhorov-uniform topology, *Electron. J. Probab.*, 22:paper no. 84, 47, 2017.

[27] Ewain Gwynne and Jason Miller, Convergence of the free Boltzmann quadrangulation with simple boundary to the Brownian disk, *Ann. Inst. Henri Poincaré Probab. Stat.*, 55(1):551–589, 2019.

[28] Nina Holden and Xin Sun, Convergence of uniform triangulations under the cardy embedding, 2020.

[29] Jean-François Le Gall, Uniqueness and universality of the Brownian map, *Ann. Probab.*, 41(4):2880–2960, 2013.

[30] Jean-François Le Gall, Random trees and applications, *Probab. Surv.*, 2:245–311 (electronic), 2005.

[31] Grégory Miermont, The Brownian map is the scaling limit of uniform random plane quadrangulations, *Acta Math.*, 210(2):319–401, 2013.

[32] Jason Miller and Scott Sheffield, Liouville quantum gravity and the Brownian map II: geodesics and continuity of the embedding, 2019.

[33] Jason Miller and Scott Sheffield, An axiomatic characterization of the brownian map, 2020, to appear in Journal de l'École Polytechnique.

[34] Jason Miller and Scott Sheffield, Liouville quantum gravity and the Brownian map I: the QLE$(8/3, 0)$ metric, *Invent. Math.*, 219(1):75–152, 2020.

[35] Cédric Villani, *Optimal transport, old and New*, volume 338 of *Grundlehren der Mathematischen Wissenschaften* [*Fundamental Principles of Mathematical Sciences*], Springer-Verlag, Berlin, 2009.

Unité de Mathématiques Pures et Appliquées
École Normale Supérieure de Lyon
46 allée d'Italie,
69364 Lyon cedex 07,
France
E-mail address: gregory.miermont@ens-lyon.fr

Advanced Studies in Pure Mathematics 87, 2021
Stochastic Analysis, Random Fields and Integrable Probability — Fukuoka 2019
pp. 199–211

Avoided points of two-dimensional random walks

Yoshihiro Abe

Abstract.

This is a review of research on the cover time and avoided points for the two-dimensional simple random walk.

§1. Introduction

In 1990s, physicists studied statistical properties of the set of avoided points by a simple random walk (SRW) on $\mathbb{Z}^d$ [15, 16, 17, 20, 36]. Brummelhuis and Hilhorst [15] investigated that while the avoided points by SRW on $\mathbb{Z}^d$ in $d \geq 3$ are essentially independent and uniform, the ones in $d = 2$ are fractal-like. In $d = 2$, they found that the avoided points tend to form clusters. The fractal nature of the two-dimensional avoided points was rigorously studied by Dembo, Peres, Rosen, and Zeitouni [22] and Okada [32]. In $d \geq 3$, Belius [5] and Miller and Sousi [30] proved the uniformity of the avoided points.

The study of the avoided points is closely related to the cover time, which is the first time at which SRW visits every vertex. Dembo, Peres, Rosen, and Zeitouni [21] established the leading term of the cover time for the two-dimensional discrete torus. Belius and Kistler [6] obtained the second-order term of the cover time for Brownian motion on the two-dimensional continuum torus.

In this paper, we review research on the avoided points and the cover time in two dimensions. In Section 2, we will state known results for the cover time. In Section 3, we will review results on the avoided points. In Section 4, we will describe idea of proof of a result on statistics of avoided points studied in [2].

Received May 15, 2020.
Revised July 30, 2020.
2010 *Mathematics Subject Classification.* Primary 60J10; Secondary 60J55, 60G57, 60G70.
Key words and phrases. Cover time, Avoided point, Discrete Gaussian free field, Random interlacement.
Partially supported by JSPS KAKENHI Grant Number 18K13429.

§2. Two-dimensional cover time

Consider a SRW on $\mathbb{Z}_N^2 := (\mathbb{Z}/N\mathbb{Z})^2$. The cover time of $\mathbb{Z}_N^2$ by the SRW is defined by

$$\tau_{\mathrm{cov}}^N := \max_{x \in \mathbb{Z}_N^2} H_x, \tag{2.1}$$

where H_x is the first hitting time of x by the SRW.

In [4, Example L9], Aldous conjectured that $\tau_{\mathrm{cov}}^N/(N^2(\log N)^2) \to 4/\pi$ in probability as $N \to \infty$. He gave heuristics of the upper bound and pointed out the difficulty of obtaining the lower bound due to strong correlations between hitting times. See also [29] for a covering problem for a disk by SRW in two dimensions and [31, 28] for numerical simulations. Aldous's conjecture was settled down by Dembo, Peres, Rosen, and Zeitouni [21].

Theorem 2.1 (First-order term [21]). *As $N \to \infty$,*

$$\frac{\tau_{cov}^N}{N^2(\log N)^2} \to \frac{4}{\pi} \quad \textit{in probability.}$$

The crucial idea of the proof of Theorem 2.1 is that they studied the number of excursions around each site rather than the first hitting time. At each point, they considered concentric annuli and analyzed the number of crossings by SRW between each annulus. Let us call it a traversal process. Then, they used a branching structure of traversal processes which enabled them to apply the second moment method to obtain the lower bound of the cover time.

In [21, Section 9], they raised a question on a limit law of τ_{cov}^N. From 2004 to 2012, almost no progress has been made in this direction. In 2012, Ding [24] made a significant step and showed that $\sqrt{\tau_{\mathrm{cov}}^N/N^2} - (2/\sqrt{\pi})\log N$ is of order $\log\log N$ with probability tending to 1 as $N \to \infty$. His method is based on a strong connection between the cover time and the discrete Gaussian free field (DGFF) which was established by Ding, Lee, and Peres [25] and Zhai [37] in a general setting. Belius and Kistler [6] refined Ding's result in the Brownian motion case. Let $\mathbb{T}^2 := (\mathbb{R}/\mathbb{Z})^2$ be the two-dimensional continuum torus. The cover time of $\mathbb{T}^2$ by the Brownian motion on $\mathbb{T}^2$ is defined by $C_{\varepsilon,\mathbb{T}^2}^* := \sup_{x\in\mathbb{T}^2} C_{x,\varepsilon}$, where $C_{x,\varepsilon}$ is the first hitting time of the ε-ball centered at x.

Theorem 2.2 (Second-order term [6]). *For any $s > 0$ and $x \in \mathbb{T}^2$,*

$$2\log\varepsilon^{-1} - (1+s)\log\log\varepsilon^{-1} \le \frac{C_{\varepsilon,\mathbb{T}^2}^*}{\frac{1}{\pi}\log\varepsilon^{-1}} \le 2\log\varepsilon^{-1} - (1-s)\log\log\varepsilon^{-1} \tag{2.2}$$

holds with P_x-probability tending to 1 *as* $\varepsilon \to 0$.

They improved the branching structure of traversal processes introduced in [21] and showed that the law of the traversal process is the same as that of a critical Galton-Watson process with a geometric offspring distribution. This enabled them to apply so-called barrier estimates (versions of the ballot theorem) developed in the study of the branching Brownian motion. See also [7] for improved barrier estimates for Galton-Watson processes. A discrete analogue of Theorem 2.2 was established by the author [1].

Further progress has been made by Belius, Rosen, and Zeitouni [8]. Let C^*_{ε,S^2} be the cover time of the two-dimensional sphere S^2 by the Brownian motion on S^2 which is defined in the same way as $C^*_{\varepsilon,\mathbb{T}^2}$. They proved the tightness of C^*_{ε,S^2}.

Theorem 2.3 (Tightness [8]).

(2.3)
$$\lim_{\lambda\to\infty} \limsup_{\varepsilon\to 0} P\left[\left|\sqrt{C^*_{\varepsilon,S^2}} - 2\sqrt{2}\left(\log\varepsilon^{-1} - \frac{1}{4}\log\log\varepsilon^{-1}\right)\right| > \lambda\right] = 0.$$

The novelty of their work is that they invented two new methods, stochastic continuity of the traversal processes and decoupling estimates which break dependence of excursions inside and outside a ball. The stochastic continuity is essential to transfer a tightness result for the traversal process to that of the cover time. The decoupling is crucial to apply the second moment method in the proof of the lower bound on the cover time. They established the decoupling estimate by constructing coupling between the original excursions and independent ones. The symmetry of S^2 plays an important role in those estimates.

We should mention that more are known for the cover time of the binary tree. A properly normalized cover time of the binary tree of depth N converges in law to a randomly-shifted Gumbel distribution as $N \to \infty$ [19, 23]. It is believed that a similar convergence result would hold for the two-dimensional cover time. We also note that in $d \geq 3$, Belius [5] showed that a normalized cover time of $\mathbb{Z}^d_N$ converges in law to a Gumbel distribution.

§3. Avoided points in two dimensions

3.1. Late points in $\mathbb{Z}^2_N$

Inspired by [15], Dembo, Peres, Rosen, and Zeitouni [22] and Okada [32] studied a fractal structure of the set of avoided points of SRW on

$\mathbb{Z}_N^2$. We use the same notation as in Section 2. Recall from Theorem 2.1 that τ_{cov}^N is approximated by $(4/\pi)N^2(\log N)^2$. In view of this fact, it is natural to define the set of avoided points

$$\mathcal{L}_N(\alpha) := \left\{ x \in \mathbb{Z}_N^2 : H_x \geq \alpha \cdot \frac{4}{\pi} N^2 (\log N)^2 \right\} \tag{3.1}$$

for $\alpha \in (0,1)$. A point in $\mathcal{L}_N(\alpha)$ is called a α-late point. They counted the number of pairs of late points. Given a set A, let $|A|$ be the cardinality of A. For a sequence of random sets $(A_N)_{N\geq 1}$ and a sequence of positive values $(a_N)_{N\geq 1}$, we will write $|A_N| \approx a_N$ to denote that $\lim_{N\to\infty} \frac{\log |A_N|}{\log a_N} = 1$ in probability.

Theorem 3.1 (Counts of pairs of late points [22, 32]). *For any $\alpha, \beta \in (0,1)$,*

$$\left|\left\{ (x,y) \in \mathcal{L}_N(\alpha) \times \mathcal{L}_N(\alpha) : d(x,y) \leq N^\beta \right\}\right| \approx N^{\rho(\alpha,\beta)}, \tag{3.2}$$

where $d(\cdot,\cdot)$ is the ℓ^2-distance on $\mathbb{Z}_N^2$ and

$$\rho(\alpha,\beta) := \begin{cases} 2 + 2\beta - \dfrac{4\alpha}{2-\beta}, & (\beta \leq 2(1-\sqrt{\alpha})), \\ 8(1-\sqrt{\alpha}) - \dfrac{4(1-\sqrt{\alpha})^2}{\beta}, & (\beta \geq 2(1-\sqrt{\alpha})). \end{cases} \tag{3.3}$$

They also showed that $|\mathcal{L}_N(\alpha)| \approx N^{2(1-\alpha)}$ and that if Y is a uniformly chosen point from $\mathcal{L}_N(\alpha)$, then $|\mathcal{L}_N(\alpha) \cap B(Y, N^\beta)| \approx N^{2\beta(1-\alpha)}$, where $B(x,r)$ is the ball of radius r centered at x. From these, one may expect that the number of pairs of α-late points is approximated by $N^{2(1-\alpha)} \cdot N^{2\beta(1-\alpha)} = N^{2+2\beta-2\alpha(\beta+1)}$. However, this guess is not correct as one can see from Theorem 3.1. In fact, when β is small enough, the number of such pairs is much larger than the one we expected. This indicates that late points form clusters.

3.2. The two-dimensional random interlacement

Comets, Popov, and Vachkovskaia [18] and Rodriguez [33] introduced the two-dimensional random interlacement to describe the local structure around late points. Roughly speaking, the model is a Poissonian soup of trajectories of SRWs conditioned on never hitting the origin. The conditioned SRW $\widehat{S} = (\widehat{S}_i, i \geq 0, \widehat{P}_x, x \in \mathbb{Z}^2\backslash\{0\})$ is a Markov chain on $\mathbb{Z}^2\backslash\{0\}$ and its transition probability from x to y is $a(y)/(4a(x))$ if x and y are neighbours on $\mathbb{Z}^2\backslash\{0\}$, where $a(\cdot)$ is the potential kernel of SRW $S = (S_i, i \geq 0, P_x^{\mathbb{Z}^2}, x \in \mathbb{Z}^2)$ on $\mathbb{Z}^2$, i.e. $a(x) := \frac{1}{4}\sum_{i=0}^{\infty}(P_0^{\mathbb{Z}^2}[S_i = 0] - P_0^{\mathbb{Z}^2}[S_i = x])$. Informally, one can construct the two-dimensional random interlacement at level α (we will write 2D RI(α)) on a given finite subset A of $\mathbb{Z}^2$ containing the origin as follows:

- Take i.i.d. random variables $(Z_i)_{i\geq 1}$ from the probability measure $e_A(\cdot)/\mathrm{cap}(A)$, where $e_A(\cdot)$ is the equilibrium measure of A defined by $e_A(x) := 4a(x)\lim_{|y|\to\infty} P_y^{\mathbb{Z}^2}[S_{H_A} = x]$, $x \in A$ (H_A is the first hitting time of A by S) and $\mathrm{cap}(A)$ is the total mass $\mathrm{cap}(A) := \sum_{x\in A} e_A(x)$.
- Let M be a Poisson random variable with mean $\pi\alpha\mathrm{cap}(A)$ which is independent of $(Z_i)_{i\geq 1}$.
- From each point Z_i ($i = 1, \ldots, M$), run two independent random walks under the law $\widehat{P}_{Z_i}$ and the law of $\widehat{P}_{Z_i}$ conditioned on never returning to A.
- 2D RI(α) that hits A consists of trajectories of those random walks.

We note that the random interlacement was originally constructed by Sznitman [34] and Teixeira [35] for transient weighted graphs. One of the main results of [18] states that the law of the set of α-late points conditioned on not hitting the origin is approximated by that of 2D RI(α).

Theorem 3.2 (Late points and 2D RI [18]). *For any $\alpha \in (0,1)$ and any finite subset $A \subset \mathbb{Z}^2$ containing the origin,*

$$\lim_{N\to\infty} P\big[\pi_N(A) \subset \mathcal{L}_N(\alpha) \mid 0 \in \mathcal{L}_N(\alpha)\big] = \mathbb{P}[A \subset \mathcal{V}^\alpha], \tag{3.4}$$

where $\mathcal{V}^\alpha$ is the set of avoided points by 2D RI(α) and $\pi_N : \mathbb{Z}^2 \to \mathbb{Z}_N^2$ is the natural projection modulo N.

Let $(w_i)_{i\in\mathbb{N}}$ be the doubly-infinite trajectories in 2D RI(α). The occupation time field of 2D RI(α) is defined by

$$\ell_\alpha^{\mathrm{RI}}(x) := \sum_{i\in\mathbb{N}} \frac{1}{4}\int_{-\infty}^{\infty} 1_{\{w_i(t)=x\}}dt, \ x \in \mathbb{Z}^2. \tag{3.5}$$

Rodriguez [33] established the Dynkin type isomorphism theorem.

Theorem 3.3 (Pinned isomorphism theorem [33]).
For any $\alpha \in (0,1)$,

(3.6)
$$\left\{\ell_\alpha^{RI}(z) + \frac{1}{2}(\phi_z)^2 : z \in \mathbb{Z}^2\right\} \overset{law}{=} \left\{\frac{1}{2}\left(\phi_z - 2\sqrt{2\pi\alpha}\ a(z)\right)^2 : z \in \mathbb{Z}^2\right\},$$

where " $\stackrel{law}{=}$ *" denote the equality in law and* $(\phi_z)_{z\in\mathbb{Z}^2}$ *is a centered Gaussian field on* $\mathbb{Z}^2$ *with covariances*

$$\mathbb{E}[\phi_z\phi_w] = a(z) + a(w) - a(z-w), \quad z, w \in \mathbb{Z}^2 \tag{3.7}$$

and $(\phi_z)_{z\in\mathbb{Z}^2}$ *is independent of* $(\ell_\alpha^{RI}(z))_{z\in\mathbb{Z}^2}$.

Remark 3.4. *The extra factors of* 2 *and* π *in* (3.6) *compared to* [33, Theorem 5.5] *are due to different normalization of the occupation time, the pinned field, the level of* 2*D RI, and the potential kernel.*

3.3. Statistics of late points

Inspired by recent progress on the extremal value theory of the two-dimensional DGFF by Biskup and Louidor [11], Biskup and the author [2] studied statistics of late points for the two-dimensional SRW with wired boundary conditions. The reader may feel that this is inconsistent with the setting of periodic boundary conditions in the previous sections. In fact, our setting is unnatural and we impose the wired boundary conditions just for technical reasons.

To describe the result, let us prepare some notation. The following definitions of continuum and lattice domains are taken from [10]:

Definition 3.5. *An admissible domain is a bounded open subset of* $\mathbb{R}^2$ *that consists of a finite number of connected components and whose boundary is composed of a finite number of connected sets each of which has a positive Euclidean diameter.*

We write $\mathfrak{D}$ to denote a family of all admissible domains and let $d_\infty(\cdot,\cdot)$ denote the ℓ^∞-distance on $\mathbb{R}^2$.

Definition 3.6. *An admissible lattice approximation of* $D \in \mathfrak{D}$ *is a sequence* $D_N \subset \mathbb{Z}^2$ *such that the following holds: There exists* $N_0 \in \mathbb{N}$ *such that for all* $N \geq N_0$ *we have*

$$D_N \subseteq \left\{x \in \mathbb{Z}^2 : d_\infty(N^{-1}x, \mathbb{R}^2 \setminus D) > N^{-1}\right\} \tag{3.8}$$

and, for any $\delta > 0$ *there exists* $N_1 \in \mathbb{N}$ *such that for all* $N \geq N_1$,

$$D_N \supseteq \left\{x \in \mathbb{Z}^2 : d_\infty(N^{-1}x, \mathbb{R}^2 \setminus D) > \delta\right\}. \tag{3.9}$$

Throughout the paper, we fix any admissible domain $D \in \mathfrak{D}$ and its admissible lattice approximation D_N of D.

Biskup and Louidor [10] showed that the discrete Green function

$$G^{D_N}(x,y) := \frac{1}{4} E_x^{\mathbb{Z}^2}\left[\sum_{i=0}^{T_{D_N}-1} 1_{\{S_i=y\}}\right], \quad x, y \in D_N \tag{3.10}$$

of SRW killed upon the exit time $T_{D_N} := \min\{i : S_i \notin D_N\}$ from D_N can be approximated by the continuum one. See also [9, Chapter 1].

Let $X = (X_t, t \geq 0, P_x, x \in D_N)$ be a continuous-time SRW on D_N with i.i.d. exponential holding times with mean 1. We impose a technical assumption on X: When X hits the outer boundary ∂D_N of D_N, then it re-enters through a uniformly-chosen boundary edge. In other words, we regard ∂D_N as a single point ρ. We need this assumption to apply known results for DGFF on D_N with zero-boundary conditions via the generalized second Ray-Knight theorem.

Next, we define our local time by

$$L_t^{D_N}(x) := \int_0^{\tau_\rho(t)} 1_{\{X_s = x\}} ds \frac{1}{\deg(x)}, \tag{3.11}$$

where

$$\tau_\rho(t) := \inf\left\{s \geq 0 : \int_0^s 1_{\{X_r = \rho\}} dr \frac{1}{\deg(\rho)} > t\right\}. \tag{3.12}$$

Let t_N be a sequence with $\frac{t_N}{\frac{1}{\pi}(\log N)^2} \overset{N\to\infty}{\longrightarrow} \alpha \in (0,1)$. We take this so that $\tau_\rho(t_N)$ is approximated by α times the cover time of D_N. To study statistics of α-late points, we define the random measure κ_N^D on $D \times [0,\infty)^{\mathbb{Z}^2}$ by

$$\kappa_N^D := \frac{1}{W_N} \sum_{x \in D_N} 1_{\{L_{t_N}^{D_N}(x) = 0\}} \delta_{\frac{x}{N}} \otimes \delta_{\{L_{t_N}^{D_N}(x+z)\ :\ z \in \mathbb{Z}^2\}}, \tag{3.13}$$

where $W_N := N^2 e^{-\frac{2t_N}{\frac{1}{\pi}\log N}}$. When we consider convergence of random measures, we endow the space of Radon measures with the vague topology and write " $\underset{N\to\infty}{\overset{\text{law}}{\longrightarrow}}$ " to denote the convergence in law. Biskup and the author [2] showed that κ_N^D converges in law to a non-degenerate random measure.

Theorem 3.7 (Statistics of late points [2]). *There exists a random Borel measure $Z_{\sqrt{\alpha}}^D$ on D such that*

$$\kappa_N^D \underset{N\to\infty}{\overset{law}{\longrightarrow}} Z_{\sqrt{\alpha}}^D(dx) \otimes \nu_\alpha^{RI}(d\phi), \tag{3.14}$$

where ν_α^{RI} is the law of the occupation time field $(\ell_\alpha^{RI}(x))_{x\in\mathbb{Z}^2}$ of 2D $RI(\alpha)$ defined in (3.5).

Remark 3.8. *The random Borel measure Z_λ^D $(\lambda \in (0,1))$ is called the Liouville quantum gravity (LQG) and constructed by Duplantier and Sheffield [26]. Let $\Pi^D(x,\cdot)$ be the harmonic measure from x in D, i.e. $\Pi^D(x,A) := P_x[W_{T_D} \in A]$, $A \subset \mathbb{R}^2$, where $(W_t)_{t\geq 0}$ is the standard Brownian motion on $\mathbb{R}^2$ and $T_D := \inf\{t \geq 0 : W_t \notin D\}$ is the exit time from D. For each Borel set $A \subset D$ and $\lambda \in (0,1)$,*

$$\mathbb{E}[Z_\lambda^D(A)] = c_\lambda \int_A r^D(x)^{2\lambda^2}\, dx, \tag{3.15}$$

where $c_\lambda > 0$ is a constant and $r^D(x) := \exp\left\{\int_{\partial D} \Pi^D(x,dz)\log|x-z|\right\}$.

Remark 3.9. *A discrete-time analogue of Theorem 3.7 was studied by Biskup, Lee, and the author [3]. As mentioned above, it is more natural to study statistics of late points for SRW on $\mathbb{Z}_N^2$ and it is an open question whether Theorem 3.7 holds for $\mathbb{Z}_N^2$ or not. See* [3, Section 2.5] *for further discussion.*

§4. Idea of proof of Theorem 3.7

In the proof of Theorem 3.7, we use two key facts. The first one is the result of Biskup and Louidor [11] on statistics of intermediate level sets for the two-dimensional DGFF. Let $h^{D_N} = (h_x^{D_N})_{x\in D_N}$ be DGFF on D_N, i.e. a centered Gaussian field whose covariances are given by the Green function $G^{D_N}(\cdot,\cdot)$ defined in (3.10). Bolthausen, Deuschel, and Giacomin [12] proved that the first-order term of $\max_{x\in D_N} h_x^{D_N}$ is given by $\sqrt{2/\pi}\log N$. (Note that more details such as the second-order term and the convergence in law are known [14, 13].) Biskup and Louidor [11] studied statistics of the set of points whose h^{D_N}-values are larger than λ times the maximal value for $\lambda \in (0,1)$ (λ-high points) and their local structures via the random measure η_N^D on $D \times \mathbb{R} \times \mathbb{R}^{\mathbb{Z}^2}$ defined by

$$\eta_N^D := \frac{1}{K_N}\sum_{x\in D_N} \delta_{\frac{x}{N}} \otimes \delta_{h_x^{D_N} - a_N} \otimes \delta_{\{h_x^{D_N} - h_{x+z}^{D_N}\,:\, z\in\mathbb{Z}^2\}}, \tag{4.1}$$

where $\frac{a_N}{\log N} \overset{N\to\infty}{\to} \lambda\sqrt{\frac{2}{\pi}}$ and $K_N := \frac{N^2}{\sqrt{\log N}} e^{-\frac{a_N^2}{\frac{1}{\pi}\log N}}$. They showed that in the limit, the spatial and local parts of η_N^D are governed by LQG and the pinned DGFF ϕ which is defined in Theorem 3.3.

Theorem 4.1 (Statistics of high points for DGFF [11])**.**

$$\eta_N^D \underset{N\to\infty}{\overset{law}{\longrightarrow}} Z_\lambda^D(dx) \otimes e^{-2\sqrt{2\pi}\lambda h} dh \otimes \nu_\lambda, \tag{4.2}$$

where Z^D_λ *is LQG as in Remark* 3.8 *and* ν_λ *is the law of the random field* $\{\phi_z + 2\sqrt{2\pi}\ \lambda a(z) : z \in \mathbb{Z}^2\}$.

The second key ingredient of the proof of Theorem 3.7 is a Dynkin-type isomorphism theorem which connects our local time with DGFF.

Theorem 4.2 (The generalized second Ray-Knight theorem [27]). *For any* $N \geq 1$, *there is a coupling of* $L^{D_N}_{t_N}$ *(sampled under* P_ρ*) and two copies of DGFF* h^{D_N} *and* $\widetilde{h}^{D_N}$ *on* D_N *such that*

$$L^{D_N}_{t_N}(x) + \frac{1}{2}(h^{D_N}_x)^2 = \frac{1}{2}\left(\widetilde{h}^{D_N}_x - \sqrt{2t_N}\right)^2 \text{ for all } x \in D_N, \tag{4.3}$$

and $L^{D_N}_{t_N}$ *and* h^{D_N} *are independent.*

We note that this theorem is usually stated as a distributional equation. The construction of the coupling in Theorem 4.2 can be found in the proof of Theorem 3.1 of [37].

Now, we give heuristics of the proof of Theorem 3.7. We will write " = " to denote approximate equality in a vague sense. First, we focus on the spatial part of κ^D_N. Let $\widetilde{\kappa}^D_N$ be the point measure obtained by extracting only the spatial part from κ^D_N:

$$\widetilde{\kappa}^D_N := \frac{1}{W_N} \sum_{x \in D_N} \delta_{\frac{x}{N}} 1_{\{L^{D_N}_{t_N}(x)=0\}}.$$

It is not difficult to show that the sequence of point measures $\widetilde{\kappa}^D_N$ is tight since $L^{D_N}_{t_N}(x)$ is the sum of a Poisson number of i.i.d. exponential random variables and one can compute the probability $P_\rho[L^{D_N}_{t_N}(x) = 0]$ explicitly. See [2, Lemma 4.4 and Corollary 4.6] for precise arguments. Let $\widetilde{\kappa}^D$ be any subsequential weak limit of $\widetilde{\kappa}^D_N$. To prove the uniqueness of $\widetilde{\kappa}^D$, we will apply Theorems 4.1 and 4.2. We take the test function $f(h) := \max\{1-nh, 0\}$ on $[0, \infty)$ where n is sufficiently large. Let $A \subset D$ be any Borel set. By Theorem 4.2, we have

$$\begin{aligned} &\frac{\sqrt{\log N}}{W_N} \sum_{\substack{x \in D_N, \\ x/N \in A}} f\left(\frac{1}{2}\left(\widetilde{h}^{D_N}_x - \sqrt{2t_N}\right)^2\right) \\ &= \frac{\sqrt{\log N}}{W_N} \sum_{\substack{x \in D_N, \\ x/N \in A}} f\left(L^{D_N}_{t_N}(x) + \frac{1}{2}(h^{D_N}_x)^2\right). \end{aligned} \tag{4.4}$$

It is technically involved, but one can show that the right of (4.4) is concentrated around the expectation with respect to h^{D_N}. By this and the fact that $\mathrm{Var}(h_x^{D_N})$ "$=$" $(2\pi)^{-1}\log N$, the right of (4.4) is approximated by

$$\frac{\sqrt{\log N}}{W_N}\sum_{\substack{x\in D_N,\\ x/N\in A}}\frac{1}{\sqrt{\log N}}\int f\left(L_{t_N}^{D_N}(x)+\frac{h^2}{2}\right)e^{-\frac{\pi h^2}{\log N}}dh \tag{4.5}$$

$$\text{“}=\text{”}\frac{1}{W_N}\sum_{\substack{x\in D_N,\\ x/N\in A}}1_{\{L_{t_N}^{D_N}(x)=0\}}\int f(h^2/2)dh,$$

where the last line is due to the fact that the support of f is the sufficiently small interval $[0,1/n]$. By (4.4), (4.5), and Theorem 4.1, as $N\to\infty$, we have

$$Z_{\sqrt{\alpha}}^D(A)\int f(h^2/2)dh\text{“}=\text{”}\widetilde{\kappa}^D(A)\int f(h^2/2)dh.$$

This implies that the spatial part of κ_N^D should be governed by $Z_{\sqrt{\alpha}}^D$.

Next, let us consider the local structure of α-late points. By Theorem 4.2, for $x\in D_N$ and $z\in\mathbb{Z}^2$, we have

$$\left\{\left(\widetilde{h}_x^{D_N}-\sqrt{2t_N}\right)+\frac{1}{2}\left(\widetilde{h}_{x+z}^{D_N}-\widetilde{h}_x^{D_N}\right)\right\}\left(\widetilde{h}_{x+z}^{D_N}-\widetilde{h}_x^{D_N}\right) \tag{4.6}$$

$$=L_{t_N}^{D_N}(x+z)-L_{t_N}^{D_N}(x)+\frac{1}{2}(h_{x+z}^{D_N})^2-\frac{1}{2}(h_x^{D_N})^2.$$

By the domain Markov property of DGFF, we have

$$\widetilde{h}_{x+z}^{D_N}\overset{\text{law}}{=}\widetilde{h}_{x+z}^{D_N\setminus\{x\}}+\frac{G^{D_N}(x,x+z)}{G^{D_N}(x,x)}\widetilde{h}_x^{D_N}, \tag{4.7}$$

where $\widetilde{h}^{D_N\setminus\{x\}}$ is DGFF on $D_N\setminus\{x\}$ which is independent of $\widetilde{h}_x^{D_N}$. Comparing covariances of $\widetilde{h}_{x+\cdot}^{D_N\setminus\{x\}}$ with those of ϕ, we have $\widetilde{h}_{x+z}^{D_N\setminus\{x\}}$ "$=$" ϕ_z. Thus, if x is a $\sqrt{\alpha}$-high point for $\widetilde{h}^{D_N}$ (i.e. $\widetilde{h}_x^{D_N}$ "$=$" $\sqrt{2t_N}$), the left of (4.6) is close to $\frac{1}{2}(\phi_z-a(z)2\sqrt{2\pi\alpha})^2$ since $G^{D_N}(x,x)$ "$=$" $(2\pi)^{-1}\log N$ and $G^{D_N}(x,x)-G^{D_N}(x,x+z)\overset{N\to\infty}{\longrightarrow}a(z)$. On the other hand, the right of (4.6) is approximated by $L_{t_N}^{D_N}(x+z)+\frac{1}{2}\phi_z^2$ since $L_{t_N}^{D_N}(x)$ "$=$" 0 and $h_x^{D_N}$ "$=$" 0 when $\widetilde{h}_x^{D_N}$ "$=$" $\sqrt{2t_N}$ and h^{D_N} has a decomposition similar to (4.7). Therefore, when x is a α-late point, we have

$$L_{t_N}^{D_N}(x+z)+\frac{1}{2}(\phi_z)^2\text{“}=\text{”}\frac{1}{2}\left(\phi_z-a(z)2\sqrt{2\pi\alpha}\right)^2. \tag{4.8}$$

Comparing (4.8) with Theorem 3.3, we have $L_{t_N}^{D_N}(x+z)$" = "$\ell_\alpha^{\mathrm{RI}}(z)$ when x is a α-late point and thus the local part of κ_N^D should be governed by the occupation time field of 2D RI(α).

Acknowledgment. The author would like to thank the referee for very helpful suggestions.

References

[1] Y. Abe, Second-order term of cover time for planar simple random walk, J. Theoret. Probab., (2020), `https://doi.org/10.1007/s10959-020-01011-2`.

[2] Y. Abe and M. Biskup, Exceptional points of two-dimensional random walks at multiples of the cover time, arXiv:1903.04045.

[3] Y. Abe, M. Biskup and S. Lee, Exceptional points of discrete-time random walks in planar domains, arXiv:1911.11810.

[4] D. Aldous, Probability approximations via the Poisson clumping heuristic, Applied Mathematical Sciences, **77**, Springer, NY, 1989.

[5] D. Belius, Gumbel fluctuations for cover times in the discrete torus, Probab. Theory Related Fields, **157** (2013), 635–689.

[6] D. Belius and N. Kistler, The subleading order of two dimensional cover times, Probab. Theory Related Fields, **167** (2017), 461–552.

[7] D. Belius, J. Rosen and O. Zeitouni, Barrier estimates for a critical Galton-Watson process and the cover time of the binary tree. Ann. Inst. Henri Poincaré, **55** (2019), 127–154.

[8] D. Belius, J. Rosen and O. Zeitouni, Tightness for the cover time of the two dimensional sphere. Probab. Theory Related Fields, **176** (2020), 1357–1437.

[9] M. Biskup, Extrema of the two-dimensional Discrete Gaussian Free Field, In: M. Barlow and G. Slade (eds.): Random graphs, phase transitions, and the Gaussian free field, Springer Proc. Math. Stat., **304** Springer, Cham, (2020), 163–407.

[10] M. Biskup and O. Louidor, Conformal symmetries in the extremal process of two-dimensional discrete Gaussian free field, Comm. Math. Phys., **375** (2020), 175–235.

[11] M. Biskup and O. Louidor, On intermediate level sets of two-dimensional discrete Gaussian free field, Ann. Inst. Henri Poincaré, **55** (2019), 1948–1987.

[12] E. Bolthausen, J.-D. Deuschel and G. Giacomin, Entropic repulsion and the maximum of the two-dimensional harmonic crystal, Ann. Probab., **29** (2001), 1670–1692.

[13] M. Bramson, J. Ding and O. Zeitouni, Convergence in law of the maximum of the two-dimensional discrete Gaussian free field, Comm. Pure Appl. Math., **69** (2016), 62–123.

[14] M. Bramson and O. Zeitouni, Tightness of the recentered maximum of the two-dimensional discrete Gaussian free field, Comm. Pure Appl. Math., **65** (2012), 1–20.

[15] M. J. A. M. Brummelhuis and H. J. Hilhorst, Covering of a finite lattice by a random walk, Phys. A., **176** (1991), 387–408.

[16] M. J. A. M. Brummelhuis and H. J. Hilhorst, How a random walk covers a finite lattice, Phys. A., **185** (1992), 35–44.

[17] S. Caser and H. J. Hilhorst, Topology of the support of the two-dimensional lattice random walk, Phys. Rev. Lett., **77** (1996), 992–995.

[18] F. Comets, S. Popov and M. Vachkovskaia, Two-dimensional random interlacements and late points for random walks, Comm. Math. Phys., **343** (2016), 129–164.

[19] A. Cortines, O. Louidor and S. Saglietti, A scaling limit for the cover time of the binary tree, arXiv:1812.10101.

[20] K. R. Coutinho, M. D. Coutinho-Filho, M. A. F. Gomes and A. M. Nemirovsky, Partial and random lattice covering times in two dimensions, Phys. Rev. Lett., **72** (1994), 3745.

[21] A. Dembo, Y. Peres, J. Rosen and O. Zeitouni, Cover times for Brownian motion and random walks in two dimensions, Ann. Math., **160** (2004), 433–464.

[22] A. Dembo, Y. Peres, J. Rosen and O. Zeitouni, Late points for random walks in two dimensions, Ann. Probab., **34** (2006), 219–263.

[23] A. Dembo, J. Rosen and O. Zeitouni, Limit law for the cover time of a random walk on a binary tree, arXiv:1906.07276.

[24] J. Ding, On cover times for 2D lattices, Electron. J. Probab., **17** no. 45 (2012).

[25] J. Ding, J. R. Lee and Y. Peres, Cover times, blanket times, and majorizing measures, Ann. of Math., **175** (2012), 1409–1471.

[26] B. Duplantier and S. Sheffield, Liouville quantum gravity and KPZ, Invent. Math., **185** (2011), 333–393.

[27] N. Eisenbaum, H. Kaspi, M. B. Marcus, J. Rosen and Z. Shi, A Ray-Knight theorem for symmetric Markov processes, Ann. Probab., **28** (2000), 1781–1796.

[28] P. Grassberger, How fast does a random walk cover a torus?, Phys. Rev. E, **96** (2017), 012115.

[29] G. F. Lawler, On the covering time of a disc by simple random walk in two dimensions, In: Seminar on Stochastic Processes, 1992, Progr. Probab., **33**, Birkhäuser Boston, Boston, MA, 1993, pp. 189–207.

[30] J. Miller and P. Sousi, Uniformity of the late points of random walk on $\mathbb{Z}_n^d$ for $d \geq 3$, Probab. Theory Related Fields, **167** (2017), 1001–1056.

[31] A. M. Nemirovsky, H. O. Mártin and M. D. Coutinho-Filho, Universality in the lattice-covering time problem, Phys. Rev. A, **41** (1990) 761–767.

[32] I. Okada, Geometric structures of late points of a two-dimensional simple random walk, Ann. Probab., **47** (2019), 2869–2893.

[33] P.-F. Rodriguez, On pinned fields, interlacements, and random walk on $(\mathbb{Z}/N\mathbb{Z})^2$, Probab. Theory Related Fields, **173** (2019), 1265–1299.

[34] A.-S. Sznitman, Vacant set of random interlacements and percolation, Ann. Math., **171** (2010), 2039–2087.
[35] A. Teixeira, Interlacement percolation on transient weighted graphs, Electron. J. Probab., **14**, (2009), 1604–1627.
[36] F. van Wijland, S. Caser and H. J. Hilhorst, Statistical properties of the set of sites visited by the two-dimensional random walk, J. Phys. A: Math. Gen., **30** (1997), 507–531.
[37] A. Zhai, Exponential concentration of cover times, Electron. J. Probab., **23** no. 32 (2018).

Department of Mathematics and Informatics,
Chiba University,
1-33 Yayoi-cho Inage-ku Chiba-shi Chiba 263-8522, Japan
E-mail address: `yosihiro@math.s.chiba-u.ac.jp`

Advanced Studies in Pure Mathematics 87, 2021
Stochastic Analysis, Random Fields and Integrable Probability — Fukuoka 2019
pp. 213–226

Note on the maximal jump size in a continuum model of directed first passage percolation

Ryoki Fukushima

Abstract.

In this note, we consider the directed first passage percolation introduced in [F. Comets, R. Fukushima, S. Nakajima and N. Yoshida: Journal of Statistical Physics, 161-(3), 577–597 (2015)]. It is proved that the shortest path from the origin to the n-th hyperplane makes a jump larger than a positive power of $\log n$. Some numerical results are also provided, which indicates that the maximal jump size is much larger in a certain parameter region.

§1. Introduction and result

We study a geometric property of the the shortest path (geodesic) in a continuum model of directed first passage percolation introduced in [2]. Specifically, we are interested in the maximal jump size in the geodesic in this model. The result in this note implies that it grows unbounded as the system size goes to infinity. In the end of this article, some numerical results are also presented.

We start by recalling the first passage percolation model in [2]. Let $(\omega, \mathbb{P})$ be the Poisson point process on $\mathbb{N} \times \mathbb{R}^d$ whose intensity is the product of the counting measure and the Lebesgue measure. It is natural to realize it as a sum of the Dirac measures but with some abuse of notation, we will often identify ω with its support. For a given sample ω, we call $\gamma = (\gamma(k))_{k=1}^n \in (\mathbb{R}^d)^n$ an open path (with length n) if $\{(k, \gamma(k))\}_{k=1}^n$ is a subset of ω and we define its passage times by

$$T_\gamma(\omega) = \sum_{k=1}^{n} |\gamma(k-1) - \gamma(k)|^\alpha.$$

Received January 16, 2020.
Revised April 6, 2020.
2010 *Mathematics Subject Classification.* Primary 60K37; Secondary 60K35, 82A51, 82D30.
Key words and phrases. directed polymer, random environment, first passage percolation, ground states, zero temperature.

The passage time from 0 to the n-th section $\{n\} \times \mathbb{R}^d$ is defined by

$$T_n(\omega) = \inf \{T_\gamma(\omega) \colon \gamma(0) = 0 \text{ and } \gamma \text{ is open with length } n\} \tag{1.1}$$

and $T_{v,w}(\omega)$ for space-time points v, w is defined similarly. This model can be regarded as a directed version of the Euclidean first passage percolation introduced and studied in [4, 5, 6].

As in other first passage percolation models (see, for example, the monograph [1]), the following two problems are of interest: (i) asymptotic behaviors of T_n and (ii) properties of the minimizing path. For the first problem (i), a direct application of the subadditive ergodic theorem shows that the so-called *time constant*

$$\mu = \lim_{n \to \infty} \frac{1}{n} T_n(\omega) \tag{1.2}$$

exists $\mathbb{P}$-almost surely. This limit is deterministic and shown to be positive in [2]. The rate of convergence is a more interesting but challenging question. The works [2, 8] contain some results in this direction but they do not seem to be optimal. The problem (ii) concerns the geodesic, denoted by $\Gamma_{n;\omega}$, which is defined by the open path γ with length n for which $T_\gamma(\omega) = T_n(\omega)$. A point-to-point geodesic $\Gamma_{v,w;\omega}$ is defined similarly. Note that these geodesics are almost surely uniquely determined in our continuum setting if $\alpha \neq 1$. In a recent work [8, Corollary 2.7], it is proved that for any $\alpha > 1$, the maximal jump size in $\Gamma_{n;\omega}$ is of order $n^{o(1)}$ as $n \to \infty$. On the other hand, a numerical experiment presented in Section 3 indicates that the maximal jump size is much larger when α is small. It is an interesting problem to determine the asymptotics of the maximal jump size. The following result is a modest step toward this direction, which in particular shows that the maximal jump size is unbounded even for $\alpha > 1$.

Theorem 1. *For any $d \in \mathbb{N}$ and $\alpha \neq 1$, there exists $\epsilon > 0$ such that $\mathbb{P}$-almost surely, the maximal jump of $\Gamma_{n;\omega}$ is larger than $(\log n)^\epsilon$ for sufficiently large n.*

Remark 1. This result is analogous to the one in [9] for the classical lattice first passage percolation, which states (as a special case) that the geodesic goes through an edge with an arbitrarily large passage time as soon as the distribution of the edge weight is unbounded. In this lattice case, a more detailed result has recently been obtained in [7].

Remark 2. The case $\alpha = 1$ is somewhat singular and not covered in this note. For instance, the geodesic is not necessarily unique in $d = 1$.

Notational convention
For a time-space point $w \in \mathbb{N} \times \mathbb{R}^d$, we write its time coordinate as $w_{\rm t}$ and space coordinate as $w_{\rm s}$.

§2. Proof of Theorem 1

The basic strategy is similar to the work [9], with the help of an idea from [3]. Fix four positive constants $c_1 > 0$, $0 < \epsilon < \beta$ and $M > 0$ whose values will be determined later. For $(k, x) \in \mathbb{N} \times 4\mathbb{Z}^d$, let us define a *face* by

$$F_n(k, x) = \{k(\log n)^\beta\} \times (x + [-2, 2)^d).$$

Definition 1. A *face* $F_n(k, x)$ is *black* if the following holds for all $v \in F_n(k, x)$:

(i) when $\alpha < 1$,

$$\inf_{w \in \{(k+1)(\log n)^\beta\} \times \mathbb{Z}^d} T_{v,w}(\omega) \geq c_1(\log n)^\beta, \tag{2.1}$$

(ii) when $\alpha > 1$, for all $w \in \{(k+1)(\log n)^\beta\} \times \mathbb{Z}^d$ satisfying $|x - w_{\rm s}| \leq 2M(\log n)^\beta$,

$$T_{v,w}(\omega) \geq \left(\left(\frac{|v_{\rm s} - w_{\rm s}|}{(\log n)^\beta}\right)^\alpha + c_1\right)(\log n)^\beta. \tag{2.2}$$

Lemma 1. *Let $\alpha \neq 1$ and $\beta > 0$. Then for all sufficiently small $c_1 > 0$,*

$$\mathbb{P}(F_n(k, x) \text{ is black}) \geq 1 - \exp\{-(\log n)^{c_1}\}.$$

Proof. We may assume $(k, x) = (0, 0)$ without loss of generality and write F_n for $F_n(0, 0)$.

Let us first deal with the case $\alpha < 1$. Note that for $k = 0$ and fixed v, the left hand side of (2.1) has the same law as $T_{(\log n)^\beta}(\omega)$. Recall the tail estimate proved in [2, Proposition 3.1]: for any $\lambda < 1/2$, there exist $C_1, C_2 > 0$ such that for all $N \in \mathbb{N}$,

$$\mathbb{P}\big(T_N(\omega) - N\mu < -N^{1-\lambda}\big) \leq C_1 \exp\{-C_2 N^\lambda\}. \tag{2.3}$$

Choosing $0 < c_1 < (\mu \wedge \beta)/4$ and $N = [(\log n)^\beta]$, one finds that

$$\mathbb{P}\big(T_{(\log n)^\beta}(\omega) < 2c_1(\log n)^\beta\big) \leq \exp\{-2(\log n)^\beta\}. \tag{2.4}$$

While this is for a fixed v, the extension to all $v \in F_n$ is straightforward: if $\inf_w T_{v,w}(\omega) < \epsilon(\log n)^\beta$ for some $v \in F_n$, then by considering the path starting at 0 and following the geodesic for $\inf_w T_{v,w}(\omega)$ after the first step, we find that

$$T_{(\log n)^\beta}(\omega) \le c_1(\log n)^\beta + 2\sqrt{d}$$

by using the triangle inequality $|x+y|^\alpha \le |x|^\alpha + |y|^\alpha$ for $\alpha < 1$. Thus this case is covered by (2.4).

Next we consider the case $\alpha > 1$. In this case, the shortest path from $v = (v_{\mathrm{t}}, v_{\mathrm{s}})$ to $w = (w_{\mathrm{t}}, w_{\mathrm{s}})$ is, if the open path constraint is dropped, the straight line with the direction $D = (w_{\mathrm{s}} - v_{\mathrm{s}})(\log n)^{-\beta}$. Suppose that

$$T_{v,w}(\omega) < (|D|^\alpha + c_1)(\log n)^\beta. \tag{2.5}$$

When $|D| < \delta$ with $\delta^\alpha + 2\epsilon < \mu/2$, then the probability of this event is controlled in the same way as before. For the case $|D| \ge \delta$, we apply an affine tilting to the configuration ω to define $\tilde{\omega}$ so that $v \mapsto (0,0)$ and $w \mapsto ((\log n)^\beta, 0)$, that is, we translate each section as $\tilde{\omega}|_{\{j\}\times\mathbb{R}^d} = \omega|_{\{j\}\times\mathbb{R}^d} - v_{\mathrm{s}} - jD$. Then, since $\{(j, \Gamma_{v,w;\omega}(j) - jD)\}_{j=1}^{(\log n)^\beta} \subset \tilde{\omega}$, we have

$$T_{(\log n)^\beta}(\tilde{\omega}) \le \sum_{j=1}^{(\log n)^\beta} \left|\Gamma_{v,w;\omega}(j-1) - \Gamma_{v,w;\omega}(j) - D\right|^\alpha.$$

We show that this is smaller than $\mu(\log n)^\beta/2$ under the assumption (2.5). Let $\Delta_j = \Gamma_{v,w;\omega}(j-1) - \Gamma_{v,w;\omega}(j) - D$. By using the Taylor expansion, we get

$$\begin{aligned} T_{v,w}(\omega) &= \sum_{j=1}^{(\log n)^\beta} \left|\Gamma_{v,w;\omega}(j-1) - \Gamma_{v,w;\omega}(j)\right|^\alpha \\ &\ge \sum_{j=1}^{(\log n)^\beta} \Big[|D|^\alpha + \langle \nabla |D|^\alpha, \Delta_j\rangle \\ &\qquad + c\big(|D|^{\alpha-2}|\Delta_j|^2 1_{\{|\Delta_j| \le |D|/2\}} + |\Delta_j|^\alpha 1_{\{|\Delta_j| > |D|/2\}}\big)\Big]. \end{aligned} \tag{2.6}$$

Note that the second term sums up to zero since $\sum_{k=1}^{(\log n)^\beta} \Delta_j = 0$. Recalling $\delta \le |D| \le M$, we have

$$\delta^{\alpha-2} \wedge M^{\alpha-2} \le |D|^{\alpha-2} \le \delta^{\alpha-2} \vee M^{\alpha-2},$$

and hence (2.5) and (2.6) imply

$$\begin{aligned}\sum_{j=1}^{(\log n)^\beta} |\Delta_j|^\alpha &\leq \sum_{j=1}^{(\log n)^\beta} \left(\delta^\alpha + \delta^{\alpha-2}|\Delta_j|^2\, 1_{\{\delta\le|\Delta_j|\le|D|/2\}} + |\Delta_j|^\alpha\, 1_{\{|\Delta_j|>|D|/2\}}\right)\\ &\leq c\left(\delta^\alpha + c_1\left(\frac{M}{\delta}\right)^{|\alpha-2|} + c_1\right)(\log n)^\beta.\end{aligned}$$

The coefficient in thc last line can be made smaller than $\mu/2$ by letting c_1 small. Therefore, for $|D|\geq\delta$, we arrive at

$$\mathbb{P}\big(T_{v,w}(\omega) < (|D|^\alpha + c_1)(\log n)^\beta\big) \leq \mathbb{P}\big(T_{(\log n)^\beta}(\tilde{\omega}) \leq \mu(\log n)^\beta/2\big).$$

Since $\tilde{\omega}$ has the same law as ω, this right hand side is again bounded by $\exp\{-(\log n)^c\}$. This bound can be extended to all lattice point w satisfying $|w|\leq 2M(\log n)^\beta$ by the union bound. In order to pass from lattice points to continuum, note that under (2.5), the first and last jumps of the geodesic must be smaller than $(M^\alpha+2c_1)^{1/\alpha}(\log n)^{\beta/\alpha}$. Then the costs to change the first and last steps are bounded by $c((\log n)^{\beta/\alpha})^{\alpha-1} = o((\log n)^\beta)$ by the Taylor expansion. Q.E.D.

Let us consider the following conditions:

$$\max_{1\leq j\leq n} |\Gamma_{n;\omega}(j-1) - \Gamma_{n;\omega}(j)| \leq (\log n)^\epsilon, \tag{2.7}$$

$$\#\{F_n(k,x)\colon \text{black and crossed by } \Gamma_{n;\omega}\} \geq (1-\epsilon)n(\log n)^{-\beta}, \tag{2.8}$$

$$\begin{aligned}\#\big\{k\colon \big|\big[\Gamma_{n;\omega}(k(\log n)^\beta)\big] - \big[\Gamma_{n;\omega}((k+1)(\log n)^\beta)\big]\big| \geq M(\log n)^\beta\big\}\\ \leq \big(1_{\{\alpha<1\}} + \epsilon^2 1_{\{\alpha>1\}}\big)n(\log n)^{-\beta}.\end{aligned} \tag{2.9}$$

The last condition is non-trivial only for $\alpha > 1$. We first check that under (2.7), we have (2.8) and (2.9) with high probability.

Lemma 2. *When M is sufficiently large depending on $\epsilon > 0$,*

$$\begin{aligned}&\mathbb{P}((2.7)\ \textit{holds and either}\ (2.8)\ \textit{or}\ (2.9)\ \textit{fails})\\ &\quad\leq \exp\{-n^{1\wedge\frac{d}{\alpha}}(\log n)^{-\beta}\}.\end{aligned}$$

Proof. Let us begin with

$$\begin{aligned}&\mathbb{P}((2.7)\text{ holds and }(2.8)\text{ fails})\\&\quad\le \sum_{v_1,\dots,v_{n(\log n)^{-\beta}}} \mathbb{P}\left(\#\{k\colon F_n(k,v_k)\text{ is black}\} < (1-\epsilon)n(\log n)^{-\beta}\right),\end{aligned}$$

where the sum runs over sequences in $4\mathbb{Z}^d$ satisfying $v_1 = 0$ and $|v_k - v_{k+1}| \le (\log n)^{(\beta+\epsilon)}$. There are at most $((\log n)^{d(\beta+\epsilon)})^{n(\log n)^{-\beta}}$ such sequences and for each of them, the number of possibilities to place less than $(1-\epsilon)n(\log n)^{-\beta}$ black faces is bounded by

$$(1-\epsilon)n(\log n)^{-\beta}\binom{n(\log n)^{-\beta}}{(1-\epsilon)n(\log n)^{-\beta}} \le \exp\{n(\log n)^{-\beta}\},$$

which is much smaller than the above $((\log n)^{d(\beta+\epsilon)})^{n(\log n)^{-\beta}}$. As the rest of faces must be white, by using Lemma 1 we get

$$\begin{aligned}&\mathbb{P}((2.7)\text{ holds and }(2.8)\text{ fails})\\&\quad\le \left((\log n)^{d\beta}\right)^{2n(\log n)^{-\beta}} \mathbb{P}(F_n\text{ is white})^{\epsilon n(\log n)^{-\beta}}\\&\quad\le \exp\{-n(\log n)^{-\beta}\}.\end{aligned}$$

Next, when $\alpha > 1$ and (2.9) fails to hold, then $T_n(\omega) > M^\alpha\epsilon^2 n$ since the passage time cannot be shorter than the one for the straight line. Thus if we choose M sufficiently large, then $\mathbb{P}((2.9)\text{ fails}) \le \exp\{-cn^{1\wedge\frac{d}{\alpha}}\}$ by Lemma 3.2 in [2]. Q.E.D.

Let us define

$$\begin{aligned}\mathcal{E} = \{\omega\colon &(2.7)\text{–}(2.9)\text{ hold and for every } 0\le k\le n(\log n)^{-\beta},\\&\omega(\{k(\log n)^\beta\}\times B(\Gamma_{n;\omega}(k(\log n)^\beta), e^{-n})) = 1\}\end{aligned}$$

and for $\omega \in \mathcal{E}$, we define a re-sampling operation as follows (see also Figure 1):

(i) fix $\theta < 1$ and choose an ordered subset $(k_1,\dots,k_{n^\theta})$ of

$$\begin{aligned}\{k\colon &F_n(k,x)\text{ is black and crossed by }\Gamma_{n;\omega}\text{ and when }\alpha>1,\\&\left|[\Gamma_{n;\omega}(k(\log n)^\beta)] - [\Gamma_{n;\omega}((k+1)(\log n)^\beta)]\right| < M(\log n)^\beta\}\end{aligned}$$

uniformly at random ((2.8) and (2.9) ensure that there are at least $(1-2\epsilon)n(\log n)^{-\beta}$ such k's),

(ii) for each $1 \leq j \leq n^\theta$, replace ω by an independent copy in the tubes with the width $(\log n)^{2\beta}$ around

$$(2.10) \qquad \bigcup_{l=0}^{(\log n)^\beta} \left\{ \left(k_j (\log n)^\beta + l, [\Gamma_{n;\omega}(k_j(\log n)^\beta)] \right) \right\}$$

when $\alpha < 1$, and around the straight lines connecting

$$(2.11) \qquad \begin{aligned} &\left(k_j(\log n)^\beta, [\Gamma_{n;\omega}(k_j(\log n)^\beta)]\right) \text{ and} \\ &\quad \left((k_j+1)(\log n)^\beta, [\Gamma_{n;\omega}((k_j+1)(\log n)^\beta)]\right) \end{aligned}$$

when $\alpha > 1$, except for the e^{-n} neighborhoods of $\Gamma_{n;\omega}(k_j(\log n)^\beta)$ and $\Gamma_{n;\omega}((k_j+1)(\log n)^\beta)$ in both cases.

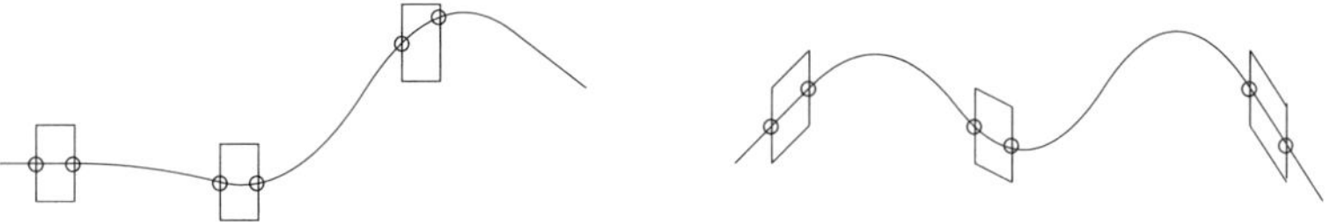

Fig. 1. Schematic figures of where re-sampling is done. Left figure is for $\alpha < 1$ and right figure is for $\alpha > 1$. Curves are $\Gamma_{n;\omega}$ and the rectangles (left) and parallelograms (right), with $\circ$'s omitted, are σ_ω.

We write σ_ω ($\omega \in \mathcal{E}$) for the collection of tubes where the re-sampling is done, $\tilde{\omega}$ for the independent copy used in (ii), and $[\omega_1, \omega_2]_S = \omega_1 1_{S^c} + \omega_2 1_S$ for $S \subset \mathbb{N} \times \mathbb{R}^d$. By using these terms, the result of the above re-sampling is written as $[\omega, \tilde{\omega}]_{\sigma_\omega}$. For a given ω, let Σ_ω denotes the set of possible choices of σ_ω and $\mathbb{Q}_\omega$ the (conditional) law of σ_ω. By choosing ϵ much smaller than θ, (2.8) and (2.9) imply

$$(2.12) \qquad \#\Sigma_\omega \geq \binom{(1-2\epsilon)n(\log n)^{-\beta}}{n^\theta} \geq \exp\left\{(1-2\theta)n^\theta \log n\right\}$$

for sufficiently large n uniformly in $\omega \in \mathcal{E}$. Finally, the joint law of $(\omega, \sigma_\omega, \tilde{\omega})$ is denoted by $\bar{\mathbb{P}}$.

Definition 2. For a tube of the form (2.10) or (2.11), let

$$D_{k_j} = \begin{cases} 0, & \alpha < 1, \\ (\log n)^{-\beta}\left(\Gamma_{n;\omega}((k_j+1)(\log n)^\beta) - \Gamma_{n;\omega}(k_j(\log n)^\beta)\right), & \alpha > 1. \end{cases}$$

The tube is *tunneling* for $\tilde{\omega}$ if the following conditions are satisfied: for every $1 \le l < (\log n)^\beta$,

$$\{k_j(\log n)^\beta + l\} \times \left(B\left(\Gamma_{n;\omega}(k_j(\log n)^\beta) + lD_{k_j} + \frac{3}{2}\mathbf{e}_1(\log n)^\epsilon 1_{\{l=(\log n)^\beta/2\}}, c_1^{2/(\alpha\wedge 1)}\right)\right)$$

contains exactly one point of $\tilde{\omega}$, and otherwise there is no point in

$$\bigcup_{l=0}^{(\log n)^\beta} \{k_j(\log n)^\beta + l\} \times \left(B([\Gamma_{n;\omega}(k_j(\log n)^\beta)] + lD_{k_j}, (\log n)^{2\beta})\right).$$

Figure 2 shows how tunneling tubes look like. A simple computation shows the following lemma.

Lemma 3. *There exist positive constants c_2 such that for sufficiently large n, for any tube of the form* (2.10) *or* (2.11),

$$\bar{\mathbb{P}}(\textit{The tube is tunneling for } \tilde{\omega}) \ge \exp\left\{-c_2(\log n)^{(2d+1)\beta}\right\}.$$

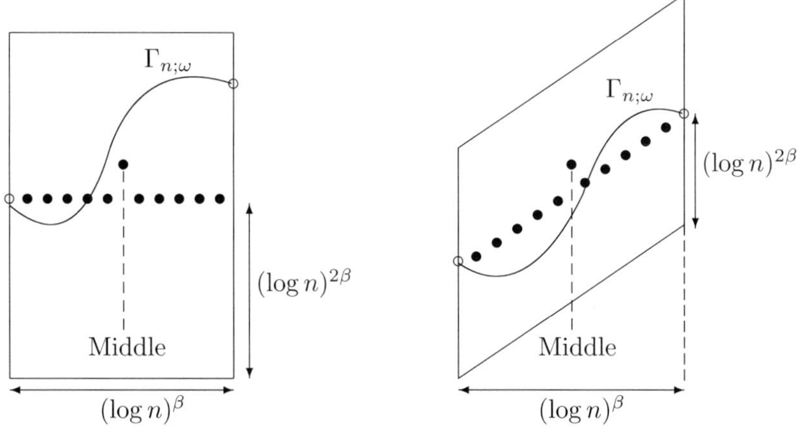

Fig. 2. Tunneling tubes for $\alpha < 1$ (left) and $\alpha > 1$ (right). Except for the middle, there are points of $\tilde{\omega}$ (indicated by $\bullet$) near the straight line at the center of the tube. The gap in the middle is close to $\frac{3}{2}(\log n)^\epsilon$.

Definition 3. Let

$$(2.13)\quad T_n^{\wedge}(\omega) = \inf\left\{T_n(\gamma)\colon \gamma(0)=0, \gamma \text{ is open and } \max_{1\le k\le n} |\gamma(k-1)-\gamma(k)| \le 2(\log n)^{\beta+\epsilon}\right\}$$

with the convention that $\inf \emptyset = \infty$. Corresponding geodesics are denoted by $\Gamma_{n;\omega}^{\wedge}$.

Now we introduce the following event:

$$\bar{\mathcal{E}} = \left\{(\omega, \sigma_\omega, \tilde{\omega})\colon \omega \in \mathcal{E}, \Gamma_{n;\omega}^{\wedge} = \Gamma_{n;[\omega,\tilde{\omega}]_{\sigma_\omega}}^{\wedge} \text{ outside } \sigma_\omega \text{ and all tubes in } \sigma_\omega \text{ are tunneling for } \tilde{\omega}\right\}.$$

Lemma 4. *For ϵ sufficiently small depending on α and β,*

$$\bar{\mathbb{P}}(\bar{\mathcal{E}}) \ge \mathbb{P}(\mathcal{E}) \exp\left\{-c_2 n^\theta (\log n)^{\beta(2d+1)}\right\}.$$

Proof. We show that the second condition in $\bar{\mathcal{E}}$ is a consequence of the others. Once this is shown, we have

$$\bar{\mathbb{P}}(\bar{\mathcal{E}}) = \bar{\mathbb{E}}\left[\bar{\mathbb{P}}(\text{All tubes in } \sigma_\omega \text{ are tunneling for } \tilde{\omega} \mid \sigma_\omega) 1_{\{\omega\in\mathcal{E}\}}\right].$$

Since σ_ω consists of n^θ tubes for any $\omega \in \mathcal{E}$, the above conditional probability is bounded from below by $\exp\{-c_2 n^\theta (\log n)^{\beta(2d+1)}\}$ by Lemma 3 and the desired bound follows.

Note first that $\Gamma_{n;\omega}^{\wedge} = \Gamma_{n;\omega}$ for $\omega \in \mathcal{E}$ due to (2.7). Suppose $\tilde{\omega}$ provides a tunneling configuration for all tubes in σ_ω. Then for each of tube in σ_ω, there is only one path satisfying the constraint in (2.13) inside, which we call *tunnel*. We show that $\Gamma_{n;[\omega,\tilde{\omega}]_{\sigma_\omega}}^{\wedge}$ is the path γ obtained by changing $\Gamma_{n;\omega}^{\wedge}$ to the tunnels in σ_ω.

Let us first check that the above operation improves the passage time, that is, $T_\gamma([\omega,\tilde{\omega}]_{\sigma_\omega}) < T_n(\omega)$. When $\alpha < 1$, a tunnel has jumps smaller than $2c_1^{2/\alpha}$ except for the middle two jumps and the last one. The middle jumps are bounded by $2(\log n)^\epsilon$. Also as $\omega \in \mathcal{E}$, endpoints of a tunnel differ at most $(\log n)^{(\beta+\epsilon)} + 2\sqrt{d}$ and hence the last jump is bounded by $2(\log n)^{(\beta+\epsilon)}$. Therefore the above path satisfies the restriction in (2.13) and, if ϵ is so small that $\alpha(\beta+\epsilon) < \beta$, its passage time is less than

$$2^\alpha\left[c_1^2(\log n)^\beta + 2(\log n)^{\alpha\epsilon} + (\log n)^{\alpha(\beta+\epsilon)}\right] < c_1(\log n)^\beta/2$$

for large n, by letting c_1 small if necessary. Since the tube in σ_ω is assumed to be black, this is strictly smaller than the original passage time. The argument for $\alpha > 1$ is similar. Indeed, the jumps in the j-th tunnel are, except for the middle two, smaller than $|D_{k_j}| + 2c_1^2$ (D_{k_j} is defined in Defninition 2). Therefore, the above path again satisfies the restriction in (2.13) and its passage time is, if $\epsilon < (\beta/\alpha) \wedge (1/2)$, less than

$$(|D_{k_j}| + 2c_1^2)^\alpha (\log n)^\beta + 2^{1+\alpha}(\log n)^{\alpha\epsilon} < (|D_{k_j}|^\alpha + c_1)(\log n)^\beta.$$

This is smaller than the original passage time by the definition of the black box.

Next we show that γ is the best among the paths satisfying the constraint in (2.13). Any path which does not use the points in $\tilde{\omega}$ has a passage time larger than $T_n(\omega)$ and by the previous step, it is not a geodesic for $[\omega, \tilde{\omega}]_{\sigma_\omega}$. On the other hand, the only way for a path satisfying the constraint in (2.13) to enter a tube $\tau \in \sigma_\omega$ is to share the left and right endpoints with $\Gamma_{n;\omega}$, which we call l_τ and r_τ respectively, since any other way forces the path to make a jump larger than $(\log n)^{2\beta}/2 \gg 2(\log n)^{\beta+\epsilon}$ (here we use the fact that the "slope" D_{k_j} of tunnels are bounded by M). Repeating this argument for the passage time from 0 to l_τ and from r_τ to $\{n\} \times \mathbb{R}^d$, we see that $\Gamma^\wedge_{n;[\omega,\tilde{\omega}]_{\sigma_\omega}}$ has to go through all the tubes in σ_ω sharing the left and right endpoints with $\Gamma_{n;\omega}$. This implies $\gamma = \Gamma^\wedge_{n;[\omega,\tilde{\omega}]_{\sigma_\omega}}$ and hence $\Gamma^\wedge_{n;\omega} = \Gamma^\wedge_{n;[\omega,\tilde{\omega}]_{\sigma_\omega}}$ outside σ_ω.

Q.E.D.

Lemma 5. *There exists $c_3 > 0$ such that*

$$\bar{\mathbb{P}}(\bar{\mathcal{E}}) \le \exp\{-c_3 n^\theta \log n\}.$$

Proof. For any $S \in \{\sigma_\omega \colon \omega \in \mathcal{E}\}$, we define

$$\mathcal{E}_S = \big\{(\omega_1, \omega_2) \colon \omega_1 \in \mathcal{E}, S \in \Sigma_{\omega_1}, \omega_1 = \omega_2 \text{ and } \Gamma^\wedge_{n;\omega_1} = \Gamma^\wedge_{n;\omega_2} \text{ outside } S, \\ \text{and all tubes in } S \text{ are tunneling for } \omega_2\big\}.$$

Then we have

$$\begin{aligned}\bar{\mathbb{P}}(\bar{\mathcal{E}}) &= \int \mathbb{P}^{\otimes 2}(\mathrm{d}\omega, \mathrm{d}\tilde{\omega}) \mathbb{Q}_\omega(\mathrm{d}\sigma) \sum_S 1_{\{\sigma = S\}} 1_{\{(\omega, [\omega,\tilde{\omega}]_S) \in \mathcal{E}_S\}} \\ &\le \frac{1}{\min_{\omega\in\mathcal{E}} \#\Sigma_\omega} \int \mathbb{P}^{\otimes 2}(\mathrm{d}\omega, \mathrm{d}\tilde{\omega}) \sum_S 1_{\{([\omega,\tilde{\omega}]_S, \omega) \in \mathcal{E}_S\}},\end{aligned}$$

where in the second inequality, we have used the fact that $(\omega, [\omega, \tilde{\omega}]_S)$ has the same law as $([\omega, \tilde{\omega}]_S, \omega)$. We shall show that the sum on the right hand side does not exceed one, that is, $\{(\omega,\tilde{\omega})\colon([\omega,\tilde{\omega}]_S,\omega)\in\mathcal{E}_S\}$ are disjoint for different S.

Suppose that $([\omega,\tilde{\omega}]_S,\omega)\in\mathcal{E}_S$ and $([\omega,\tilde{\omega}]_{S'},\omega)\in\mathcal{E}_{S'}$ for $S\neq S'$. Then $S\in\Sigma_{[\omega,\tilde{\omega}]_S}$ and thus S consists of n^θ tubes crossed by $\Gamma^\wedge_{n;[\omega,\tilde{\omega}]_S}$ as shown in Figure 1. Noting that $\Gamma^\wedge_{n;[\omega,\tilde{\omega}]_S}=\Gamma^\wedge_{n;\omega}$ outside S, it follows that $\Gamma^\wedge_{n;\omega}$ enters and exits the tubes in S as if $S\in\Sigma_\omega$. Since these observations apply for S' as well, it follows that S' contains a tube disjoint with S located along $\Gamma^\wedge_{n;\omega}$ (since both S and S' consist of the same number of tubes located along $\Gamma^\wedge_{n;\omega}$). However this contradicts $[\omega,\tilde{\omega}]_S\in\mathcal{E}$ as follows: note that $\Gamma^\wedge_{n;[\omega,\tilde{\omega}]_S}=\Gamma^\wedge_{n;\omega}$ in the tube found above ($\notin S$) which is assumed to be tunneling for ω. Then $\Gamma^\wedge_{n;[\omega,\tilde{\omega}]_S}$ must follow the tunnel (see the proof of Lemma 4 for the terminology), and hence it makes a jump larger than $(\log n)^\epsilon$ there.

Recalling (2.12), we arrive at

$$\bar{\mathbb{P}}(\bar{\mathcal{E}})\le\frac{1}{\min_{\omega\in\mathcal{E}}\#\Sigma_\omega}\le\exp\left\{-(1-2\theta)n^\theta\log n\right\}.$$

Q.E.D.

Proof of Theorem 1. Combining Lemmas 4 and 5, we obtain for $\beta<(2d+1)^{-1}$ that

$$\mathbb{P}(\mathcal{E})\le\exp\{-n^\theta\}.$$

Further invoking Lemma 2, we arrive at

$$\begin{aligned}\mathbb{P}((2.7))&\le\mathbb{P}(\mathcal{E})+\mathbb{P}((2.7)\text{ holds and either (2.8) or (2.9) fails})\\&\quad+\mathbb{P}\left(\exists x,y\in\omega\cap[0,n]\times\left[-n^{1+\epsilon},n^{1+\epsilon}\right]\text{ with }|x-y|<e^{-n}\right)\\&\le\exp\{-n^\theta\}+\exp\left\{-n^{1\wedge\frac{d}{\alpha}}(\log n)^{-\beta}\right\}+e^{-cn}.\end{aligned}$$

This and the Borel–Cantelli lemma complete the proof. Q.E.D.

§3. Numerical experiments

In this section, some results from a numerical experiment are presented. We first see how the geodesics look like for $\alpha<1$ and $\alpha>1$. Keeping track of geodesics requires a large memory and it is done only for 1024 steps. Then we proceed to statistical studies. For this purpose, about 1300 ($n=4096$), 1000 ($n=8192$), 700 ($n=16382$), and 400 ($n=32764$) samples for nine values of α ranging from 0.5 to 1.3 are generated numerically. Using these data, we see how the maximal jump size, the maximal displacement, and the variance of the passage time behave as functions of n.

Figure 3 shows the geodesics in 1024 steps with $\alpha = 0.6$ and $\alpha = 1.2$. It makes a large jump at the beginning when $\alpha = 0.6$ while it is much smoother when $\alpha = 1.2$. The reason can be explained as follows. The motivation for the geodesic to make a large displacement is to go to a "good point" from which the passage time is atypically small. Since the fluctuation of the passage time grows in time, there is a better chance to find a good point earlier than later. Now if there is a good point (k, x) for relatively small k, then x is typically far from the origin. The shortest path to go from $(0, 0)$ to such a point (k, x) tries to make one large jump and otherwise mostly horizontal when $\alpha < 1$, and tries to stay close to the straight line between $(0, 0)$ and (k, x) when $\alpha > 1$.

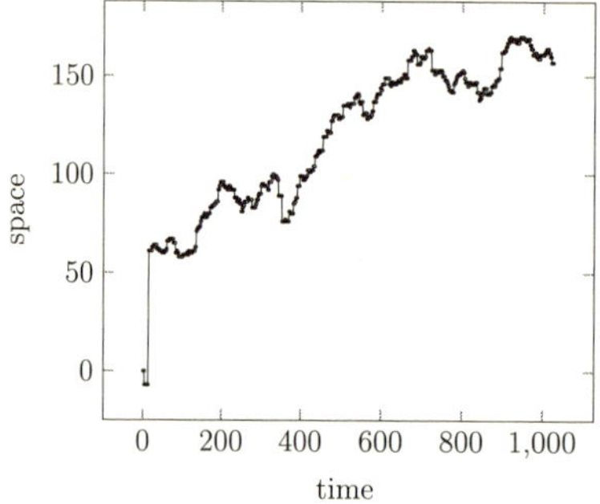

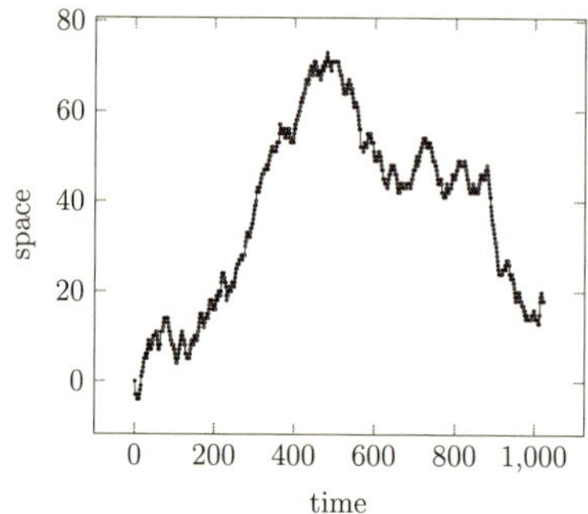

Fig. 3. The geodesics in 1024 steps with $\alpha = 0.6$ (left), whose maximal jump size is 68, and with $\alpha = 1.2$ (right) whose maximal jump size is 4.

Table 1 shows the maximal jump sizes of geodesic. The size decreases rapidly as α approaches to 1. Assuming that the maximal jump size obeys a power law, their exponents are computed. We know from [8, Corollary 2.7] that this is not the case for $\alpha > 1$. But the interesting regime here is $\alpha < 1$, where the exponents are larger. This suggest that $(\log n)^{\epsilon}$ in Theorem 1 can be improved to n^{ϵ}. We leave this as a conjecture.

Tables 2 and 3 show the maximal displacement and the variance, respectively. The maximal displacement decreases in α while the variance increases. This is natural since smaller α allows the path to explore better environments in a wider area, and it should cause a stronger self-averaging. The so-called fluctuation exponent χ and the wondering exponent ξ are computed, again assuming the power law behaviors. The results are not far from the Kardar–Parisi–Zhang prediction $2\chi = \xi = 2/3$, at least when $\alpha > 1$.

α	$n = 4096$	$n = 8192$	$n = 16382$	$n = 32764$	Exponent
0.5	128	218	314	434	0.586
0.6	46.7	57.6	74.5	89.1	0.311
0.7	19.6	23.1	28.1	31.7	0.232
0.8	12.2	13.3	14.9	16.8	0.155
0.9	9.02	9.88	10.7	11.5	0.116
1.0	5.59	5.97	6.44	6.76	0.091
1.1	3.57	3.82	4.14	4.45	0.106
1.2	3.49	3.69	4.03	4.24	0.094
1.3	3.27	3.45	3.73	3.95	0.091

Table 1. Empirical means of the maximal jump size. The exponents in the rightmost column are read from the slopes in log-log plot.

It would be desirable to do a numerical experiment in a larger scale to make sharper predictions.

α	$n = 4096$	$n = 8192$	$n = 16382$	$n = 32764$	ξ
0.5	432	739	1200	1870	0.704
0.6	277	448	713	1110	0.666
0.7	216	350	553	869	0.669
0.8	193	303	479	767	0.664
0.9	179	280	447	703	0.658
1.0	171	264	411	662	0.651
1.1	169	260	408	634	0.646
1.2	164	264	423	652	0.664
1.3	166	265	410	662	0.665

Table 2. Empirical means of the maximal displacement. The wondering exponents ξ are read from the slopes in log-log plot.

Acknowledgments. The author is indebted to Hugo Duminil-Copin for explaining the content of the paper [3] prior to its publication. He also thanks Nobuo Yoshida for carefully reading an early version of this article. Part of this work has been done during the author's stay at ENS Paris in 2016 October. He thanks ENS for hospitality and Mitsubishi Heavy Industries for the financial support. This work was also supported by JSPS KAKENHI Grant Number JP24740055 and ISHIZUE 2019 of Kyoto University Research Development Program.

α	$n = 4096$	$n = 8192$	$n = 16382$	$n = 32764$	2χ
0.5	43.9	67.1	95.0	158	0.615
0.6	58	102	162	255	0.712
0.7	77	123	204	354	0.733
0.8	104	158	269	394	0.641
0.9	113	201	317	490	0.705
1.0	154	252	374	641	0.685
1.1	164	241	438	624	0.624
1.2	183	267	459	721	0.659
1.3	176	287	467	747	0.695

Table 3. The variance of the passage time. The fluctuation exponents χ are read from the slopes in log-log plot.

References

[1] A. Auffinger, M. Damron and J. Hanson, 50 *years of first-passage percolation*, volume 68 of *University Lecture Series*, American Mathematical Society, Providence, RI, 2017.

[2] F. Comets, R. Fukushima, S. Nakajima and N. Yoshida, Limiting results for the free energy of directed polymers in random environment with unbounded jumps, *J. Stat. Phys.*, 161(3):577–597, 2015.

[3] H. Duminil-Copin, H. Kesten, F. Nazarov, Y. Peres and V. Sidoravicius, On the number of maximal paths in directed last-passage percolation, *Ann. Probab.*, 48(5):2176–2188, 2020.

[4] C. D. Howard and C. M. Newman, Euclidean models of first-passage percolation, *Probab. Theory Related Fields*, 108(2):153–170, 1997.

[5] C. D. Howard and C. M. Newman, From greedy lattice animals to Euclidean first-passage percolation, In: *Perplexing problems in probability*, volume 44 of *Progr. Probab.*, pages 107–119, Birkhäuser Boston, Boston, MA, 1999.

[6] C. D. Howard and C. M. Newman, Geodesics and spanning trees for Euclidean first-passage percolation, Ann. Probab., 29(2):577–623, 2001.

[7] S. Nakajima, Maximal edge-traversal time in first passage percolation, 2016, *preprint*, arXiv:1605.04787.

[8] S. Nakajima, Concentration results for directed polymer with unbounded jumps, *ALEA Lat. Am. J. Probab. Math. Stat.*, 15(1):1–20, 2018.

[9] J. van den Berg and H. Kesten, Inequalities for the time constant in first-passage percolation, *Ann. Appl. Probab.*, 3(1):56–80, 1993.

Research Institute for Mathematical Sciences, Kyoto University.
Current address: Institute of Mathematics, University of Tsukuba, 1-1-1 Tennodai, Tsukuba, Ibaraki 305–8571, Japan
E-mail address: `ryoki@math.tsukuba.ac.jp`

Advanced Studies in Pure Mathematics 87, 2021
Stochastic Analysis, Random Fields and Integrable Probability — Fukuoka 2019
pp. 227–238

Large deviation for dynamic model of three dimensional Young diagrams

Tadahisa Funaki

Abstract.

In this notes, we study the large deviation principle for the dynamic model of three dimensional Young diagrams, having the jump rate originally given by [10]. We employ the level-set function approach discussed in [8]. Our argument is rather heuristic by relying on the assumption of local equilibrium and the local ergodicity for slope variables. We are planning to give rigorous proofs in [4] based on a different formulation due to height functions and dimer-cover picture as in [9].

§1. Introduction

Three dimensional (3D) Young diagram is a collection of unit cubes being piled up in the first octant of $\mathbb{R}^3$, given usual x, y, z coordinates, in such a way that its height $h(x, y)$ measured from the x-y plane to the surface of 3D Young diagram (i.e., top located cube) at the position (x, y) is non-increasing in x, y, see Fig. 1 (or Figure 1 of [8]). Several different and essentially equivalent ways, other than the height function $h(x, y)$, are known to represent 3D Young diagrams, including descriptions due to lozenge tiling and dimer cover of honeycomb lattice. Here, we adopt the level-set function approach as in Section 3.2 of [8] and discuss the large deviation principle corresponding to the hydrodynamic limit, which is at the level of the law of large numbers, for the dynamic model of 3D Young diagrams. Note that, at the static level, the law of large numbers and the large deviation principle are shown by [1], see also Section 2.3.1 of [2].

Received January 5, 2020.
Revised July 6, 2020.
2010 *Mathematics Subject Classification.* 60K35, 60F10, 82C22, 74A50.
Key words and phrases. Large deviation, 3D Young diagram, Hydrodynamic limit.
The author is supported in part by JSPS KAKENHI, Grant-in-Aid for Scientific Researches (S) 16H06338, (A) 18H03672.

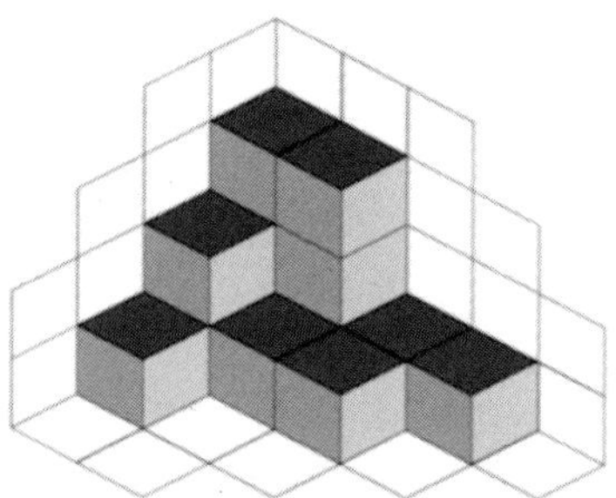

Fig. 1. 3D Young diagram (taken from [2], [1])

In Section 2, for a given macroscopic domain V in $\mathbb{R}^2$, we introduce the microscopic model by describing its generator and explain the (microscopic) level-set function $\hat{h}(v)$, $v = (v_1, v_2) \in V_L$, where $V_L := LV \cap (\mathbb{Z} \times (\mathbb{Z} + \frac{1}{2}))$ with large $L > 0$ is a microscopic domain corresponding to V. In Section 3, we study the behavior of the diffusively scaled (macroscopic) level-set function $\hat{\psi}^L(t, u)$, $u = (u_1, u_2) \in V$, defined in (3.5) or (3.2) with the time-evolution $\hat{h}_t(v)$ and derive the hydrodynamic equation (3.6) in the limit, under the assumption of local ergodicity. The large deviation rate function $I(h)$ is stated in Section 4. Section 5 explains how one can derive $I(h)$.

There is a small gap between two approaches due to level-set functions $\hat{h}(v)$ and height functions $h(x, y)$. Indeed, when we study the hydrodynamic limit, it turns out to be convenient to rely on the height functions, see [9]. The reason is that it is usually hard to deal with a specific boundary condition or consider on an infinite region in a study of the hydrodynamic limit. The formulation of [9] is convenient to make the system periodic and to avoid the boundary condition. In Section 6, we show that two rate functions of the large deviation principle derived in different descriptions due to $\hat{h}(v)$ and $h(x, y)$ are equivalent.

§2. Model, generator and level-set function

In the approach explained in Section 3.2 of [8], each unit cube $\mathbf{C}$ located on the surface of 3D Young diagram is described by the level-set function $\hat{h}(v), v = (v_1, v_2) \in \mathbb{Z} \times (\mathbb{Z} + \frac{1}{2})$, see Figure 7 of [8]. Here, (v_1, v_2) is the coordinate of $\mathbf{C}$ obtained by projecting it on the plane $P_a \equiv P_{110} := \{x + y = 0\}$ in $\mathbb{R}^3$, i.e., v_1 is the signed distance of $\mathbf{C}$ from the plane $P_b = \{x = y\}$ in $\mathbb{R}^3$ and v_2 is the height of $\mathbf{C}$ (of the middle point of **C**) measured from the x-y plane $P_c = \{z = 0\}$, and $\hat{h}(v)$ is the negatively signed distance of **C** measured from the plane P_a.

More precisely, let a smooth bounded domain V in $\mathbb{R}^2$ be given and let $V_L := LV \cap (\mathbb{Z} \times (\mathbb{Z} + \frac{1}{2}))$ be a discrete domain with side-length of order L. Let $\hat{h}(v)$ be the level-set function of 3D Young diagram at $v \in V_L$ defined as above. The configuration space of microscopic level-set functions is denoted by $\Omega_L = \{\hat{h}\}$. Then, the generator of the dynamics of 3D Young diagram is given by

$$\mathcal{L}F(\hat{h}) = \sum_{v \in V_L} \frac{|\varepsilon(\hat{h}, v)|}{2k(\hat{h}, v)} \left(F(\sigma_{v,k(\hat{h},v)}\hat{h}) - F(\hat{h}) \right), \tag{2.1}$$

for any function F on Ω_L, where $\sigma_{v,k}\hat{h}$ denotes the configuration $\hat{h}$ after a transition occurred initiated at v with k flips at $v + (0, n)\varepsilon(\hat{h}, v)$, $0 \leq n \leq k - 1$, (up or below v due to $\varepsilon(\hat{h}, v) = +1$ or -1, respectively). Here, $\varepsilon(\hat{h}, v)$ describes the local situation of the surface $\hat{h}$ at v (it is determined only from the slopes of $\hat{h}$ around v), i.e., if $\varepsilon(\hat{h}, v) = +1$, one can remove at most $k(\hat{h}, v)$ unit cubes keeping the monotone property of the surface, while, if $\varepsilon(\hat{h}, v) = -1$, one can add at most $k(\hat{h}, v)$ cubes. Note that $k(\hat{h}, v)$ counts how many cubes upward or downward have the same state as that at v.

The jump rate of $\hat{h}$ is $\frac{1}{2k(\hat{h},v)}$ and 0 if $\varepsilon(\hat{h}, v) = 0$, so that we put $|\varepsilon(\hat{h}, v)|$ in the jump rate. This choice of rate is due to [10]. To study the hydrodynamic limit, this choice turns out to be convenient, since the model becomes of gradient type and provides a good cancellation. The natural choice of the jump rate is to remove or add a single unit cube located at the surface keeping the monotone property, but this choice makes the model of non-gradient type, which is hard to analyze, see [2], p. 76.

Our goal is to study the behavior of the Markov process $\hat{h}_t = \hat{h}_t(v)$ on Ω_L with generator $L^2\mathcal{L}$, in which the time is already speeded up in diffusive scale L^2.

§3. Hydrodynamic limit

For a macroscopic test function $f \in C^\infty(V)$, we set

$$\langle \hat{\psi}^L, f \rangle := \frac{1}{L^2} \sum_{v \in V_L} f\left(\frac{v}{L}\right) \frac{\hat{h}(v)}{L}, \tag{3.1}$$

so that $\langle \hat{\psi}^L, f\rangle$ is the macroscopically scaled empirical measure of the microscopic level-set function $\hat{h}$ on V acted to f. We sometimes abuse the notation $\hat{\psi}^L$ to describe the step function approximation of the macroscopically scaled level-set function:

$$\hat{\psi}^L(u) = \sum_{v\in V_L} \frac{\hat{h}(v)}{L} 1_{B(\frac{v}{L},\frac{1}{L})}(u), \quad u \in V, \tag{3.2}$$

where $B(\frac{v}{L},\frac{1}{L}) = \{u=(u_1,u_2)\in V; |u_1-\frac{v_1}{L}|\le \frac{1}{2L}, |u_2-\frac{v_2}{L}|\le\frac{1}{2L}\}$.

Then, as in (5.24) of [8], we have

$$\begin{aligned} L^2\mathcal{L}\langle\hat{\psi}^L, f\rangle = \frac{L}{L^2}\sum_{v\in V_L} \frac{|\varepsilon(\hat{h},v)|}{k(\hat{h},v)}\varepsilon(\hat{h},v) \\ \times \sum_{n=0}^{k(\hat{h},v)-1} f\left(\frac{(v_1, v_2+n\varepsilon(\hat{h},v))}{L}\right), \end{aligned} \tag{3.3}$$

since

$$\sigma_{v,k(\hat{h},v)}(\hat{h}(v')) = \begin{cases} \hat{h}(v')+2\varepsilon(\hat{h},v), & \text{if } v' = v+(0,n)\varepsilon(\hat{h},v), \\ & \quad 0\le n\le k(\hat{h},v)-1, \\ \hat{h}(v'), & \text{otherwise.} \end{cases}$$

As in (5.25), (5.26) of [8], one can rewrite the right hand side of (3.3) as

$$\begin{aligned} &\frac{1}{2L^2}\sum_{v\in V_L}\frac{\hat{h}(v)}{L}\partial_{u_1}^2 f\left(\frac{v}{L}\right) \\ &+ \frac{1}{L^2}\sum_{v\in V_L} |\varepsilon(\hat{h},v)| \frac{k(\hat{h},v)-1}{2}\partial_{u_2} f\left(\frac{v}{L}\right) + O\left(\frac{1}{L}\right). \end{aligned} \tag{3.4}$$

Indeed, by Taylor expansion,

$$f\left(\frac{(v_1,v_2+n\varepsilon(\hat{h},v))}{L}\right) = f\left(\frac{v}{L}\right) + \frac{n\varepsilon(\hat{h},v)}{L}\partial_{u_2} f\left(\frac{v}{L}\right) + O\left(\frac{1}{L^2}\right).$$

For the term coming from $f(\frac{v}{L})$, $\frac{1}{k(\hat{h},v)}$ cancels with the $k(\hat{h},v)$ from the sum and then, noting the gradient property: $2\varepsilon(\hat{h},v) = \Delta_{v_1}\hat{h}(v)$ (cf. (5.22) of [8]), we can apply the summation by parts formula, which kills the diverging factor L^2 due to the C^2-property of f. On the other hand, for the term with $\partial_{u_2} f(\frac{v}{L})$, note that $\sum_{n=0}^{k-1} n = \frac{k(k-1)}{2}$ and

$\varepsilon^2(\hat{h}, v) = |\varepsilon(\hat{h}, v)|$. The number $k(\hat{h}, v)$ of unit cubes, that one can remove or add keeping the monotone property of the surface, is $O(L)$ especially if a steep bluff exists, but it actually behaves as $O(1)$ under the Gibbs measures π_ρ as we will see below or from Proposition 5.3 of [9]. Moreover, at dynamic level, a certain uniform bound is given for $k(\hat{h}, v)$ in Proposition 4.8 of [9] in the height function approach.

Recall that $\hat{h}_t = \hat{h}_t(v)$ is the process with generator $L^2\mathcal{L}$. Let us assume that the scaled level-set function $\hat{\psi}^L(t, u)$ of $\hat{h}_t$ as in (3.2) has a macroscopic limit $\hat{\psi}(t, u), (t, u) \in [0, T] \times V$:

$$\hat{\psi}^L\left(t, \frac{v}{L}\right) \equiv \frac{\hat{h}_t(v)}{L} \underset{L\to\infty}{\longrightarrow} \hat{\psi}\left(t, \frac{v}{L}\right), \quad \text{(in probability)}. \tag{3.5}$$

Then, the first term of (3.4) (with $\hat{h}$ replaced by $\hat{h}_t$) behaves as

$$\frac{1}{2}\int_V \hat{\psi}(t, u)\partial_{u_1}^2 f(u)du.$$

On the other hand, by the local ergodicity (which could be realized by showing one and two blocks' estimates), the second term of (3.4) can be replaced by (5.31) of [8], i.e.,

$$\int_V E_{\pi_{\nabla\hat{\psi}(t,u)}}\left[|\varepsilon(\hat{h}, 0)|\frac{k(\hat{h}, 0) - 1}{2}\right] \partial_{u_2} f(u)du,$$

where $0 = (0, \frac{1}{2}) \in \mathbb{Z} \times (\mathbb{Z} + \frac{1}{2})$ and π_ρ is the translation-invariant (on the space of slopes) ergodic Gibbs measure with average slope ρ, see [5], [11]. By the computation (5.32)–(5.35) of [8], this can be further rewritten as

$$\int_V \left(1 + \frac{1}{2}\partial_{u_2}\hat{\psi}(t, u)\right) E_{\pi_{\nabla\hat{\psi}(t,u)}}[A]\, \partial_{u_2} f(u)du,$$

where $A = A(\hat{h}) := \frac{1}{2}1_{X(0)}|\varepsilon(\hat{h}, 0)|(k(\hat{h}, 0) - 1)$ is defined below (5.19) with $X(0)$ given in Definition 6 of [8]. Thus, we obtain the hydrodynamic equation (3.32) in [8] for the limit function $\hat{\psi}(t, u)$:

$$\partial_t\hat{\psi}(t, u) = \frac{1}{2}\partial_{u_1}^2\hat{\psi}(t, u) - \partial_{u_2}\left\{\left(1 + \frac{1}{2}\partial_{u_2}\hat{\psi}(t, u)\right) E_{\pi_{\nabla\hat{\psi}(t,u)}}[A]\right\}. \tag{3.6}$$

In the above arguments, we have ignored the boundary effect at ∂V.

§4. The rate function

The limit $\hat{\psi}(t,u)$ given by (3.6) is non-random, so that the hydrodynamic limit is probabilistically at the level of the law of large numbers. Therefore, one of the natural next problems is to study the large deviation from the limit $\hat{\psi}(t,u)$.

The dynamic part of the large deviation rate function is expected to be

$$(4.1)\quad I(h) = \int_0^T dt \int_V \frac{du}{4E_{\pi_{\nabla h(t,u)}}[B]} \Big(\partial_t h(t,u) - \frac{1}{2}\partial_{u_1}^2 h(t,u) \\ + \partial_{u_2} \Big\{ \Big(1 + \frac{1}{2}\partial_{u_2} h(t,u)\Big) E_{\pi_{\nabla h(t,u)}}[A] \Big\} \Big)^2,$$

for every macroscopic function $h = h(t,u) \in C^{1,2}([0,T]\times V)$, where

$$(4.2)\qquad B = |\varepsilon(\hat{h},0)| k(\hat{h},0).$$

The full large deviation rate function should be the sum of $I(h)$ and the contribution $I_0(h(0))$ of the initial distribution. In the next section, we derive the rate function $I(h)$ given in (4.1).

§5. Weak perturbation and derivation of $I(h)$

5.1. Hydrodynamic limit for the perturbed process

We introduce a weak perturbation of $\mathcal{L}$ in order to find the rate function $I(h)$, cf. [7], Chapter 10 of [6], [3]. Take a macroscopic function $G = G(t,u)$ on $[0,T]\times V$ and perturb the generator $\mathcal{L}$ to $\mathcal{L}^G$ defined by

$$\mathcal{L}^G F(\hat{h}) \equiv \mathcal{L}_t^G F(\hat{h}) \\ = \sum_{v\in V_L} \frac{|\varepsilon(\hat{h},v)|}{2k(\hat{h},v)} e^{\mathbb{F}(t,\sigma_{v,k(\hat{h},v)}\hat{h}) - \mathbb{F}(t,\hat{h})} \Big(F(\sigma_{v,k(\hat{h},v)}\hat{h}) - F(\hat{h}) \Big),$$

for every function F on Ω_L, where

$$(5.1)\qquad \mathbb{F}(t,\hat{h}) := \sum_{v\in V_L} G\Big(t, \frac{v}{L}\Big) \frac{\hat{h}(v)}{L} = L^2 \langle \hat{\psi}^L, G(t,\cdot)\rangle.$$

Since one can compute as

$$e^{\mathbb{F}(t,\sigma_{v,k(\hat{h},v)}\hat{h}) - \mathbb{F}(t,\hat{h})} = \exp\Big\{ \frac{2\varepsilon(\hat{h},v)}{L} \sum_{n=0}^{k(\hat{h},v)-1} G\Big(t, \frac{(v_1, v_2 + n\varepsilon(\hat{h},v))}{L}\Big) \Big\}$$

$$= 1 + \frac{2\varepsilon(\hat{h},v)k(\hat{h},v)}{L} G\left(t, \frac{v}{L}\right) + O\left(\frac{1}{L^2}\right),$$

by Taylor expansion for the second line, we see

$$L^2 \mathcal{L}^G F(\hat{h}) = L^2 \mathcal{L} F(\hat{h}) + L \sum_{v \in V_L} \left(\varepsilon(\hat{h},v) G\left(t, \frac{v}{L}\right) + O\left(\frac{1}{L}\right) \right) \times \left(F(\sigma_{v,k(\hat{h},v)}\hat{h}) - F(\hat{h}) \right).$$

Taking $F(\hat{h}) = \langle \hat{\psi}^L, f \rangle$ with $f \in C^\infty(V)$, as we saw in (3.3), we have

$$\begin{aligned} F(\sigma_{v,k(\hat{h},v)}\hat{h}) - F(\hat{h}) &= \frac{2\varepsilon(\hat{h},v)}{L^3} \sum_{n=0}^{k(\hat{h},v)-1} f\left(\frac{(v_1, v_2 + n\varepsilon(\hat{h},v))}{L} \right) \\ &= \frac{2\varepsilon(\hat{h},v)k(\hat{h},v)}{L^3} f\left(\frac{v}{L}\right) + O\left(\frac{1}{L^4}\right). \end{aligned}$$

Therefore, noting $\varepsilon^2(\hat{h},v) = |\varepsilon(\hat{h},v)|$, we have

$$\begin{aligned} L^2 \mathcal{L}^G \langle \hat{\psi}^L, f \rangle &= L^2 \mathcal{L} \langle \hat{\psi}^L, f \rangle \\ &\quad + \frac{2}{L^2} \sum_{v \in V_L} |\varepsilon(\hat{h},v)| k(\hat{h},v) G\left(t, \frac{v}{L}\right) f\left(\frac{v}{L}\right) + O\left(\frac{1}{L}\right). \end{aligned}$$

The limit of the first term was discussed in Section 3. By the local ergodicity, the second term behaves as

$$2 \int_V E_{\pi_{\nabla \hat{\psi}(t,u)}}[B] G(t,u) f(u) du,$$

where B is the function defined in (4.2). Therefore, the hydrodynamic equation for the perturbed system is

$$\begin{aligned} \partial_t \hat{\psi}(t,u) = \frac{1}{2}\partial_{u_1}^2 \hat{\psi}(t,u) - \partial_{u_2} \left\{ \left(1 + \frac{1}{2}\partial_{u_2}\hat{\psi}(t,u) \right) E_{\pi_{\nabla \hat{\psi}(t,u)}}[A] \right\} \\ + 2E_{\pi_{\nabla \hat{\psi}(t,u)}}[B] G(t,u). \end{aligned} \tag{5.2}$$

5.2. Radon-Nikodým density and derivation of $I(h)$

To derive the rate function $I(h)$, we will compute the Radon-Nikodým density M_T^G of the distribution of the perturbed process P^G with respect to that of the original process P on $[0,T]$. Once this is done, we can rewrite

$$P\left(\hat{\psi}^L(t,u) \sim h(t,u), t \in [0,T] \right)$$

$$= E^{P^G}\left[(M_T^G)^{-1}, \hat{\psi}^L(t,u) \sim h(t,u), t \in [0,T]\right]$$

for any given $h(t,u), t \in [0,T], u \in V$, where "$\hat{\psi}^L(t,u) \sim h(t,u), t \in [0,T]$" means the set that these two functions are close as paths on $[0,T]$ in certain topology. We choose $G(t,u)$ in such a way that $P^G(\hat{\psi}^L(t,u) \sim h(t,u), t \in [0,T]) \to 1$ holds as $L \to \infty$. This is possible from (5.2) under the choice of

(5.3)
$$G(t,u) = \frac{\partial_t h(t,u) - \frac{1}{2}\partial_{u_1}^2 h(t,u) + \partial_{u_2}\{(1+\frac{1}{2}\partial_{u_2}h(t,u))E_{\pi_{\nabla h(t,u)}}[A]\}}{2E_{\pi_{\nabla h(t,u)}}[B]},$$

to make $h(t,u)$ the limit for the perturbed system P^G.

The Radon-Nikodým density M_T^G, especially for jump Markov processes, is given by

$$M_T^G = H(T,\hat{h}_T)e^{\int_0^T V(t,\hat{h}_t)dt},$$

where

$$H(T,\hat{h}_T) = e^{L^2\left(\langle\hat{\psi}^L(T,\cdot),G(T,\cdot)\rangle - \langle\hat{\psi}^L(0,\cdot),G(0,\cdot)\rangle\right)},$$
$$V(t,\hat{h}) = -e^{-\mathbb{F}(t,\hat{h})}(\partial_t + L^2\mathcal{L})e^{\mathbb{F}(t,\hat{h})},$$

see Proposition 7.3 in Appendix 1.7 of [6] and also [7]. Here, to compute $V(t,\hat{h})$,

$$\begin{aligned} L^2\mathcal{L}e^{\mathbb{F}(t,\hat{h})} = L^2 \sum_{v\in V_L} \frac{|\varepsilon(\hat{h},v)|}{2k(\hat{h},v)} \\ \times \left[\exp\left\{\mathbb{F}(t,\hat{h}) + \frac{2\varepsilon(\hat{h},v)}{L}\sum_{n=0}^{k(\hat{h},v)-1} G\left(t, \frac{(v_1, v_2+n\varepsilon(\hat{h},v))}{L}\right)\right\}\right. \\ \left. - \exp\{\mathbb{F}(t,\hat{h})\}\right], \end{aligned}$$

$$\partial_t e^{\mathbb{F}(t,\hat{h})} = \sum_{v\in V_L} \partial_t G\left(t,\frac{v}{L}\right)\frac{\hat{h}(v)}{L}e^{\mathbb{F}(t,\hat{h})}.$$

Therefore, we have

$$V(t,\hat{h}) = -L^2\langle\hat{\psi}^L(t,\cdot),\partial_t G(t,\cdot)\rangle - L^2\sum_{v\in V_L}\frac{|\varepsilon(\hat{h},v)|}{2k(\hat{h},v)}$$

$$\times\left[\exp\left\{\frac{2\varepsilon(\hat h,v)}{L}\sum_{n=0}^{k(\hat h,v)-1}G\left(t,\frac{(v_1,v_2+n\varepsilon(\hat h,v))}{L}\right)\right\}-1\right].$$

By Taylor expansion, the term in the brackets $[\cdots]$ above is equal to

$$\frac{2\varepsilon(\hat h,v)}{L}\sum_{n=0}^{k(\hat h,v)-1}G\left(t,\frac{(v_1,v_2+n\varepsilon(\hat h,v))}{L}\right)\\+\frac{1}{2}\left(\frac{2\varepsilon(\hat h,v)}{L}\sum_{n=0}^{k(\hat h,v)-1}G\left(t,\frac{(v_1,v_2+n\varepsilon(\hat h,v))}{L}\right)\right)^2+O\left(\frac{1}{L^3}\right).$$

Thus,

$$V(t,\hat h)=-L^2\left(\langle\hat\psi^L(t,\cdot),\partial_t G(t,\cdot)\rangle+I_2+I_3\right)+O(L),$$

where

$$I_2=\frac{L}{L^2}\sum_{v\in V_L}\frac{|\varepsilon(\hat h,v)|}{2k(\hat h,v)}2\varepsilon(\hat h,v)\sum_{n=0}^{k(\hat h,v)-1}G\left(t,\frac{(v_1,v_2+n\varepsilon(\hat h,v))}{L}\right),$$

$$I_3=\frac{1}{L^2}\sum_{v\in V_L}\frac{|\varepsilon(\hat h,v)|}{2k(\hat h,v)}\frac{4}{2}\varepsilon^2(\hat h,v)\left(\sum_{n=0}^{k(\hat h,v)-1}G\left(t,\frac{(v_1,v_2+n\varepsilon(\hat h,v))}{L}\right)\right)^2\\=\frac{1}{L^2}\sum_{v\in V_L}|\varepsilon(\hat h,v)|k(\hat h,v)G^2\left(t,\frac{v}{L}\right)+O\left(\frac{1}{L}\right).$$

For $I_1:=\langle\hat\psi^L(t,\cdot),\partial_t G(t,\cdot)\rangle$, we have

$$I_1\sim\int_V h(t,u)\partial_t G(t,u)du=:I_1(h,G,t),$$

on the set $\hat\psi^L(t,u)\sim h(t,u)$. As we saw in Section 3,

$$I_2\sim\frac{1}{2}\int_V\left\{h(t,u)\partial_{u_1}^2G(t,u)\right.\\\left.+\left(1+\frac{1}{2}\partial_{u_2}h(t,u)\right)E_{\pi_{\nabla h(t,u)}}[A]\partial_{u_2}G(t,u)\right\}du\\=:I_2(h,G,t),$$

on the set $\hat{\psi}^L(t,u) \sim h(t,u)$. Similarly, recalling $B = |\varepsilon(\hat{h},0)|k(\hat{h},0)$, we have

$$I_3 \sim \int_V E_{\pi_{\nabla h(t,u)}}[B] G^2(t,u)du =: I_3(h,G,t),$$

on the set $\hat{\psi}^L(t,u) \sim h(t,u)$. Finally, on the set $\hat{\psi}^L(t,u) \sim h(t,u)$, we have

$$\begin{aligned}\frac{1}{L^2}\log H(T,\hat{h}_T) &\sim \int_V h(T,u)G(T,u)du - \int_V h(0,u)G(0,u)du \\ &=: I_0(h,G,T).\end{aligned}$$

Summarizing all these computations, on the set $\hat{\psi}^L(t,u) \sim h(t,u)$, we have

$$\begin{aligned}&\frac{1}{L^2}\log(M_T^G)^{-1} \\ &\sim -I_0(h,G,T) + \int_0^T \{I_1(h,G,t) + I_2(h,G,t) + I_3(h,G,t)\}\,dt \\ &= -\int_0^T \int_V \Big(\partial_t h(t,u) - \frac{1}{2}\partial_{u_1}^2 h(t,u) \\ &\qquad + \partial_{u_2}\Big\{\Big(1 + \frac{1}{2}\partial_{u_2} h(t,u)\Big) E_{\pi_{\nabla h(t,u)}}[A]\Big\}\Big) \cdot G(t,u)dtdu \\ &\quad + \int_0^T \int_V E_{\pi_{\nabla h(t,u)}}[B] G(t,u)^2 dtdu \\ &= -\int_0^T \int_V \Big(\partial_t h(t,u) - \frac{1}{2}\partial_{u_1}^2 h(t,u) \\ &\qquad + \partial_{u_2}\Big\{\Big(1 + \frac{1}{2}\partial_{u_2} h(t,u)\Big) E_{\pi_{\nabla h(t,u)}}[A]\Big\}\Big)^2 \cdot \frac{dtdu}{4E_{\pi_{\nabla h(t,u)}}[B]},\end{aligned}$$

by noting

$$-I_0(h,G,T) + \int_0^T I_1(h,G,t)dt = -\int_0^T \int_V \partial_t h(t,u)G(t,u)dtdu,$$

and by the choice of G in (5.3) determined from $h = h(t,u)$. Thus, the speed of the large deviation principle is L^2 and the dynamic part of the rate function $I(h)$ is given by (4.1).

§6. Identification of two rate functions

In [4], taking V as two dimensional torus $\mathbb{T}^2$, we will derive another form of the rate function $\bar{I}(h)$ based on the representation of the (microscopic) height function $h(x,y)$ explained in Section 1 as follows:

$$(6.1)\quad \bar{I}(h) = \int_0^T dt \int_{\mathbb{T}^2} \frac{du}{E_{\pi_{\nabla h(t,u)}}[B]} \bigg(\partial_t h(t,u) - \frac{1}{4}\Delta h(t,u) \\ - \frac{1}{4}(\partial_{u_1} + \partial_{u_2}) E_{\pi_{\nabla h(t,u)}}[\bar{A}] \bigg)^2,$$

where $u = (u_1, u_2) \in \mathbb{T}^2$ is the macroscopic spatial variable corresponding to the microscopic variable (x,y), $\Delta = \partial_{u_1}^2 + \partial_{u_2}^2$, B is the same function as in (4.2) and $\bar{A}$ is a certain function of (microscopic) height function $(h(x,y))$. It is not important here, but it is written as

$$\bar{A} = |\varepsilon(\eta, 0)|(k(\eta, 0) - 1) - |\eta_{b_1(0)} - \eta_{b_2(0)}|$$

in the (microscopic) slope function η with suitably defined bonds $b_i(0)$, $i = 1, 2$. See Definition 2.2 of [9] for the relation between η and the microscopic height function $H = H_\eta$, which is denoted by $(h(x,y))$ in the present setting. Two equations appearing in the parentheses $(\cdots)$ in (4.1) and (6.1) are exactly the hydrodynamic equations represented in each setting. These equations are called (3.32) (or (3.41)) and (3.3) (with (3.12)), respectively, in [8] and it is shown that these are the same. Therefore, the difference of $I(h)$ and $\bar{I}(h)$ is only the factor $\frac{1}{4}$ in (4.1), but this is caused by the fact that the height h changes by ± 1, while $\hat{h}$ changes by ± 2, see Remark 13 of [8]. Thus, we see that $I(h)$ and $\bar{I}(h)$ are essentially the same.

References

[1] R. Cerf and R. Kenyon, *The low-temperature expansion of the Wulff crystal in the* 3D *Ising model*, Comm. Math. Phys., **222** (2001), 147–179.

[2] T. Funaki, *Lectures on Random Interfaces*, SpringerBriefs in Probability and Mathematical Statistics, Springer, 2016, xii+138 pp.

[3] T. Funaki and T. Nishikawa, *Large deviations for the Ginzburg-Landau $\nabla\phi$ interface model*, Probab. Theory Related Fields, **120** (2001), 535–568.

[4] T. Funaki and F. Toninelli, *Large deviation for lozenge tiling dynamics*, in preparation.

[5] R. Kenyon, A. Okounkov and S. Sheffield, *Dimers and amoebae*, Ann. Math., **163** (2006), 1019–1056.

[6] C. Kipnis and C. Landim, *Scaling Limits of Interacting Particle Systems*, Springer, 1999.

[7] C. Kipnis, S. Olla and S. R. S. Varadhan, *Hydrodynamics and large deviation for simple exclusion processes*, Comm. Pure Appl. Math., **42** (1989), 115–137.

[8] B. Laslier and F. Toninelli, *Hydrodynamic limit equation for a lozenge tiling Glauber dynamics*, Ann. Henri Poincaré, **18** (2017), 2007–2043.
[9] B. Laslier and F. Toninelli, *Lozenge tiling dynamics and convergence to the hydrodynamic equation*, Comm. Math. Phys., **358** (2018), 1117–1149.
[10] M. Luby, D. Randall and A. Sinclair, *Markov chain algorithms for planar lattice structures*, SIAM J. Comput., **31** (2001), 167–192.
[11] S. Sheffield, *Random surfaces*, Astérisque, **304** (2005), vi+175 pp.

Department of Mathematics,
School of Fundamental Science and Engineering,
Waseda University,
3-4-1 Okubo, Shinjuku-ku, Tokyo 169-8555, Japan
E-mail address: funaki@ms.u-tokyo.ac.jp

Advanced Studies in Pure Mathematics 87, 2021
Stochastic Analysis, Random Fields and Integrable Probability — Fukuoka 2019
pp. 239–259

Iterated paraproducts and iterated commutator estimates in Besov spaces

Masato Hoshino

Abstract.

In the previous study [6], the author provided an algebraic proof of the multicomponent commutator estimate in Besov spaces $C^\alpha = B^\alpha_{\infty,\infty}$ with $0 < \alpha < 1$. In this paper, we extend that result to general Besov spaces $B^\alpha_{p,q}$ with $p, q \in [1, \infty]$ and $0 < \alpha < 1$.

§1. Introduction

It is well known that, the definition of Besov space in Euclidean space $\mathbb{R}^d$ by Littlewood-Paley theory is equivalent to the one based on the estimate of Taylor remainder, when the regularity parameter is positive. See [1, Theorem 2.36] for instance. Especially, Besov space $B^\alpha_{\infty,\infty}(\mathbb{R}^d)$ is the same as Hölder space $C^\alpha(\mathbb{R}^d)$ if α is a positive noninteger.

In [6], the author showed a similar equivalence result for *Bony's paraproduct* and its iterated versions. For any distributions $f, g \in \mathcal{S}'(\mathbb{R}^d)$, the paraproduct $f \prec g$ is defined via Littlewood-Paley theory, so this is not a local operator. Nevertheless, in the case $f \in C^\alpha(\mathbb{R}^d)$ and $g \in C^\beta(\mathbb{R}^d)$ with $0 < \alpha, \beta$ and $\alpha + \beta < 1$, the previous result [6, Theorem 3.1] implies

$$(1) \qquad (f \prec g)(y) = (f \prec g)(x) + f(x)(g(y) - g(x)) + O(|y - x|^{\alpha+\beta}).$$

Conversely, we can show that the function h of such a local behavior is essentially the same as $f \prec g$. In [6], the author studied a generalized version of (1) for the iterated paraproducts, and as a consequence, provided an algebraic proof of the *commutator estimate* [4, Lemma 2.4], which has an important role in the theory of paracontrolled calculus [4].

Received January 17, 2020.
Revised May 18, 2020.
2010 *Mathematics Subject Classification.* 35S50, 60H15.
Key words and phrases. Besov space, paraproduct, paracontrolled calculus, commutator estimate.

In this paper, we consider the Besov type extension of the results in [6]. First we show the estimate like (1), see Theorem 3.1 below. The result is no longer a uniform bound on $\mathbb{R}^d$, but an L^pL^q type estimate of Taylor remainder. As a consequence, we also show the commutator estimate in Besov spaces, stated as below. Commutators discussed in this paper is defined as follows.

Definition 1.1. *For any functions $\xi, f_1, f_2, \dots$ in $\mathcal{S}(\mathbb{R}^d)$, define*

$$\begin{aligned} \mathsf{C}(f_1,\xi) &:= f_1 \succeq \xi \ (:= f_1\xi - f_1 \prec \xi), \\ \mathsf{C}(f_1,f_2,\xi) &:= \mathsf{C}(f_1 \prec f_2, \xi) - f_1\mathsf{C}(f_2,\xi), \\ \mathsf{C}(f_1,\dots,f_n,\xi) &:= \mathsf{C}(f_1 \prec f_2, f_3,\dots,f_n,\xi) - f_1\mathsf{C}(f_2,f_3,\dots,f_n,\xi). \end{aligned}$$

We denote by $B_{p,q}^{\alpha,0}$ the closure of $\mathcal{S}(\mathbb{R}^d)$ in the space $B_{p,q}^{\alpha}(\mathbb{R}^d)$. The following theorem is a generalization of [6, Theorem 4.2] onto Besov norms.

Theorem 1.1. *Let $\alpha_1,\dots,\alpha_n \in (0,1)$ and $\alpha_\circ < 0$ be such that*

$$\begin{aligned} &\alpha_1 + \cdots + \alpha_n < 1, \\ &\alpha_2 + \cdots + \alpha_n + \alpha_\circ < 0 < \alpha_1 + \cdots + \alpha_n + \alpha_\circ, \end{aligned}$$

and let $\alpha := \alpha_1 + \cdots + \alpha_n + \alpha_\circ$. Let $p_1,\dots,p_n,p_\circ,q_1,\dots,q_n,q_\circ \in [1,\infty]$ be such that

$$\frac{1}{p} := \frac{1}{p_1} + \cdots + \frac{1}{p_n} + \frac{1}{p_\circ} \le 1, \quad \frac{1}{q} := \frac{1}{q_1} + \cdots + \frac{1}{q_n} + \frac{1}{q_\circ} \le 1.$$

Then there exists a unique multilinear continuous operator

$$\tilde{\mathsf{C}} : B_{p_1,q_1}^{\alpha_1,0} \times \cdots \times B_{p_n,q_n}^{\alpha_n,0} \times B_{p_\circ,q_\circ}^{\alpha_\circ,0} \to B_{p,q}^{\alpha,0}$$

such that,

$$\tilde{\mathsf{C}}(f_1,\dots,f_n,\xi) = \mathsf{C}(f_1,\dots,f_n,\xi)$$

for any smooth inputs $(f_1,\dots,f_n,\xi)$.

To show the main theorem, we introduce a regularity structure suitable for our context, and impose $B_{p,q}$ type bounds on models and modelled distributions. Thus in our case, each basis vector τ of the model space has three homogeneity parameters $(\alpha_\tau, p_\tau, q_\tau)$, which is slightly different from the original setting [5]. See [7, 8, 9] for relevant studies. In [7, 8], the authors defined $B_{p,q}$ type modelled distributions and proved a generalized reconstruction theorem, while $B_{\infty,\infty}$ type bounds are imposed on models. In [9], the authors defined $B_{p,p}$ (Sobolev) type models and modelled distributions to consider the Sobolev type rough paths.

This paper is organized as follows. In Section 2, we define some important notions used in this paper; Besov type norms, paraproducts, and the word Hopf algebra. In Section 3, we show the Besov type estimates of Taylor remainders of iterated paraproducts. In Section 4, we show the Besov type commutator estimates.

§2. Preliminaries

We introduce some important notions used throughout this paper.

2.1. Besov type norms

In this paper, we often use a sequence $\{a_j\}_{j=-1}^{\infty}$ of numbers, functions, or operators. We use simplifying notations for partial sums as follows.

$$a_{<j} := \sum_{i<j} a_i, \quad a_{\geq j} := \sum_{i\geq j} a_i.$$

Lemma 2.1. *Let $q \in [1,\infty]$ and let $\{c_j\}_{j=-1}^{\infty}$ be a sequence of nonnegative numbers. If $\alpha > 0$, then we have*

$$\big\|\{2^{j\alpha}c_{\geq j}\}_j\big\|_{\ell^q} \lesssim \big\|\{2^{j\alpha}c_j\}_j\big\|_{\ell^q}, \tag{2}$$

$$\big\|\{2^{-j\alpha}c_{\leq j}\}_j\big\|_{\ell^q} \lesssim \big\|\{2^{-j\alpha}c_j\}_j\big\|_{\ell^q}. \tag{3}$$

Proof. We extend $\{c_j\}_{j=-1}^{\infty}$ into a sequence $\{c_j\}_{j\in\mathbb{Z}}$ by setting $c_j = 0$ if $j \leq -2$. For (2), by using Young's inequality on the group $\mathbb{Z}$,

$$\begin{aligned}\big\|\{2^{J\alpha}c_{\geq J}\}_J\big\|_{\ell^q} &= \Big\|\Big\{\sum_{j\geq J} 2^{(J-j)\alpha}2^{j\alpha}c_j\Big\}_J\Big\|_{\ell^q} \\ &\leq \sum_{-\infty<i\leq 0} 2^{i\alpha}\big\|\{2^{j\alpha}c_j\}_j\big\|_{\ell^q} \lesssim \big\|\{2^{j\alpha}c_j\}_j\big\|_{\ell^q}.\end{aligned}$$

The proof of (2) is just an analogue. Q.E.D.

Denote by $\mathcal{S} = \mathcal{S}(\mathbb{R}^d)$ the space of Schwartz functions, and by $\mathcal{S}'$ its dual space. Fix smooth radial functions χ and ρ such that,

- $\mathrm{supp}(\chi) \subset \{x\,;|x| < \frac{4}{3}\}$ and $\mathrm{supp}(\rho) \subset \{x\,;\frac{3}{4} < |x| < \frac{8}{3}\}$,
- $\chi(x) + \sum_{j=0}^{\infty}\rho(2^{-j}x) = 1$ for any $x \in \mathbb{R}^d$.

Set $\rho_{-1} := \chi$ and $\rho_j := \rho(2^{-j}\cdot)$ for $j \geq 0$. We define the Littlewood-Paley blocks

$$\Delta_j f := \mathcal{F}^{-1}(\rho_j\mathcal{F}f)$$

for $f \in \mathcal{S}'$, where $\mathcal{F}$ is the Fourier transform and $\mathcal{F}^{-1}$ is its inverse. It is useful to write

$$\Delta_j f(x) = \int_{\mathbb{R}^d} Q_j(x,y) f(y) dy,$$

where $Q_j(x,y) = \mathcal{F}^{-1}(\rho_j)(x-y)$. We also write $Q_j(h) = \mathcal{F}^{-1}(\rho_j)(h)$.

Definition 2.1. *For any $\alpha \in \mathbb{R}$ and $p,q \in [1,\infty]$, we define the (nonhomogeneous) Besov space $B^\alpha_{p,q}$ by the space of all $f \in \mathcal{S}'$ such that*

$$\|f\|_{B^\alpha_{p,q}} := \left\| \{2^{j\alpha} \|\Delta_j f\|_{L^p}\}_{j \geq -1} \right\|_{\ell^q} < \infty.$$

As stated in [1, Theorem 2.36], it is possible to define Besov norms without Littlewood-Paley theory. The aim of this paper is to study the following norm for two parameter functions.

Definition 2.2. *Let $\alpha \in \mathbb{R}$ and $p,q \in [1,\infty]$. For any two parameter measurable function $\omega(x,y)$ on $\mathbb{R}^d \times \mathbb{R}^d$, define*

$$\|\omega\|_{D^\alpha_{p,q}} := \left\| |h|^{-\alpha} \|\omega(x,x+h)\|_{L^p(dx)} \right\|_{L^q(dh/|h|^d)}.$$

The following well-known result provides an alternative definition of Besov norm. A self-contained proof appears in the next subsection.

Proposition 2.2 ([1, Theorem 2.36])**.** *If $\alpha \in (0,1)$, then*

$$\|f\|_{B^\alpha_{p,q}} \simeq \|f\|_{L^p} + \|\omega_f\|_{D^\alpha_{p,q}},$$

where $\omega_f(x,y) = f(y) - f(x)$.

2.2. Technical lemmas

We prove some technical lemmas used throughout this paper.

Definition 2.3. *Let $\alpha \in \mathbb{R}$ and $p,q \in [1,\infty]$. For any sequence $\{f_j(x)\}_{j=-1}^\infty$ of measurable functions on $\mathbb{R}^d$, define*

$$\|\{f_j\}_j\|_{\mathbb{B}^\alpha_{p,q}} := \left\| \{2^{j\alpha} \|f_j\|_{L^p}\}_j \right\|_{\ell^q}.$$

By definition, $\|f\|_{B^\alpha_{p,q}} = \|\{\Delta_j f\}_j\|_{\mathbb{B}^\alpha_{p,q}}$. We often emphasize the variables j and x and write

$$\|f_j(x)\|_{\mathbb{B}^\alpha_{p,q}} = \left\| \{2^{j\alpha} \|f_j(x)\|_{L^p(dx)}\}_j \right\|_{\ell^q},$$

by an abuse of notation.

Lemma 2.3. *Let $\alpha > 0$ and $q \in [1, \infty]$. Let F be a nonnegative function on $\mathbb{R}^d$ such that*

$$\big\| |h|^{-\alpha} F(h) \big\|_{L^q(dh/|h|^d)} \leq C.$$

Then for any nonnegative function $\varphi \in \mathcal{S}$, one has

$$\left\| 2^{j\alpha} \int_{\mathbb{R}^d} 2^{jd} \varphi(2^j h) F(h) dh \right\|_{\ell^q} \lesssim C.$$

Proof. The proof is essentially contained in the latter half part of the proof of [1, Theorem 2.36]. There $\alpha \in (0,1)$ is assumed, but we can see that Lemma 2.3 holds for any $\alpha > 0$. The only point to be modified is the integration over $2^j|h| > 1$ when $q < \infty$. Indeed, for any $\varepsilon > 0$,

$$\begin{aligned}
&2^{j\alpha} \left| \int_{2^j|h|>1} 2^{jd} \varphi(2^j h) F(h) dh \right| \\
&\leq 2^{-j\varepsilon} \int_{2^j|h|>1} |2^j h|^{d+\alpha+\varepsilon} |\varphi(2^j h)| \frac{F(h)}{|h|^{\alpha+\varepsilon}} \frac{dh}{|h|^d} \\
&\lesssim 2^{-j\varepsilon} \left(\int_{2^j|h|>1} \frac{F(h)^q}{|h|^{(\alpha+\varepsilon)q}} \frac{dh}{|h|^d} \right)^{1/q}
\end{aligned}$$

by Hölder's inequality for the measure $dh/|h|^d$, and we have

$$\left\| 2^{-j\varepsilon} \left(\int_{2^j|h|>1} \frac{F(h)^q}{|h|^{(\alpha+\varepsilon)q}} \frac{dh}{|h|^d} \right)^{1/q} \right\|_{\ell^q} \lesssim \big\| |h|^{-\alpha} F(h) \big\|_{L^q(dh/|h|^d)},$$

since the sum of $2^{-j\varepsilon q}$ over j such that $2^j|h| > 1$ is bounded by $|h|^{\varepsilon q}$.

Q.E.D.

Lemma 2.4. *Let $\alpha > 0$. For any $\omega \in D^\alpha_{p,q}$, one has the bound*

$$\big\| \Delta_{<j}(\omega(x, \cdot))(x) \big\|_{\mathbb{B}^\alpha_{p,q}} \lesssim \|\omega\|_{D^\alpha_{p,q}}.$$

Proof. Since

$$\begin{aligned}
\Delta_{<j}(\omega(x, \cdot))(x) &= \int_{\mathbb{R}^d} Q_{<j}(x, y) \omega(x, y) dy \\
&= \int_{\mathbb{R}^d} Q_{<j}(-h) \omega(x, x+h) dh,
\end{aligned}$$

by using Minkowski's inequality and Lemma 2.3 we have

$$\begin{aligned}\big\|\Delta_{<j}(\omega(x,\cdot))(x)\big\|_{L^p} &\lesssim \int_{\mathbb{R}^d} |Q_{<j}(-h)|\,\|\omega(x,x+h)\|_{L^p(dx)}dh\\ &\lesssim \|\omega\|_{D^\alpha_{p,q}} 2^{-j\alpha}\mathbf{1}^q_j,\end{aligned}$$

where $\mathbf{1}^q_j$ denotes a sequence belonging to the unit sphere of ℓ^q. Q.E.D.

Through this paper, we often use the notation $\mathbf{1}^q_j$ without notice.

Lemma 2.5. *Let $\{\omega_j(x,y)\}_{j=-1}^\infty$ be a sequence of two parameter functions. Assume that for some $C>0$ and $\alpha>0$, the bound*

$$\begin{aligned}(4)\qquad \|\omega_j(x+h,x)\|_{\mathbb{B}^{\alpha-\theta}_{p,q}} &:= \Big\|\big\{2^{j(\alpha-\theta)}\|\omega_j(x+h,x)\|_{L^p(dx)}\big\}_{j\ge -1}\Big\|_{\ell^q}\\ &\le C|h|^\theta\end{aligned}$$

holds for any $h\in\mathbb{R}^d$ and any θ in a neighborhood of α. Then $\omega=\sum_{j\ge -1}\omega_j$ converges in $D^\alpha_{p,q}$ and one has the bound

$$\|\omega\|_{D^\alpha_{p,q}} \lesssim C.$$

Proof. We follow the proof of [1, Theorem 2.36]. Since the case $q=\infty$ is the same as [6, Lemma 3.7], we consider $q<\infty$. Assume $C\le 1$ without loss of generality. Let

$$\begin{aligned}A_N &= \big\{h\in\mathbb{R}^d;\, 2^{-N-1}\le |h| < 2^{-N}\big\}\quad (N\ge 0),\\ A_{-1} &= \big\{h\in\mathbb{R}^d;\, 1\le |h|\big\}.\end{aligned}$$

Fix a small $\varepsilon>0$ such that (4) holds for $\theta=\alpha\pm\varepsilon$. If $h\in A_N$ with $N\ge 0$,

$$\begin{aligned}|h|^{-\alpha}\|\omega(x+h,x)\|_{L^p(dx)} &\lesssim \sum_j |h|^{-\alpha}\|\omega_j(x+h,x)\|_{L^p(dx)}\\ &\lesssim \sum_{-1\le j<N}\mathbf{1}^q_j 2^{j\varepsilon}|h|^\varepsilon + \sum_{j\ge N}\mathbf{1}^q_j 2^{-j\epsilon}|h|^{-\varepsilon}.\end{aligned}$$

By Hölder's inequality with the weight $2^{\pm j\varepsilon}$, we have

$$\begin{aligned}\Big(\sum_{j<N}\mathbf{1}^q_j 2^{j\varepsilon}|h|^\varepsilon\Big)^q &\lesssim |h|^{\varepsilon q}\Big(\sum_{j<N}2^{j\varepsilon}\Big)^{q-1}\sum_{j<N}\mathbf{1}^1_j 2^{j\varepsilon}\\ &\lesssim |h|^{\varepsilon q}2^{N\varepsilon(q-1)}\sum_{j<N}\mathbf{1}^1_j 2^{j\varepsilon},\end{aligned}$$

and
$$\Big(\sum_{j\geq N}\mathbf{1}_j^q 2^{-j\varepsilon}|h|^{-\varepsilon}\Big)^q\lesssim |h|^{-\varepsilon q}2^{-N\varepsilon(q-1)}\sum_{j\geq N}\mathbf{1}_j^1 2^{-j\varepsilon}.$$

Since $|h|\sim 2^{-N}$ on A_N, we have
$$\begin{aligned}\int_{A_N}&\Big(|h|^{-\alpha}\|\omega(x+h,x)\|_{L^p(dx)}\Big)^q\frac{dh}{|h|^d}\\&\lesssim 2^{-N\varepsilon}\sum_{j<N}\mathbf{1}_j^1 2^{j\varepsilon}+2^{N\varepsilon}\sum_{j\geq N}\mathbf{1}_j^1 2^{-j\varepsilon}\lesssim\sum_{j\geq -1}2^{-|j-N|\varepsilon}\mathbf{1}_j^1.\end{aligned}$$

Summing them over $N\geq 0$, by Young's inequality we have
$$\sum_{N\geq 0}\sum_{j\geq -1}2^{-|j-N|\varepsilon}\mathbf{1}_j^1<\infty.$$

If $h\in A_{-1}$, similarly to above,
$$\begin{aligned}\int_{A_{-1}}&\Big(|h|^{-\alpha}\|\omega(x+h,x)\|_{L^p(dx)}\Big)^q\frac{dh}{|h|^d}\\&\lesssim\sum_{j\geq -1}\mathbf{1}_j^1 2^{-j\varepsilon}\int_{|h|\geq 1}|h|^{-\varepsilon q-d}dh\lesssim 1,\end{aligned}$$

which completes the proof. Q.E.D.

Using above lemmas, we can prove Proposition 2.2.

Proof of Proposition 2.2. Note that $\|\Delta_{-1}f\|_{L^p}\lesssim\|f\|_{L^p}$. For $j\geq 0$, since $\int Q_j=0$ we have
$$\Delta_j f(x)=\Delta_j(\omega_f(x,\cdot))(x).$$

Lemma 2.4 yields $\|f\|_{B_{p,q}^\alpha}\lesssim\|f\|_{L^p}+\|\omega_f\|_{D_{p,q}^\alpha}$. To show the converse, let $\omega_j(x,y)=\Delta_j f(y)-\Delta_j f(x)$ and apply Lemma 2.5. Obviously,
$$\|\omega_j(x,x+h)\|_{L^p(dx)}\lesssim\|\Delta_j f\|_{L^p}\lesssim 2^{-j\alpha}\mathbf{1}_j^q\|f\|_{B_{p,q}^\alpha}.$$

By Minkowski's inequality and the continuity of the differentiation $B_{p,q}^\alpha\ni f\mapsto\nabla f\in(B_{p,q}^{\alpha-1})^d$ (see [1, Proposition 2.78]), we have
$$\begin{aligned}\|\omega_j(x,x+h)\|_{L^p(dx)}&\leq\left\|h\cdot\int_0^1\nabla\Delta_j f(x+\theta h)d\theta\right\|_{L^p(dx)}\\&\leq|h|\,\|\Delta_j(\nabla f)\|_{L^p}\lesssim|h|2^{j(1-\alpha)}\mathbf{1}_j^q\|f\|_{B_{p,q}^\alpha}.\end{aligned}$$

By an interpolation, for any $\theta \in [0,1]$ we have

$$\|\omega_j(x+h,x)\|_{L^p(dx)} \lesssim |h|^\theta 2^{j(\theta-\alpha)} \mathbf{1}_j^q \|f\|_{B^\alpha_{p,q}},$$

so $\|\omega_f\|_{D^\alpha_{p,q}} \lesssim \|f\|_{B^\alpha_{p,q}}$ by Lemma 2.5. Q.E.D.

2.3. Paraproduct

For any smooth functions f, g, we can decompose the product fg by

$$\begin{aligned} fg = \sum_{j,k \geq -1} \Delta_j f \Delta_k g &= \Big(\sum_{j<k-1} + \sum_{|j-k|\leq 1} + \sum_{j+1<k} \Big) \Delta_j f \Delta_k g \\ &=: f \prec g + f \circ g + f \succ g. \end{aligned}$$

$f \prec g = g \succ f$ is called a *paraproduct*, and $f \circ g$ is called a *resonant*.

As in [2], we often use the two parameter extension of the paraproduct. For any measurable function $\omega(x,y)$, we define

$$\mathsf{P}_j(\omega)(z) := \iint_{\mathbb{R}^d \times \mathbb{R}^d} Q_{<j-1}(z,x) Q_j(z,y) \omega(x,y) dx dy$$

and $\mathsf{P}(\omega) := \sum_j \mathsf{P}_j(\omega)$. Obviously, for the case $\omega(x,y) = f(x)g(y)$ we have $\mathsf{P}(\omega) = f \prec g$.

Lemma 2.6. *Let $\alpha > 0$. If $\omega \in D^\alpha_{p,q}$, then $\mathsf{P}(\omega) \in B^\alpha_{p,q}$. The mapping $\omega \mapsto \mathsf{P}(\omega)$ is continuous.*

Proof. In view of [2, Proposition 8],

$$\|\mathsf{P}(\omega)\|_{B^\alpha_{p,q}} \lesssim \|\mathsf{P}_j(\omega)\|_{\mathbb{B}^\alpha_{p,q}}.$$

For the right hand side, exchanging variables $y = z+h$ and $x = z+h+k$ and using Minkowski's inequality, we have

$$\begin{aligned} &\|\mathsf{P}_j(\omega)\|_{L^p} \\ &\lesssim \iint |Q_{<j-1}(-h)|\,|Q_j(-h-k)|\,\|\omega(z+h, z+h+k)\|_{L^p(dz)} dh dk \\ &\lesssim \int 2^{jd} \varphi(2^j k) \|\omega(z, z+k)\|_{L^p(dz)} dk, \end{aligned}$$

for some $\varphi \in \mathcal{S}(\mathbb{R}^d)$. Thus Lemma 2.3 completes the proof. Q.E.D.

2.4. Word Hopf algebra

We introduce a specific regularity structure. Fix an integer n. For any integers $1 \leq k \leq \ell \leq n$, denote by $(k \dots \ell)$ the sequence from k to ℓ, which is called a *word* throughout this paper. We discuss the algebras made from the set W of all such words. Let $\mathrm{Alg}(W)$ be the commutative algebra freely generated by W with unit $\mathbf{1}$. We regard $\mathbf{1}$ as an empty word and consider the extended set $\overline{W} = W \cup \{\mathbf{1}\}$ of words. For any nonempty words $\sigma = (k \dots \ell)$ and $\eta = ((\ell+1) \dots m)$ in W, we define $\sigma \sqcup \eta = (k \dots m)$. We also define $\mathbf{1} \sqcup \tau = \tau \sqcup \mathbf{1} = \tau$.

Definition 2.4. *Define the linear map* $\Delta : \mathrm{Alg}(W) \to \mathrm{Alg}(W) \otimes \mathrm{Alg}(W)$ *by*

$$\Delta\tau = \sum_{\sigma,\eta \in \overline{W},\, \sigma \sqcup \eta = \tau} \sigma \otimes \eta$$

for any $\tau \in \overline{W}$.

The map Δ is coassociative; $(\Delta \otimes \mathrm{id})\Delta = (\mathrm{id} \otimes \Delta)\Delta$. It is easy to show the existence of the algebra map $A : \mathrm{Alg}(W) \to \mathrm{Alg}(W)$ such that

$$\begin{aligned} &A\mathbf{1} = \mathbf{1}, \\ &M(A \otimes \mathrm{id})\Delta\tau = M(\mathrm{id} \otimes A)\Delta\tau = 0 \quad (\tau \in W), \end{aligned}$$

where $M : \mathrm{Alg}(W) \otimes \mathrm{Alg}(W) \to \mathrm{Alg}(W)$ is the product map. Such A is called an antipode. In other words, $\mathrm{Alg}(W)$ is a *Hopf algebra*. The existence of A yields that, the set G of all algebra maps $\gamma : \mathrm{Alg}(W) \to \mathbb{R}$ forms a group by the product

$$(\gamma_1 * \gamma_2)(\tau) = (\gamma_1 \otimes \gamma_2)\Delta\tau.$$

The inverse of $\gamma \in G$ is given by $\gamma^{-1} = \gamma \circ A$.

In Section 3, we study the family $\{f_\tau = f_\tau(x)\}_{\tau \in W}$ of functions on $\mathbb{R}^d$, indexed by words. We regard $f(x) \in G$ by extending the map $\tau \mapsto f_\tau(x)$ algebraically. Then we define the G-valued two parameter function by

$$\omega(x, y) = f(x)^{-1} * f(y).$$

In other words, we have a family $\{\omega_\tau(x, y) := \omega(x, y)(\tau)\}_{\tau \in W}$ of two parameter functions, indexed by words. The following relationships between f and ω are useful in Section 3.

Lemma 2.7. *For any $1 \le k \le \ell \le n$, one has*

$$\omega_{k\ldots\ell}(x,y) = f_{k\ldots\ell}(y) - f_{k\ldots\ell}(x) - \sum_{m=k}^{\ell-1} f_{k\ldots m}(x)\omega_{(m+1)\ldots\ell}(x,y), \tag{5}$$

$$\begin{aligned}\omega_{k\ldots\ell}(x,z) &= \omega_{k\ldots\ell}(x,y) + \omega_{k\ldots\ell}(y,z) \\ &\quad + \sum_{m=k}^{\ell-1} \omega_{k\ldots m}(x,y)\omega_{(m+1)\ldots\ell}(y,z).\end{aligned} \tag{6}$$

Proof. Immediate consequences of the simple formulas. (5): $f(y) = f(x) * \omega(x,y)$, (6): $\omega(x,z) = \omega(x,y) * \omega(y,z)$. Q.E.D.

§3. Taylor remainders of iterated paraproducts

For a given sequence $f_1, f_2, \ldots$ of functions, we define the *iterated paraproducts*

$$(f_1)^{\prec} := f_1, \quad (f_1, \ldots, f_n)^{\prec} := (f_1, \ldots, f_{n-1})^{\prec} \prec f_n.$$

The aim of this section is to show the following Besov type estimate, which is an extension of [6, Theorem 3.1]. We write $f_{k\ldots\ell}^{\prec} := (f_k, \ldots, f_\ell)^{\prec}$.

Theorem 3.1. *For any measurable functions $f_1, \ldots, f_n$, we define the family*

$$\{\omega_{k\ldots\ell}^{\prec}(x,y)\}_{1 \le k \le \ell \le n}$$

of two parameter functions by the recursive formula (5) with $f_{k\ldots\ell}$ replaced by $f_{k\ldots\ell}^{\prec}$. Let $\alpha_1, \ldots, \alpha_n \in (0,1)$, $p_1, \ldots, p_n, q_1, \ldots, q_n \in [1, \infty]$, and $f_i \in B_{p_i,q_i}^{\alpha_i}$ for each i. If $\alpha := \alpha_1 + \cdots + \alpha_n < 1$, $\frac{1}{p} = \frac{1}{p_1} + \cdots + \frac{1}{p_n} \le 1$, and $\frac{1}{q} = \frac{1}{q_1} + \cdots + \frac{1}{q_n} \le 1$, then we have $\omega_{1\ldots n}^{\prec} \in D_{p,q}^{\alpha}$ and

$$\|\omega_{1\ldots n}^{\prec}\|_{D_{p,q}^{\alpha}} \lesssim \|f_1\|_{B_{p_1,q_1}^{\alpha_1}} \cdots \|f_n\|_{B_{p_n,q_n}^{\alpha_n}}. \tag{7}$$

3.1. Simplified iterated paraproducts

Fix the parameters and the functions as in Theorem 3.1. For any $1 \le k \le \ell \le n$, we use the following simplifying notations.

$$\alpha_{k\ldots\ell} := \alpha_k + \cdots + \alpha_\ell, \quad \frac{1}{p_{k\ldots\ell}} := \frac{1}{p_k} + \cdots + \frac{1}{p_\ell}, \quad \frac{1}{q_{k\ldots\ell}} := \frac{1}{q_k} + \cdots + \frac{1}{q_\ell}.$$

First we show the existence of the family $\{\tilde{f}_{k\ldots\ell}\}_{1 \le k \le \ell \le n}$ such that the corresponding $\{\tilde{\omega}_{k\ldots\ell}\}_{1 \le k \le \ell \le n}$ satisfies the bound (7).

Definition 3.1. *For any $j \geq -1$, we define*

$$(\tilde{f}_k)_j := \Delta_j f_k, \quad (\tilde{f}_{k\dots\ell})_j := (\tilde{f}_{k\dots(\ell-1)})_{<j-1}(\tilde{f}_\ell)_j,$$

(the latter definition has a meaning only if $j \geq 1$) and set

$$\tilde{f}_{k\dots\ell} = \sum_j (\tilde{f}_{k\dots\ell})_j.$$

Moreover, we define the family $\{\tilde{\omega}_{k\dots\ell}\}_{1\leq k\leq\ell\leq n}$ by the recursive formula (5) with $f_{k\dots\ell}$ replaced by $\tilde{f}_{k\dots\ell}$.

We consider the decomposition $\tilde{\omega}_{k\dots\ell} = \sum_j (\tilde{\omega}_{k\dots\ell})_j$ as follows. The proof of this lemma is left to the reader.

Lemma 3.2. *Define $(\tilde{\omega}_{k\dots\ell})_j$ recursively by*

$$(\tilde{\omega}_{k\dots\ell})_j(x,y) = (\tilde{f}_{k\dots\ell})_j(y) - (\tilde{f}_{k\dots\ell})_j(x) - \sum_{m=k}^{\ell-1} \tilde{f}_{k\dots m}(x)(\tilde{\omega}_{(m+1)\dots\ell})_j(x,y).$$

Then one has the following formulas.

(1) $(\tilde{\omega}_k)_j(x,y) = \Delta_j f_k(y) - \Delta_j f_k(x)$.

(2) *If $k < \ell$,*

$$(\tilde{\omega}_{k\dots\ell})_j(x,y)$$
$$= (\tilde{\omega}_{k\dots(\ell-1)})_{<j-1}(x,y)(\tilde{f}_\ell)_j(y) - (C_{k\dots(\ell-1)})_{\geq j-1}(x)(\tilde{\omega}_\ell)_j(x,y),$$

where $(C_{k\dots\ell})_{1\leq k\leq\ell\leq n}$ is recursively defined by $(C_k)_j(x) = \Delta_j f_k(x)$ and

$$(C_{k\dots\ell})_j(x) = (\tilde{f}_{k\dots\ell})_j(x) - \sum_{m=k}^{\ell-1} \tilde{f}_{k\dots m}(x)(C_{(m+1)\dots\ell})_j(x).$$

(3) *If $k < \ell$,*

$$(C_{k\dots\ell})_j(x) = -(C_{k\dots(\ell-1)})_{\geq j-1}(x)(\tilde{f}_\ell)_j(x).$$

Proof of the bound (7) *for $\tilde{\omega}$.* Without loss of generality, we assume $\|f_i\|_{B^{\alpha_i}_{p_i,q_i}} \leq 1$ for any i. To apply Lemma 2.5, we show the bound

$$\big\|(\tilde{\omega}_{1\dots n})_j(x,x+h)\big\|_{\mathbb{B}^{\alpha_{1\dots n}-\theta}_{p_{1\dots n},q_{1\dots n}}} \lesssim |h|^\theta \tag{8}$$

uniformly over θ in a neighborhood of $\alpha_{1\dots n} < 1$. The case $n = 1$ is already proved in the proof of Proposition 2.2, in Section 2.2. Let $n \geq 2$. By Lemma 3.2-(3), we inductively have

$$\begin{aligned}&\left\|(C_{1\dots n})_j\right\|_{\mathbb{B}^{\alpha_{1\dots n}}_{p_{1\dots n},q_{1\dots n}}}\\&\leq \left\|(C_{1\dots(n-1)})_{\geq j-1}\right\|_{\mathbb{B}^{\alpha_{1\dots(n-1)}}_{p_{1\dots(n-1)},q_{1\dots(n-1)}}}\left\|(\tilde{f}_n)_j\right\|_{\mathbb{B}^{\alpha_n}_{p_n,q_n}} \lesssim 1,\end{aligned}$$

where we use Lemma 2.1-(2) for the bound of $(C_{1\dots(n-1)})_{\geq j-1}$. Assume (8) holds for the word $(1\dots(n-1))$, uniformly over $\theta \in (\alpha_{1\dots(n-2)}, 1]$. Then by Lemma 2.1-(3), we have

$$\left\|(\tilde{\omega}_{1\dots(n-1)})_{<j-1}(x, x+h)\right\|_{\mathbb{B}^{\alpha_{1\dots(n-1)}-\theta}_{p_{1\dots(n-1)},q_{1\dots(n-1)}}} \lesssim |h|^\theta$$

for any $\theta \in (\alpha_{1\dots(n-1)}, 1]$. For such θ, by Lemma 3.2-(2),

$$\begin{aligned}&\left\|(\tilde{\omega}_{1\dots n})_j(x, x+h)\right\|_{\mathbb{B}^{\alpha_{1\dots n}-\theta}_{p_{1\dots n},q_{1\dots n}}}\\&\leq \left\|(\tilde{\omega}_{1\dots(n-1)})_{<j-1}(x, x+h)\right\|_{\mathbb{B}^{\alpha_{1\dots(n-1)}-\theta}_{p_{1\dots(n-1)},q_{1\dots(n-1)}}}\left\|(\tilde{f}_n)_j\right\|_{\mathbb{B}^{\alpha_n}_{p_n,q_n}}\\&\quad+\left\|(C_{1\dots(n-1)})_{\geq j-1}\right\|_{\mathbb{B}^{\alpha_{1\dots(n-1)}}_{p_{1\dots(n-1)},q_{1\dots(n-1)}}}\left\|(\tilde{\omega}_n)_j\right\|_{\mathbb{B}^{\alpha_n-\theta}_{p_n,q_n}} \lesssim |h|^\theta.\end{aligned}$$

Thus we have the required bound by an induction on n. Q.E.D.

3.2. Proof of Theorem 3.1

We show the bound (7) for $\omega^{\prec}$, which is really required. For any word $\tau = (k\dots\ell)$, denote by $\Pi(\tau)$ the set of all *partitions* of τ, that is, we write

$$\{\tau_1, \dots, \tau_m\} \in \Pi(\tau)$$

if $\tau_1, \dots, \tau_m$ are nonempty words of the form $\tau_j = (k_j \dots \ell_j)$ for each j, where $k_1 = k$, $\ell_m = \ell$, and $\ell_j + 1 = k_{j+1}$ for any j. Recall the definitions of $\alpha_\tau = \alpha_{k\dots\ell}$, p_τ, and q_τ as before.

Lemma 3.3. *For any word $\tau = (k\dots\ell)$, there exists a function $[\tilde{f}]_\tau \in B^{\alpha_\tau}_{p_\tau,q_\tau}$ continuously depending on $f_k, \dots, f_\ell$, such that, one has the formula*

$$\tilde{f}_\tau = \sum_{\{\sigma,\eta\}\in\Pi(\tau)} \tilde{f}_\sigma \prec [\tilde{f}]_\eta + [\tilde{f}]_\tau. \tag{9}$$

Moreover, one has the atomic decomposition

$$\tilde{f}_\tau = \sum_{m=1}^{\infty} \sum_{\{\tau_1,\dots,\tau_m\}\in\Pi(\tau)} ([\tilde{f}]_{\tau_1}, \dots, [\tilde{f}]_{\tau_m})^{\prec}. \tag{10}$$

Proof. Second formula (10) is an immediate consequence of the first one (9). The proof of (9) is essentially the same as [2, Proposition 12]. The point is that we use Besov norms $B^{\alpha}_{p,q}$, while in [2] the particular case $p = q = \infty$ is considered.

Here we give a proof of (9). Write $\omega_f(x,y) = f(y) - f(x)$ for simplicity. Expanding $\tilde{\omega}_\tau$ by repeating (5), we have

(11)

$$\begin{aligned}
&\tilde{\omega}_\tau(x,y)\\
&= \omega_{\tilde{f}_\tau}(x,y) - \sum_{\{\tau_1,\tau_2\}\in\Pi(\tau)} \tilde{f}_{\tau_1}(x)\tilde{\omega}_{\tau_2}(x,y)\\
&= \cdots\\
&= \omega_{\tilde{f}_\tau}(x,y) - \sum_{m=2}^{\infty}(-1)^m \sum_{\{\tau_1,\dots,\tau_m\}\in\Pi(\tau)} (\tilde{f}_{\tau_1}\cdots\tilde{f}_{\tau_{m-1}})(x)\omega_{\tilde{f}_{\tau_m}}(x,y).
\end{aligned}$$

Applying the two parameter operator P to both sides, we have

$$\mathsf{P}(\tilde{\omega}_\tau) = 1 \prec \tilde{f}_\tau - \sum_{m=2}^{\infty}(-1)^m \sum_{\{\tau_1,\dots,\tau_m\}\in\Pi(\tau)} (\tilde{f}_{\tau_1}\cdots\tilde{f}_{\tau_{m-1}}) \prec \tilde{f}_{\tau_m}.$$

By Lemma 2.6, $\mathsf{P}(\tilde{\omega}_\tau)$ belongs to $B^{\alpha_\tau}_{p_\tau,q_\tau}$ and continuously depends on $f_k,\dots,f_\ell$. If $\tilde{f}_{\tau_m}$ has a decomposition (9),

$$\begin{aligned}
&\sum_{m=2}^{\infty}(-1)^m \sum_{\{\tau_1,\dots,\tau_m\}\in\Pi(\tau)} (\tilde{f}_{\tau_1}\cdots\tilde{f}_{\tau_{m-1}}) \prec \tilde{f}_{\tau_m}\\
&= \sum_{m=2}^{\infty}(-1)^m \sum_{\{\tau_1,\dots,\tau_m\}\in\Pi(\tau)} (\tilde{f}_{\tau_1}\cdots\tilde{f}_{\tau_{m-1}}) \prec [\tilde{f}]_{\tau_m}\\
&\quad + \sum_{m=2}^{\infty}(-1)^m \sum_{\{\tau_1,\dots,\tau_m,\tau_{m+1}\}\in\Pi(\tau)} (\tilde{f}_{\tau_1}\cdots\tilde{f}_{\tau_{m-1}}) \prec (\tilde{f}_{\tau_m} \prec [\tilde{f}]_{\tau_{m+1}})\\
&= \sum_{\{\tau_1,\tau_2\}\in\Pi(\tau)} \tilde{f}_{\tau_1} \prec [\tilde{f}]_{\tau_2}\\
&\quad + \sum_{m=2}^{\infty}(-1)^m \sum_{\{\tau_1,\dots,\tau_m,\tau_{m+1}\}\in\Pi(\tau)} \mathsf{R}(\tilde{f}_{\tau_1}\cdots\tilde{f}_{\tau_{m-1}}, \tilde{f}_{\tau_m}, [\tilde{f}]_{\tau_{m+1}}),
\end{aligned}$$

where R is the correcting operator defined by

$$\mathsf{R}(a,b,c) := a \prec (b \prec c) - (ab) \prec c.$$

The sum of all R terms belongs to $B^{\alpha_\tau}_{p_\tau,q_\tau}$ and continuously depends on $f_k,\dots,f_\ell$. Its proof is left to Lemma 3.4 below. Then we obtain the formula (9) since

$$\|\tilde{f}_\tau - 1 \prec \tilde{f}_\tau\|_{B^r_{p_\tau,q_\tau}} = \|\Delta_{\le 0}\tilde{f}_\tau\|_{B^r_{p_\tau,q_\tau}} \lesssim \|\tilde{f}_\tau\|_{B^{\alpha_\ell}_{p_\tau,q_\tau}}$$

for any $r > 0$. Q.E.D.

Lemma 3.4. *Let* $\sigma = (k\dots\ell)$, $\alpha' > 0$, $p',q' \in [1,\infty]$, *and* $g \in B^{\alpha'}_{p',q'}$. *Assume that* $\alpha = \alpha_\sigma + \alpha' < 1$, $1/p = 1/p_\sigma + 1/p' \le 1$, *and* $1/q = 1/q_\sigma + 1/q' \le 1$. *Then one has the bound*

$$\left\| \sum_{m=2}^{\infty} (-1)^m \sum_{\{\tau_1,\dots,\tau_m\}\in\Pi(\sigma)} \mathsf{R}(\tilde{f}_{\tau_1}\dots\tilde{f}_{\tau_{m-1}}, \tilde{f}_{\tau_m}, g) \right\|_{B^\alpha_{p,q}}$$
$$\lesssim \|f_k\|_{B^{\alpha_k}_{p_k,q_k}} \cdots \|f_\ell\|_{B^{\alpha_\ell}_{p_\ell,q_\ell}} \|g\|_{B^{\alpha'}_{p',q'}}.$$

Proof. Just an analogue of [2, Proposition 10], so see it for details. In view of the formula (11), it is sufficient to show that

$$\big\|\mathsf{P}_j\big((\tilde{\omega}_\sigma(x,\cdot) \prec g)(y)\big)\big\|_{\mathbb{B}^\alpha_{p,q}} < \infty,$$

where we write $\mathsf{P}_j(\Omega) = \mathsf{P}_j(\Omega(x,y))$ as an abuse of notation. Since the integral $\int Q_j(z,y)Q_{<i-1}(y,u)Q_i(y,v)dy$ vanishes if $|i-j| \ge N$ for some constant N,

$$\int Q_j(z,y)\big((\tilde{\omega}_\sigma(x,\cdot) \prec g)(y)\big)dy$$
$$= \sum_{i;|i-j|<N} \int Q_j(z,y)\Delta_{<i-1}(\tilde{\omega}_\sigma(x,\cdot))(y)\Delta_i g(y)dy.$$

By the formula (6),

$$\Delta_{<i-1}(\tilde{\omega}_\sigma(x,\cdot))(y) = \tilde{\omega}_\sigma(x,y) + \Delta_{<i-1}(\tilde{\omega}_\sigma(y,\cdot))(y)$$
$$+ \sum_{\{\eta,\zeta\}\in\Pi(\sigma)} \tilde{\omega}_\eta(x,y)\Delta_{<i-1}(\tilde{\omega}_\zeta(y,\cdot))(y).$$

Hence, by using Lemma 2.4, we see that $\mathsf{P}_j\big((\tilde{\omega}_\sigma(x,\cdot) \prec g)(y)\big)(z)$ is a sum of the integrals of the form

$$\sum_{i;|i-j|<N} \iint Q_{<j-1}(z,x)Q_j(z,y)A(x,y)B_i(y)C_i(y)dxdy, \tag{12}$$

where $A \in D^{\alpha_A}_{p_A,q_A}$, $B \in \mathbb{B}^{\alpha_B}_{p_B,q_B}$, $C \in \mathbb{B}^{\alpha_C}_{p_C,q_C}$, and parameters are such that $\alpha = \alpha_A + \alpha_B + \alpha_C$, $1/p = 1/p_A + 1/p_B + 1/p_C$, and $1/q = 1/q_A + 1/q_B + 1/q_C$. Exchanging variables $y = z + h$ and $x = z + h + k$, we see that the $L^p(dz)$ bound of such an integral is as follows.

$$\sum_{i;|i-j|<N} \iint |Q_{<j-1}(-h-k)||Q_j(-h)| \\ \times \|A(z+k,z)\|_{L^{p_A}} \|B_i\|_{L^{p_B}} \|C_i\|_{L^{p_C}} dhdk$$
$$\lesssim \int 2^{jd} K(2^j k) \|A(z+k,z)\|_{L^{p_A}} dk \; 2^{-j(\alpha_B+\alpha_C)} c_j,$$

where $K \in \mathcal{S}(\mathbb{R}^d)$ and $\{c_j\} \in \ell^{q_{BC}}$ with $1/q_{BC} = 1/q_B + 1/q_C$. By Lemma 2.3, we have that the above integral is bounded by $2^{-j\alpha} d_j$ with $\{d_j\} \in \ell^q$, which completes the proof. Q.E.D.

For any partition $\{\tau_1, \ldots, \tau_m\} \in \sqcap(\tau)$, we define

$$[\tilde{f}]^{\prec}_{\tau_1 \ldots \tau_m} = ([\tilde{f}]_{\tau_1}, \ldots, [\tilde{f}]_{\tau_m})^{\prec}, \tag{13}$$
$$[\tilde{\omega}]^{\prec}_{\tau_1 \ldots \tau_m}(x,y) = [\tilde{f}]^{\prec}_{\tau_1 \ldots \tau_m}(y) - [\tilde{f}]^{\prec}_{\tau_1 \ldots \tau_m}(x) \\ - \sum_{j=1}^{m-1} [\tilde{f}]^{\prec}_{\tau_1 \ldots \tau_j}(x) [\tilde{\omega}]^{\prec}_{\tau_{j+1} \ldots \tau_m}(x,y). \tag{14}$$

Summing (14) over all $\{\tau_1, \ldots, \tau_m\} \in \sqcap(\tau)$, we can inductively obtain

$$\tilde{\omega}_\tau = \sum_{m=1}^{\infty} \sum_{\{\tau_1,\ldots,\tau_m\} \in \sqcap(\tau)} [\tilde{\omega}]^{\prec}_{\tau_1 \ldots \tau_m} =: \sum_{\Xi \in \sqcap(\tau)} [\tilde{\omega}]^{\prec}_{\Xi}. \tag{15}$$

Proof of Theorem 3.1. We emphasize the dependence of $\omega^{\prec}_{k \ldots \ell}$ on $f_k \ldots, f_\ell$ by writing

$$\omega^{\prec}_{k \ldots \ell} = \omega^{\prec}(f_k, \ldots, k_\ell).$$

We prove the result by an induction on the number of the components of $\omega^{\prec}$. By the formula (9), $[\tilde{f}]_{(k)} = f_k$ for a word with only one letter. Hence if $\Xi \in \sqcap(\tau)$ has the same cardinality as the length of τ (denoted by $|\Xi| = |\tau|$), we have $[\tilde{f}]^{\prec}_{\Xi} = f^{\prec}_\tau$ and $[\tilde{\omega}]^{\prec}_{\Xi} = \omega^{\prec}_\tau$. Hence by (15),

$$\omega^{\prec}_\tau = \tilde{\omega}_\tau - \sum_{\Xi \in \sqcap(\tau), |\Xi| < |\tau|} [\tilde{\omega}]^{\prec}_{\Xi}.$$

The bound for $\tilde{\omega}_\tau$ was already obtained. By an assumption of the induction, $\omega^{\prec}$ is continuous as a less than $|\tau|$-component operator. Hence

$$\begin{aligned}\left\|[\tilde{\omega}]^{\prec}_{\tau_1\dots\tau_m}\right\|_{D^{\alpha_\tau}_{p_\tau,q_\tau}} &= \left\|\omega^{\prec}([\tilde{f}]_{\tau_1},\dots,[\tilde{f}]_{\tau_m})\right\|_{D^{\alpha_\tau}_{p_\tau,q_\tau}} \\ &\lesssim \|[\tilde{f}]_{\tau_1}\|_{B^{\alpha_{\tau_1}}_{p_{\tau_1},q_{\tau_1}}}\cdots\|[\tilde{f}]_{\tau_m}\|_{B^{\alpha_{\tau_m}}_{p_{\tau_m},q_{\tau_m}}} \\ &\lesssim \|f_k\|_{B^{\alpha_k}_{p_k,q_k}}\cdots\|f_\ell\|_{B^{\alpha_\ell}_{p_\ell,q_\ell}},\end{aligned}$$

where we use the continuity of $(f_{k_j},\dots,f_{\ell_j})\mapsto[\tilde{f}]_{\tau_j}$ (Lemma 3.3). As a result, we obtain the continuity of $\omega^{\prec}_\tau$ with respect to $f_k,\dots,f_\ell$. Q.E.D.

§4. Besov type regularity structure and commutator estimates

We prove Theorem 1.1 in the rest of this paper. We show only the existence of the continuous map $\tilde{\mathsf{C}}$. The uniqueness of $\tilde{\mathsf{C}}$ and its multilinearity follows from the denseness argument.

4.1. Besov type regularity structure

We return to the Hopf algebra $\mathrm{Alg}(W)$ with the character group G. We consider a subset

$$V=\{\mathbf{1}\}\cup\{(k\dots n)\ ;\ k=1,\dots,n\},$$

of $\overline{W}$ and a linear subspace $T=\langle V\rangle$. Since $\Delta T\subset\mathrm{Alg}(W)\otimes T$, for any $\gamma\in G$ we can define the linear map $\Gamma_\gamma:T\to T$ by

$$\Gamma_\gamma=(\gamma\otimes\mathrm{id})\Delta.$$

The pair (T,G) is an example of the *regularity structure*.

Remark 4.1. *Note that the position of γ is opposite to the original definition [5]. Because of it, in the mapping $\gamma\mapsto\Gamma_\gamma$, the order of multiplication is turned over as follows.*

$$\Gamma_{\gamma_1}\Gamma_{\gamma_2}=\Gamma_{\gamma_2*\gamma_1}. \tag{16}$$

We define a model (Π,Γ) on the regularity structure (T,G). Fix the parameters and the functions satisfying the assumptions in Theorem 1.1. For any $1\le k\le n$, we define

$$\begin{aligned}&\alpha'_{k\dots n}=\alpha_k+\cdots+\alpha_n+\alpha_\circ,\\ &\frac{1}{p'_{k\dots n}}=\frac{1}{p_k}+\cdots+\frac{1}{p_n}+\frac{1}{p_\circ},\qquad \frac{1}{q'_{k\dots n}}=\frac{1}{q_k}+\cdots+\frac{1}{q_n}+\frac{1}{q_\circ}.\end{aligned}$$

Moreover, set $\alpha'_{\mathbf{1}}=\alpha_\circ$, $p'_{\mathbf{1}}=p_\circ$, and $q'_{\mathbf{1}}=q_\circ$.

Definition 4.1. *Let $f_{k\dots\ell}^{\prec} = (f_k, \dots, f_\ell)^{\prec}$ be the iterated paraproduct. We regard $f^{\prec}(x) \in G$ by extending the map $\tau \mapsto f_\tau^{\prec}(x)$ algebraically, and define*

$$\omega^{\prec}(x,y) = f^{\prec}(x)^{-1} * f^{\prec}(y), \quad \Gamma_{xy} = \Gamma_{\omega^{\prec}(x,y)}.$$

Definition 4.2. *For any linear map $\Pi : T \to \mathcal{S}'$, define*

$$\Pi_x \tau = (f^{\prec}(x)^{-1} \otimes \Pi)\Delta\tau.$$

Denote by $\mathcal{M}$ the set of all maps Π such that

$$\|\Pi\|_{\mathcal{M}} := \sup_{\tau \in V} \left\| \Delta_{<j}(\Pi_x \tau)(x) \right\|_{\mathbb{B}^{\alpha'_\tau}_{p'_\tau, q'_\tau}} < \infty.$$

It is easy to show the following formulas.

$$\Gamma_{yx}\Gamma_{zy} = \Gamma_{zx}, \quad \Pi_y \Gamma_{xy} = \Pi_x. \tag{17}$$

The pair (Π, Γ) is called a *model* on the regularity structure (T, G). Note that these formulas are slightly different from the original ones [5], like the formula (16).

As an analogue of [6, 3], we can show that the space $\mathcal{M}$ has a simple topological structure. Let

$$V^- = \{\mathbf{1}\} \cup \{(k \dots n)\,;\, k = 2, \dots, n\}.$$

Note that $\alpha'_\tau < 0$ for any $\tau \in V^-$ by assumption.

Theorem 4.2. *For any $\Pi \in \mathcal{M}$, define the linear map $[\Pi] : T \to \mathcal{S}'$ by*

$$\Pi\tau = \sum_{\sigma \sqcup \eta = \tau,\, \sigma \neq \mathbf{1}} f_\sigma^{\prec} \prec [\Pi]\eta + [\Pi]\tau. \tag{18}$$

Then $[\Pi]\tau \in B^{\alpha'_\tau}_{p'_\tau, q'_\tau}$ and the mapping

$$(f_1, \dots, f_n, \Pi) \mapsto [\Pi]\tau \in B^{\alpha'_\tau}_{p'_\tau, q'_\tau}$$

is continuous. Conversely, for any given family

$$\{[\Pi]\tau\}_{\tau \in V^-} \in \prod_{\tau \in V^-} B^{\alpha'_\tau}_{p'_\tau, q'_\tau},$$

there exists a unique element $\Pi \in \mathcal{M}$ satisfying (18). *Moreover, the map $\{[\Pi]\tau\}_{\tau \in V^-} \mapsto \Pi$ is continuous.*

The above theorem is just an analogue of [2, Theorem 14 and Corollary 15], so we leave the details to the reader. The only modification is that we have to use the Besov type reconstruction theorem in Appendix.

4.2. Proof of Theorem 1.1

Now we show the iterated commutator estimates. This part is strictly an analogue of [6, Section 4].

Proof of Theorem 1.1. For any given $\xi \in B^{\alpha_\circ}_{p_\circ,q_\circ}$, we can define $\Pi^\xi \in \mathcal{M}$ by

$$[\Pi^\xi]\mathbf{1} = \xi, \quad [\Pi^\xi](k\dots n) = 0 \quad (2 \le k \le n).$$

Note that

$$\begin{aligned} \Pi^\xi \mathbf{1} = \xi, \quad \Pi^\xi(k\dots n) &= f^{\prec}_{k\dots n} \prec \xi \quad (2 \le k \le n), \\ \Pi^\xi(1\dots n) &= f^{\prec}_{1\dots n} \prec \xi + [\Pi^\xi](1\dots n), \end{aligned} \tag{19}$$

by the formula (18). Then the map

$$(f_1, \dots, f_n, \xi) \mapsto [\Pi^\xi](1\dots n)$$

is continuous, which turns out to be the required map $\tilde{\mathsf{C}}$. It remains to show that

$$[\Pi^\xi](1\dots n) = \mathsf{C}(f_1, \dots, f_n, \xi) \tag{20}$$

if all inputs $(f_1, \dots, f_n, \xi)$ are in $\mathcal{S}(\mathbb{R}^d)$. Since $\Pi^\xi = (f^{\prec}(x) \otimes \Pi^\xi_x)\Delta$,

$$\begin{aligned} \Pi^\xi(k\dots n)(x) = \Pi^\xi_x(k\dots n)(x) &+ f^{\prec}_{k\dots n}(x)\xi(x) \\ &+ \sum_{\ell=k}^{n-1} f^{\prec}_{k\dots\ell}(x)\big(\Pi^\xi_x((\ell+1)\dots n)\big)(x), \end{aligned} \tag{21}$$

for any $1 \le k \le n$. By using it and (19), we can inductively show that

$$\Pi^\xi_x(k\dots n)(x) = -\mathsf{C}(f_k, \dots, f_n, \xi)(x)$$

for $2 \le k \le n$. Then letting $k = 1$ in (21) and using

$$\Pi^\xi_x(1\dots n)(x) = \lim_{j\to\infty} \Delta_{<j}(\Pi^\xi_x(1\dots n))(x) = 0$$

because $\Delta_{<j}(\Pi^\xi_x(1\dots n))(x) \in \mathbb{B}^{\alpha'_{1\dots n}}_{p'_{1\dots n},q'_{1\dots n}}$ and $\alpha'_{1\dots n} > 0$, we have (20) by the definition of C. Q.E.D.

§Appendix A. Besov type reconstruction theorem

We define Besov type modelled distributions. Recall that $V = \{\mathbf{1}\} \cup \{(k \dots n)\ ;\ k = 1, \dots, n\}$, and $T = \langle V \rangle$.

Definition A.1. *For any function $\boldsymbol{g} : \mathbb{R}^d \to T$, define*

$$\omega^{\boldsymbol{g}}(x, y) = \boldsymbol{g}(y) - \Gamma_{xy}\boldsymbol{g}(x)$$

and denote by $\omega^{\boldsymbol{g}}_\tau(y, x)$ its τ-component. Let k be the smallest integer such that $\omega^{\boldsymbol{g}}_{k\dots n}(y, x)$ does not vanish, and let $\alpha > \alpha'_{k\dots n}$, $p \in [1, p'_{k\dots n}]$, and $q \in [1, q'_{k\dots n}]$. For such parameters, we define

$$\|\boldsymbol{g}\|_{\mathcal{D}^\alpha_{p,q}} := \sup_\tau \big\|\omega^{\boldsymbol{g}}_\tau\big\|_{D^{\alpha - \alpha'_\tau}_{p \backslash p'_\tau, q \backslash q'_\tau}},$$

where $\frac{1}{p \backslash p'_\tau} = \frac{1}{p} - \frac{1}{p'_\tau}$, $\frac{1}{q \backslash q'_\tau} = \frac{1}{q} - \frac{1}{q'_\tau}$. Let $\mathcal{D}^\alpha_{p,q}$ be the set of functions $\boldsymbol{g} : \mathbb{R}^d \to T$ such that $\|\boldsymbol{g}\|_{\mathcal{D}^\alpha_{p,q}} < \infty$.

Such $\boldsymbol{g}$ is called a *modelled distribution* controlled by Γ. We show the Besov type reconstruction theorem.

Proposition A.1. *For any $\boldsymbol{g} \in \mathcal{D}^\alpha_{p,q}$ and $\Pi \in \mathcal{M}$, we define*

$$\mathcal{P}\boldsymbol{g}(z) = \sum_j \iint_{\mathbb{R}^d \times \mathbb{R}^d} P_j(z, x) Q_j(z, y) \Pi_x(\boldsymbol{g}(x))(y)\, dx\, dy.$$

(1) *If $\alpha > 0$, there exists a unique continuous bilinear map $\mathcal{Q} : \mathcal{D}^\alpha_{p,q} \times \mathcal{M} \to B^\alpha_{p,q}$ such that*

$$\big\|\Delta_{<j}(\mathcal{P}\boldsymbol{g} + \mathcal{Q}\boldsymbol{g} - (\Pi_x \boldsymbol{g}(x))(x))\big\|_{\mathbb{B}^\alpha_{p,q}} < \infty.$$

(2) *If $\alpha < 0$,*

$$\big\|\Delta_{<j}(\mathcal{P}\boldsymbol{g} - (\Pi_x \boldsymbol{g}(x))(x))\big\|_{\mathbb{B}^\alpha_{p,q}} < \infty.$$

(The operator $\mathcal{R}$ defined by $\mathcal{R}\boldsymbol{g} = \mathcal{P}\boldsymbol{g} + \mathcal{Q}\boldsymbol{g}$ if $\alpha > 0$ and $\mathcal{R}\boldsymbol{g} = \mathcal{P}\boldsymbol{g}$ if $\alpha < 0$ is called a reconstruction operator.*)*

Proof. The proof is almost the same as [2, Proposition 9]. In view of it, here it is sufficient to show the bound

$$\big\|\Delta_j(\mathcal{P}\boldsymbol{g} - \Pi_x \boldsymbol{g}(x))(x)\big\|_{\mathbb{B}^\alpha_{p,q}} < \infty.$$

We have only to consider $j \geq 1$. For such j,

$$\begin{aligned}
&\Delta_j(\mathcal{P}\boldsymbol{g} - \Pi_x\boldsymbol{g}(x))(x) \\
&= \sum_{i;i\sim j} \iiint Q_j(x,y)Q_{<i-1}(y,u)Q_i(y,v)(\Pi_u\boldsymbol{g}(u) - \Pi_x\boldsymbol{g}(x))(v)dydudv,
\end{aligned}$$

where $i \sim j$ means that $|i-j| \leq N$ for some constant N. By (17),

$$\Pi_u\boldsymbol{g}(u) - \Pi_x\boldsymbol{g}(x) = \Pi_x(\Gamma_{ux}\boldsymbol{g}(u) - \boldsymbol{g}(x)) = -\sum_{\tau\in V}\omega_\tau^{\boldsymbol{g}}(u,x)\Pi_x\tau.$$

Hence the above integral is equal to

$$-\sum_{i;i\sim j}\sum_{\tau}\int Q_j(x,y)\Delta_{<i-1}(\omega_\tau^{\boldsymbol{g}}(\cdot,x))(y)\Delta_i(\Pi_x\tau)(y)dy.$$

For the $\omega^{\boldsymbol{g}}$ part, since

$$\begin{aligned}
\omega^{\boldsymbol{g}}(u,x) = \boldsymbol{g}(x) - \Gamma_{ux}\boldsymbol{g}(u) &= \boldsymbol{g}(x) - \Gamma_{yx}\boldsymbol{g}(y) + \Gamma_{yx}(\boldsymbol{g}(y) - \Gamma_{uy}\boldsymbol{g}(u)) \\
&= \omega^{\boldsymbol{g}}(y,x) + \Gamma_{yx}\omega^{\boldsymbol{g}}(u,y),
\end{aligned}$$

we have

$$\Delta_{<i-1}(\omega_\tau^{\boldsymbol{g}}(\cdot,x))(y) = \omega_\tau^{\boldsymbol{g}}(y,x) + \sum_\sigma \omega_\sigma^{\prec}(y,x)\Delta_{<i-1}(\omega_{\sigma\sqcup\tau}^{\boldsymbol{g}}(\cdot,y))(y).$$

Similarly, for the Π part,

$$\Delta_i(\Pi_x\tau)(y) = \Delta_i(\Pi_y\Gamma_{xy}\tau)(y) = \sum_{\sigma\sqcup\eta=\tau}\omega_\sigma^{\prec}(x,y)\Delta_i(\Pi_y\eta)(y).$$

Hence it turns out that $\Delta_j(\mathcal{P}\boldsymbol{g} - \Pi_x\boldsymbol{g}(x))(x)$ is a sum of the integrals of the form

$$\sum_{i;i\sim j}\int Q_j(x,y)A(x,y)B_i(y)C_i(y)dy,$$

where $A \in D_{p_A,q_A}^{\alpha_A}$, $B \in \mathbb{B}_{p_B,q_B}^{\alpha_B}$, $C \in \mathbb{B}_{p_C,q_C}^{\alpha_C}$, and parameters are such that $\alpha = \alpha_A + \alpha_B + \alpha_C$, $1/p = 1/p_A + 1/p_B + 1/p_C$, and $1/q = 1/q_A + 1/q_B + 1/q_C$. Since we are in exactly the same situation as the previous one (12), we can complete the proof by a similar way to Lemma 3.4. Q.E.D.

Acknowledgements. The author is supported by JSPS KAKENHI Early-Career Scientists 19K14556. The author thanks the anonymous referee for reading the paper carefully and providing helpful comments.

References

[1] H. Bahouri, J.-Y. Chemin and R. Danchin, Fourier Analysis and Nonlinear Partial Differential Equations, Springer, 2011.

[2] I. Bailleul and M. Hoshino, Paracontrolled calculus and regularity structures I, J. Math. Soc. Japan, **73** (2021), 553–595.

[3] I. Bailleul and M. Hoshino, Paracontrolled calculus and regularity structures II, J. Éc. polytech. Math., **8** (2021), 1275–1328.

[4] M. Gubinelli, P. Imkeller and N. Perkowski, Paracontrolled distributions and singular PDEs, Forum Math. Pi, **3** (2015), e6, 75pp.

[5] M. Hairer, A theory of regularity structures, Invent. Math., **198** (2014), no. 2, 269–504.

[6] M. Hoshino, Commutator estimates from a viewpoint of regularity structures, RIMS Kôkyûroku Bessatsu B79 (2020), 179–197.

[7] M. Hairer and C. Labbé, The reconstruction theorem in Besov spaces, J. Funct. Anal., **273** (2017), no. 8, 2578–2618.

[8] C. Liu, D. J. Prömel and J. Teichmann, Stochastic Analysis with Modelled Distributions, Stoch. Partial Differ. Equ. Anal. Comp., **9** (2021), 343–379. (doi:10.1007/s40072-020-00166-7)

[9] C. Liu, D. J. Prömel and J. Teichmann, Optimal extension to Sobolev rough paths, arXiv:1811.05173.

Graduate School of Engineering Science,
Osaka University,
1-3, Machikaneyama, Toyonaka, Osaka 560-8531, Japan.
E-mail address: hoshino@sigmath.es.osaka-u.ac.jp

Advanced Studies in Pure Mathematics 87, 2021
Stochastic Analysis, Random Fields and Integrable Probability — Fukuoka 2019
pp. 261–291

Determinantal structures in the q-Whittaker measure

Takashi Imamura, Matteo Mucciconi and Tomohiro Sasamoto

Abstract.

The q-Whittaker measure is a probability measure on the set of partitions. We show a Fredholm determinant formula for an expectation value with respect to this measure, which is the q-Laplace transform of the marginal distribution on the last element of the partition. Contrary to the typical approaches in integrable probability, our method does not focus on the q-moment generating function but explains the origin of its determinantal structure using Ramanujan's summation formula and the Frobenius determinant. This method can be applied to analyze on the fluctuations in the stationary situations in q-TASEP, higher spin exclusion process, and the directed polymer in the beta-distributed random environment.

§1. Introduction

The study on the stochastic interacting particle systems with some algebraic structures have attracted much interest recently. Utilizing these structures, one can analyze some interesting quantities in detail including their asymptotic behavior. Recently such a research field has been called "integrable probability".

One of the most basic models in integrable probaiblity is the Schur measure [30]. It is a probability measure on the set of partitions $\mathcal{S}_N$

$$\mathcal{S}_N := \{\lambda = (\lambda_1, \lambda_2, \dots, \lambda_N) \in \mathbb{Z}^N \mid \lambda_1 \geq \cdots \geq \lambda_N \geq 0\}. \tag{1}$$

For $\lambda \in \mathcal{S}_N$, it is expressed as

$$\frac{1}{\Pi_0(a;b)} s_\lambda(a_1, \dots, a_N) s_\lambda(b_1, \dots, b_M), \tag{2}$$

Received May 22, 2020.
Revised October 5, 2020.
2010 *Mathematics Subject Classification.* 82C22, 60K35, 33D52.
Key words and phrases. Integrable probability, Macdonald polynomials, KPZ class.

where $s_\lambda(a_1,\dots,a_N)$ is the Schur symmetric polynomial [26] and $\Pi_0(a;b) = 1/\prod_{i=1}^N \prod_{j=1}^M (1-a_ib_j)$. Since the late 1990's, it has been found that the Schur measure (6) and some of its valiants have some relation to stochastic processes belonging to the Kardar-Parisi-Zhang (KPZ) class such as the totally asymmetric simple exclusion process (TASEP) [22] through the RSK correspondence. We can analyze the fluctuation property of the TASEP using the Schur measure.

In [30], it has been shown that the Schur measure is a determinantal point process and its correlation kernel has been obtained explicitly. Such a determinantal structure can be explained from the Jacobi-Trudi formula which says that the Schur function is written as a single determinant

$$s_\lambda(a_1,\dots,a_N) = \det\big(h_{\lambda_i+j-i}(a_1,\dots,a_N)\big), \tag{3}$$

where $h_k(a_1,\dots,a_N)$ is the kth order complete symmetric polynomial. Using this with the antisymmetry with respect to the exchange of $\lambda_i - i$, we see that for $N \le M$,

$$\begin{aligned}\text{Prob}(s \le x_N) = &\frac{1}{\Pi_0(a;b)} \sum_{x_1=s}^{\infty} \cdots \sum_{x_N=s}^{\infty} \\ &\times \det\big(h_{x_i+j}(a_1,\dots,a_N)\big) \det\big(h_{x_i+j}(b_1,\dots,b_M)\big),\end{aligned} \tag{4}$$

where $x_k = \lambda_k - k,\ k = 1,\dots,N$. This structure, i.e. the summations of a product of two determinants, appears typically in the random matrix theory, non-intersecting random walk/Brownian motion, Free fermions etc. Following the procedures discussed in [33, 23] one can easily obtain a Fredholm determinant

$$\text{Prob}(s \le x_N) = \det(1+K)_{\ell^2(\{s,s+1,\dots\})}, \tag{5}$$

where the kernel $K(m_1, m_2)$ can be expressed explicitly as a double integral form.

The q-Whittaker measure, which has been introduced in [6], is also a probability measure on $\mathcal{S}_N$ having the following form,

$$\frac{1}{\Pi(a;b)} P_\lambda(a_1,\dots,a_N) Q_\lambda(b_1,\dots,b_M), \tag{6}$$

where P_λ, Q_λ are the q-Whittaker functions. They are a special case of the Macdonald polynomials i.e. the case $t = 0$ in two parameters (q,t) of the Macdonald polynomials and when $q = 0$ both of them go to the Schur polynomial. Their combinatorial definitions are stated in

Section 2.1. The normalization constant $\Pi(a;b)$ is written as

$$\Pi(a;b) = \prod_{i=1}^{N}\prod_{j=1}^{M}\frac{1}{(a_i b_j;q)_\infty}, \tag{7}$$

where $(a;q)_\infty = \prod_{j=1}^{\infty}(1-aq^{j-1})$ is the q-Pochhammer symbol.

The q-Whittaker measure is related to the q-totally asymmetric simple exclusion process (q-TASEP) [6]. It is a version of exclusion process where the hopping rate of each particle depends on the "gap" (the number of empty sites) between the particle and the closest particle in the traveling direction as $1-q^{\mathrm{gap}}$ with parameter $0 \le q < 1$. One immediately sees that the q-TASEP becomes the usual TASEP in the case $q=0$. In addition it has been known that it goes to the directed polymer models with finite temperature in the $q \to 1$ scaling limits: the convergence to the O'Connell-Yor polymer model has been discussed in [6] and more recently in [15], the higher spin vertex model, which includes the q-TASEP as a special case, has been shown to converge to the stochastic heat equation (SHE) which is equivalent to the KPZ equation. In particular the latter is a sign to the weak universality in the KPZ class.

A big problem of the q-Whittaker measure is that for the q-Whittaker functions $P_\lambda(a)$, $Q_\lambda(b)$, determinantal formulas have not been known except the case $q=0$ (3) so it is not clear if one can generalize the relation (5) to the case $0<q<1$ keeping the determinantal structure in the rhs.

More recently, such a generalized relations have been found out. In the case of the q-Whittaker measure (6), one focuses on the q-moment of the marginal distribution on λ_N,

$$\mathbb{E}\left[q^{k\lambda_N}\right],\ k=1,2,\dots, \tag{8}$$

and has obtained a representation in term of $k\times k$ determinant. This determinantal expression has been first obtained by using the Macdonald difference operator [6]. Since then, other interesting techniques such as stochastic duality [8], Yang-Baxter equation [14, 9] has been developed. Then they have got the Fredholm determinant representation for the q-moment generating function,

$$\sum_{n=0}^{\infty}\frac{\zeta^n}{(q;q)_n}\mathbb{E}\left[q^{n\lambda_N}\right] = \mathbb{E}\left[\frac{1}{(\zeta q^{\lambda_N};q)_\infty}\right], \tag{9}$$

which can be obtained by the special case ($a=0$) of the q-binomial theorem,

$$\sum_{k=0}^{\infty} x^k \frac{(a;q)_k}{(q;q)_k} = \frac{(ax;q)_\infty}{(x;q)_\infty}. \tag{10}$$

This will be stated in Theorem 1 below.

In this note, on the other hand, we review our result of [20], which is another Fredholm determinant formula for the same quantity, the rhs of (9). It will be written as Theorem 2 below. The main difference from the usual approaches in the integrable probability is that we do not focus on the q-moments (8) but on directly the rhs of (9). By using the three relations, Cauchy's identity, Ramanujan's summation formula, and the Frobenius determinant, we obtain a similar determinantal structure to the rhs of (4), i.e. multiple integral of a product of two determinants. This relation will be stated in (63) below. An advantage of this approach is that it can be applicable to the stationary situation of the q-TASEP [20] and the higer spin vertex model [18], where in the usual approach the q-moments (8) are not well-defined for large ks and thus we need some regularization developed in [1].

This note is organized as follows. In Section 2, we define the q-Whittaker function, q-Whittaker measure and its relation to the q-TASEP. In Section 3, we introduce two Fredholm determinant formuals for the q-Laplace transform of the marginal distribution, the first one is obtained by the usual approaches (Theorem 1) and the next one is obtained by our approach (Theorem 2). We also give an outline of derivation of Theorem 2. In the final section, we give applications of our approach to the stationary q-TASEP and the stationary higher spin vertex model.

§2. q-Whittaker measure

In this section, we define the q-Whittaker measure and state its relation to the q-TASEP.

2.1. q-Whittaker functions

Let $\mathcal{S}_N$ be the set of partition with length N defined by (1) and $\mathbb{G}_N$ be the Gelfand-Tsetlin cone,

$$\begin{aligned}\mathbb{G}_N := \Big\{\underline{\lambda}_N = (\lambda^{(1)}, \lambda^{(2)}, \ldots, \lambda^{(N)}) \mid \lambda^{(n)} \in \mathcal{S}_n, 1 \le n \le N, \\ \lambda^{(m+1)}_{\ell+1} \le \lambda^{(m)}_\ell \le \lambda^{(m+1)}_\ell, 1 \le \ell \le m \le N-1\Big\}.\end{aligned} \tag{11}$$

The q-Whittaker functions are special case $t = 0$ in the two parameters q, t of the Macdonald symmetric functions. They are defined as follows.

First, for $\lambda^{(n)} \in S_n$ and $\lambda^{(n-1)} \in \mathcal{S}_{n-1}$ with $\lambda_\ell^{(n)} \le \lambda_{\ell-1}^{(n-1)} \le \lambda_{\ell-1}^{(n)}$, $1 \le \ell \le n$, we define the skew q-Whittaker function with one variable,

$$P_{\lambda^{(n)}/\lambda^{(n-1)}}(a) = a^{|\lambda^{(n)}|-|\lambda^{(n-1)}|} \prod_{i=1}^{n-1} \binom{\lambda_i^{(n)} - \lambda_{i+1}^{(n)}}{\lambda_i^{(n-1)} - \lambda_{i+1}^{(n)}}_q \tag{12}$$

where for $\lambda \in \mathcal{S}_N$, $|\lambda| = \lambda_1 + \cdots + \lambda_N$ and

$$\binom{n}{m}_q := \frac{(q;q)_n}{(q;q)_m (q;q)_{n-m}} \quad \text{with} \quad (a;q)_n := \prod_{j=1}^{n} (1 - aq^{j-1}). \tag{13}$$

Here LHSs of the first and second equations are called the q-binomial coefficient and q-Pochhammer symbol respectively and they are common objects in q-calculus. For details see Section 10 in [2].

Next using (12), we define the q-Whittaker functions P_λ, Q_λ with $\lambda \in \mathcal{S}_N$ and multi variables $a_1, \dots, a_N$.

$$P_\lambda(a_1, \dots, a_N) = \sum_{\substack{\underline{\lambda}_N \in \mathbb{G}_N \\ \lambda^{(N)} = \lambda}} \prod_{j=1}^{N} P_{\lambda^{(j)}/\lambda^{(j-1)}}(a_j), \tag{14}$$

$$Q_\lambda(a_1, \dots, a_N) = \prod_{i=1}^{N} \frac{1}{(q;q)_{\lambda_i - \lambda_{i+1}}} \cdot P_\lambda(a_1, \dots, a_N). \tag{15}$$

Here we summarize some properties of the q-Whittaker functions which are useful in this paper.

(i) **Schur function:** The q-Whittaker functions are one parameter q-generalization of the Schur polynomial $s_\lambda(a_1, \dots, a_N)$, i.e. in the case $q = 0$ both (14) and (15) correspond to the combinatorial definition of the Schur polynomial

$$s_\lambda(a_1, \dots, a_N) = \sum_{\substack{\underline{\lambda}_N \in \mathbb{G}_N \\ \lambda^{(N)} = \lambda}} \prod_{j=1}^{N} a_j^{|\lambda^{(j)}|-|\lambda^{(j-1)}|}. \tag{16}$$

(ii) **Torus scalar product and the specialization ρ:** $Q_\lambda(a_1, \dots, a_N)$ has the following integral representation (called the torus scalar product)

$$Q_\lambda(a) = \prod_{i=1}^{N-1} (q^{\lambda_i - \lambda_{i+1} + 1}; q)_\infty \int_{\mathbb{T}^N} \prod_{i=1}^{N} \frac{dz_i}{z_i} \cdot P_\lambda\left(\frac{1}{z}\right) \Pi(z;a) m_N^q(z), \tag{17}$$

where we used the shorthand notations $a = (a_1, \ldots, a_N)$, $z = (z_1, \ldots, z_N)$ and $1/z = (1/z_1, \ldots, 1/z_N)$ and $\Pi(z; a)$ is defined by (7) and $m_N^q(z)$ is

$$m_N^q(z) = \frac{1}{(2\pi i)^N N!} \prod_{1 \le i < j \le N} (z_i/z_j; q)_\infty (z_j/z_i; q)_\infty. \tag{18}$$

In this paper we define $Q_\lambda(\rho)$ with specialization ρ by the representation (17) with $\Pi(z; a)$ replaced by $\Pi(z; \rho)$,

$$\Pi(z; \rho) = \prod_{i=1}^{N} f_q(z_i; \rho), \quad \text{with} \quad f_q(z; \rho) = e^{\gamma z} \prod_{i=1}^{\infty} \frac{(1 + \beta_i z)}{(\alpha_i z; q)_\infty}. \tag{19}$$

The specialization ρ has three sequences of parameters $\alpha = (\alpha_j)_{j \in \mathbb{Z}_{\ge 1}}$, $\beta = (\beta_j)_{j \in \mathbb{Z}_{\ge 1}}$ and γ. Recently in [27], it has been showed that $P_\lambda(\rho) \ge 0$ if and only if $\alpha_i, \beta_i \ge 0,\ i = 1, 2, \ldots$ such that $\sum_i (\alpha_i + \beta_i) < \infty$ and $\gamma \in \mathbb{R}$. (Actually in [27], the positivity result was proved for the more general Macdonald functions.) Note that the case (17) can be realized as the following choice of the parameters in ρ,

$$\alpha_i = \begin{cases} a_i, & i = 1, 2, \ldots, N, \\ 0, & i \ge N + 1, \end{cases} \quad \beta_j = \gamma = 0 \ \text{ for } \ j \in \mathbb{Z}_{\ge 0}. \tag{20}$$

(iii) **Cauchy identity:** For $P_\lambda(a_1, \ldots, a_N)$ and $Q_\lambda(b_1, b_2, \ldots, b_M)$, we have the following relation called the Cauchy identity

$$\sum_\lambda P_\lambda(a_1, a_2, \ldots, a_N) Q_\lambda(b_1, \ldots, b_N) = \Pi(a; b), \tag{21}$$

where $\Pi(a; b)$ is defined by (7). This relation is generalized to the case of $Q_\lambda(\rho)$ with specialization ρ,

$$\sum_\lambda P_\lambda(a_1, a_2, \ldots, a_N) Q_\lambda(\rho) = \Pi(a; \rho), \tag{22}$$

where $\Pi(a; \rho)$ is defined by (19).

2.2. q-Whittaker measures and processes

The q-Whittaker measure and processes are introduced by Borodin and Corwin [6]. The q-Whittaker measure is a probability measure on the set of partition $\mathcal{S}_N$ (1). For $\lambda \in \mathcal{S}_N$, it is defined as

$$\mathbb{M}_{a;\rho}(\lambda) = \frac{1}{\Pi(a; \rho)} P_\lambda(a) Q_\lambda(\rho). \tag{23}$$

The (ascending) q-Whittaker process is a probability measure on the Gelfand-Tsetlin $\mathbb{G}_N$ (11). For $\underline{\lambda}_N = (\lambda^{(1)}, \dots, \lambda^{(N)}) \in \mathbb{G}_N$, we define

(24)
$$\mathbb{P}_{a;\rho}(\underline{\lambda}_N) = \frac{1}{\Pi(a;\rho)} P_{\lambda^{(1)}}(a_1) P_{\lambda^{(2)}/\lambda^{(1)}}(a_2) \cdots P_{\lambda^{(N)}/\lambda^{(N-1)}}(a_N) \cdot Q_{\lambda^{(N)}}(\rho).$$

From (14), we see that the above two measures are related as

$$\sum_{\substack{\lambda^{(1)},\dots,\lambda^{(N-1)} \\ \underline{\lambda}_N \in \mathbb{G}_N}} \mathbb{P}_{a;\rho}(\underline{\lambda}_N) = \mathbb{M}_{a;\rho}(\lambda^{(N)}), \tag{25}$$

i.e. $\mathbb{M}_{a;\rho}(\lambda^{(N)})$ appears as the marginal measure of $\mathbb{P}_{a;\rho}(\underline{\lambda}_N)$ on the top layer $\lambda^{(N)}$.

2.3. q-TASEP

The q-TASEP is a model of an interacting particle process introduced in [6]. Throughout this paper, we assume the total number of particles is finite N. Let $X_j(t) \in \mathbb{Z}$, $j = 1, 2, \dots, N$ with $X_N(t) < X_{N-1}(t) < \cdots < X_1(t)$ be the position of the jth particle at time t. The jth particle has its own exponential clock with the rate

$$a_j(1 - q^{\mathrm{gap}_j(t)}), \tag{26}$$

where $0 \le q < 1$, $a_j > 0$, $j = 1, \dots, N$ and

$$\mathrm{gap}_j(t) = \begin{cases} \infty, & j = 1, \\ X_{j-1}(t) - X_j(t) - 1, & j = 2, 3, \dots, N \end{cases} \tag{27}$$

represents the gap (the number of empty sites) between the $j-1$th and jth particles at time t. That is the clock of the jth particle waits an exponentially distributed time (with rate (26)) before ringing. If the clock rings, the jth particle moves forward by 1. Note that it cannot move if $\mathrm{gap}_j(t) = 0$ since the rate vanishes. This property represents the exclusion interaction.

In [6], Borodin and Corwin found out a connection between the N-particle q-TASEP and the q-Whittaker measure with the specialization ρ_t defined by (19) with $f_q(z;\rho_t) = e^{tz}$, i.e. the following choice of the parameters in ρ,

$$\alpha_j = \beta_j = 0 \ \text{ for } j = 1, 2, \dots \ \text{ and } \ \gamma = t > 0, \tag{28}$$

where we write t in place of γ since it corresponds to the time in the q-TASEP. To see this, we consider the following Markov chain. Let

$\underline{\lambda}_N(t) = (\lambda^{(1)}(t), \ldots, \lambda^{(N)}(t))$ be a $\mathbb{G}_N$-valued random variable with time parameter $t > 0$. Note that it has $N(N+1)/2$ degrees of freedom in total. We apply the following rule.

Suppose that the positions of all elements at t are fixed as $\lambda_k^{(j)}(t) = \lambda_k^{(j)}$. Each $\lambda_i^{(j)}(t)$, $(1 \le i \le j \le N)$ has its own independent exponential clock with rate

$$a_j \frac{(1-q^{\lambda_{i-1}^{(j-1)}-\lambda_i^{(j)}})(1-q^{\lambda_i^{(j)}-\lambda_{i+1}^{(j)}+1})}{1-q^{\lambda_i^{(j)}-\lambda_i^{(j-1)}+1}}, \tag{29}$$

where we set $\lambda_0^{(j-1)} = \infty$, $\lambda_{j+1}^{(j)} = -\infty$. If the clock of the element $\lambda_i^{(j)}(t)$ rings, then it moves forward by 1 (i.e. it increases by 1). In addition all elements $\lambda_i^{(j+\ell)}(t), \ell = 1, 2, \ldots, \ell_c$ such that $\lambda_i^{(j)}(t) = \lambda_i^{(j+1)}(t) = \cdots = \lambda_i^{(j+\ell_c)}(t) < \lambda_i^{(j+\ell_c+1)}(t)$ are pushed to move forward by 1. We interpret it to mean that $\lambda_i^{(j)}(t)$ *pushes* the sequence $\lambda_i^{(j)}(t)$, $\lambda_i^{(j+1)}(t), \ldots, \lambda_i^{(j+\ell_c)}(t)$. Note that when $\lambda_i^{(j)} = \lambda_{i-1}^{(j-1)}$, the rate (29) vanishes and thus $\lambda_i^{(j)}(t)$ cannot move, (i.e. it has to keep its value $\lambda_i^{(j)}$). This means that $\lambda_{i-1}^{(j-1)}(t)$ *blocks* $\lambda_i^{(j)}(t)$.

When we regard each element $\lambda_i^{(j)}(t)$ as a position of particle, the dynamics described by the above rule can be interpreted as $N(N+1)/2$ particle random walk model with push and block of interactions. Note that this dynamics does not change the order $\lambda_{\ell+1}^{(m+1)} \le \lambda_\ell^{(m)} \le \lambda_\ell^{(m+1)}$, $1 \le \ell \le m \le N-1$ in $\mathbb{G}_N$.

Let us focus on the marginal $(\lambda_j^{(j)}(t))_{j=1,\ldots,N}$. Note that each element is independent of the ones other than this marginal since by definition, $\lambda_j^{(j)}(t)$ is not pushed by any elements and is blocked only by $\lambda_{j-1}^{(j-1)}(t)$. Furthermore, the rate of each element $\lambda_j^{(j)}$ becomes $a_j(1 - q^{\lambda_{j-1}^{(j-1)}-\lambda_j^{(j)}})$. Thus comparing the dynamics with the one in the q-TASEP, one immediately sees

$$(\lambda_k^{(k)}(t) - k)_{k=1,2,\ldots,N} \stackrel{\mathrm{d}}{=} (X_k(t))_{k=1,2,\ldots,N} \tag{30}$$

where "$\stackrel{\mathrm{d}}{=}$" means an equality in distribution and $X_k(t)$ is the position of the kth particle in the q-TASEP.

In [6], on the other hand, it has been shown that when at $t = 0$, $\underline{\lambda}_N(0) = \phi$, i.e. $\lambda_i^{(j)} = 0$ for $1 \le i \le j \le N$, the transition probability of this Markov chain is given by the q-Whittaker process,

Proposition 1 ([6]).

$$\mathrm{Prob}\big(\underline{\lambda}_N(t) = \underline{\lambda}_N \mid \underline{\lambda}_N(0) = \phi\big) = \mathbb{P}_{a;\rho_t}(\underline{\lambda}_N), \tag{31}$$

where the rhs is the q-Whittaker process (24) *with the specialization* ρ_t *defined by* (28).

Combining (31) with (25), the distribution of the "top" marginal $\lambda^{(N)}(t) = (\lambda_1^{(N)}(t), \lambda_2^{(N)}(t), \dots, \lambda_N^{(N)}(t))$ is given by the q-Whittaker measure,

$$\text{Prob}\big(\lambda^{(N)}(t) = \lambda^{(N)} \mid \lambda^{(N)} = \phi\big) = \mathbb{M}_{a;\rho_t}(\lambda^{(N)}). \tag{32}$$

In q-TASEP, we are interested in quantities such that $\mathbb{E}_{q\mathrm{T}}[f(X_N(t))]$, the expectation value over the dynamics of the q-TASEP with the step initial condition $X_k(0) = -k$, $k = 1, 2, \dots, N$ with some arbitrary function f. From the discussion above we see that

$$\mathbb{E}_{q\mathrm{T}}\big[f(X_N(t))\big] = \mathbb{E}_{q\mathrm{P}}^{a;\rho_t}\big[f(\lambda_N^{(N)}(t) - N)\big] = \mathbb{E}_{q\mathrm{M}}^{a;\rho_t}\big[f(\lambda_N(t) - N)\big] \tag{33}$$

where $\mathbb{E}_{q\mathrm{P}}^{a;\rho_t}$ and $\mathbb{E}_{q\mathrm{M}}^{a;\rho_t}$ represent the expectation with respect to $\mathbb{P}_{a;\rho_t}$ and $\mathbb{M}_{a;\rho_t}$. In Section 3, we focus on the expectation with respect to the q-Whittaker measure.

2.4. Analytic continuations on the parameters

Here we give two types of analytic continuations of the parameters. Both of them will be used in Section 4.2. Let ρ_J for $J \in \mathbb{Z}_{>0}$ be the specialization ρ defined by (19) the parameters $\beta_j, j = 1, 2, \dots$ are set to

$$\beta_j = \begin{cases} q^{j-1}v, & j = 1, 2, \dots, J, \\ \beta_{j-J+1}, & j = J+1, J+2, \dots, \end{cases} \tag{34}$$

with $v > 0$ and other parameters α_j, $j = 1, 2, \dots$ and γ remain unchanged. That is $f_q(z; \rho_J)$ in (19) is written as

$$f_q(z; \rho_J) = e^{\gamma z} \prod_{i=1}^{\infty} \frac{(1 + \beta_i z)}{(\alpha_i z; q)_\infty} \cdot (-vz; q)_J. \tag{35}$$

Using the representation (17) in terms of the torus scaler product, the q-Whittaker measure $\mathbb{M}_{a;\rho_J}(\lambda)$ can be expressed as

$$\begin{aligned} \mathbb{M}_{a;\rho_J}(\lambda) = &\prod_{i=1}^{N-1} (q^{\lambda_i - \lambda_{i+1} + 1}; q)_\infty \\ &\times \int_{\mathbb{T}^N} \prod_{i=1}^{N} \frac{dz_i}{z_i} \frac{f_q(z_i; \rho_J)}{f_q(a_i; \rho_J)} \cdot P_\lambda(a) P_\lambda(1/z) m_N^q(z) \end{aligned} \tag{36}$$

where $f_q(z; \rho_J)$ is defined by (35) above.

Now we regard (36) as a function of $J \in \mathbb{Z}_{\geq 0}$ and consider the following two types of analytic continuation.

type 1 Set $p \equiv -q^{-J}$, do the analytic continuation to $p \in \mathbb{C}$, then $p \to 0$

type 2 Set $d \equiv -q^J$, do the analytic continuation to $d \in \mathbb{C}$.

We have the following

Proposition 2.

(i) *Let* $\lambda - J^N := (\lambda_1 - J, \lambda_2 - J, \dots, \lambda_N - J)$. *By performing the type* 1 *above, the* q*-Whittaker measure* $\mathbb{M}_{a;\rho_J}(\lambda - J^N)$ *is transformed to the double-sided* q*-Whittaker measure* $\mathbb{M}_{a;\rho^v}(\lambda)$ *with the specialization* ρ^v *defined by*

$$f_q(z;\rho^v) = e^{\gamma z} \prod_{i=1}^{\infty} \frac{(1+\beta_i z)}{(\alpha_i z;q)_\infty} \cdot \frac{1}{(q/vz;q)_\infty}. \tag{37}$$

(ii) *By performing the type* 2 *above, the* q*-Whittaker measure* $\mathbb{M}_{a;\rho_J}(\lambda)$ *is transformed to* $\mathbb{M}_{a;\rho_d}(\lambda)$ *with the specialization* ρ_d *defined by*

$$f_q(z;\rho_d) = e^{\gamma z} \prod_{i=1}^{\infty} \frac{(1+\beta_i z)}{(\alpha_i z;q)_\infty} \cdot \frac{(-vz;q)_\infty}{(-dvz;q)_\infty}. \tag{38}$$

Outline of the proof. For proving the first relation, we use the relation

$$P_{\lambda - J^N}(x_1,\dots,x_N) = \prod_{j=1}^{N} x_j^{-J} \cdot P_\lambda(x_1,\dots,x_N) \tag{39}$$

which is obtained from the definition of P_λ (14). We also note the part in (36) $(-vz_j;q)_J/(-va_j;q)_J$ in $\mathbb{M}_{a;\rho_J}$ can be rewritten as

$$\begin{aligned} \frac{(-vz;q)_J}{(-va;q)_J} &= \frac{(1+vz)(1+vqz)\cdots(1+vq^{J-1}z)}{(1+va)(1+vqa)\cdots(1+vq^{J-1}a)} \\ &= \left(\frac{z}{a}\right)^J \frac{(-q^{1-J}/vz;q)_J}{(-q^{1-J}/va;q)_J} = \left(\frac{z}{a}\right)^J \frac{(qp/vz;q)_\infty}{(q/vz;q)_\infty} \frac{(q/va;q)_\infty}{(qp/va;q)_\infty} \end{aligned} \tag{40}$$

where in the last equality we set $p \equiv q^{-J}$. The result is obtained by substituting (39) and (40) into (36) and then perform the analytic continuation of type 1.

For showing the second relation, we rewrite $(-vz_j;q)_J$ as follows,

$$(-vz;q)_J = \frac{(-vz;q)_\infty}{(dvz;q)_\infty}, \tag{41}$$

where we set $\mathscr{A} \equiv -q^J$. Then we extend the domain to $\mathbb{C}$. Q.E.D.

Remark. The double-sided q-Whittaker measure and process The functions $P_\lambda(a_1, \dots, a_N)$ (14) and $Q_\lambda(\rho)$ defined by (17) with (19) are originally defined on $\mathcal{S}_N$ (1). We see that the domain can be extended to the space $\tilde{\mathcal{S}}_N$,

$$\tilde{\mathcal{S}}_N := \{\lambda = (\lambda_1, \lambda_2, \dots, \lambda_N) \in \mathbb{Z}^N \mid \lambda_1 \geq \cdots \geq \lambda_N\}, \tag{42}$$

where the nonnegativity condition in $\mathcal{S}_N$ is eliminated. $P_\lambda(a_1, \dots, a_N)$, $\lambda \in \tilde{\mathcal{S}}_N$ can be defined by (14) with $\mathbb{G}_N$ in rhs replaced by $\tilde{\mathbb{G}}_N$,

$$\tilde{\mathbb{G}}_N := \Big\{\underline{\lambda}_N = (\lambda^{(1)}, \lambda^{(2)}, \cdots, \lambda^{(N)}) \mid \lambda^{(n)} \in \tilde{\mathcal{S}}_n, 1 \leq n \leq N, \\ \lambda_{\ell+1}^{(m+1)} \leq \lambda_\ell^{(m)} \leq \lambda_\ell^{(m+1)}, 1 \leq \ell \leq m \leq N-1\Big\}. \tag{43}$$

Also $Q_\lambda(\rho)$, $\lambda \in \tilde{\mathcal{S}}_N$ can be defined by (17) and (19) with $P_\lambda(1/z)$ in (17) defined as above.

Thus the q-Whittaker measure (23) and process (24) can also be defined on larger spaces $\tilde{\mathcal{S}}_N$ (42) and $\tilde{\mathbb{G}}_N$ (43) respectively. However, we easily find $Q_\lambda(\rho) = 0$ when $0 > \lambda_k > \cdots > \lambda_N$ for some k and thus as long as we consider the specialization ρ defined by (19), the supports of the q-Whittaker measure and process are still limited to $\mathcal{S}_N$ (1) and $\mathbb{G}_N$ (19) respectively.

This situation changes in the specialization ρ^v appearing in Proposition 2(i). We see that due to the factor $1/(q/vz;q)_\infty$, $Q_\lambda(\rho^v)$ can take nonzero even when $0 > \lambda_k > \cdots > \lambda_N$ and the supports of $\mathbb{M}_{a;\rho^v}(\lambda)$ and $\mathbb{P}_{a;\rho}(\underline{\lambda}_N)$ extend. In this paper we call them the double-sided (or two-sided) q-Whittaker measure and process respectively. They play an important role in the analyses of the stationary q-TASEP and higher spin exclusion process.

§3. Determinantal structures

3.1. Fredholm determinant formulas

In the recent development of the integrable probability, we have Fredholm determinant formulas for the quantity like the rhs of (33) when we choose the function f in some special form. In this note, we provide the determinant formulas on the case $f(x) = 1/(\zeta q^x; q)_\infty$, where $\zeta \in \mathbb{C} \setminus \mathbb{R}$ and $(x;q)_\infty = (1-x)(1-xq)(1-xq^2)\cdots$.

The Fredholm determinant formula was first obtained in [6].

Theorem 1 ([6]).

$$\mathbb{E}^{a;\rho}_{q\mathrm{M}}\left[\frac{1}{(\zeta q^{\lambda_N};q)_\infty}\right] = \det(1+K_\zeta)_{L^2(\gamma_a)}. \tag{44}$$

Here γ_a is a contour enclosing all a_is positively and the kernel $K_\zeta(w_1,w_2)$ is

$$K_\zeta(w_1,w_2) = \frac{1}{2\pi i}\int_{C_{1,2,\ldots}} ds \frac{\pi s}{\sin \pi s}\frac{(-\zeta)^s}{q^s w_1 - w_2}\frac{F_q(q^s w_1;a,\rho)}{F_q(w_1;a,\rho)}, \tag{45}$$

$$F_q(w;a,\rho) = \prod_{m=1}^{N}(w/a_m;q)_\infty \cdot f_q(w;\rho),$$

where $C_{1,2,\ldots}$ encloses the pole $1,2,\ldots$ negatively and $f_q(w,\rho)$ is defined in (19).

The lhs of (44) is a q-deformation of the Laplace transform of the marginal distribution of the q-Whittaker measure on λ_N. Thus one can obtain the distribution by inverse Laplace like transform. Furthermore, we can obtain the limiting distribution of $\lambda_N(t)$ by considering the limiting behavior of the Fredholm determinant of the rhs of (44). The problem is reduced to see an asymptotic behavior of kernel (45), which one can find by applying the saddle point analysis to (45).

This theorem was first obtained by [6] using the Macdonald difference operators. (In [6], the special case of $\rho = \rho_t$ (28) has been treated. The above result on the full Macdonald positive specialization ρ has been obtained in [10]). Since then similar types of the formula have been obtained by various techniques such as Markov duality [8] and Yang-Baxter equations [14, 9, 31].

In spite of the diversity of the techniques, they are based on one common idea. They focus on the fact that the lhs of (44) is expressed as the generating function of the q-moments $\mathbb{E}^{\rho_t}_{q\mathrm{W}}[q^{n\lambda_N}]$, $n=0,1,2,\ldots$, i.e.

$$\mathbb{E}^{a;\rho}_{q\mathrm{M}}\left[\frac{1}{(\zeta q^{\lambda_N};q)_\infty}\right] = \sum_{n=0}^{\infty}\frac{\zeta^n}{(q;q)_n}\mathbb{E}^{a;\rho}_{q\mathrm{M}}[q^{n\lambda_N}], \tag{46}$$

which can be obtained by the special case ($a=0$) of the q-binomial theorem (10). Furthermore by using the techniques above, one can get nice determinantal forms for the q-moments. Combining these, one eventually obtain Fredholm determinant formulas such as (44).

We obtain another Fredholm determinant representation without using the q-moment (46).

Theorem 2 ([20]). *For $\zeta \in \mathbb{C} \setminus q^n$, n, $m \in \mathbb{Z}_{\geq 0}$, we have*

$$\mathbb{E}^{a;\rho}_{q\mathrm{M}}\left[\frac{1}{(\zeta q^{\lambda_N};q)_\infty}\right] = \det(1+fK_{a;\rho})_{\ell^2(\mathbb{Z}_{\geq 0})}, \tag{47}$$

$$f(n) = \frac{\zeta q^n}{1-\zeta q^n},$$

$$K_{a;\rho}(n,m) = \int_D dv \int_C \frac{dz}{z} \frac{1}{z-v} \frac{v^n}{z^m} \frac{F_q(z;a,\rho)}{F_q(v;a,\rho)}$$

where $F_q(z;a,\rho)$is defined in (45) *and the contour C encloses the origin with small radius anticlockwise while D encloses $a_1,\ldots,a_N$ anticlockwise.*

This double contour integral representation is similar to the kernel of the Schur measure.

3.2. Cauchy, Ramanujan, and Frobenius

Before giving the proof of our main Theorem 2, we state the relations by Cauchy, Ramanujan, and Frobenius, which are key to prove it.

Cauchy's identity. This is already given by (21) and (22). In the proof we will use a slightly deformed version of the former one (21). By the definition (14), we note the fact that $P_\lambda(x_1,\ldots,x_N)$ can be factorized as the parts depending on λ_N and on $\ell_k = \lambda_k - \lambda_{k+1}$, $k=1,\ldots,N-1$,

$$P_\lambda(x_1,x_2,\ldots,x_N) = X^{\lambda_N} R_\ell(x_1,x_2,\ldots,x_N), \tag{48}$$

where $X = x_1x_2\cdots x_N$. (21) can be translated into the relation on $R_\ell(x_1,x_2,\ldots,x_N)$,

$$\sum_{\ell_1,\ldots,\ell_{N-1}=0}^{\infty} \frac{R_\ell(x)R_\ell(y)}{\prod_{j=1}^{N-1}(q;q)_{\ell_j}} = \frac{(XY;q)_\infty}{\prod_{ij}(x_iy_j;q)_\infty}. \tag{49}$$

For more details, see Appendix B in [20].

Ramanujan's summation formula. For $|q|<1, |b/a|<|z|<1$, we have

$$\sum_{n=-\infty}^{\infty} \frac{(bq^n;q)_\infty}{(aq^n;q)_\infty} z^n = \frac{(az;q)_\infty(\frac{q}{az};q)_\infty(q;q)_\infty(\frac{b}{a};q)_\infty}{(a;q)_\infty(\frac{q}{a};q)_\infty(z;q)_\infty(\frac{b}{az};q)_\infty}. \tag{50}$$

For some properties, see e.g. Section 10.5 in [2].

Frobenius determinant. This is the determinantal formula on the theta functions. Let

$$\tilde{\theta}(z) = \sqrt{z}(1/z;q)_\infty(zq;q)_\infty. \tag{51}$$

The function has a symmetry $\tilde{\theta}(z) = -\tilde{\theta}(1/z)$ is related to the Jacobi theta functions $\theta_1(z;q)$ and $\theta_3(z;q)$,

$$\tilde{\theta}(z) = -iq^{-\frac{1}{8}}(q;q)_\infty^{-1}\theta_1(z^{-1};q) = (q;q)_\infty^{-1}\sqrt{z}\theta_3(-q^{-1/2}z^{-1};q) \tag{52}$$

where $\theta_1(z;q)$ and $\theta_3(z;q)$ are defined by

$$\theta_1(z;q) = i^{-1}\sum_{n\in\mathbb{Z}}(-1)^n q^{(n+1/2)^2/2}z^{n+1/2}, \quad \theta_3(z;q) = \sum_{n\mathbb{Z}} q^{n^2/2}z^n. \tag{53}$$

For $\tilde{\theta}(z)$, we have the following determinantal formula,

$$\frac{\prod_{i<j}^N \tilde{\theta}(a_i/a_j)\tilde{\theta}(z_i/z_j)}{\prod_{i,j=1}^N \tilde{\theta}(a_i/z_j)} \cdot \frac{\tilde{\theta}(\zeta A/Z)}{\tilde{\theta}(\zeta)} = \det\left[\frac{\tilde{\theta}(\zeta a_i/z_j)}{\tilde{\theta}(\zeta)\tilde{\theta}(a_i/z_j)}\right]_{i,j=1}^N, \tag{54}$$

where $A = a_1 \cdots a_N$, $Z = z_1 \cdots z_N$. This has been obtained by Frobenius in [17], so we call it a Frobenius determinant. Recently (54) has been further studied in [24].

The Frobenius determinant (54) is a generalization of the Cauchy's determinant:

$$\frac{\prod_{i<j}^N (a_i - a_j)(w_i - w_j)}{\prod_{i,j=1}^N (a_i - w_j)} = \det\left[\frac{1}{a_i - w_j}\right]_{i,j=1}^N. \tag{55}$$

This is obtained from (54) by setting $q = 0$ and taking the limit $\zeta \to \infty$.

3.3. Outline of the proof of Thorem 2

Now we give an outline of the proof of Theorem 2. As we mentioned above Theorem 2, we do not utilize (46). Rather than focusing on the q-moments, we will deal with the q-Laplace transform itself,

$$\begin{aligned} \mathbb{E}_{q\mathrm{M}}^{a;\rho}\left[\frac{1}{(\zeta q^{\lambda_N};q)_\infty}\right] &= \sum_{\lambda\in\mathcal{S}_N} \frac{1}{(\zeta q^{\lambda_N};q)_\infty}\mathbb{M}_{a;\rho}(\lambda) \\ &= \sum_{\lambda_N=0}^{\infty} \frac{1}{(\zeta q^{\lambda_N};q)_\infty} \sum_{\substack{\lambda_1,\dots,\lambda_{N-1} \\ \lambda\in\mathcal{S}_N}} \mathbb{M}_{a;\rho}(\lambda), \end{aligned} \tag{56}$$

where $\mathcal{S}_N$ is defined by (1). Our strategy is to rewrite the rhs of (56) in the form of the integral of a product of two determinants. Actually at the end we will obtain (63) below. For this purpose, in (56) we divide the sum $\sum_{\lambda\in\mathcal{S}_N}$ into two parts $\sum_{\lambda_1,\dots,\lambda_{N-1}}$ and $\sum_{\lambda_N}$. We will use the results by Cauchy (49) and Ramanujan (50) for taking each summation, then we will find out the determinants with the help of Frobenius (54).

First step: taking the sum $\sum_{\lambda_1,\dots,\lambda_{N-1}}$

Let $P(\lambda_N)$ be the marginal distribution of the q-Whittaker measure $\mathbb{M}_{a;\rho}(\lambda)$ on the last element λ_N. We see that

$$P(\lambda_N) = \sum_{\substack{\lambda_1,\dots,\lambda_{N-1} \\ \lambda \in \mathcal{S}_N}} \mathbb{M}_{a;\rho}(\lambda) = \sum_{\ell_1,\dots,\ell_{N-1}=0}^{\infty} \mathbb{M}_{a;\rho}(\lambda), \tag{57}$$

where $\ell_k = \lambda_k - \lambda_{k+1}$, $k = 1, \dots, N = 1$. By using the representation (17) with the specialization ρ (19) and (48), one has an integral representation of $\mathbb{M}_{a;\rho}(\lambda)$,

$$\frac{\mathbb{M}_{a;\rho}(\lambda)}{(q;q)_\infty^{N-1}} = \int_{\mathbb{T}^N} \prod_{i=1}^{N} \frac{dz_i}{z_i} \cdot \frac{R_\ell(a) R_\ell(1/z)}{\prod_{k=1}^{N-1} (q;q)_{\ell_k}} \frac{\Pi(z;\rho)}{\Pi(a;\rho)} m_N^q(z) \left(\frac{A}{Z}\right)^{\lambda_N}, \tag{58}$$

where $A = \prod_{j=1}^N a_j$, $Z = \prod_{j=1}^N z_j$ and we used the shorthand notations $R_\ell(a) = R_\ell(a_1, \dots, a_N)$ and $R_\ell(1/z) = R_\ell(1/z_1, \dots, 1/z_N)$. Using (58) with (49), we have

$$\begin{aligned} P(\lambda_N) &= \frac{(q;q)_\infty^{N-1}}{N!} \int_{\mathbb{T}^N} \prod_{j=1}^{N} \frac{dz_j}{2\pi i z_j} \cdot \prod_{i \neq j}^{N} \left(\frac{z_i}{z_j}; q\right)_\infty \cdot \frac{\Pi(z;\rho)}{\Pi(a;\rho)} \\ &\quad \times \prod_{i,j=1}^{N} \frac{1}{(\frac{a_i}{z_j}; q)_\infty} \cdot \left(\frac{A}{Z}; q\right)_\infty \left(\frac{A}{Z}\right)^{\lambda_N}. \end{aligned} \tag{59}$$

Second step: taking the sum $\sum_{\lambda_N=0}^{\infty}$

From (56), (57) and (59) we find

$$\begin{aligned} \mathbb{E}_{q\mathrm{M}}^{a;\rho}\left[\frac{1}{(\zeta q^{\lambda_N}; q)_\infty}\right] &= \sum_{\lambda_N=0}^{\infty} \frac{P(\lambda_N)}{(\zeta q^{\lambda_N}; q)_\infty} = \sum_{\lambda_N=-\infty}^{\infty} \frac{P(\lambda_N)}{(\zeta q^{\lambda_N}; q)_\infty} \\ &= \frac{1}{N!} \int_{\mathbb{T}^N} \prod_{j=1}^{N} \frac{dz_j}{2\pi i z_j} \cdot \prod_{i \neq j}^{N} \left(\frac{z_i}{z_j}; q\right)_\infty \cdot \frac{\Pi(z;\rho)}{\Pi(a;\rho)} \\ &\quad \times \prod_{i,j=1}^{N} \frac{1}{(\frac{a_i}{z_j}; q)_\infty} \cdot \left(\frac{A}{Z}; q\right)_\infty \sum_{\lambda_N=-\infty}^{\infty} \frac{(q;q)_\infty^{N-1}}{(\zeta q^{\lambda_N}; q)_\infty} \left(\frac{A}{Z}\right)^{\lambda_N}, \end{aligned} \tag{60}$$

where in the second equality we used the fact $P(\lambda_N) = 0$ for $\lambda_N \leq -1$. Using Ramanujan summation formula (50) with $a = \zeta, b = 0$, we have

$$\sum_{\lambda_N=-\infty}^{\infty} \frac{(q;q)_\infty^{N-1}}{(\zeta q^{\lambda_N}; q)_\infty} \left(\frac{A}{Z}\right)^{\lambda_N} = \frac{(\zeta A/Z; q)_\infty (qZ/\zeta A; q)_\infty (q;q)_\infty^N}{(\zeta; q)_\infty (q/\zeta; q)_\infty (A/Z; q)_\infty}. \tag{61}$$

Substituting (61) into (60), we find

$$\begin{aligned}(62)\quad &\mathbb{E}_{q\mathrm{M}}^{a;\rho}\left[\frac{1}{(\zeta q^{\lambda_N};q)_\infty}\right]\\ &=\frac{(q;q)_\infty^N}{N!}\int_{\mathbb{T}^N}\prod_{i=1}^N\frac{dz_i}{\sqrt{z_ia_i}}\cdot\prod_{i\neq j}^N\left(\frac{z_i}{z_j};q\right)_\infty\cdot\prod_{i,j=1}^N\frac{1}{(\frac{a_i}{z_j};q)_\infty}\cdot\frac{\tilde{\theta}(\frac{\zeta A}{Z})}{\tilde{\theta}(\zeta)}\frac{\Pi(z;\rho)}{\Pi(a;\rho)}.\end{aligned}$$

Here $\tilde{\theta}(x)$ is defined by (52).

Final step: extracting a determinantal structure

Applying the determinant formulas (54) and (55) to (62), we arrive at the determinantal representation,

$$\begin{aligned}(63)\quad &\mathbb{E}_{q\mathrm{M}}^{a;\rho}\left[\frac{1}{(\zeta q^{\lambda_N};q)_\infty}\right]\\ &=\frac{1}{N!}\int_{\mathbb{T}^N}\prod_{i=1}^N\frac{dz_i}{2\pi i z_i}\cdot\prod_{i\neq j}^N\left(\frac{a_i}{a_j};q\right)_\infty\prod_{i,j=1}^N\left(\frac{z_i}{a_j};q\right)_\infty\frac{\Pi(z;\rho)}{\Pi(a;\rho)}\\ &\qquad\times\det\left[\frac{a_i}{a_i-z_j}\right]_{i,j=1}^N\det\left[\frac{\tilde{\theta}(\zeta a_i/z_j)}{\tilde{\theta}(\zeta)\tilde{\theta}(a_i/z_j)}\right]_{i,j=1}^N.\end{aligned}$$

Note that this representation has a similar structure to the rhs of (4). In both cases, the integrands are a product of two determinants. From (63), one can readily obtain the determinantal formula (47) applying the techniques in [23, 33].

§4. Applications

4.1. Stationary q-TASEP

In this subsection, we consider the stationary situation of the q-TASEP. Contrary to Section 2.3, now we consider infinitely many particles on $\mathbb{Z}$. We set the particle which is initially at the closest positive site to the origin to be the "0 th" particle and label all the other particles from left to right. In the stationary q-TASEP, $\mathrm{gap}_k(t) := X_{k-1}(t)-X_k(t)-1$, $k\in\mathbb{Z}$ are the i.i.d. random variables with the q-Poisson distribution $q\mathrm{Po}(\alpha)$ with parameter $\alpha\in[0,1]$ defined by

$$(64)\qquad \mathbb{P}\big(q\mathrm{Po}(\alpha)=n\big)=(\alpha;q)_\infty\frac{\alpha^n}{(q;q)_n},\quad n\in\mathbb{Z}_{\geq 0}.$$

Based on this, we consider the initial condition $\mathrm{gap}_k(0)\stackrel{\mathrm{d}}{=}q\mathrm{Po}(\alpha)$ and set $X_1(0)=0$. This initial condition illustrated in Figure 1(a) and we call it the stationary initial condition. Under this initial condition, we would like to find the fluctuation property of the Nth particle $X_N(t)$ especially its limiting behavior for large N and t under the initial condition.

In order to study this problem, we introduce finite particle q-TASEP with two different initial conditions as depicted in Figure 1(b) and (c). First we apply the Burke's theorem to the infinitely many q-TASEP particles with the stationary initial condition. This theorem says that in the stationary situation, the marginal dynamics of each $X_k(t)$ behaves as the Poisson random walk with rate α, i.e. $X_k(t) - X_k(0) \overset{\mathrm{d}}{=} \mathrm{Po}(\alpha)$. This means that as long as we focus on the fluctuation property of $X_N(t)$, the contributions of the whole particles with nonnegative label can be represented as only one particle with rate α starting at the origin. Considering this fact, we introduce the following N-particle q-TASEP: Let $Y_k^0(t)$, $k = 1, \ldots, N$ be the position of the kth q-TASEP particle with rate $a_k(1 - q^{\mathrm{gap}_k(t)})$ where $a_k > 0$ and we set the initial condition

$$(65) \quad Y_1^0(0) = 0, \ (\mathrm{gap}_k(0))_{k=2,\ldots,N}, \text{ i.i.d. with } \mathrm{gap}_k(0) \overset{\mathrm{d}}{=} q\mathrm{Po}(\alpha/a_k),$$

which is illustrated in Figure 1(b). Note that by Burke's thoerem, one sees when $a_1 = \alpha$, $a_2 \cdots = a_N = 1$,

$$(66) \qquad X_N(t) \overset{\mathrm{d}}{=} Y_N^0(t).$$

Although the above model $Y_k^0(t)$, $k = 1, \ldots, N$ is already a nice reduction to the finite particle system, the model which we actually analyze is the slightly modified one $Y_k(t)$, $k = 1, \ldots, N$, which is more tractable to our approach. In this model, the hopping rates and initial conditions are the same as $Y_k^0(t)$'s except that the initial position of the first particle is also distributed as

$$(67) \qquad -Y_1(0) \overset{\mathrm{d}}{=} q\mathrm{Po}(\alpha/a_1).$$

We call this initial condition the half-stationary initial condition, which is shown in Figure 1(c). Note that for $N \geq 2$, $Y_N(t)$ is related to $Y_N^0(t)$ as

$$(68) \qquad Y_N(t) \overset{\mathrm{d}}{=} Y_N^0(t) - \chi,$$

where χ is the random variable independent of $Y_N^0(t)$ with distributed as $\chi \overset{\mathrm{d}}{=} q\mathrm{Po}(\alpha/a_1)$. Thus combining (66) with (68), one can obtain the distribution of $X_N(t)$ from that of $Y_N(t)$.

There is a difficulty when we apply the conventional approaches developed in [6, 8, 14, 9, 31] to the q-TASEP with the half stationary

(a) 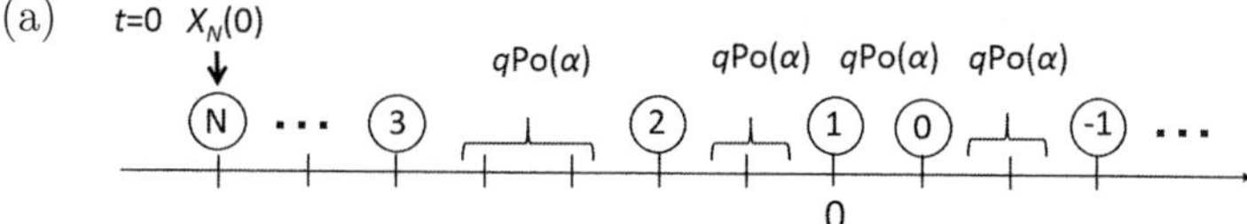

(b) 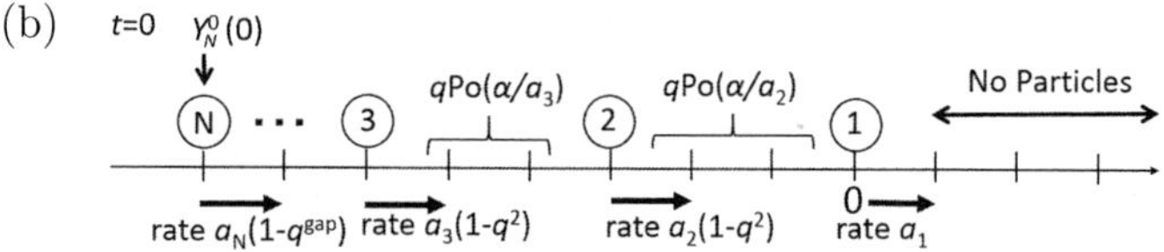

(c) 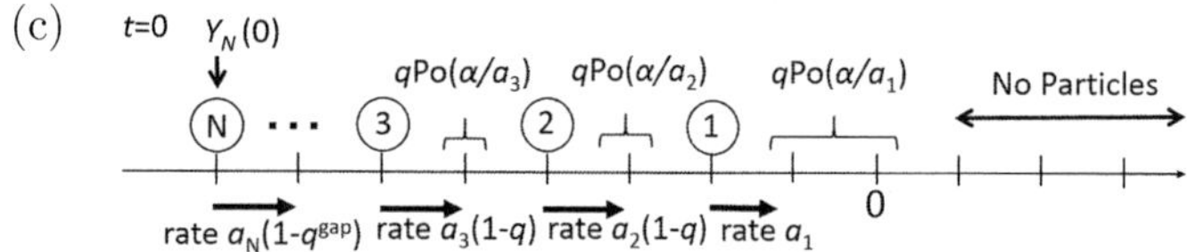

Fig. 1. Initial configurations for the three models about the stationary q-TASEP, (a) $X_k(t),\ k \in \mathbb{Z}$ (b) $Y_k^0(t),\ k = 1, \ldots, N$ and (c) $Y_k(t),\ k = 1, \ldots, N$.

initial condition. In these approaches one focuses on the q-moments $\langle q^{n(Y_N(t)+N)} \rangle$, $n = 1, 2, \ldots$, where $\langle \cdot \rangle$ represents the average over the q-TASEP with half-stationary initial condition. But they diverge for sufficiently large n and thus the moment generating function such as (46) for $Y_N(t)$ is not well defined. This can be easily seen as follows: From (67), we have

$$\langle q^{n(Y_N(0)+N)} \rangle = \left\langle \prod_{k=0}^{N-1} q^{-n\mathrm{gap}_k(0)} \right\rangle = \prod_{k=0}^{N-1} \left\langle q^{-n\mathrm{gap}_k(0)} \right\rangle, \tag{69}$$

where we set $\mathrm{gap}_0(0) = -Y_1(0)$. From (64) we find for $n > \log_q(\alpha/a_k)$

$$\langle q^{-n\mathrm{gap}_k(0)} \rangle = \left(\frac{\alpha}{a_k}; q \right)_\infty \sum_{y=0}^{\infty} \frac{\left(\frac{\alpha}{q^n a_k} \right)^y}{(q;q)_y} = \infty. \tag{70}$$

Thus the q-moment generating function is not well-defined.

Similar divergence appears in the study of the stationary state in the ASEP. In [1], Aggarwal has developed a method of regularization in terms of the fusion technique in quantum integrable system and showed that the distribution of the particle current converges to the Baik-Rains distribution [3] in the KPZ scaling limit.

In our approach, we do not need to use the fusion technique. In fact our approach discussed in Section 2.3 and Section 3 can be generalized to the half stationary initial condition (67). As in Section 2.3, we consider the stochastic process $\underline{\lambda}_N(t)$ on $\tilde{\mathbb{G}}_N$ (43) where the initial distribution is

$$\text{Prob}(\underline{\lambda}_N(0) = \underline{\lambda}_N) = P_0(\underline{\lambda}_N),$$
$$P_0(\underline{\lambda}_N) = \prod_{j=1}^{N} (\alpha/a_j;q)_\infty \frac{(\alpha/a_j)^{\lambda_{j-1}^{(j-1)}-\lambda_j^{(j)}}}{(q;q)_{\lambda_{j-1}^{(j-1)}-\lambda_j^{(j)}}} \cdot \prod_{k=1}^{N}\prod_{j=1}^{k-1} \delta_{\lambda_j^{(k)},0}. \tag{71}$$

Here we set $\lambda_0^{(0)} \equiv 0$ and $(a;q)_n \equiv 0$ for $n < 0$. Note that when $\alpha = 0$, $\underline{\lambda}_N(0) = \phi$, i.e. $\lambda_i^{(j)} = 0,\ 1 \le i \le j \le N$ and the marginal distribution on $(\lambda_k^{(k)})_{k=1,\dots,N}$ is given by

$$\begin{aligned}\text{Prob}\big(\{\lambda_k^{(k)}(t) = \lambda_k^{(k)}\}_{k=1,\dots,N}\big) &= \prod_{j=1}^{N} (\alpha/a_j;q)_\infty \frac{(\alpha/a_j)^{\lambda_{j-1}^{(j-1)}-\lambda_j^{(j)}}}{(q;q)_{\lambda_{j-1}^{(j-1)}-\lambda_j^{(j)}}} \\ &= \prod_{k=1}^{N} \text{Prob}\left(\text{gap}_k(0) = \lambda_{j-1}^{(j-1)} - \lambda_j^{(j)}\right),\end{aligned} \tag{72}$$

where $\text{gap}_k(t) := Y_{k-1} - Y_k - 1$. Thus we find that $\lambda_k^{(k)}(t) - k \overset{\text{d}}{=} Y_k(t)$, $k = 1, \dots, N$. Furthermore in [20], we obtain a relation to the q-Whittaker process, which is a generalization to (31).

Proposition 3 ([20]).

$$\sum_{\underline{\mu}_N \in \tilde{\mathbb{G}}_N} \text{Prob}\left(\underline{\lambda}_N(t) = \underline{\lambda}_N \mid \underline{\lambda}_N(0) = \underline{\mu}_N\right) P_0(\underline{\mu}_N) = \mathbb{P}_{a;\rho_{t,\alpha}}(\underline{\lambda}_N), \tag{73}$$

where the rhs is the double-sided q-Whittaker process (24) *(see the remark below Proposition* 2) *with the specialization $\rho_{t,\alpha}$ defined by* (19) *with $f_q(z;\rho)$ replaced by*

$$f_q(z;\rho_{t,\alpha}) = \frac{e^{tz_j}}{(\alpha/z_j;q)_\infty}. \tag{74}$$

Note that our approach in Section 2.3 and Section 3 does not rely on the q-moments and their generating function. Thus we can simply apply our approach in Section 2.3 and Section 3 to this case just changing ρ to $\rho_{t,\alpha}$. As a result we have the following

Theorem 3 ([20]).

$$\mathbb{E}\left[\frac{1}{(\zeta q^{Y_N(t)+N};q)_\infty}\right] = \det(1+fK_{a;\rho_{t,\alpha}})_{\ell^2(\mathbb{Z})}. \tag{75}$$

Here $Y_N(t)$ represents the position of the Nth particle in the q-TASEP with the half-stationary initial condition defined around (67), $f(n)$ is the second equation in (47) and the kernel $K_{a;\rho_{t,\alpha}}(n,m)$ is defined by the third equation of (47) with ρ replaced by $\rho_{t,\alpha}$ defined by (74).

Another Fredholm determinant formula, which is similar to Theorem 1 was also obtained in [21].

Combining the single Fredholm determinant formula with (66) and (68) we finally get the limiting distribution of $X_N(t)$, the position of the Nth particle with the stationary q-TASEP discussed below (64). We obtain

Theorem 4 ([20]).

$$\lim_{N\to\infty} \mathbb{P}\left(\frac{Y_N(\kappa N)-\eta N}{\gamma N^{1/3}} > -s\right) = F_\omega(s), \tag{76}$$

where κ, η, and γ are defined by

(77)

$$\kappa = \sum_{n=0}^{\infty} \frac{q^n}{(1-\alpha q^n)^2},\quad \eta = \sum_{n=0}^{\infty} \frac{\alpha^2 q^{2n}}{(1-\alpha q^n)^2},\quad \gamma = \left(\sum_{n=0}^{\infty} \frac{\alpha^2 q^{2n}}{(1-\alpha q^n)^3}\right)^{\frac{1}{3}},$$

and $F_\omega(s)$ is the Baik-Rains distribution.

The Baik-Rains distribution is first introduced by [3] by using the solution to the Painlevé equation. Later the other representations in terms of Fredholm determinants have been found out [16, 19].

4.2. Higher spin exclusion process (HSEP)

4.2.1. *HSEP with step initial condition.* The higher spin vertex model is introduced in [14] as a generalization of the stochastic six vertex model. In this note we describe it in the language of the exclusion process and call it the higher spin exclusion process (HSEP). The HSEP is a discrete time exclusion process with sequential update. Let $X_k(t)$ be the position of the kth particle and $\mathrm{dis}_k(t) := X_k(t+1) - X_k(t)$ be the hopping distance during the update between time t to $t+1$. As before we denote the gap between kth and $k-1$th particle as $\mathrm{gap}_k(t) = X_{k-1}(t) - X_k(t) - 1$. The hopping probability in the HSEP is defined by

$$\text{Prob}\big(\text{dis}_k(t) = m \mid \text{dis}_{k-1}(t) = \ell,\ \text{gap}_k(t-1) = g\big) \\ = \begin{cases} \dfrac{1}{1+a_k\beta_t} + \dfrac{q^g a_k}{1+a_k\beta_t}(\beta_t\delta_{\ell,0} - \alpha_k\delta_{\ell,1}), & m = 0, \\ \dfrac{a_k\beta_t}{1+a_k\beta_t} - \dfrac{q^g a_k}{1+a_k\beta_t}(\beta_t\delta_{\ell,0} - \alpha_k\delta_{\ell,1}), & m = 1. \end{cases} \tag{78}$$

Like the continuous time q-TASEP, the hopping probability (78) depends on the value of $\text{gap}_k(t)$. When $\alpha_k = 0$, (78) corresponds to the hopping probability of the discrete-time Bernoulli q-TASEP introduced in [7], and in the case $q = 0$, determinantal structure on the probability measure has been found out [5, 25]. When both $\alpha = q = 0$, it reduces to the discrete time TASEP with sequential update.

For later purpose of the analysis on the q-Hahn TASEP (Section 4.2.3), we introduce a generalization of the HSEP. We call it the fused higher spin exclusion process (fHSEP) since it is associated with the fusion of the stochastic vertex model [9]. For $\boldsymbol{a} = (a_1, \ldots, a_m) \in \mathbb{R}^m$ and $\boldsymbol{b} = (b_1, \ldots, b_n) \in \mathbb{R}^n$, we define $\boldsymbol{a} \sqcup \boldsymbol{b} \in \mathbb{R}^{m+n}$ by

$$\boldsymbol{a} \sqcup \boldsymbol{b} = (a_1, \ldots, a_m, b_1, \ldots, b_n). \tag{79}$$

Using this notation, we define the fHSEP as follows. We fix $J \in \{1, 2, 3, \ldots\}$ and set the sequence of parameters $(\beta_t)_{t=1,2,\ldots}$ in (78) to be

$$(\beta_t)_{t=1,2,\ldots} = \hat{\boldsymbol{\beta}}_J^{(1)} \sqcup \hat{\boldsymbol{\beta}}_J^{(2)} \sqcup \cdots \sqcup \hat{\boldsymbol{\beta}}_J^{(\tau)} \sqcup \cdots, \tag{80}$$

where $\hat{\boldsymbol{\beta}}_J^{(\tau)} = (\hat{\beta}_\tau, q\hat{\beta}_\tau, \ldots, q^{J-1}\hat{\beta}_\tau)$. Then in the HSEP with the parametrization (80), we regard τ as new time variable, i.e. a single time step $\tau \to \tau + 1$ in the new model contains J time steps $J\tau + 1 \to J\tau + J$ in the original HSEP. Note that the case $J = 1$ (and $\tau = t$) is reduced to the HSEP. The hopping probability in the fHSEP is expressed as

(81)

$$\text{Prob}\big(\text{dis}_k(\tau) = m \mid \text{dis}_{k-1}(\tau) = \ell,\ \text{gap}_k(\tau - 1) = g\big) = L^{(J)}_{a_k,\alpha_k,\beta_\tau}(m|\ell, g)$$

$$L^{(J)}_{a_k,\alpha_k,\beta_\tau}(m|\ell, g) = \frac{q^{\frac{1}{2}g(g+2\ell-1)}\hat{\beta}_\tau^{\ g}(a_k\alpha_k)^{g/2+\ell}(-\hat{\beta}_\tau/(a_k\alpha_k)^{1/2}; q)_{m-g}}{(q;q)_{m-g-\ell}(-\hat{\beta}_\tau(a_k\alpha_k); q)_{g+\ell}(q^{J+1-\ell}; q)_{\ell-m}} \\ \times {}_4\bar{\phi}_3\left(\begin{matrix} q^{g+\ell-m}, & q^{-g}, & -(a_k\alpha_k)^{1/2}\hat{\beta}_t q^J, & -q(a_k\alpha_k)^{1/2}/\hat{\beta}_t \\ & a_k\alpha_k, & q^{1+m-g}, & q^{J+1-g-\ell} \end{matrix} \middle| q,\ q \right),$$

where ${}_{r+1}\bar{\phi}_r$ is the regularized terminating q-hypergeometric function defined as

(82)

$$ {}_{r+1}\bar{\phi}_r\left(\begin{matrix} q^{-n}, a_1, \dots, a_r \\ b_1,\ b_2, \dots, b_r \end{matrix} \middle| q, z\right) = \sum_{k=0}^{n} z^k \frac{(q^{-n};q)_k}{(q;q)_k} \prod_{j=1}^{r} (a_j;q)_k (q^k b_j; q)_{n-k}. $$

In [31], Orr and Petrov found a correspondence between the HSEP with step initial conditions $X_k(0) = -k,\ k = 1, 2, \dots, N$ and the q-Whittaker measure.

Proposition 4 ([31]). *Let $X_k(t),\ k = 1, \dots, N$ be the positions of the N-particle the HSEP with the step initial condition $X_k(0) = -k,\ k = 1, 2, \dots, N$. We have*

$$ X_N(t) + N \overset{d}{=} \lambda_N(t), \tag{83} $$

where $\lambda_N(t)$ is the last element of the random partition $\lambda(t) = (\lambda_1(t), \dots, \lambda_N(t))$ distributed by the q-Whittaker measure (23) *with the specialization $\rho_{N,t}$ defined by* (19) *with*

$$ f_q(z; \rho_{N,t}) = \prod_{i=1}^{N-1} \frac{1}{(\alpha_i z; q)_\infty} \cdot \prod_{j=1}^{t} (1 + \beta_j z). \tag{84} $$

That is, among the three sequence of parameters in ρ (19), *we set $\alpha_i = \beta_j = \gamma = 0$ for $i \geq N, j \geq t+1$.*

For later discussions in Section 4.2.2 and Section 4.2.3, we introduce another model in which the parametrization (80) is slightly deformed. We fix $J, K \in \{1, 2, 3 \cdots\}$ and set the sequence of parameters $(\beta_t)_{t=1,2,\dots}$ in (78) to be

$$ (\beta_t)_{t=1,2,\dots} = \hat{\boldsymbol{\alpha}}_K \sqcup \hat{\boldsymbol{\beta}}_J^{(1)} \sqcup \hat{\boldsymbol{\beta}}_J^{(2)} \sqcup \cdots \sqcup \hat{\boldsymbol{\beta}}_J^{(\tau)} \sqcup \cdots, \tag{85} $$

where $\hat{\boldsymbol{\alpha}}_K = (q/\hat{\alpha}, q^2/\hat{\alpha}, \dots, q^K/\hat{\alpha})$ and $\hat{\boldsymbol{\beta}}_J^{(\tau)}$ is defined below (80).

Let $\hat{X}_k(\tau)$ $(k = 1, \dots, N)$ be the position of the kth particle in the deformed fHSEP defined by (85). From Proposition 4, we have

Corollary 1.

$$ \hat{X}_N(\tau) + N \overset{d}{=} \hat{\lambda}_N(\tau), \tag{86} $$

where $\hat{\lambda}_N(\tau)$ is the last element of the random partition $\hat{\lambda}(\tau) = (\hat{\lambda}_1(t), \dots, \hat{\lambda}_N(\tau))$ distributed by the q-Whittaker measure (23) *with the specialization $\hat{\rho}_{N,\tau}$ defined by* (19) *with*

$$(87)\qquad f_q(z;\hat{\rho}_{N,\tau}) = \prod_{i=1}^{N-1} \frac{1}{(\alpha_i z;q)_\infty} \cdot \prod_{j=1}^{\tau} (-\hat{\beta}_j z;q)_J \cdot (-qz/\hat{\alpha};q)_K.$$

Thus just applying Theorem 2, we have the following

Theorem 5 ([18]). *For the N-paticle deformed* fHSEP *with the step initial condition we obtain*

$$(88)\qquad \mathbb{E}\left[\frac{1}{(\zeta q^{\hat{X}_N(\tau)+N};q)_\infty}\right] = \det(1+fK_{a;\rho_{N,J,\tau}})_{\ell^2(\mathbb{Z}_{\geq 0})},$$

where f, $K_{a;\rho}(m,n)$ are defined in (47) *and $\hat{\rho}_{N,\tau}$ is defined by* (87).

4.2.2. *Stationary fHSEP.* For discussing the stationary state of the fHSEP, we consider the infinitely many particle system described by $X_k(t)$, $k \in \mathbb{Z}$. We also set $a_k = a$, $\alpha_k = \alpha$ for $k \in \mathbb{Z}$ in (81) to realize the uniform hopping probability of each particle. In [18], it has been shown that in the stationary state, the sequence of the gaps $\mathrm{gap}_k(t) = X_{k-1}(t) - X_k(t) - 1$, $k \in \mathbb{Z}$ is independently distributed as

$$(89)\qquad \mathrm{gap}_k(t) \overset{\mathrm{d}}{=} q\mathrm{NB}(a\alpha, \hat{\alpha}/a),$$

where $q\mathrm{NB}(b,p)$ is the negative binomial distribution defined by

$$(90)\qquad \mathbb{P}(X=n) = p^n \frac{(b;q)_n}{(q;q)_n} \frac{(p;q)_\infty}{(pb;q)_\infty}, \qquad \text{for all } n \in \mathbb{Z}_{\geq 0}.$$

Note that $q\mathrm{NB}(0,p) = q\mathrm{Po}(p)$ where $q\mathrm{Po}(p)$ is defined by (64).

We are interested in the position of the Nth particle $X_N(t)$ when we set $X_1(0) = 0$. Similarly to Section 4.1, one can apply Burke's theorem to this system, which says that the marginal distribution on each particle $X_k(t)$ is distributed as $q\mathrm{NB}(q^{-J}, -q^J\hat{\alpha}\beta_\tau)$. Thus as long as we focus on $X_N(t)$, the system with infinitely many particle is equivalent to the N-particle system where the hopping probability of the first particle at the time step $\tau \to \tau+1$ is $q\mathrm{NB}(q^{-J}, -q^J\hat{\alpha}\beta_\tau)$. It corresponds to Figure 1(b) in the continuous time q-TASEP.

Now let us define the fHSEP with the half stationary initial condition corresponding to Figure 1(c) in the q-TASEP. Let $Y_k(\tau)$, $k = 1, 2, \ldots, N$ be the position of the kth particle in the fHSEP at time τ. The half stationary initial condition is defined as follows.

(91)

$$-Y_1(0) \sim q\mathrm{Po}(\hat{\alpha}/a_1),\ \mathrm{gap}_k(0) := Y_{k-1}(0) - Y_k(0) - 1 \sim q\mathrm{NB}(a_k\alpha_k, \hat{\alpha}/a_k)$$

with $k = 2, \dots, N$.

We would like to get a determinantal formula for $Y_N(\tau)$ using the result on the q-Whittaker measure discussed in Section 3. However the problem is that Proposition 4, which states the coupling with the q-Whittaker measure, can be applicable only to the step initial condition.

In fact, we can construct the half stationary initial condition (91) from the step initial condition by applying the analytic continuation of type 1 stated above Proposition 2. The idea is based on [1] in the study on the stationary ASEP.

Proposition 5 ([18]). *Let $\hat{X}_k(\tau)$, $k = 1, 2, \dots, N$ be the position of the kth particle in the deformed fHSEP defined by* (85). *Then we have*

$$
\begin{aligned}
(92) \qquad & \mathrm{Prob}(\hat{X}_k(0) - K = m_k, k = 1, 2, \dots, N) \\
\overset{\text{type 1}}{\longrightarrow}\ & \mathrm{Prob}(Y_k(0) = m_k, k = 1, \dots, N) \\
&= \left(\frac{\hat{\alpha}}{a_1}\right)^{-m_1} \frac{(\hat{\alpha}/a_1; q)_\infty}{(q;q)_{-m_1}} \prod_{k=2}^{N} \left(\frac{\hat{\alpha}}{a_k}\right)^{n_k} \frac{(a_k \alpha_k; q)_{n_k}}{(q;q)_{n_k}} \frac{(\hat{\alpha}/a_k; q)_\infty}{(\alpha_k \hat{\alpha}; q)_\infty},
\end{aligned}
$$

where $n_k = m_{k-1} - m_k - 1$ and "$\overset{\text{type 1}}{\longrightarrow}$" represents the analytic continuation of type 1 *defined above Proposition* 2 *with $J \equiv K$.*

Combining this with Theorem 5 and Proposition 2(i), we obtain a determinantal formula for the fHSEP with the half stationary initial condition.

Theorem 6. *For N-paticle* fHSEP *with the half stationary initial condition* (91), *we obtain*

$$
(93) \qquad \mathbb{E}\left[\frac{1}{(\zeta q^{Y_N(\tau)+N}; q)_\infty}\right] = \det\left(1 + f K_{a;\rho^{\hat{\alpha}}_{N,J,\tau}}\right)_{\ell^2(\mathbb{Z})},
$$

where f, $K_{a;\rho}(m,n)$ are defined in (47) *and the specialization $\rho^{\hat{\alpha}}_{N,J,\tau}$ is defined by*

$$
(94) \qquad f_q(z; \rho^{\hat{\alpha}}_{N,J,\tau}) = \prod_{i=1}^{N-1} \frac{1}{(\alpha_i z; q)_\infty} \cdot \prod_{j=1}^{\tau} (-\beta_j z; q)_J \cdot \frac{1}{(\hat{\alpha}/z; q)_\infty}.
$$

Outline of the proof. From Corollary 1, we have

$$
(95) \qquad \mathrm{Prob}\left(\hat{X}_N(\tau) = m\right) = \mathrm{Prob}\left(\hat{\lambda}_N(\tau) - N = m\right).
$$

Now we apply the analytic continuation of type 1 stated above Proposition 1 with $J \equiv K$ to both hand sides of (95). We see that from

Proposition 5, the lhs goes to $\mathrm{Prob}(Y_N(\tau) = m)$, while from Proposition 2(i) the rhs goes to $\mathrm{Prob}(\lambda_N(\tau) = m)$, which is the marginal distribution of the q-Whittaker measure on $\lambda_N(\tau)$ with the specialization $\rho^{\hat{\alpha}}_{N,J,\tau}$. Applying our approach discussed in Section 3 to this q-Whittaker measure, we get the result (93). Q.E.D.

Recently in [11] different Fredholm determinant formulas having a similar form to (44) in Theorem 1 were obtained for the stationary higher spin vertex model by considering the Yang-Baxter random field associated with the spin q-Whittaker function.

4.2.3. *Staionary q-Hahn TASEP.* The q-Hahn TASEP has been introduced in [32]. For $g = 0, 1, 2, \ldots$ and $m = 0, \ldots, g$, the hopping probability of this process is given by

$$\mathrm{Prob}\big(\mathrm{dis}_k(t) = m \mid \mathrm{gap}_k(t-1) = g\big) = \varphi_{q,\mu,\nu}(m|g) \tag{96}$$

where for $0 \le \nu < \mu < 1$ and $0 \le q < 1$, $\varphi_{q,\mu,\nu}(m|g)$ is defined by

$$\varphi_{q,\mu,\nu}(m|g) = \mu^m \frac{(\nu/\mu; q)_m (\mu; q)_{g-m}}{(\nu; q)_g} \frac{(q; q)_g}{(q; q)_m (q; q)_{g-m}}. \tag{97}$$

In the stationary state of the q-Hahn TASEP, one sees that for $k \in \mathbb{Z}$,

$$\mathrm{gap}_k(t) \overset{\mathrm{d}}{=} q\mathrm{NB}(\nu, \mathscr{d}). \tag{98}$$

Similarly to the case of q-TASEP and fHSEP, one can use Burke's theorem to reduce the problem of infinitely many particles to that with N particles. Let $Y_k(\tau)$, $k = 1, 2, \ldots, N$ be the position of the kth particle in the q-Hahn TASEP with the half stationary initial condition defined as follows. First the hopping probability of each particle is given by $\varphi_{q,d\mu,\mathscr{d}_+\nu}(m|\infty) = q\mathrm{NB}(\nu/\mu, \mathscr{d}_+\mu)$ with $0 < \mathscr{d}_+ < 1$ for $Y_1(\tau)$ and $\varphi_{q,d\mu,d\nu}(m|g)$ for $Y_k(\tau)$, $k = 2, \ldots, N$. Next the initial condition is defined by

$$-Y_1(0) \sim q\mathrm{Po}(\mathscr{d}/\mathscr{d}_+), \ \mathrm{gap}_k(0) \sim q\mathrm{NB}(\nu, \mathscr{d}), \tag{99}$$

with $0 < \mathscr{d} < \mathscr{d}_+ < 1$.

In [5], the hopping probability of the q-Hahn TASEP $\varphi_{q,\mu,\nu}(m|g)$ (97) can be obtained from that of fHSEP (81) by applying the analytic continuation of type 2 stated above Proposition 2.

Proposition 6 ([5]). *In the hopping probability $L^{(J)}_{a_k,\alpha_k,\beta_\tau}(m|\ell, g)$ (81) of the fHSEP, we specialize the parameter $a_k, \alpha_k, \beta_\tau$ as*

$$a_k = 1, \ \alpha_k = \beta_\tau = \nu. \tag{100}$$

We have

$$L^{(J)}_{1,\nu,\nu}(m|\ell,g) \overset{\text{type 2}}{\longrightarrow} \varphi_{q,\mu,\nu}(m|g) \tag{101}$$

where "$\overset{\text{type 2}}{\longrightarrow}$" *represents the analytic continuation of type 2 defined above Proposition 2 with* $d \equiv \mu/\nu$.

In addition, in the special case $g = \infty$ in (81), we can easily perform the analytic continuation since the hopping probability in the HSEP (78) is reduced to the Bernoulli distribution. We obtain the following result [18]

$$a_k = d_+, \ \alpha_k = \beta_\tau = \nu. \tag{102}$$

with $0 < d_+ < 1$. Then under the same type 2 analytic continuation above, we have

$$L^{(J)}_{d_+,\nu,\nu}(m|\ell,\infty) \overset{\text{type 2}}{\longrightarrow} q\mathrm{NB}(\nu/\mu, d_+\mu). \tag{103}$$

Thus using Theorems 5, Propositions 2, 6 and (103), we have the Fredholm determinant formula for the stationary q-Hahn TASEP.

Theorem 7 ([18]). *Let* $Y_N(\tau)$ *be the position of the* N*th particle in* q*-Hahn TASEP with the half stationary initial condition defined around* (99). *We obtain*

$$\mathbb{E}^{\mathrm{hs}}_{q\mathrm{Hahn}}\left[\frac{1}{(\zeta q^{Y_N(\tau)+N};q)_\infty}\right] = \det\left(1 + fK_{a_{d_+};\rho_{\mu,\nu,d}}\right)_{\ell^2(\mathbb{Z})}. \tag{104}$$

Here f, $K_{a;\rho}(m,n)$ *are defined in* (47) *where*

$$a_{d_+} = (d_+, 1, 1, \ldots, 1) \tag{105}$$

and the specialization $\rho_{\mu,\nu,d}$ *is defined by*

$$f_q(z;\rho) = \left(\frac{(\nu z;q)_\infty}{(\mu z;q)_\infty}\right)^t \left(\frac{(z;q)_\infty}{(\nu z;q)_\infty}\right)^{N-1} \frac{1}{(qd/z;q)_\infty}. \tag{106}$$

Outline of the proof. We take a similar approach to the proof of Theorem 6. The starting point is (95) with specific choice of parameters (105) and $\alpha_k = \hat{\beta}_\tau = \nu$ for $k = 1, \ldots, N$, and $\tau = 1, 2, \ldots$. We apply both type 1 with $J \equiv K$ and type 2 with $h \equiv \mu/\nu$ stated above Proposition 2 to the both hand sides of this equation. From Proposition 5, (101) and (103), the lhs goes to $\mathrm{Prob}(Y_N(\tau) = m)$, while from Proposition 2(i) (ii) the rhs goes to $\mathrm{Prob}(\lambda_N(\tau) = m)$, which is the marginal distribution of the q-Whittaker measure on the last element $\lambda_N(\tau)$ with the specialization $\rho_{\mu,\nu,d}$. Applying our approach discussed in Section 3 to this q-Whittaker measure, we get the result (104). Q.E.D.

The proof of this theorem relies on Proposition 4 obtained in [31], which says the coupling between the HSEP and the q-Whittaker measure. Recently another type of coupling has been found out: In [28], the spin q-Whittaker process was introduced and it was shown that the q-Hahn TASEP was realized as a marginal of this process.

4.2.4. *Stationary beta-polymer.* The beta-polymer is an exactly solvable model of a directed polymer in beta-distributed random environment. It is introduced in [4] and further studied in [12, 13, 28, 29].

Let $\pi = \{e_1, e_2, \dots, e_t\}$ be a directed path on a two-dimensional lattice from $(1,0)$ to (n,t). The element e_j $(j = 1, 2, \dots, t)$ represents the directed edge and is often denoted as $e_j : s_{j-1} \to s_j$ in terms of the starting and end points s_{j-1} and s_j respectively. For each e_j, two types are allowed: the vertical edge $s_j - s_{j-1} = (0,1)$ and the diagonal edge $s_j - s_{j-1} = (1,1)$. Depending on the types, we assign the following weight w_{e_j},

$$w_e = \begin{cases} B_{\mu,\nu-\mu} & \text{if } e = (i, j-1) \to (i,j), \\ 1 - B_{\mu,\nu-\mu} & \text{if } e = (i-1, j-1) \to (i,j), \\ B_{d+\mu,\nu-\mu} & \text{if } e = (1,j) \to (1, j+1), \\ B^{-1}_{d,\nu} & \text{if } e = (i, i-1) \to (i+1, i), \end{cases} \tag{107}$$

where $B_{\alpha,\beta} \overset{\mathrm{d}}{=} \mathrm{Beta}(\alpha,\beta)$ and B^{-1} is an inverse beta random variable. $\mathrm{Beta}(\alpha,\beta)$ represents the beta distribution with parameters α, β whose pdf is

$$x^{\alpha-1}(1-x)^{\beta-1}\frac{\Gamma(\alpha+\beta)}{\Gamma(\alpha)\Gamma(\beta)} \tag{108}$$

with $0 \le x \le 1$. The partition function is defined as

$$Z(N,t) = \sum_{\pi:(0,1)\to(t,n)} \prod_{i=1}^{t} w_{e_i}. \tag{109}$$

We derive an integral formula for the probability density of the partition function of the stationary Beta polymer model using the q-Whittaker measure formalism.

It was proven in [4] that in case there are no boundary sources the Beta polymer model is obtained as a $q \to 1$ scaling limit of the q-Hahn TASEP. The same result holds in presence of sources,

$$q^{Y_N^0(t)+N} \xrightarrow[q=e^{-\epsilon},\ \epsilon\to 0]{} Z(N,t), \tag{110}$$

where $Y_N^0(\tau)$ is the same definition as $Y_N(\tau)$ defined around (99) except that the first equation in (99) is replaced by $Y_N^0(0) = 0$ and parameters are changed to $\mu \to q^\mu, \nu \to q^\nu$ and $d,\ d_+ \to q^d$. The proof of this fact is an adaptation of the argument in [4] and we do not reproduce it here. Thus applying the $q \to 1$ scaling limit to Theorem 7, one could get a Fredholm determinant formula for the stationary beta-polymer model. However to do this one has to figure out the limiting behaviors of the poles in the integrand of the kernel $K_{a;\rho^{\hat{\alpha}}_{N,J,\tau}}$ in (104). This problem is an interesting future task. In addition, there are some different approaches which would be useful to analyze the beta polymer: e.g. another Fredholm determinant formula for the stationary higher spin vertex model in [11] and the spin Whittaker process introduced in [28], whose marginal describes the beta polymer.

Here we focus instead on how to produce exact formulas to describe the law of $Z(n,t)$ using results connecting the q-Whittaker measure and the q-Hahn TASEP. The result is contained in the following

Proposition 7. *The density $\rho(x)dx$ of the partition function $Z(N,t)$ is given by*

(111)

$$\rho(x) = \frac{1}{N!} \int_{(1+\epsilon+\mathrm{i}\mathbb{R})^N} \prod_{i=1}^{N} \frac{dy_i}{2\pi \mathrm{i}}$$
$$\times \prod_{1 \le i \neq j \le N} \frac{\Gamma(-y_j)\Gamma(\nu + y_j)}{\Gamma(y_i - y_j)\Gamma(\nu)} x^{d - Y - 1} \frac{\prod_{i=1}^{N} \Gamma(d - y_i)^2}{\Gamma(d - Y)^2}$$
$$\times \left(\frac{\Gamma(\nu)}{\Gamma(d)\Gamma(d+\nu)} \right)^{N-1} \left(\frac{\Gamma(d+\nu)}{\Gamma(d+\mu)} \frac{\Gamma(\nu)^{N-1}}{\Gamma(\mu)^{N-1}} \prod_{i=1}^{n} \frac{\Gamma(Z_i + \mu)}{\Gamma(Z_i + \nu)} \right)^t$$

where $Y = y_1 + \cdots + y_N$ and $\rho(x)$ has support on $\mathbb{R}_{>0}$. The value ϵ in the expression of the contour is small enough so that the poles larger than 1 stay on the outside.

Outline of the proof. The starting point is the integral expression for the probability mass function of $Y_N(t)$, a tagged particle in q-Hahn TASEP with the half-stationary initial condition defined around (99). Combining (59) with the discussion in Section 4.2.3, we have

$$
\begin{aligned}
(112)\quad \operatorname{Prob}(Y_N(\tau)=\ell) &= \operatorname{Prob}\big(Y_N^0(\tau;\text{đ},\text{đ}_+)+N-\chi=\ell\big) \\
&= \frac{1}{N!}\int_{(1+\epsilon)\mathbb{T}^N}\prod_{i=1}^{N}\frac{dz_i}{2\pi \mathrm{i} z_i}\prod_{1\le i\ne j\le N}\frac{(z_i/z_j,\nu;q)_\infty}{(1/z_j,\nu z_j;q)_\infty}\left(\frac{\text{đ}_+}{Z}\right)^{\ell} \\
&\times(\text{đ}/\text{đ}_+;q)_\infty\frac{(q;q)_\infty^{N-1}(\text{đ};q)_\infty^{N-1}(\text{đ}_+/Z;q)_\infty}{\prod_{i=1}^{N}(\text{đ}_+/z_i;q)_\infty(\text{đ}/z_i;q)_\infty} \\
&\times\left(\frac{(\text{đ}\mu;q)_\infty}{(\text{đ}\nu;q)_\infty}\frac{(\mu;q)_\infty^{N-1}}{(\nu;q)_\infty^{N-1}}\prod_{i=1}^{N}\frac{(z_i\nu;q)_\infty}{(z_i\mu;q)_\infty}\right)^{\tau}.
\end{aligned}
$$

Here $Z=z_1\cdots z_N$ and $Y_N^0(\tau;\text{đ},\text{đ}_+)$ is defined by the relation

$$
(113)\qquad Y_N^0(\tau;\text{đ},\text{đ}_+)-\chi=Y_N(\tau)
$$

where χ is a random variable independent of $Y_N^0(\tau;\text{đ},\text{đ}_+)$ and $\chi\overset{\mathrm{d}}{=}q\mathrm{Po}(\text{đ}/\text{đ}_-)$ Note that by definition $Y_N^0(0;\text{đ},\text{đ}_+)=0$ and $Y_N^0(\tau;\text{đ},\text{đ})=Y_N^0(\tau)$. The random shift χ can be removed by applying the relation given in Lemma 5.1 in [20],

$$
f(z)=\frac{1}{(\text{đ}/\text{đ}_+;q)_\infty}\sum_{k\ge 0}\left(\frac{\text{đ}}{\text{đ}_+}\right)^k\frac{(-1)^k q^{\binom{k}{2}}}{(q;q)_k}\mathbb{E}_\chi\left[f(z-\chi-k)\right]
$$

to the function $f: z\mapsto \operatorname{Prob}(Y_N^0(\tau;\text{đ},\text{đ}_+)+N+z=\ell)$. On expression (112) such transformation has the effect of transforming the ℓ depending term as

$$
\left(\frac{\text{đ}_+}{Z}\right)^{\ell}\to\left(\frac{\text{đ}_+}{Z}\right)^{\ell}\frac{(\text{đ}_-/Z)}{(\text{đ}_-/\text{đ}_+;q)_\infty}.
$$

The singularities caused by the q-Pochhammer symbols $(\text{đ}/\text{đ}_+;q)_\infty$ cancel out and we can obtain expressions for the stationary q-Hahn TASEP setting $\text{đ}_-=\text{đ}_+=\text{đ}$. The resulting formulas are also well suited for a $q\to 1$ limit after rescaling all quantities as

$$
z_i\to q^{y_i},\qquad \nu\to q^{\nu},\qquad \text{đ}\to q^{\text{đ}},\qquad \mu\to q^{\mu},\qquad \ell\to\log_q(x)
$$

so that (111) follows. Q.E.D.

Acknowledgements. We thank the referee for reading this paper carefully and useful comments. The work of TI has been supported by JSPS KAKENHI Grants No.16K05192, No.19H01793, and No.20K03626. The work of TS has been supported by JSPS KAKENHI Grants No.15K05203, No.16H06338, No.18H01141, No.18H03672, No.19L03665, No.21H04432.

References

[1] A. Aggarwal, Current fluctuations of the stationary ASEP and six-vertex model, Duke Math. J., **167** (2018), 269–384.

[2] G. E. Andrews, R. Askey and R. Roy, Special Functions, Encyclopedia of Mathematics and its Applications, **71**, Cambridge University Press, 2000.

[3] J. Baik and E. M. Rains, Limiting Distributions for a Polynuclear Growth Model with External Sources, J. Stat. Phys., **100** (2000), 523–541.

[4] G. Barraquand and I. Corwin, Random-walk in Beta-distributed random environment, Probab. Theory Related Fields, **167** (2017), 1057–1116.

[5] A. Borodin, On a family of symmetric rational functions, Adv. Math., **306** (2017), 973–1018.

[6] A. Borodin and I. Corwin, Macdonald processes, Probab. Theory Related Fields, **158** (2014), 225–400.

[7] A. Borodin and I. Corwin, Discrete time q-TASEPs, Int. Math. Res. Not. IMRN, **2015** (2015), 499–537.

[8] A. Borodin, I. Corwin and T. Sasamoto, From duality to determinants for q-TASEP and ASEP, Ann. Probab., **42** (2014), 2314–2382.

[9] A. Borodin and L. Petrov, Higher spin six vertex model and symmetric rational functions, Selecta Math. (N.S.), **24** (2018), 751–874.

[10] A. Borodin, I. Corwin, V. Gorin and S. Shakirov, Observables of Macdonald processes, Trans. Amer. Math. Soc., **368** (2016), 1517–1558.

[11] A. Bufetov, M. Mucciconi and L. Petrov, Yang-Baxter random fields and stochastic vertex models, to appear in Adv. Math., arXiv:1905.06815.

[12] H. Chaumont and C. Noack, Characterizing stationary $1+1$ dimensional lattice polymer models, Electron. J. Probab., **23** (2018), 1–19.

[13] H. Chaumont and C. Noack, Fluctuation exponents for stationary exactly solvable lattice polymer models via a Mellin transform framework, ALEA Lat. Ann. J. Probab. Math. Stat., **15** (2018), 509–547.

[14] I. Corwin and L. Petrov, Stochastic higher spin vertex models on the line, Comm. Math. Phys., **343** (2016), 651–700.

[15] I. Corwin and L.-C. Tsai, KPZ equation limit of higher-spin exclusion processes, Ann. Probab., **45** (2017), 1771–1798.

[16] P. L. Ferrari and H. Spohn, Scaling Limit for the Space-Time Covariance of the Stationary Totally Asymmetric Simple Exclusion Process, Comm. Math. Phys., **265** (2006), 45–46.

[17] G. Frobenius, Ueber die elliptischen Funktionen zweiter Art, J. Reine Angew. Math., **93** (1882), 53–68.

[18] T. Imamura, M. Mucciconi and T. Sasamoto, Stationary Higher Spin Six Vertex Model and q-Whittaker measure, Probab. Theory Related Fields, **177** (2020), 923–1042.

[19] T. Imamura and T. Sasamoto, Stationary Correlations for the 1D KPZ Equation, J. Stat. Phys., **150** (2013), 908–939.

[20] T. Imamura and T. Sasamoto, Fluctuations for stationary q-TASEP, Probab. Theory Related Fields, **174** (2019), 647–730.

[21] T. Imamura and T. Sasamoto, The q-TASEP with a Random Initial Condition, Theoret. and Math. Phys., **198** (2019), 69–88.
[22] K. Johansson, Shape Fluctuations and Random Matrices, Comm. Math. Phys., **209** (2000), 437–476.
[23] K. Johansson, Discrete Polynuclear Growth and Determinantal Processes, Comm. Math. Phys., **242** (2003), 277–329.
[24] Y. Kajihara and M. Noumi, Multiple elliptic hypergeometric series. An approach from the Cauchy determinant, Indag. Math., **14** (2003), 395–421.
[25] A. Knizel, L. Petrov and A. Saenz, Generalizations of TASEP in Discrete and Continuous Inhomogeneous Space, Comm. Math. Phys., **372** (2019), 797–864.
[26] I. G. Macdonald, Symmetric Functions and Hall Polynomials, Oxford classic texts in the physical sciences. Clarendon Press, 1998.
[27] K. Matveev, Macdonald-positive specializations of the algebra of symmetric functions: Proof of the Kerov conjecture, Ann. Math., **189** (2019), 277–316.
[28] M. Mucciconi and L. Petrov, Spin q-Whittaker polynomials and deformed quantum Toda, arXiv:2003.14260.
[29] C. Noack and P. Sosoe, Concentration for integrable directed polymer models, arXiv:2005.00126.
[30] A. Okounkov, Infinite wedge and random partitions, Sel. Math. New Ser., **7** (2001), 57–81.
[31] D. Orr and L. Petrov, Stochastic higher spin six vertex model and q-TASEPs, Adv. Math., **317** (2017), 473–525.
[32] A. M. Povolotsky, On the integrability of zero-range chipping models with factorized steady states, J. Phys. A, **46** (2013), 465205.
[33] C. Tracy and H. Widom, Correlation Functions, Cluster Functions, and Spacing Distributions for Random Matrices, J. Stat. Phys., **92** (1998), 809–835.

Takashi Imamura:
Department of Mathematics and Informatics,
Chiba University
E-mail address: imamura@math.s.chiba-u.ac.jp

Matteo Mucciconi:
Department of Physics,
Tokyo Institute of Technology
E-mail address: matteomucciconi@gmail.com

Tomohiro Sasamoto:
Department of Physics,
Tokyo Institute of Technology
E-mail address: sasamoto@phys.titech.ac.jp

Advanced Studies in Pure Mathematics 87, 2021
Stochastic Analysis, Random Fields and Integrable Probability — Fukuoka 2019
pp. 293–314

The Laplacian on some self-conformal fractals and Weyl's asymptotics for its eigenvalues: A survey of the analytic aspects

Naotaka Kajino

Abstract.

This article surveys the analytic aspects of the author's recent studies on the construction and analysis of a "*geometrically canonical*" *Laplacian* on circle packing fractals invariant with respect to certain Kleinian groups (i.e., discrete groups of Möbius transformations on the Riemann sphere $\widehat{\mathbb{C}} = \mathbb{C} \cup \{\infty\}$), including the classical *Apollonian gasket* and some *round Sierpiński carpets*. The main result on Weyl's asymptotics for its eigenvalues is of the same form as that by Oh and Shah [*Invent. Math.* **187** (2012), 1–35, Theorem 1.4] on the asymptotic distribution of the circles in a very large class of such fractals.

§1. Introduction

This article, which is a considerable expansion of [12], concerns the author's recent studies in [11, 14, 15, 16] on Weyl's eigenvalue asymptotics for a "geometrically canonical" Laplacian defined by the author on circle packing fractals which are invariant with respect to certain Kleinian groups (i.e., discrete groups of Möbius transformations on $\widehat{\mathbb{C}} := \mathbb{C} \cup \{\infty\}$), including the classical *Apollonian gasket* (Figure 1) and some *round Sierpiński carpets* (Figure 5). Here we focus on sketching the construction of the Laplacian, the proof of its uniqueness and basic properties, and the analytic aspects of the proof of the eigenvalue asymptotics;

Received January 20, 2020.
Revised August 31, 2020.
2010 *Mathematics Subject Classification.* Primary 28A80, 35P20, 53C23; Secondary 31C25, 37B10, 60J35.
Key words and phrases. Apollonian gasket, Kleinian groups, round Sierpiński carpets, Dirichlet forms, Laplacian, Weyl's eigenvalue asymptotics.
This work was supported by JSPS KAKENHI Grant Numbers JP25887038, JP15K17554, JP18K18720 and by the Research Institute for Mathematical Sciences, an International Joint Usage/Research Center located in Kyoto University.

the reader is referred to [13] for a survey of the ergodic-theoretic aspects of the proof of the eigenvalue asymptotics.

This article is organized as follows. First in §2 we introduce the Apollonian gasket $K(\mathcal{D})$ and recall its basic geometric properties. In §3, after a brief summary of how the Laplacian on $K(\mathcal{D})$ was discovered by Teplyaev in [34], we give its definition and sketch the proof of the result in [14] that it is the infinitesimal generator of the *unique* strongly local, regular symmetric Dirichlet form over $K(\mathcal{D})$ with respect to which the inclusion map $K(\mathcal{D}) \hookrightarrow \mathbb{C}$ is *harmonic* on the complement of the three outmost vertices. In §4, we state the principal result in [14] that the Laplacian on $K(\mathcal{D})$ satisfies Weyl's eigenvalue asymptotics of the same form as the asymptotic distribution of the circles in $K(\mathcal{D})$ by Oh and Shah in [30, Corollary 1.8], and sketch the proof of certain estimates on the eigenvalues required to conclude Weyl's asymptotics by applying the ergodic-theoretic result explained in [13]. Finally, in §5 we present a partial extension of these results to the case of round Sierpiński carpets which are invariant with respect to certain concrete Kleinian groups.

Notation. We use the following notation throughout this article.

(0) The symbols $\subset$ and $\supset$ for set inclusion *allow* the case of the equality.
(1) $\mathbb{N} := \{n \in \mathbb{Z} \mid n > 0\}$, i.e., $0 \notin \mathbb{N}$.
(2) $\widehat{\mathbb{C}} := \mathbb{C} \cup \{\infty\}$ denotes the Riemann sphere.
(3) $i := \sqrt{-1}$ denotes the imaginary unit. The real and imaginary parts of $z \in \mathbb{C}$ are denoted by $\operatorname{Re} z$ and $\operatorname{Im} z$, respectively.
(4) The cardinality (number of elements) of a set A is denoted by $\#A$.
(5) Let E be a non-empty set. We define $\mathrm{id}_E : E \to E$ by $\mathrm{id}_E(x) := x$. For $x \in E$, we define $\mathbf{1}_x = \mathbf{1}_x^E \in \mathbb{R}^E$ by $\mathbf{1}_x(y) := \mathbf{1}_x^E(y) := \left\{\begin{smallmatrix} 1 \text{ if } y = x, \\ 0 \text{ if } y \neq x. \end{smallmatrix}\right.$ For $u : E \to [-\infty, +\infty]$ we set $\|u\|_{\sup} := \|u\|_{\sup,E} := \sup_{x \in E} |u(x)|$.
(6) Let E be a topological space. The Borel σ-field of E is denoted by $\mathcal{B}(E)$. For $A \subset E$, its interior, closure and boundary in E are denoted by $\mathrm{int}_E A$, $\overline{A}^E$ and $\partial_E A$, respectively, and when $E = \mathbb{C}$ they are simply denoted by $\mathrm{int}\, A$, $\overline{A}$ and ∂A, respectively. We set $\mathcal{C}(E) := \{u \mid u : E \to \mathbb{R},\ u \text{ is continuous}\}$, $\mathrm{supp}_E[u] := \overline{u^{-1}(\mathbb{R} \setminus \{0\})}^E$ for $u \in \mathcal{C}(E)$, and $\mathcal{C}_{\mathrm{c}}(E) := \{u \in \mathcal{C}(E) \mid \mathrm{supp}_E[u] \text{ is compact}\}$.
(7) Let $n \in \mathbb{N}$. The Lebesgue measure on $(\mathbb{R}^n, \mathcal{B}(\mathbb{R}^n))$ is denoted by vol_n. The Euclidean inner product and norm on $\mathbb{R}^n$ are denoted by $\langle \cdot, \cdot \rangle$ and $|\cdot|$, respectively. For $A \subset \mathbb{R}^n$ and $f : A \to \mathbb{C}$ we set $\mathbf{Lip}_A f := \sup_{x,y \in A,\, x \neq y} \frac{|f(x)-f(y)|}{|x-y|}$ $(\sup \emptyset := 0)$. For a non-empty open subset U of $\mathbb{R}^n$ and $u : U \to \mathbb{R}$ with $\mathbf{Lip}_U u < +\infty$, the first-order partial derivatives of u, which exist vol_n-a.e. on U, are denoted by $\partial_1 u, \ldots, \partial_n u$, and we set $\nabla u := (\partial_1 u, \ldots, \partial_n u)$.

§2. The Apollonian gasket and its fractal geometry

In this section, we introduce the Apollonian gasket and state its geometric properties needed for our purpose. The same framework is presented also in [13, Section 2], but we repeat it here for the reader's convenience. The following definition and proposition form the basis of the construction and further detailed studies of the Apollonian gasket.

Definition 2.1 (tangential disk triple). (0) We set $S := \{1, 2, 3\}$.

(1) Let $D_1, D_2, D_3 \subset \mathbb{C}$ be either three open disks or two open disks and an open half-plane. The triple $\mathcal{D} := (D_1, D_2, D_3)$ of such sets is called a *tangential disk triple* if and only if $\#(\overline{D_j} \cap \overline{D_k}) = 1$ (i.e., D_j and D_k are *externally* tangent) for any $j, k \in S$ with $j \neq k$. If $\mathcal{D}$ is such a triple consisting of three disks, then the open triangle in $\mathbb{C}$ with vertices the centers of D_1, D_2, D_3 is denoted by $\triangle(\mathcal{D})$.

(2) Let $\mathcal{D} = (D_1, D_2, D_3)$ be a tangential disk triple. The open subset $\mathbb{C} \setminus \bigcup_{j \in S} \overline{D_j}$ of $\mathbb{C}$ is then easily seen to have a unique bounded connected component, which is denoted by $T(\mathcal{D})$ and called the *ideal triangle* associated with $\mathcal{D}$. We also set $\{q_j(\mathcal{D})\} := \overline{D_k} \cap \overline{D_l}$ for each $(j, k, l) \in \{(1, 2, 3), (2, 3, 1), (3, 1, 2)\}$ and $V_0(\mathcal{D}) := \{q_j(\mathcal{D}) \mid j \in S\}$.

(3) A tangential disk triple $\mathcal{D} = (D_1, D_2, D_3)$ is called *positively oriented* if and only if its associated ideal triangle $T(\mathcal{D})$ is to the left of $\partial T(\mathcal{D})$ when $\partial T(\mathcal{D})$ is oriented so as to have $\{q_j(\mathcal{D})\}_{j=1}^{3}$ in this order.

Finally, we define

$$\mathsf{TDT}^{+} := \{\mathcal{D} \mid \mathcal{D} \text{ is a positively oriented tangential disk triple}\},$$
$$\mathsf{TDT}^{\oplus} := \{\mathcal{D} \mid \mathcal{D} = (D_1, D_2, D_3) \in \mathsf{TDT}^{+},\ D_1, D_2, D_3 \text{ are disks}\}.$$

The following proposition is classical and can be shown by some elementary (though lengthy) Euclidean-geometric arguments. We set $\operatorname{rad}(D) := r$ and $\operatorname{curv}(D) := r^{-1}$ for each open disk $D \subset \mathbb{C}$ of radius $r \in (0, +\infty)$ and $\operatorname{curv}(D) := 0$ for each open half-plane $D \subset \mathbb{C}$.

Proposition 2.2. *Let* $\mathcal{D} = (D_1, D_2, D_3) \in \mathsf{TDT}^{+}$, *set* $(\alpha, \beta, \gamma) := \big(\operatorname{curv}(D_1), \operatorname{curv}(D_2), \operatorname{curv}(D_3)\big)$ *and set* $\kappa := \kappa(\mathcal{D}) := \sqrt{\beta\gamma + \gamma\alpha + \alpha\beta}$.

(1) *Let* $D_{\mathrm{cir}}(\mathcal{D}) \subset \mathbb{C}$ *denote the* circumscribed disk *of* $T(\mathcal{D})$, *i.e., the unique open disk with* $\{q_1(\mathcal{D}), q_2(\mathcal{D}), q_3(\mathcal{D})\} \subset \partial D_{\mathrm{cir}}(\mathcal{D})$. *Then* $\overline{T(\mathcal{D})} \setminus \{q_1(\mathcal{D}), q_2(\mathcal{D}), q_3(\mathcal{D})\} \subset D_{\mathrm{cir}}(\mathcal{D})$, $\partial D_{\mathrm{cir}}(\mathcal{D})$ *is orthogonal to* ∂D_j *for any* $j \in S$, *and* $\operatorname{curv}(D_{\mathrm{cir}}(\mathcal{D})) = \kappa$.

(2) *There exists a unique* inscribed disk $D_{\mathrm{in}}(\mathcal{D})$ *of* $T(\mathcal{D})$, *i.e., a unique open disk* $D_{\mathrm{in}}(\mathcal{D}) \subset \mathbb{C}$ *such that* $D_{\mathrm{in}}(\mathcal{D}) \subset T(\mathcal{D})$ *and* $\#(\overline{D_{\mathrm{in}}(\mathcal{D})} \cap \overline{D_j}) = 1$ *for any* $j \in S$. *Moreover,* $\operatorname{curv}(D_{\mathrm{in}}(\mathcal{D})) = \alpha + \beta + \gamma + 2\kappa$.

The following notation is standard in studying self-similar sets.

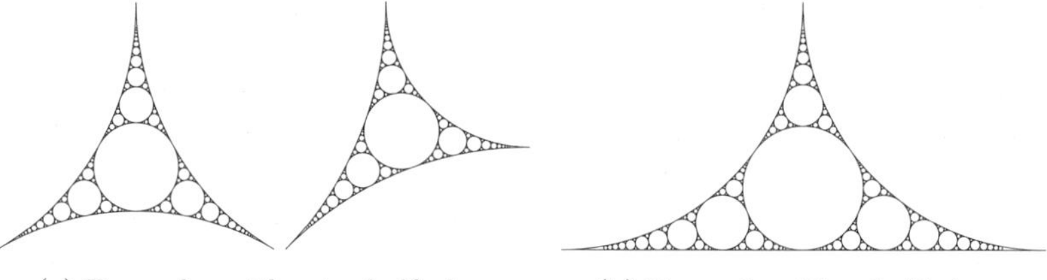

(a) Examples without a half-plane (b) Example with a half-plane

Figure 1. The Apollonian gaskets $K(\mathcal{D})$ associated with $\mathcal{D} \in \mathsf{TDT}^+$.

Definition 2.3. (1) We set $W_0 := \{\emptyset\}$, where $\emptyset$ is an element called the *empty word*, $W_m := S^m$ for $m \in \mathbb{N}$ and $W_* := \bigcup_{m \in \mathbb{N} \cup \{0\}} W_m$. For $w \in W_*$, the unique $m \in \mathbb{N} \cup \{0\}$ satisfying $w \in W_m$ is denoted by $|w|$ and called the *length* of w.

(2) Let $w, v \in W_*$, $w = w_1 \ldots w_m$, $v = v_1 \ldots v_n$. We define $wv \in W_*$ by $wv := w_1 \ldots w_m v_1 \ldots v_n$ ($w\emptyset := w$, $\emptyset v := v$). We also define $w^{(1)} \ldots w^{(k)}$ for $k \geq 3$ and $w^{(1)}, \ldots, w^{(k)} \in W_*$ inductively by $w^{(1)} \ldots w^{(k)} := (w^{(1)} \ldots w^{(k-1)}) w^{(k)}$. For $w \in W_*$ and $n \in \mathbb{N} \cup \{0\}$ we set $w^n := w \ldots w \in W_{n|w|}$. We write $w \leq v$ if and only if $w = v\tau$ for some $\tau \in W_*$, and write $w \not\asymp v$ if and only if neither $w \leq v$ nor $v \leq w$ holds.

Proposition 2.2-(2) enables us to define natural "contraction maps" $\Phi_w : \mathsf{TDT}^+ \to \mathsf{TDT}^+$ for each $w \in W_*$, which in turn is used to define the Apollonian gasket $K(\mathcal{D})$ associated with $\mathcal{D} \in \mathsf{TDT}^+$, as follows.

Definition 2.4. We define maps $\Phi_1, \Phi_2, \Phi_3 : \mathsf{TDT}^+ \to \mathsf{TDT}^+$ by

$$\begin{cases} \Phi_1(\mathcal{D}) := (D_{\mathrm{in}}(\mathcal{D}), D_2, D_3), \\ \Phi_2(\mathcal{D}) := (D_1, D_{\mathrm{in}}(\mathcal{D}), D_3), \quad \mathcal{D} = (D_1, D_2, D_3) \in \mathsf{TDT}^+. \\ \Phi_3(\mathcal{D}) := (D_1, D_2, D_{\mathrm{in}}(\mathcal{D})), \end{cases} \tag{2.1}$$

We also set $\Phi_w := \Phi_{w_m} \circ \cdots \circ \Phi_{w_1}$ ($\Phi_\emptyset := \mathrm{id}_{\mathsf{TDT}^+}$) and $\mathcal{D}_w := \Phi_w(\mathcal{D})$ for $w = w_1 \ldots w_m \in W_*$ and $\mathcal{D} \in \mathsf{TDT}^+$.

Definition 2.5 (Apollonian gasket)**.** Let $\mathcal{D} \in \mathsf{TDT}^+$. We define the *Apollonian gasket* $K(\mathcal{D})$ associated with $\mathcal{D}$ (see Figure 1) by

$$K(\mathcal{D}) := \overline{T(\mathcal{D})} \setminus \bigcup_{w \in W_*} D_{\mathrm{in}}(\mathcal{D}_w) = \bigcap_{m \in \mathbb{N}} \bigcup_{w \in W_m} \overline{T(\mathcal{D}_w)}. \tag{2.2}$$

The curvatures of the disks involved in (2.2) admit the following simple expression.

Definition 2.6. We define 4×4 real matrices M_1, M_2, M_3 by

$$(2.3)\quad M_1 := \begin{pmatrix} 1 & 0 & 0 & 0 \\ 1 & 1 & 0 & 1 \\ 1 & 0 & 1 & 1 \\ 2 & 0 & 0 & 1 \end{pmatrix}, \quad M_2 := \begin{pmatrix} 1 & 1 & 0 & 1 \\ 0 & 1 & 0 & 0 \\ 0 & 1 & 1 & 1 \\ 0 & 2 & 0 & 1 \end{pmatrix}, \quad M_3 := \begin{pmatrix} 1 & 0 & 1 & 1 \\ 0 & 1 & 1 & 1 \\ 0 & 0 & 1 & 0 \\ 0 & 0 & 2 & 1 \end{pmatrix}$$

and set $M_w := M_{w_1} \cdots M_{w_m}$ for $w = w_1 \ldots w_m \in W_*$ ($M_\emptyset := \mathrm{id}_{4\times 4}$). Note that then for any $n \in \mathbb{N} \cup \{0\}$ we easily obtain

(2.4)

$$M_{1^n} = \begin{pmatrix} 1 & 0 & 0 & 0 \\ n^2 & 1 & 0 & n \\ n^2 & 0 & 1 & n \\ 2n & 0 & 0 & 1 \end{pmatrix}, \quad M_{2^n} = \begin{pmatrix} 1 & n^2 & 0 & n \\ 0 & 1 & 0 & 0 \\ 0 & n^2 & 1 & n \\ 0 & 2n & 0 & 1 \end{pmatrix}, \quad M_{3^n} = \begin{pmatrix} 1 & 0 & n^2 & n \\ 0 & 1 & n^2 & n \\ 0 & 0 & 1 & 0 \\ 0 & 0 & 2n & 1 \end{pmatrix}.$$

Proposition 2.7. *Let $\mathcal{D} = (D_1, D_2, D_3) \in \mathsf{TDT}^+$, let $\alpha, \beta, \gamma, \kappa$ be as in Proposition 2.2, let $w \in W_*$ and $(D_{w,1}, D_{w,2}, D_{w,3}) := \mathcal{D}_w$. Then*

$$(2.5)\quad \big(\mathrm{curv}(D_{w,1}), \mathrm{curv}(D_{w,2}), \mathrm{curv}(D_{w,3}), \kappa(\mathcal{D}_w)\big) = (\alpha, \beta, \gamma, \kappa) M_w.$$

Proof. This follows by an induction in $|w|$ using Proposition 2.2-(2) and Definition 2.4. Q.E.D.

We next collect basic facts regarding the Hausdorff dimension and measure of $K(\mathcal{D})$. For each $s \in (0, +\infty)$ let $\mathcal{H}^s : 2^{\mathbb{C}} \to [0, +\infty]$ denote the s-dimensional Hausdorff (outer) measure on $\mathbb{C}$ with respect to the Euclidean metric, and for each $A \subset \mathbb{C}$ let $\dim_{\mathrm{H}} A$ denote its Hausdorff dimension; see, e.g., [25, Chapters 4–7] for details. As is well known, it easily follows from the definition of $\mathcal{H}^s$ that the image $f(A)$ of $A \subset \mathbb{C}$ by $f : A \to \mathbb{C}$ with $\mathbf{Lip}_A f < +\infty$ satisfies $\mathcal{H}^s(f(A)) \leq (\mathbf{Lip}_A f)^s \mathcal{H}^s(A)$ for any $s \in (0, +\infty)$ and hence in particular $\dim_{\mathrm{H}} f(A) \leq \dim_{\mathrm{H}} A$. On the basis of this observation, we easily get the following lemma.

Lemma 2.8. *Let $\mathcal{D}, \mathcal{D}' \in \mathsf{TDT}^+$. Then there exists $c \in (0, +\infty)$ such that $\mathcal{H}^s(K(\mathcal{D})) \leq c^s \mathcal{H}^s(K(\mathcal{D}'))$ for any $s \in (0, +\infty)$. In particular, $\dim_{\mathrm{H}} K(\mathcal{D}) = \dim_{\mathrm{H}} K(\mathcal{D}')$.*

Proof. Let $f_{\mathcal{D}',\mathcal{D}}$ denote the unique orientation-preserving Möbius transformation on $\widehat{\mathbb{C}}$ such that $f_{\mathcal{D}',\mathcal{D}}(q_j(\mathcal{D}')) = q_j(\mathcal{D})$ for any $j \in S$. Then $f_{\mathcal{D}',\mathcal{D}}(K(\mathcal{D}')) = K(\mathcal{D})$, since a Möbius transformation on $\widehat{\mathbb{C}}$ maps any open disk in $\widehat{\mathbb{C}}$ onto another. Now the assertion follows from the observation in the last paragraph and $\mathbf{Lip}_{\overline{D_{\mathrm{cir}}(\mathcal{D}')}} f_{\mathcal{D}',\mathcal{D}} < +\infty$. Q.E.D.

Definition 2.9. Noting Lemma 2.8, we define

$$(2.6)\qquad d_{\mathsf{AG}} := \dim_{\mathrm{H}} K(\mathcal{D}), \qquad \text{where } \mathcal{D} \in \mathsf{TDT}^+ \text{ is arbitrary.}$$

Theorem 2.10 (Boyd [2]; see also [7, 26, 27]).

$$1.300197 < d_{\mathsf{AG}} < 1.314534. \tag{2.7}$$

Moreover, for the d_{AG}-dimensional Hausdorff measure $\mathcal{H}^{d_{\mathsf{AG}}}(K(\mathcal{D}))$ of $K(\mathcal{D})$ we have the following theorem, which was proved first by Sullivan [33] through considerations on the isometric action of Möbius transformations on the three-dimensional hyperbolic space, and later by Mauldin and Urbański [26] through purely two-dimensional arguments.

Theorem 2.11 ([33, Theorem 2], [26, Theorem 2.6]).

$$0 < \mathcal{H}^{d_{\mathsf{AG}}}(K(\mathcal{D})) < +\infty \qquad \text{for any } \mathcal{D} \in \mathsf{TDT}^{+}. \tag{2.8}$$

Remark 2.12. The self-conformality of $K(\mathcal{D})$ is required most crucially in the proof of Theorem 2.11, and is heavily used further to obtain certain equicontinuity properties of $\{\mathcal{H}^{d_{\mathsf{AG}}}(K(\mathcal{D}_w))\}_{w\in W_*}$ as a family of functions of $\bigl(\mathrm{curv}(D_1), \mathrm{curv}(D_2), \mathrm{curv}(D_3)\bigr)$, where $(D_1, D_2, D_3) := \mathcal{D}$. This equicontinuity is the key to verifying the ergodic-theoretic assumptions of Kesten's renewal theorem [19, Theorem 2], which is then applied to conclude Theorem 4.4 below.

§3. The canonical Dirichlet form on the Apollonian gasket

In this section, we introduce the canonical Dirichlet form on the Apollonian gasket $K(\mathcal{D})$, whose infinitesimal generator is our Laplacian on $K(\mathcal{D})$, and state its properties established by the author in [14]; see [6, 4] for the basics of the theory of regular symmetric Dirichlet forms.

Before giving its actual definition, we briefly summarize how it has been discovered. The initial idea for its construction was suggested by the theory of analysis on the *harmonic Sierpiński gasket* $K_{\mathcal{H}}$ (Figure 2, right) due to Kigami [20, 22]. This is a compact subset of $\mathbb{C}$ defined as the image of a *harmonic map* $\Phi : K \to \mathbb{C}$ from the *Sierpiński gasket* K (Figure 2, left) to $\mathbb{C}$. More precisely, let $V_0 = \{q_1, q_2, q_3\}$ be the set of the three outmost vertices of K, let $(\mathcal{E}, \mathcal{F})$ be the (self-similar) *standard Dirichlet form* on K (so that $\mathcal{F}$ is known to be a dense subalgebra of $(\mathcal{C}(K), \|\cdot\|_{\sup})$), and let $h_1^K, h_2^K \in \mathcal{F}$ be $\mathcal{E}$*-harmonic* on $K \setminus V_0$ and satisfy $\mathcal{E}(h_j^K, h_k^K) = \delta_{jk}$ for any $j, k \in \{1, 2\}$ (see [10, Sections 2 and 3] and the references therein for details). Then we can define a continuous map $\Phi : K \to \mathbb{C}$ by $\Phi(x) := \bigl(h_1^K(x), h_2^K(x)\bigr)$, and its image $K_{\mathcal{H}} := \Phi(K)$ is called the harmonic Sierpiński gasket. In fact, Kigami has proved in [20, Theorem 3.6] that $\Phi : K \to K_{\mathcal{H}}$ is injective and hence a homeomorphism, and further in [20, Theorem 4.1] that a one-dimensional, measure-theoretic "Riemannian structure" can be defined on K through

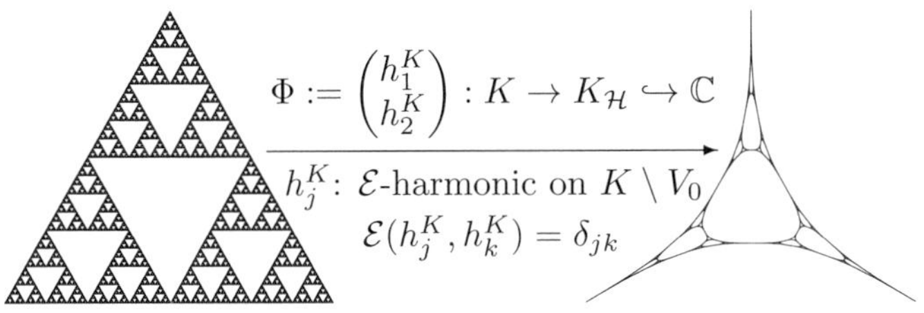

Figure 2. Sierpiński gasket K and harmonic Sierpiński gasket $K_{\mathcal{H}}$.

the embedding Φ and the $\mathcal{E}$-*energy measure* μ[1] of Φ, which plays the role of the "Riemannian volume measure" and is given by

$$\mu := \mu_{\langle h_1^K \rangle} + \mu_{\langle h_2^K \rangle} = \text{“}|\nabla \Phi|^2 \, d\,\mathrm{vol}\text{”}; \tag{3.1}$$

here $\mu_{\langle u \rangle}$ denotes the $\mathcal{E}$-energy measure of $u \in \mathcal{F}$ playing the role of "$|\nabla u|^2 \, d\,\mathrm{vol}$" and defined as the unique Borel measure on K such that

$$\int_K f \, d\mu_{\langle u \rangle} = \mathcal{E}(fu, u) - \frac{1}{2}\mathcal{E}(f, u^2) \qquad \text{for any } f \in \mathcal{F}. \tag{3.2}$$

Kigami has also proved in [22, Theorem 6.3] that the heat kernel of $(K, \mu, \mathcal{E}, \mathcal{F})$ satisfies the two-sided *Gaussian* estimate of the same form as for Riemannian manifolds, and further detailed studies of $(K, \mu, \mathcal{E}, \mathcal{F})$ have been done in [9, 23, 10]; see [10] and the references therein for details.

As observed from Figures 1 and 2, the overall geometric structure of the Apollonian gasket $K(\mathcal{D})$ resembles that of the harmonic Sierpiński gasket $K_{\mathcal{H}}$, and then it is natural to expect that the above-mentioned framework of the measurable Riemannian structure on K induced by the embedding $\Phi : K \to K_{\mathcal{H}}$ can be adapted to the setting of $K(\mathcal{D})$ for $\mathcal{D} \in \mathsf{TDT}^{\oplus}$ to construct a "geometrically canonical" Dirichlet form on $K(\mathcal{D})$. Namely, it is expected that there exists a non-zero strongly local regular symmetric Dirichlet form $(\mathcal{E}^{\mathcal{D}}, \mathcal{F}_{\mathcal{D}})$ over $K(\mathcal{D})$ *with respect to which the coordinate functions* $\mathrm{Re}(\cdot)|_{K(\mathcal{D})}, \mathrm{Im}(\cdot)|_{K(\mathcal{D})}$ *are harmonic on* $K(\mathcal{D}) \setminus V_0(\mathcal{D})$. The possibility of such a construction was first noted by Teplyaev in [34, Theorem 5.17], and in [14] the author has completed the construction of $(\mathcal{E}^{\mathcal{D}}, \mathcal{F}_{\mathcal{D}})$ and further proved its uniqueness and concrete identification, summarized as follows. We start with some definitions.

Definition 3.1. (1) A subset C of $\mathbb{C}$ is called a *circular arc* if and only if $C = \{z_0 + re^{i\theta} \mid \theta \in [\alpha, \beta]\}$ for some $z_0 \in \mathbb{C}$, $r \in (0, +\infty)$ and

[1] μ was first introduced in [24] and is called the *Kusuoka measure* on K.

$\alpha, \beta \in \mathbb{R}$ with $\alpha < \beta$. In this case we set $\mathrm{cent}(C) := z_0$, $\mathrm{rad}(C) := r$ and $D_C := \mathrm{int}\{(1-t)\,\mathrm{cent}(C) + tz \mid z \in C,\, t \in [0,1]\}$.

(2) For a circular arc C, the length measure on $(C, \mathcal{B}(C))$ is denoted by $\mathcal{H}^1_C$, the gradient vector along C at $x \in C$ of a function $u : C \to \mathbb{R}$ is denoted by $\nabla_C u(x)$ provided u is differentiable at x, and we set $W^{1,2}(C) := \{u \in \mathbb{R}^C \mid u \text{ is a.c. on } C,\, |\nabla_C u| \in L^2(C, \mathcal{H}^1_C)\}$, where "a.c." is an abbreviation of "absolutely continuous".

(3) We define $h_1, h_2 : \mathbb{C} \to \mathbb{R}$ by $h_1(z) := \mathrm{Re}\, z$ and $h_2(z) := \mathrm{Im}\, z$.

Definition 3.2. Let $\mathcal{D} = (D_1, D_2, D_3) \in \mathsf{TDT}^{\oplus}$. We define

$$\mathcal{A}_{\mathcal{D}} := \{\overline{T(\mathcal{D})} \cap \partial D_j \mid j \in S\} \cup \{\partial D_{\mathrm{in}}(\mathcal{D}_w) \mid w \in W_*\} \tag{3.3}$$

and set $K^0(\mathcal{D}) := \bigcup_{C \in \mathcal{A}_{\mathcal{D}}} C$, so that each $C \in \mathcal{A}_{\mathcal{D}}$ is a circular arc, $\bigcup_{C \in \mathcal{A}_{\mathcal{D}}} D_C = \triangle(\mathcal{D}) \setminus K(\mathcal{D})$, $\bigcup_{C, A \in \mathcal{A}_{\mathcal{D}}, C \neq A} (C \cap A) = \bigcup_{w \in W_*} V_0(\mathcal{D}_w)$, and an induction in $|w|$ easily shows that for any $w \in W_*$,

$$\mathcal{A}_{\mathcal{D}_w} = \{C \cap K(\mathcal{D}_w) \mid C \in \mathcal{A}_{\mathcal{D}}\} \setminus \{\emptyset\}. \tag{3.4}$$

The canonical Dirichlet form $(\mathcal{E}^{\mathcal{D}}, \mathcal{F}_{\mathcal{D}})$ on $K(\mathcal{D})$ and the associated "Riemannian volume measure" similar to (3.1) turn out to be expressed explicitly in terms of the circle packing structure of $K(\mathcal{D})$, as follows.

Definition 3.3 (cf. [14, Theorems 5.11 and 5.13])**.** Let $\mathcal{D} \in \mathsf{TDT}^{\oplus}$.

(1) We define a Borel measure $\mu^{\mathcal{D}}$ on $K(\mathcal{D})$ by

$$\mu^{\mathcal{D}} := \sum\nolimits_{C \in \mathcal{A}_{\mathcal{D}}} \mathrm{rad}(C) \mathcal{H}^1_C(\cdot \cap C), \tag{3.5}$$

so that for any $w \in W_*$ we have $\mu^{\mathcal{D}}(K(\mathcal{D}_w)) = 2\,\mathrm{vol}_2(\triangle(\mathcal{D}_w))$ by (3.4), $\bigcup_{C \in \mathcal{A}_{\mathcal{D}_w}} D_C = \triangle(\mathcal{D}_w) \setminus K(\mathcal{D}_w)$ and $\mathrm{vol}_2(K(\mathcal{D}_w)) = 0$.

(2) For each $u \in \mathbb{R}^{K^0(\mathcal{D})}$ with $u|_C$ a.c. on C for any $C \in \mathcal{A}_{\mathcal{D}}$, we define a $\mu^{\mathcal{D}}$-a.e. defined, $\mathbb{R}^2$-valued Borel measurable map $\nabla_{\mathcal{D}} u$ by $(\nabla_{\mathcal{D}} u)|_C := \nabla_C(u|_C)$ for each $C \in \mathcal{A}_{\mathcal{D}}$, so that $|\nabla_{\mathcal{D}} u|^2\, d\mu^{\mathcal{D}} = \sum_{C \in \mathcal{A}_{\mathcal{D}}} |\nabla_C(u|_C)|^2\, \mathrm{rad}(C)\, d\mathcal{H}^1_C$. Then we further define

$$\mathcal{F}_{\mathcal{D}} := W^{1,2}_{\mathcal{D}} := \left\{ u \in \mathbb{R}^{K^0(\mathcal{D})} \;\middle|\; \begin{array}{l} u|_C \in W^{1,2}(C) \text{ for any } C \in \mathcal{A}_{\mathcal{D}}, \\ |\nabla_{\mathcal{D}} u| \in L^2(K(\mathcal{D}), \mu^{\mathcal{D}}) \end{array} \right\} \tag{3.6}$$

and set $\mathcal{C}_{\mathcal{D}} := \{u \in \mathcal{C}(K(\mathcal{D})) \mid u|_{K^0(\mathcal{D})} \in \mathcal{F}_{\mathcal{D}}\}$ and $\mathcal{C}^{\mathrm{lip}}_{\mathcal{D}} := \{u \in \mathcal{C}(K(\mathcal{D})) \mid \mathbf{Lip}_{K(\mathcal{D})} u < +\infty\}$, which are considered as linear subspaces of $\mathcal{F}_{\mathcal{D}}$ through the linear injection $\mathcal{C}(K(\mathcal{D})) \ni u \mapsto u|_{K^0(\mathcal{D})} \in \mathbb{R}^{K^0(\mathcal{D})}$. Noting that $\langle \nabla_{\mathcal{D}} u, \nabla_{\mathcal{D}} v \rangle \in L^1(K(\mathcal{D}), \mu^{\mathcal{D}})$ for any $u, v \in$

$\mathcal{F}_{\mathcal{D}}$, we also define a bilinear form $\mathcal{E}^{\mathcal{D}} : \mathcal{F}_{\mathcal{D}} \times \mathcal{F}_{\mathcal{D}} \to \mathbb{R}$ on $\mathcal{F}_{\mathcal{D}}$ by

$$\begin{aligned} \mathcal{E}^{\mathcal{D}}(u,v) &:= \int_{K(\mathcal{D})} \langle \nabla_{\mathcal{D}} u, \nabla_{\mathcal{D}} v \rangle \, d\mu^{\mathcal{D}} \\ &= \sum\nolimits_{C \in \mathcal{A}_{\mathcal{D}}} \int_C \langle \nabla_C(u|_C), \nabla_C(v|_C) \rangle \operatorname{rad}(C) \, d\mathcal{H}^1_C. \end{aligned} \tag{3.7}$$

In particular, setting $d\mu^{\mathcal{D}}_{\langle u \rangle} := |\nabla_{\mathcal{D}} u|^2 \, d\mu^{\mathcal{D}}$ for each $u \in \mathcal{F}_{\mathcal{D}}$, we have $\mu^{\mathcal{D}} = \mu^{\mathcal{D}}_{\langle h_1|_{K(\mathcal{D})} \rangle} + \mu^{\mathcal{D}}_{\langle h_2|_{K(\mathcal{D})} \rangle}$ as the counterpart of (3.1) for $K(\mathcal{D})$.

Theorem 3.4 ([14, Theorem 5.18]). *Let $\mathcal{D} \in \mathsf{TDT}^{\oplus}$ and set $\mathcal{F}^0_{\mathcal{D},0} := \{u \in \mathcal{F}_{\mathcal{D}} \mid u|_{V_0(\mathcal{D})} = 0\}$. Then $(\mathcal{E}^{\mathcal{D}}, \mathcal{F}_{\mathcal{D}})$ is an irreducible, strongly local, regular symmetric Dirichlet form on $L^2(K(\mathcal{D}), \mu^{\mathcal{D}})$ with a core $\mathcal{C}^{\mathrm{lip}}_{\mathcal{D}}$, and*

$$\int_{K(\mathcal{D})} u^2 \, d\mu^{\mathcal{D}} \leq 40 \kappa(\mathcal{D})^{-2} \mathcal{E}^{\mathcal{D}}(u,u) \qquad \text{for any } u \in \mathcal{F}^0_{\mathcal{D},0}. \tag{3.8}$$

Moreover, the inclusion map $\mathcal{F}_{\mathcal{D}} \hookrightarrow L^2(K(\mathcal{D}), \mu^{\mathcal{D}})$ is a compact linear operator under the norm $\|u\|_{\mathcal{F}_{\mathcal{D}}} := (\mathcal{E}^{\mathcal{D}}(u,u) + \int_{K(\mathcal{D})} u^2 \, d\mu^{\mathcal{D}})^{1/2}$ on $\mathcal{F}_{\mathcal{D}}$.

Theorem 3.5 ([14, Theorem 5.23]). *Let $\mathcal{D} \in \mathsf{TDT}^{\oplus}$, let μ' be a finite Borel measure on $K(\mathcal{D})$ with $\mu'(U) > 0$ for any non-empty open subset U of $K(\mathcal{D})$, and let $(\mathcal{E}', \mathcal{F}')$ be a strongly local, regular symmetric Dirichlet form on $L^2(K(\mathcal{D}), \mu')$ with $\mathcal{E}'(u,u) > 0$ for some $u \in \mathcal{F}'$. Then the following two conditions are equivalent:*

1. *Any $h \in \{h_1|_{K(\mathcal{D})}, h_2|_{K(\mathcal{D})}\}$ is in $\mathcal{F}'$ and is $\mathcal{E}'$-*harmonic *on $K(\mathcal{D}) \setminus V_0(\mathcal{D})$, i.e., $\mathcal{E}'(h,v) = 0$ for any $v \in \mathcal{F}' \cap \mathcal{C}(K(\mathcal{D}))$ with $v|_{V_0(\mathcal{D})} = 0$.*
2. *$\mathcal{F}' \cap \mathcal{C}(K(\mathcal{D})) = \mathcal{C}_{\mathcal{D}}$ and $\mathcal{E}'|_{\mathcal{C}_{\mathcal{D}} \times \mathcal{C}_{\mathcal{D}}} = c\mathcal{E}^{\mathcal{D}}|_{\mathcal{C}_{\mathcal{D}} \times \mathcal{C}_{\mathcal{D}}}$ for some $c \in \mathbb{R}$.*

Remark 3.6. In contrast to the case of $K(\mathcal{D})$ described in Definition 3.3, Theorems 3.4 and 3.5, the standard Dirichlet form $(\mathcal{E}, \mathcal{F})$ on the Sierpiński gasket K satisfies $\mu_{\langle u \rangle}(K^0) = 0$ for any $u \in \mathcal{F}$ by [10, Lemma 8.26] and [8, Lemma 5.7], where K^0 denotes the union of the boundaries of the equilateral triangles constituting K. In particular, $(\mathcal{E}, \mathcal{F})$ cannot be expressed as the sum of any weighted one-dimensional Dirichlet forms on $\Phi(K^0) \subset K_{\mathcal{H}}$ similar to (3.7). The author does not have a good explanation of the reason for this difference, and it would be very nice to give one. A naive guess could be that some sufficient smoothness of the relevant curves might be required for the validity of an expression like (3.7) of a non-zero strongly local regular symmetric Dirichlet form satisfying the analog of Theorem 3.5-(1); indeed, the curves constituting $\Phi(K^0)$ are $\mathcal{C}^1$ but not $\mathcal{C}^2$ by [22, Theorem 5.4-(2)], whereas the corresponding curves $C \in \mathcal{A}_{\mathcal{D}}$ in $K(\mathcal{D})$ are circular arcs and therefore real analytic. While this guess itself might well be correct, it would be still unclear how smooth the relevant curves should need to be.

The rest of this section is devoted to a brief sketch of the proof of Theorems 3.4 and 3.5, which is rather long and occupies the whole of [14, Sections 4 and 5]. It starts with identifying what the *trace* $\mathcal{E}^{\mathcal{D}}|_{V_m(\mathcal{D})}$,

$$\mathcal{E}^{\mathcal{D}}|_{V_m(\mathcal{D})}(u,u) := \inf_{v\in\mathcal{F}_{\mathcal{D}},\, v|_{V_m(\mathcal{D})}=u} \mathcal{E}^{\mathcal{D}}(v,v), \quad u\in\mathbb{R}^{V_m(\mathcal{D})}, \tag{3.9}$$

of $(\mathcal{E}^{\mathcal{D}},\mathcal{F}_{\mathcal{D}})$ to $V_m(\mathcal{D}) := \bigcup_{w\in W_m} V_0(\mathcal{D}_w)$ *should* be for any $m\in\mathbb{N}\cup\{0\}$. In view of the desired properties of $(\mathcal{E}^{\mathcal{D}},\mathcal{F}_{\mathcal{D}})$ in Theorem 3.5, the forms $\{\mathcal{E}^{\mathcal{D}}|_{V_m(\mathcal{D})}\}_{m\in\mathbb{N}\cup\{0\}}$ should have the properties in the following theorem.

Theorem 3.7 ([34, Theorem 5.17]). *Let $\mathcal{D}\in\mathsf{TDT}^{\oplus}$. Then there exists $\{\mathcal{E}^{\mathcal{D}}_m\}_{m\in\mathbb{N}\cup\{0\}}$ such that the following hold for any $m\in\mathbb{N}\cup\{0\}$:*
(1) *$\mathcal{E}^{\mathcal{D}}_m$ is a symmetric Dirichlet form on $\ell^2(V_m(\mathcal{D}))$. $\mathcal{E}^{\mathcal{D}}_m(\mathbf{1}_x,\mathbf{1}_y)=0=\mathcal{E}^{\mathcal{D}}_m(\mathbf{1}_x,\mathbf{1})$ for any $x,y\in V_m(\mathcal{D})$ with $\{\tau\in W_m \mid x,y\in V_0(\mathcal{D}_\tau)\}=\emptyset$.*
(2) *Both $h_1|_{V_m(\mathcal{D})}$ and $h_2|_{V_m(\mathcal{D})}$ are $\mathcal{E}^{\mathcal{D}}_m$-harmonic on $V_m(\mathcal{D})\setminus V_0(\mathcal{D})$.*
(3) *$\mathcal{E}^{\mathcal{D}}_m(u,u)=\min_{v\in\mathbb{R}^{V_{m+1}(\mathcal{D})},\, v|_{V_m(\mathcal{D})}=u}\mathcal{E}^{\mathcal{D}}_{m+1}(v,v)$ for any $u\in\mathbb{R}^{V_m(\mathcal{D})}$.*
(4) *$\mathcal{E}^{\mathcal{D}}_m(h_1|_{V_m(\mathcal{D})},h_1|_{V_m(\mathcal{D})})+\mathcal{E}^{\mathcal{D}}_m(h_2|_{V_m(\mathcal{D})},h_2|_{V_m(\mathcal{D})})=2\operatorname{vol}_2(\triangle(\mathcal{D}))$.*

Teplyaev's proof of Theorem 3.7 in [34] is purely Euclidean-geometric and provides no further information on $\{\mathcal{E}^{\mathcal{D}}_m\}_{m\in\mathbb{N}\cup\{0\}}$. The author has identified it as follows, by applying a refinement of [28, Corollary 4.2].

Theorem 3.8 ([14, Theorem 4.18]). *For each $\mathcal{D}=(D_1,D_2,D_3)\in\mathsf{TDT}^{\oplus}$, a sequence $\{\mathcal{E}^{\mathcal{D}}_m\}_{m\in\mathbb{N}\cup\{0\}}$ as in Theorem 3.7 is unique, and*

$$\mathcal{E}^{\mathcal{D}}_0(u,u)=\sum_{j\in S}\frac{\kappa(\mathcal{D})^2+\operatorname{curv}(D_j)^2}{2\kappa(\mathcal{D})\operatorname{curv}(D_j)}\bigl(u(q_{j+1}(\mathcal{D}))-u(q_{j+2}(\mathcal{D}))\bigr)^2 \tag{3.10}$$

for any $u\in\mathbb{R}^{V_0(\mathcal{D})}$, where $q_{j+3}(\mathcal{D}):=q_j(\mathcal{D})$ for $j\in S$. Moreover, for any $\mathcal{D}\in\mathsf{TDT}^{\oplus}$, any $m\in\mathbb{N}\cup\{0\}$ and any $u\in\mathbb{R}^{V_m(\mathcal{D})}$,

$$\mathcal{E}^{\mathcal{D}}_m(u,u)=\sum\nolimits_{w\in W_m}\mathcal{E}^{\mathcal{D}_w}_0(u|_{V_0(\mathcal{D}_w)},u|_{V_0(\mathcal{D}_w)}). \tag{3.11}$$

Let $\mathcal{D}\in\mathsf{TDT}^{\oplus}$. Theorem 3.7-(3) allows us to apply to $\{\mathcal{E}^{\mathcal{D}}_m\}_{m\in\mathbb{N}\cup\{0\}}$ the general theory from [21, Chapter 2] of constructing a Dirichlet form by taking the "*inductive limit*" of Dirichlet forms on finite sets. Namely, setting $V_*(\mathcal{D}):=\bigcup_{m\in\mathbb{N}\cup\{0\}}V_m(\mathcal{D})$, we can define a linear subspace $\mathcal{F}'_{\mathcal{D}}$ of $\mathbb{R}^{V_*(\mathcal{D})}$ and a bilinear form $\mathcal{E}'^{\mathcal{D}}:\mathcal{F}'_{\mathcal{D}}\times\mathcal{F}'_{\mathcal{D}}\to\mathbb{R}$ on $\mathcal{F}'_{\mathcal{D}}$ by

$$\mathcal{F}'_{\mathcal{D}}:=\bigl\{u\in\mathbb{R}^{V_*(\mathcal{D})}\bigm|\lim\nolimits_{m\to\infty}\mathcal{E}^{\mathcal{D}}_m(u|_{V_m(\mathcal{D})},u|_{V_m(\mathcal{D})})<+\infty\bigr\}, \tag{3.12}$$

$$(3.13)\quad \mathcal{E}'^{\mathcal{D}}(u,v) := \lim_{m\to\infty} \mathcal{E}^{\mathcal{D}}_m(u|_{V_m(\mathcal{D})}, v|_{V_m(\mathcal{D})}) \in \mathbb{R}, \quad u,v \in \mathcal{F}'_{\mathcal{D}}.$$

The next step of the proof of Theorems 3.4 and 3.5 is the following identification of $(\mathcal{E}'^{\mathcal{D}}, \mathcal{F}'_{\mathcal{D}})$ as $(\mathcal{E}^{\mathcal{D}}, \mathcal{F}_{\mathcal{D}})$, i.e., as given by (3.6) and (3.7).

Theorem 3.9 ([14, Theorem 5.13]). *Let $\mathcal{D} \in \mathsf{TDT}^{\oplus}$. Then $\mathcal{F}'_{\mathcal{D}} = \{u|_{V_*(\mathcal{D})} \mid u \in \mathcal{F}_{\mathcal{D}}\}$, the mapping $\mathcal{F}_{\mathcal{D}} \ni u \mapsto u|_{V_*(\mathcal{D})} \in \mathcal{F}'_{\mathcal{D}}$ is a linear isomorphism, and $\mathcal{E}'^{\mathcal{D}}(u|_{V_*(\mathcal{D})}, v|_{V_*(\mathcal{D})}) = \mathcal{E}^{\mathcal{D}}(u,v)$ for any $u,v \in \mathcal{F}_{\mathcal{D}}$.*

Sketch of the proof. By Theorem 3.7-(2),(3) and (3.12) we have $h_1|_{V_*(\mathcal{D})}, h_2|_{V_*(\mathcal{D})} \in \mathcal{F}'_{\mathcal{D}}$, which together with (3.12) implies that $\mathcal{C}'_{\mathcal{D}} := \{u \in \mathcal{C}(K(\mathcal{D})) \mid u|_{V_*(\mathcal{D})} \in \mathcal{F}'_{\mathcal{D}}\}$ is a dense subalgebra of $(\mathcal{C}(K(\mathcal{D})), \|\cdot\|_{\mathrm{sup}})$ with $h_1|_{K(\mathcal{D})}, h_2|_{K(\mathcal{D})} \in \mathcal{C}^{\mathrm{lip}}_{\mathcal{D}} \subset \mathcal{C}'_{\mathcal{D}}$. Hence at this stage we can already define the $\mathcal{E}'^{\mathcal{D}}$-energy measure $\mu'^{\mathcal{D}}_{\langle u\rangle}$ of $u \in \mathcal{C}'_{\mathcal{D}}$ by (3.2) with $K(\mathcal{D}), \mathcal{E}'^{\mathcal{D}}, \mathcal{C}'_{\mathcal{D}}$ in place of $K, \mathcal{E}, \mathcal{F}$, and the analog of (3.1) by $\mu'^{\mathcal{D}} := \mu'^{\mathcal{D}}_{\langle h_1|_{K(\mathcal{D})}\rangle} + \mu'^{\mathcal{D}}_{\langle h_2|_{K(\mathcal{D})}\rangle}$. Then it follows from Theorem 3.7-(4) and (3.11) that $\mu'^{\mathcal{D}}(K(\mathcal{D}_w)) = 2\,\mathrm{vol}_2(\triangle(\mathcal{D}_w)) = \mu^{\mathcal{D}}(K(\mathcal{D}_w))$ for any $w \in W_*$, whence $\mu'^{\mathcal{D}} = \mu^{\mathcal{D}}$.

Now that $\mu'^{\mathcal{D}}$ has been identified as $\mu^{\mathcal{D}}$ given by (3.5), it is natural to guess[2] that $\mathcal{F}'_{\mathcal{D}} \subset \{u|_{V_*(\mathcal{D})} \mid u \in \mathcal{F}_{\mathcal{D}}\}$ and that $\mathcal{E}'^{\mathcal{D}}(u|_{V_*(\mathcal{D})}, u|_{V_*(\mathcal{D})}) = \mathcal{E}^{\mathcal{D}}(u,u)$ for any $u \in \mathcal{F}_{\mathcal{D}}$ with $u|_{V_*(\mathcal{D})} \in \mathcal{F}'_{\mathcal{D}}$. This guess is not difficult to verify, first for any piecewise linear $u \in \mathcal{F}_{\mathcal{D}}$ by direct calculations based on Theorem 3.7-(2), (3.10), (3.11) and (3.13), and then for any $u \in \mathcal{F}_{\mathcal{D}}$ with $u|_{V_*(\mathcal{D})} \in \mathcal{F}'_{\mathcal{D}}$ by using the canonical approximation of u by piecewise linear functions; here $u \in \mathcal{F}_{\mathcal{D}}$ is called *m-piecewise linear*, where $m \in \mathbb{N} \cup \{0\}$, if and only if $u|_{K^0(\mathcal{D}_w)}$ is a linear combination of $h_1|_{K^0(\mathcal{D}_w)}, h_2|_{K^0(\mathcal{D}_w)}, \mathbf{1}_{K^0(\mathcal{D}_w)}$ for any $w \in W_m$, and *piecewise linear* if and only if u is m-piecewise linear for some $m \in \mathbb{N} \cup \{0\}$.

Finally, for any $u \in \mathcal{F}_{\mathcal{D}}$, some direct calculations using (3.10), (3.7) and (3.4) show that $\mathcal{E}^{\mathcal{D}_w}_0(u|_{V_0(\mathcal{D}_w)}, u|_{V_0(\mathcal{D}_w)}) \leq 7\int_{K(\mathcal{D}_w)} |\nabla_{\mathcal{D}} u|^2\, d\mu^{\mathcal{D}}$ for any $w \in W_*$, which together with (3.11) yields $\mathcal{E}^{\mathcal{D}}_m(u|_{V_m(\mathcal{D})}, u|_{V_m(\mathcal{D})}) \leq 7\mathcal{E}^{\mathcal{D}}(u,u)$ for any $m \in \mathbb{N} \cup \{0\}$, whence $u|_{V_*(\mathcal{D})} \in \mathcal{F}'_{\mathcal{D}}$ by (3.12). Q.E.D.

The last main step of the proof of Theorem 3.4 is to prove (3.8), which is based mainly on (3.5), (3.7) and the following lemma.

Lemma 3.10 ([14, Lemma 5.19]). *Let $C \subset \mathbb{C}$ be a circular arc, let $u \in \mathbb{R}^C$ satisfy $\mathbf{Lip}_C\, u < +\infty$, and for $a \in \mathbb{R}$ define $\mathcal{I}^a_C u : \overline{D_C} \to \mathbb{R}$ by*

$$(3.14)\quad \mathcal{I}^a_C u((1-t)\,\mathrm{cent}(C) + tz) := (1-t)a + tu(z), \quad (t,z) \in [0,1] \times C.$$

---

[2]This is how the author first came up with the expressions (3.6) and (3.7).

Then for any $a \in [\min_C u, \max_C u]$, $\mathbf{Lip}_{\overline{D_C}} \mathcal{I}_C^a u \leq \sqrt{5}\,\mathbf{Lip}_C u$ *and*

(3.15)
$$\frac{2}{21}\int_{D_C} |\nabla \mathcal{I}_C^a u|^2 \, d\,\mathrm{vol}_2 \leq \int_C |\nabla_C u|^2 \,\mathrm{rad}(C)\, d\mathcal{H}_C^1 \leq 2\int_{D_C} |\nabla \mathcal{I}_C^a u|^2 \, d\,\mathrm{vol}_2 .$$

Further, with $\overline{u}^C := \mathcal{H}_C^1(C)^{-1} \int_C u \, d\mathcal{H}_C^1$, *for any* $a \in \{0, \overline{u}^C\}$,

$$2\int_{D_C} |\mathcal{I}_C^a u|^2 \, d\,\mathrm{vol}_2 \leq \int_C u^2 \,\mathrm{rad}(C)\, d\mathcal{H}_C^1 \leq 4\int_{D_C} |\mathcal{I}_C^a u|^2 \, d\,\mathrm{vol}_2 . \tag{3.16}$$

Combining Lemma 3.10 with (3.5) and (3.7), we obtain the following.

Lemma 3.11 ([14, Lemma 5.21]). *Let* $\mathcal{D} \in \mathsf{TDT}^{\oplus}$ *and* $u \in \mathcal{C}_{\mathcal{D}}^{\mathrm{lip}}$. *Noting* $\overline{\triangle(\mathcal{D})} \setminus (K(\mathcal{D}) \setminus K^0(\mathcal{D})) = \bigcup_{C \in \mathcal{A}_{\mathcal{D}}} \overline{D_C}$, *define* $\mathcal{I}_{\mathcal{D}}^0 u \in \mathbb{R}^{\overline{\triangle(\mathcal{D})}}$ *by*

(3.17)
$$\mathcal{I}_{\mathcal{D}}^0 u|_{K(\mathcal{D})} := u, \quad \mathcal{I}_{\mathcal{D}}^0 u|_{\overline{D_C}} := \begin{cases} \mathcal{I}_C^0(u|_C) & \text{if } C \subset \partial T(\mathcal{D}), \\ \mathcal{I}_C^{\overline{u}^C}(u|_C) & \text{if } C \not\subset \partial T(\mathcal{D}), \end{cases} \quad C \in \mathcal{A}_{\mathcal{D}}.$$

If also $u|_{V_0(\mathcal{D})} = 0$, *then* $\mathcal{I}_{\mathcal{D}}^0 u|_{\partial\triangle(\mathcal{D})} = 0$, $\mathbf{Lip}_{\overline{\triangle(\mathcal{D})}} \mathcal{I}_{\mathcal{D}}^0 u \leq \sqrt{5}\,\mathbf{Lip}_{K(\mathcal{D})} u$,

$$\frac{2}{21}\int_{\triangle(\mathcal{D})} |\nabla \mathcal{I}_{\mathcal{D}}^0 u|^2 \, d\,\mathrm{vol}_2 \leq \mathcal{E}^{\mathcal{D}}(u,u) \leq 2\int_{\triangle(\mathcal{D})} |\nabla \mathcal{I}_{\mathcal{D}}^0 u|^2 \, d\,\mathrm{vol}_2, \tag{3.18}$$

$$2\int_{\triangle(\mathcal{D})} |\mathcal{I}_{\mathcal{D}}^0 u|^2 \, d\,\mathrm{vol}_2 \leq \int_{K(\mathcal{D})} u^2 \, d\mu^{\mathcal{D}} \leq 4\int_{\triangle(\mathcal{D})} |\mathcal{I}_{\mathcal{D}}^0 u|^2 \, d\,\mathrm{vol}_2 . \tag{3.19}$$

Sketch of the proof of Theorem 3.4. Recall the following classical fact implied by [5, Lemma 6.2.1, Theorems 4.5.1, 4.5.3 and 6.1.6]: if Q is an open rectangle in $\mathbb{C}$ whose smaller side length is $\delta \in (0, +\infty)$, then

$$\int_Q u^2 \, d\,\mathrm{vol}_2 \leq \frac{\delta^2}{\pi^2}\int_Q |\nabla u|^2 \, d\,\mathrm{vol}_2 \tag{3.20}$$

for any $u \in \mathbb{R}^{\overline{Q}}$ with $\mathbf{Lip}_{\overline{Q}} u < +\infty$ and $u|_{\partial Q} = 0$. Since $\triangle(\mathcal{D}) \subset Q$ for some such Q with $\delta = 3\kappa(\mathcal{D})^{-1}$ and then each $u \in \mathbb{R}^{\overline{\triangle(\mathcal{D})}}$ with $\mathbf{Lip}_{\overline{\triangle(\mathcal{D})}} u < +\infty$ and $u|_{\partial\triangle(\mathcal{D})} = 0$ can be extended to $\overline{Q}$ by setting $u|_{\overline{Q}\setminus\triangle(\mathcal{D})} := 0$ so as to satisfy $\mathbf{Lip}_{\overline{Q}} u < +\infty$ and $u|_{\partial Q} = 0$, we easily see from Lemma 3.11 and (3.20) that (3.8) holds for any $u \in \mathcal{F}_{\mathcal{D},0}^0 \cap \mathcal{C}_{\mathcal{D}}^{\mathrm{lip}}$.

Now, by utilizing the canonical approximation of each $u \in \mathcal{F}_{\mathcal{D}}$ by piecewise linear functions as in the sketch of the proof of Theorem 3.9

above, we can show that (3.8) extends to any $u \in \mathcal{F}^0_{\mathcal{D},0}$, which implies $\mathcal{F}_{\mathcal{D}} \subset L^2(K(\mathcal{D}), \mu^{\mathcal{D}})$, and that the inclusion map $\mathcal{F}_{\mathcal{D}} \hookrightarrow L^2(K(\mathcal{D}), \mu^{\mathcal{D}})$ is the limit in operator norm of finite-rank linear operators and hence compact. The rest of the proof is straightforward. Q.E.D.

Sketch of the proof of Theorem 3.5. The implication from (2) to (1) is immediate from Theorem 3.9 and Theorem 3.7-(2),(3). That from (1) to (2) can be shown by defining the trace $\mathcal{E}|_{V_m(\mathcal{D})}$ of $(\mathcal{E}, \mathcal{F})$ to $V_m(\mathcal{D})$ for $m \in \mathbb{N} \cup \{0\}$ in essentially the same way as (3.9), proving that $\{\mathcal{E}|_{V_m(\mathcal{D})}\}_{m \in \mathbb{N}\cup\{0\}}$ satisfies Theorem 3.7-(1),(2),(3) by the assumption of (1) and then applying Theorem 3.8 to conclude that $\{\mathcal{E}|_{V_m(\mathcal{D})}\}_{m\in\mathbb{N}\cup\{0\}} = \{c\mathcal{E}^{\mathcal{D}}_m\}_{m\in\mathbb{N}\cup\{0\}}$ for some $c \in \mathbb{R}$, which is easily seen to imply (2). Q.E.D.

§4. Weyl's eigenvalue asymptotics for the Apollonian gasket

The following proposition is an easy consequence of Theorem 3.4; see also [5, Exercise 4.2, Corollary 4.2.3, Theorems 4.5.1 and 4.5.3].

Proposition 4.1. *Let $\mathcal{D} \in \mathsf{TDT}^{\oplus}$, let V be a finite subset of $V_*(\mathcal{D})$ and set $\mathcal{F}^0_{\mathcal{D},V} := \{u \in \mathcal{F}_{\mathcal{D}} \mid u|_V = 0\}$. Then $(\mathcal{E}^{\mathcal{D}}|_{\mathcal{F}^0_{\mathcal{D},V}\times\mathcal{F}^0_{\mathcal{D},V}}, \mathcal{F}^0_{\mathcal{D},V})$ is a strongly local, regular symmetric Dirichlet form on $L^2(K(\mathcal{D}) \setminus V, \mu^{\mathcal{D}})$, and there exists a unique non-decreasing sequence $\{\lambda^{\mathcal{D},V}_n\}_{n\in\mathbb{N}} \subset [0,+\infty)$ such that $-\mathcal{L}_{\mathcal{D},V}\varphi^{\mathcal{D},V}_n = \lambda^{\mathcal{D},V}_n\varphi^{\mathcal{D},V}_n$ for any $n \in \mathbb{N}$ for some complete orthonormal system $\{\varphi^{\mathcal{D},V}_n\}_{n\in\mathbb{N}} \subset \mathcal{D}(\mathcal{L}_{\mathcal{D},V})$ of $L^2(K(\mathcal{D}) \setminus V, \mu^{\mathcal{D}})$; here $\mathcal{L}_{\mathcal{D},V} : \mathcal{D}(\mathcal{L}_{\mathcal{D},V}) \to L^2(K(\mathcal{D}) \setminus V, \mu^{\mathcal{D}})$ denotes the* Laplacian, *i.e., the non-positive self-adjoint operator on $L^2(K(\mathcal{D}) \setminus V, \mu^{\mathcal{D}})$, associated with $(\mathcal{E}^{\mathcal{D}}|_{\mathcal{F}^0_{\mathcal{D},V}\times\mathcal{F}^0_{\mathcal{D},V}}, \mathcal{F}^0_{\mathcal{D},V})$. Also, $\lim_{n\to\infty}\lambda^{\mathcal{D},V}_n = +\infty$, and for any $n \in \mathbb{N}$,*

$$(4.1)\qquad \lambda^{\mathcal{D},V}_n = \min\left\{ \max_{u\in L\setminus\{0\}} \frac{\mathcal{E}^{\mathcal{D}}(u,u)}{\int_{K(\mathcal{D})} u^2\, d\mu^{\mathcal{D}}} \;\middle|\; \begin{array}{l} L \text{ is a linear subspace} \\ \text{of } \mathcal{F}^0_{\mathcal{D},V},\ \dim L = n \end{array} \right\}.$$

The proof of the following theorem is the principal aim of [14].

Theorem 4.2 ([14, Theorem 7.1]). *There exists $c_{\mathsf{AG}} \in (0,+\infty)$ such that for any $\mathcal{D} \in \mathsf{TDT}^{\oplus}$ and any finite subset V of $V_*(\mathcal{D})$,*

$$(4.2)\qquad \lim_{\lambda\to+\infty} \frac{\#\{n \in \mathbb{N} \mid \lambda^{\mathcal{D},V}_n \le \lambda\}}{\lambda^{d_{\mathsf{AG}}/2}} = c_{\mathsf{AG}}\mathcal{H}^{d_{\mathsf{AG}}}(K(\mathcal{D})).$$

The rest of this section outlines the analytic aspects of the proof of Theorem 4.2. It can be deduced from the following theorem applicable to more general counting functions, including the classical one given by $\#\{w \in W_* \mid \mathrm{curv}(D_{\mathrm{in}}(\mathcal{D}_w)) \le \lambda\}$, whose asymptotic behavior analogous to (4.2) has been obtained first by Oh and Shah in [30, Corollary 1.8].

Definition 4.3. (1) We define $I := \{j^n k \mid j,k \in S,\, j \neq k,\, n \in \mathbb{N}\}$, so that $I \subset W_* \setminus \{\emptyset\}$, $\tau \not\leq \upsilon$ for any $\tau, \upsilon \in I$ with $\tau \neq \upsilon$ and

$$K(\mathcal{D}) \setminus V_0(\mathcal{D}) = \bigcup_{\tau \in I} K(\mathcal{D}_\tau) \qquad \text{for any } \mathcal{D} \in \mathsf{TDT}^+. \tag{4.3}$$

(2) We define $\Gamma \subset [0,+\infty)^4$ by $\Gamma := \{(g, \kappa(g)) \mid g \in [0,+\infty)^3, \kappa(g) > 0\}$, where $\kappa(g) := \sqrt{\beta\gamma + \gamma\alpha + \alpha\beta}$ for $g = (\alpha, \beta, \gamma) \in [0,+\infty)^3$, and set $\Gamma^\circ := \Gamma \cap (0,+\infty)^4$, which is an open subset of Γ; recall Propositions 2.2 and 2.7 and note that $gM_w \in \Gamma$ for any $g \in \Gamma$ and any $w \in W_*$.

(3) Recalling Theorem 2.11, we set $\mathcal{H}_\Gamma(g) := \mathcal{H}^{d_{\mathsf{AG}}}(K(\mathcal{D}))$ for each $g = (\alpha, \beta, \gamma, \kappa) \in \Gamma$, where we take any $\mathcal{D} = (D_1, D_2, D_3) \in \mathsf{TDT}^+$ with $(\mathrm{curv}(D_1), \mathrm{curv}(D_2), \mathrm{curv}(D_3)) = (\alpha, \beta, \gamma)$, which is easily seen to exist. Note that $\mathcal{H}_\Gamma(g) = s^{d_{\mathsf{AG}}} \mathcal{H}_\Gamma(sg)$ for any $(g, s) \in \Gamma \times (0,+\infty)$.

Theorem 4.4 ([14]). *Let Γ' denote either of Γ and Γ°, and for each $n \in \mathbb{N}$ let $\lambda_n : \Gamma' \to (0,+\infty)$ be continuous and satisfy $\lambda_n(sg) = s\lambda_n(g)$ for any $(g,s) \in \Gamma' \times (0,+\infty)$. Suppose that $\lambda_1(g) = \min_{n \in \mathbb{N}} \lambda_n(g)$ and $\lim_{n \to \infty} \lambda_n(g) = +\infty$ for any $g \in \Gamma'$, set $\mathcal{N}(g, \lambda) := \#\{n \in \mathbb{N} \mid \lambda_n(g) \leq \lambda\}$ for $(g, \lambda) \in \Gamma' \times [0,+\infty)$, and suppose that there exist $\eta \in [0, d_{\mathsf{AG}})$ and $c \in (0,+\infty)$ such that for any $g = (\alpha, \beta, \gamma, \kappa) \in \Gamma'$ and any $\lambda \in (0,+\infty)$,*

$$\begin{aligned} &\sum_{\tau \in I} \mathcal{N}(gM_\tau, \lambda) \leq \mathcal{N}(g, \lambda) \\ &\qquad \leq \sum_{\tau \in I} \mathcal{N}(gM_\tau, \lambda) + c(\min\{\beta+\gamma, \gamma+\alpha, \alpha+\beta\})^{-\eta} \lambda^\eta + c. \end{aligned} \tag{4.4}$$

Then there exists $c_0 \in (0,+\infty)$ such that for any $g \in \Gamma'$,

$$\lim_{\lambda \to +\infty} \frac{\mathcal{N}(g, \lambda)}{\lambda^{d_{\mathsf{AG}}}} = c_0 \mathcal{H}_\Gamma(g). \tag{4.5}$$

Theorem 4.4 is proved by applying Kesten's renewal theorem [19, Theorem 2] to the Markov chain on $\widetilde{\Gamma} := \{g \in \Gamma \mid \mathcal{H}_\Gamma(g) = 1\}$, the "*space of Euclidean shapes of* $\{K(\mathcal{D})\}_{\mathcal{D} \in \mathsf{TDT}^+}$", with transition function $\mathcal{P}(g, \cdot) := \sum_{\tau \in I} \mathcal{H}_\Gamma(gM_\tau) \delta_{[gM_\tau]_\Gamma}$, where for each $g \in \Gamma$ we set $[g]_\Gamma := \mathcal{H}_\Gamma(g)^{1/d_{\mathsf{AG}}} g \in \widetilde{\Gamma}$ and $\delta_{[g]_\Gamma}$ denotes the Borel probability measure on $\widetilde{\Gamma}$ with $\delta_{[g]_\Gamma}(\{[g]_\Gamma\}) = 1$; a brief sketch of the proof of Theorem 4.4 can be found in [13], and the full details will appear in [14, Sections 3 and 7].

Sketch of the proof of Theorem 4.2 under Theorem 4.4. We define $\mathcal{N}_{\mathcal{D},V}(\lambda) := \#\{n \in \mathbb{N} \mid \lambda_n^{\mathcal{D},V} \leq \lambda\}$, $\mathcal{N}_{\mathcal{D}}(\lambda) := \mathcal{N}_{\mathcal{D},\emptyset}(\lambda)$ and $\mathcal{N}_{\mathcal{D},0}(\lambda) := \mathcal{N}_{\mathcal{D},V_0(\mathcal{D})}(\lambda)$ for $\mathcal{D} \in \mathsf{TDT}^\oplus$, each finite subset V of $V_*(\mathcal{D})$ and $\lambda \in [0,+\infty)$. Then for any such $\mathcal{D}, V, \lambda$, as noted in [21, Theorem 4.1.7 and

Corollary 4.1.8], we easily see from $\dim \mathcal{F}_{\mathcal{D}}/\mathcal{F}^0_{\mathcal{D},V} = \#V$ and (4.1) that $\lambda^{\mathcal{D},\emptyset}_n \leq \lambda^{\mathcal{D},V}_n \leq \lambda^{\mathcal{D},\emptyset}_{n+\#V}$ for any $n \in \mathbb{N}$ and thereby that

$$\mathcal{N}_{\mathcal{D},V}(\lambda) \leq \mathcal{N}_{\mathcal{D}}(\lambda) \leq \mathcal{N}_{\mathcal{D},V}(\lambda) + \#V, \tag{4.6}$$

so that it suffices to prove (4.2) for $V = V_0(\mathcal{D})$, i.e., for $\mathcal{N}_{\mathcal{D},0}(\lambda)$.

To apply Theorem 4.4, for each $n \in \mathbb{N}$ and each $g = (\alpha, \beta, \gamma, \kappa) \in \Gamma^{\circ}$ we set $\lambda_n(g) := (\lambda^{\mathcal{D},V_0(\mathcal{D})}_n)^{1/2}$, where we take any $\mathcal{D} = (D_1, D_2, D_3) \in \mathsf{TDT}^{\oplus}$ with $(\mathrm{curv}(D_1), \mathrm{curv}(D_2), \mathrm{curv}(D_3)) = (\alpha, \beta, \gamma)$, so that $\lambda_n(g) \geq \lambda_1(g) > 0$ by $\{u \in \mathcal{F}_{\mathcal{D}} \mid \mathcal{E}^{\mathcal{D}}(u, u) = 0\} = \mathbb{R}\mathbf{1}$ and $\lim_{n\to\infty} \lambda_n(g) = +\infty$ by Proposition 4.1. We also easily see from Proposition 2.7, (3.5), (3.7) and (4.1) that for any $n \in \mathbb{N}$, $\lambda_n : \Gamma^{\circ} \to (0, +\infty)$ is continuous and satisfies $\lambda_n(sg) = s\lambda_n(g)$ for any $(g, s) \in \Gamma^{\circ} \times (0, +\infty)$.

It remains to verify that $\{\lambda_n\}_{n\in\mathbb{N}}$ satisfies (4.4). To this end, let $\mathcal{D} = (D_1, D_2, D_3) \in \mathsf{TDT}^{\oplus}$, $(\alpha, \beta, \gamma, \kappa) =: g$ be as in Proposition 2.2 and $\lambda \in (0, +\infty)$. Then since $\#\{n \in \mathbb{N} \mid \lambda_n(gM_w) \leq \lambda^{1/2}\} = \mathcal{N}_{\mathcal{D}_w,0}(\lambda)$ for any $w \in W_*$ by Proposition 2.7, (4.4) for $\{\lambda_n\}_{n\in\mathbb{N}}$ can be rephrased as

$$\begin{aligned}\sum\nolimits_{\tau\in I} \mathcal{N}_{\mathcal{D}_\tau,0}(\lambda) &\leq \mathcal{N}_{\mathcal{D},0}(\lambda)\\ &\leq \sum\nolimits_{\tau\in I} \mathcal{N}_{\mathcal{D}_\tau,0}(\lambda) + c(\min\{\beta+\gamma, \gamma+\alpha, \alpha+\beta\})^{-\eta}\lambda^{\eta/2} + c,\end{aligned} \tag{4.7}$$

which can be shown with $\eta = 1 < d_{\mathsf{AG}}$ (recall Theorem 2.10) as follows. Set $c_g := \min\{\beta+\gamma, \gamma+\alpha, \alpha+\beta\}$ and $n_\lambda := \min\{n \in \mathbb{N} \mid c_g^2 n^2 \geq 40\lambda\}$. Then for any $n \in \mathbb{N}$ with $n \geq n_\lambda$, any $j \in S$ and any $\tau \in I \cup \{j^{n_\lambda}\}$ with $\tau \leq j^{n_\lambda}$, from (3.8), (4.1) and (2.4) we obtain

$$\mathcal{N}_{\mathcal{D}_\tau,0}(\lambda) = 0 \text{ by } \lambda^{\mathcal{D}_\tau,V_0(\mathcal{D}_\tau)}_1 \geq \frac{\kappa(\mathcal{D}_\tau)^2}{40} \geq \frac{\kappa(\mathcal{D}_{j^{n_\lambda}})^2}{40} > \frac{c_g^2 n_\lambda^2}{40} \geq \lambda. \tag{4.8}$$

On the other hand, setting $I_\lambda := \{\tau \in I \mid |\tau| \leq n_\lambda\} \cup \{j^{n_\lambda} \mid j \in S\}$ and $V_\lambda := \bigcup_{\tau\in I_\lambda} V_0(\mathcal{D}_\tau)$, we have $K(\mathcal{D}) \setminus V_\lambda = \bigcup_{\tau\in I_\lambda}(K(\mathcal{D}_\tau) \setminus V_0(\mathcal{D}_\tau))$ with the union disjoint, which together with (4.1) and (4.8) easily implies that

$$\mathcal{N}_{\mathcal{D},V_\lambda}(\lambda) = \sum\nolimits_{\tau\in I_\lambda} \mathcal{N}_{\mathcal{D}_\tau,0}(\lambda) = \sum\nolimits_{\tau\in I} \mathcal{N}_{\mathcal{D}_\tau,0}(\lambda). \tag{4.9}$$

Now (4.7) follows from (4.9), $\#V_\lambda = 9n_\lambda - 3$ and the fact that $\mathcal{N}_{\mathcal{D},V_\lambda}(\lambda) \leq \mathcal{N}_{\mathcal{D},0}(\lambda) \leq \mathcal{N}_{\mathcal{D},V_\lambda}(\lambda) + \#V_\lambda - 3$ by the same proof as (4.6). Theorem 4.4 is thus applicable to $\{\lambda_n\}_{n\in\mathbb{N}}$ and yields (4.5), which means (4.2). Q.E.D.

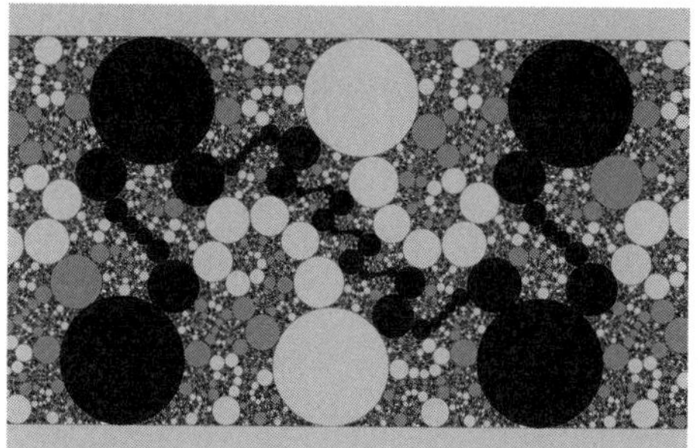

Figure 3. Limit set of $\frac{7}{43}$ double cusp group.

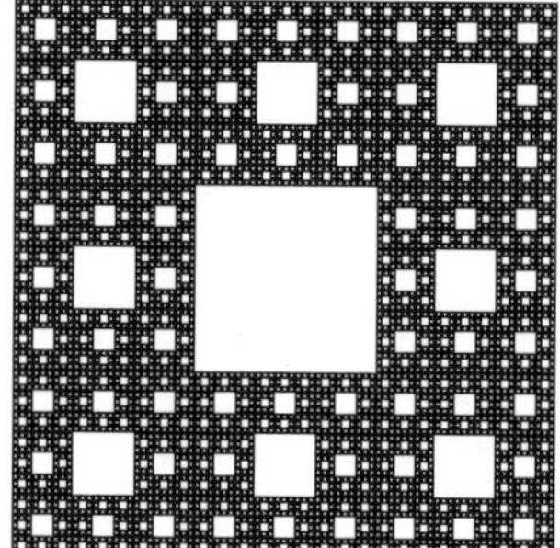

Figure 4. Sierpiński carpet.

§5. Kleinian groups with limit sets round Sierpiński carpets

In this last section, we illustrate the possibility of extending the results in §3 and §4 to other circle packing fractals, by presenting the results of the author's recent study in [16] obtained as the initial step toward developing a rich theory of construction and analysis of "geometrically canonical" Laplacians on more general self-conformal fractals.

Let $\mathrm{Möb}(\widehat{\mathbb{C}})$ denote the group of (orientation preserving or reversing) Möbius transformations on $\widehat{\mathbb{C}}$. A discrete subgroup G of $\mathrm{Möb}(\widehat{\mathbb{C}})$ is called a *Kleinian group*[3], and the smallest closed subset $\partial_\infty G$ of $\widehat{\mathbb{C}}$ invariant with respect to the action of G is called the *limit set* of G. It is known in the theory of Kleinian groups (see, e.g., [3, 17, 18, 36]) that the limit sets of certain classes of Kleinian groups are circle packing fractals, and typical examples of such circle packing fractals are provided in the book [29] together with a number of beautiful pictures of them.

Since the expressions (3.5) of $\mu^{\mathcal{D}}$ and (3.7) of the unique canonical Dirichlet form $(\mathcal{E}^{\mathcal{D}}, \mathcal{F}_{\mathcal{D}})$ on $K(\mathcal{D})$ makes sense on a general circle packing fractal, (a candidate of) a "geometrically canonical" Laplacian on it can be defined by (3.5) and (3.7), and it is natural to expect Weyl's eigenvalue asymptotics to hold when the fractal has some nice self-conformal structure. The author has recently verified this expectation in [15, 16] for the circle packing fractals arising as the limit sets of two specific classes of Kleinian groups, one of which studied in [15] is the *double cusp groups* on the boundary of Maskit's embedding of the Teichmüller space of the once-punctured torus treated in detail in [18, 29, 36]. In this case, the

[3]Kleinian groups are usually assumed to consist only of orientation preserving elements, but here we allow them to contain orientation reversing ones.

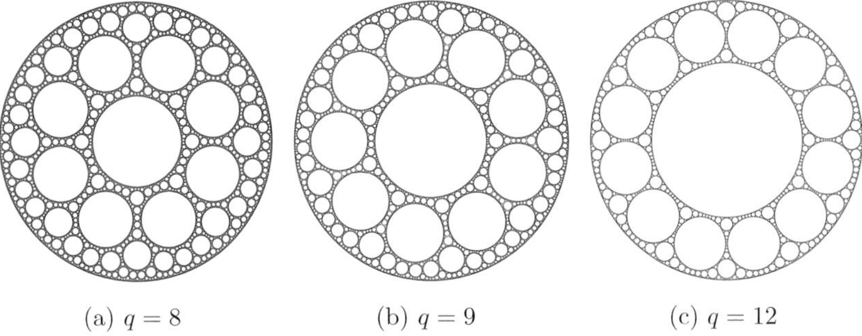

(a) $q = 8$ (b) $q = 9$ (c) $q = 12$

Figure 5. The limit sets $\partial_\infty G_q$ of the Kleinian groups G_q.

limit sets (Figure 3) can be shown to admit a self-conformal cellular decomposition similar to (4.3) which is *finitely ramified* in the sense that any cell intersects the others only on boundedly many points, and this property makes the proof of Weyl's asymptotics largely analogous to that of Theorem 4.2; a brief presentation of the precise statements of the results can be found in [11], and the full details will be given in [15].

On the other hand, each Kleinian group in the other class, which has been studied in [16], has as its limit set a *round Sierpiński carpet* (Figure 5), i.e., a subset of $\widehat{\mathbb{C}}$ homeomorphic to the standard Sierpiński carpet (Figure 4) whose complement in $\widehat{\mathbb{C}}$ consists of disjoint open disks in $\widehat{\mathbb{C}}$. In particular, this limit set is *infinitely ramified*, i.e., is not finitely ramified regardless of the choice of a cellular decomposition, which prevents the method of the above proof of (4.7) from applying to it and thereby makes the proof of Weyl's asymptotics for this case considerably more difficult.

The rest of this section is devoted to a brief summary of the results in [16] for the latter class of Kleinian groups, which are defined as follows. Let $q \in \mathbb{N}$ satisfy $q > 6$. It is a well-known fact from hyperbolic geometry (see, e.g., [31, Theorem 3.5.6]) that by $\frac{\pi}{2} + \frac{\pi}{3} + \frac{\pi}{q} < \pi$ there exists a geodesic triangle with inner angles $\frac{\pi}{2}, \frac{\pi}{3}, \frac{\pi}{q}$, unique up to hyperbolic isometry, in the Poincaré disk model $\mathbb{D} := \{z \in \mathbb{C} \mid |z| < 1\}$ of the hyperbolic plane; here we make the following specific choice of such one. The following construction is a slight modification of that given in [3].

Definition 5.1. (1) Set $\ell_1 := \mathbb{R}$, $\ell_3 := \{te^{i\pi/q} \mid t \in \mathbb{R}\}$ and choose $t_q, s_q \in (0, +\infty)$ so that $\ell_2 := \{z \in \mathbb{C} \mid |z - t_q e^{i\pi/q}| = s_q\}$ is orthogonal to $\partial\mathbb{D}$ and intersects ℓ_1 with angle $\frac{\pi}{3}$; there is a unique such choice of t_q, s_q by virtue of $\frac{\pi}{2} + \frac{\pi}{3} + \frac{\pi}{q} < \pi$. The closed geodesic triangle in $\mathbb{D}$ formed by ℓ_1, ℓ_2, ℓ_3 is denoted by $\triangle_q$, and the subgroup

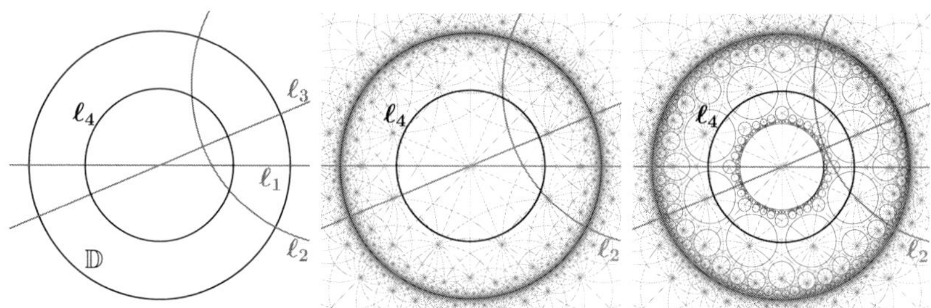

(a) Inversion circles $\{\ell_k\}_k$ (b) Tessellation by Γ_8 (c) Construction of $\partial_\infty G_8$

Figure 6. Illustration of Definition 5.1 and Proposition 5.2: $\Gamma_8, \partial_\infty G_8$.

of $\mathrm{Möb}(\widehat{\mathbb{C}})$ generated by $\{\mathrm{Inv}_{\ell_k}\}_{k=1}^3$ is denoted by Γ_q, where Inv_ℓ denotes the inversion (reflection) in a circle or a straight line $\ell \subset \mathbb{C}$.

(2) Choose $r_q \in (0,1)$ so that $\ell_4 := \{z \in \mathbb{C} \mid |z| = r_q\}$ intersects ℓ_2 with angle $\frac{\pi}{3}$; it is easy to see that there is a unique such choice of r_q. The subgroup of $\mathrm{Möb}(\widehat{\mathbb{C}})$ generated by $\{\mathrm{Inv}_{\ell_k}\}_{k=1}^4$ is denoted by G_q.

Proposition 5.2. (1) $\mathbb{D} = \bigcup_{\tau\in\Gamma_q} \tau(\triangle_q)$ *and* $\tau(\mathrm{int}\,\triangle_q) \cap \upsilon(\triangle_q) = \emptyset$ *for any* $\tau, \upsilon \in \Gamma_q$ *with* $\tau \neq \upsilon$.

(2) G_q *is a Kleinian group,* $\partial_\infty G_q = \overline{\bigcup_{\tau\in G_q} \tau(\partial\mathbb{D})} = \widehat{\mathbb{C}} \setminus \bigcup_{\tau\in G_q} \tau(\widehat{\mathbb{C}} \setminus \overline{\mathbb{D}})$, $D_1 \cap D_2 = \emptyset$ *for any* $D_1, D_2 \in \{\tau(\widehat{\mathbb{C}} \setminus \mathbb{D}) \mid \tau \in G_q\}$ *with* $D_1 \neq D_2$, *and* $\mathrm{int}\,\partial_\infty G_q = \emptyset$. *In particular,* $\partial_\infty G_q$ *is a round Sierpiński carpet.*

Proof. (1) is immediate from *Poincaré's polygon theorem* (see, e.g., [31, Theorem 7.1.3]), which applies to $\triangle_q$ since any of its inner angles is a *submultiple* of π, i.e., of the form π/n for some $n \in \mathbb{N} \cup \{+\infty\}$.

For (2), recall (see, e.g., [31, Sections 4.4–4.6]) that $\mathrm{Möb}(\widehat{\mathbb{C}})$ is canonically isomorphic to the group of isometries of the upper half-space model $\mathbb{H}^3 := \mathbb{C} \times (0,+\infty)$ of the three-dimensional hyperbolic space, where the inversion Inv_ℓ in a circle or a straight line $\ell \subset \mathbb{C}$ corresponds to the inversion in the sphere or the plane $\widetilde{\ell}$ intersecting $\mathbb{C}$ orthogonally on ℓ. Then since the closed polyhedron $\triangle_q^3$ in $\mathbb{H}^3$ formed by $\{\widetilde{\ell}_k\}_{k=1}^4$, defined as the part of $\{re^{i\theta} \mid (r,\theta) \in [0,+\infty) \times [0,\frac{\pi}{q}]\} \times (0,+\infty)$ above $\widetilde{\ell}_2$ and $\widetilde{\ell}_4$, has only submultiples of π as the dihedral angles between its faces, by *Poincaré's polyhedron theorem* (see, e.g., [31, Theorem 13.5.2]) applied to $\triangle_q^3$ we have $\mathbb{H}^3 = \bigcup_{\tau\in G_q} \tau(\triangle_q^3)$ and $\tau(\mathrm{int}_{\mathbb{H}^3}\,\triangle_q^3) \cap \upsilon(\triangle_q^3) = \emptyset$ for any $\tau, \upsilon \in G_q$ with $\tau \neq \upsilon$. Now we can obtain the first three assertions from this fact, $\mathrm{int}\,\partial_\infty G_q = \emptyset$ from [31, Theorem 12.2.7], and the last one from the topological characterization of the Sierpiński carpet in [35]. Q.E.D.

Even though in Definition 5.1 we have specifically chosen the unit disk $\mathbb{D}$ and the geodesic triangle $\triangle_q$, a particular choice of a disk D in $\mathbb{C}$ and a geodesic triangle in D should not matter for the desired Laplacian eigenvalue asymptotics. We should note also that the expressions (3.5) and (3.7) do not make perfect sense for the family $\{\tau(\partial\mathbb{D}) \mid \tau \in G_q\}$ of circles constituting $\partial_\infty G_q$, since $\partial\mathbb{D}$ should be treated together with the part $\widehat{\mathbb{C}} \setminus \overline{\mathbb{D}}$ of $\widehat{\mathbb{C}} \setminus \partial_\infty G_q$ enclosed by $\partial\mathbb{D}$ and thereby considered to be of *infinite* area and radius, which is incompatible with (3.5) and (3.7). To take care of these issues, we introduce the following definition.

Definition 5.3. We define $\mathcal{G} := \{g \in \text{Möb}(\widehat{\mathbb{C}}) \mid g^{-1}(\infty) \in \widehat{\mathbb{C}} \setminus \overline{\mathbb{D}}\}$, and for each $g \in \mathcal{G}$ we set $\mathcal{D}_g := \{g\tau(\widehat{\mathbb{C}} \setminus \overline{\mathbb{D}}) \mid \tau \in G_q\} \setminus \{g(\widehat{\mathbb{C}} \setminus \overline{\mathbb{D}})\}$ and $K_g := g(\mathbb{D} \cap \partial_\infty G_q) = g(\mathbb{D}) \setminus \bigcup_{D \in \mathcal{D}_g} D$, so that $\mathcal{D}_g$ is a family of open disks in $\mathbb{C}$ and $\overline{D_1} \subset g(\mathbb{D}) \setminus \overline{D_2}$ for any $D_1, D_2 \in \mathcal{D}_g$ with $D_1 \neq D_2$.

Definition 5.4 ([16]). Let $g \in \mathcal{G}$. We define a linear subspace $\mathcal{C}_g$ of $\mathcal{C}_c(K_g)$ by $\mathcal{C}_g := \{u \in \mathcal{C}_c(K_g) \mid \mathbf{Lip}_{K_g} u < +\infty\}$, and also define a *finite* Borel measure μ^g on K_g and a bilinear form $\mathcal{E}^g : \mathcal{C}_g \times \mathcal{C}_g \to \mathbb{R}$ on $\mathcal{C}_g$ by

$$\mu^g := \sum_{D \in \mathcal{D}_g} \text{rad}(D)\mathcal{H}^1_{\partial D}(\cdot \cap \partial D), \tag{5.1}$$

$$\mathcal{E}^g(u, v) := \sum_{D \in \mathcal{D}_g} \int_{\partial D} \langle \nabla_{\partial D}(u|_{\partial D}), \nabla_{\partial D}(v|_{\partial D}) \rangle \, \text{rad}(D) \, d\mathcal{H}^1_{\partial D}. \tag{5.2}$$

Proposition 5.5 ([16]). *Let $g \in \mathcal{G}$. Then $(\mathcal{E}^g, \mathcal{C}_g)$ is closable in $L^2(K_g, \mu^g)$ and its smallest closed extension $(\mathcal{E}^g, \mathcal{F}_g)$ in $L^2(K_g, \mu^g)$ is a strongly local, regular symmetric Dirichlet form on $L^2(K_g, \mu^g)$. Further, the inclusion map $\mathcal{F}_g \hookrightarrow L^2(K_g, \mu^g)$ is a compact linear operator under the norm $\|u\|_{\mathcal{F}_g} := (\mathcal{E}^g(u, u) + \int_{K_g} u^2 \, d\mu^g)^{1/2}$ on $\mathcal{F}_g$.*

Proposition 5.6 ([16]). *Let $g \in \mathcal{G}$. Then any $h \in \{h_1|_{K_g}, h_2|_{K_g}\}$ is $\mathcal{E}^g$-harmonic on K_g, i.e., $\mathcal{E}^g(h, v) = 0$ for any $v \in \mathcal{C}_g$, with $\mathcal{E}^g(h, v)$ still defined by (5.2).*

Proof. This follows easily by explicit calculations using the Gauss–Green theorem and the fact that ∂D is a circle for any $D \in \mathcal{D}_g$. Q.E.D.

The following is the main result of [16]. Note that for any $g \in \mathcal{G}$ and any non-empty open subset U of K_g, $d_q := \dim_{\text{H}} \partial_\infty G_q = \dim_{\text{H}} K_g \in (1, 2)$ and $\mathcal{H}^{d_q}(U) \in (0, +\infty)$ by [32, Theorem 7] and $\mathbf{Lip}_{\overline{\mathbb{D}}}\, g < +\infty$, and Proposition 5.5 implies the analog of Proposition 4.1 for $(\mathcal{E}^{g,U}, \mathcal{F}^0_{g,U})$ on $L^2(U, \mu^g|_U)$, where $\mu^g|_U := \mu^g|_{\mathcal{B}(U)}$, $\mathcal{F}^0_{g,U} := \overline{\{u \in \mathcal{C}_g \mid \text{supp}_{K_g}[u] \subset U\}}^{\mathcal{F}_g}$ and $\mathcal{E}^{g,U} := \mathcal{E}^g|_{\mathcal{F}^0_{g,U} \times \mathcal{F}^0_{g,U}}$.

Theorem 5.7 ([16]). *There exists* $c_q \in (0, +\infty)$ *such that for any* $g \in \mathcal{G}$ *and any non-empty open subset* U *of* K_g *with* $\mathcal{H}^{d_q}(\partial_{K_g} U) = 0$ *and* $\overline{U} \subset g(\mathbb{D})$, *the eigenvalues* $\{\lambda_n^{g,U}\}_{n \in \mathbb{N}}$ *(repeated according to multiplicity) of the Laplacian on* $L^2(U, \mu^g|_U)$ *associated with* $(\mathcal{E}^{g,U}, \mathcal{F}^0_{g,U})$ *satisfy*

$$\lim_{\lambda \to +\infty} \frac{\#\{n \in \mathbb{N} \mid \lambda_n^{g,U} \leq \lambda\}}{\lambda^{d_q/2}} = c_q \mathcal{H}^{d_q}(U). \tag{5.3}$$

The ergodic-theoretic aspects of the proof of Theorem 5.7 are largely analogous to those of the proof of Theorem 4.2, and in particular the roles played by the self-conformality of K_g are similar to those described in Remark 2.12. The most difficult part of the proof of Theorem 5.7 is that of an analog of (4.7), which is achieved by heavy use of heat kernel estimates in combination with the property of $\{\tau(\partial\mathbb{D}) \mid \tau \in G_q\}$ that they are *uniformly relatively separated* in the following sense (see [1]):

$$\inf_{(x,y) \in C_1 \times C_2} |x - y| \geq \varepsilon_q \min\{\mathrm{rad}(C_1), \mathrm{rad}(C_2)\} \tag{5.4}$$

for any $C_1, C_2 \in \{\tau(\partial\mathbb{D}) \mid \tau \in G_q\}$ with $C_1 \neq C_2$ for some $\varepsilon_q \in (0, +\infty)$. The full details of the proof of Theorem 5.7 will appear in [16].

References

[1] M. Bonk, Uniformization of Sierpiński carpets in the plane, *Invent. Math.*, **186** (2011), 559–665.

[2] D. W. Boyd, The residual set dimension of the Apollonian packing, *Mathematika*, **20** (1973), 170–174.

[3] S. Bullett and G. Mantica, Group theory of hyperbolic circle packings, *Nonlinearity*, **5** (1992), 1085–1109.

[4] Z.-Q. Chen and M. Fukushima, *Symmetric Markov Processes, Time Change, and Boundary Theory*, London Math. Soc. Monogr., **35**, Princeton Univ. Press, Princeton, NJ, 2012.

[5] E. B. Davies, *Spectral Theory and Differential Operators*, Cambridge Stud. Adv. Math., **42**, Cambridge Univ. Press, Cambridge, 1995.

[6] M. Fukushima, Y. Oshima and M. Takeda, *Dirichlet Forms and Symmetric Markov Processes*, 2nd ed., de Gruyter Stud. Math., **19**, Walter de Gruyter, Berlin, 2011.

[7] K. E. Hirst, The Apollonian packing of circles, *J. London Math. Soc.*, **42** (1967), 281–291.

[8] M. Hino, Energy measures and indices of Dirichlet forms, with applications to derivatives on some fractals, *Proc. London Math. Soc.*, **100** (2010), 269–302.

[9] N. Kajino, Heat kernel asymptotics for the measurable Riemannian structure on the Sierpinski gasket, *Potential Anal.*, **36** (2012), 67–115.

[10] N. Kajino, Analysis and geometry of the measurable Riemannian structure on the Sierpiński gasket, In: D. Carfì, M. L. Lapidus, E. P. J. Pearse and M. van Frankenhuijsen (eds.), *Fractal Geometry and Dynamical Systems in Pure and Applied Mathematics I: Fractals in Pure Mathematics*, Contemp. Math., **600**, Amer. Math. Soc., Providence, RI, 2013, pp. 91–133.

[11] N. Kajino, Weyl's eigenvalue asymptotics for the Laplacian on circle packing limit sets of certain Kleinian groups, In: *Heat Kernels, Stochastic Processes and Functional Inequalities*, Oberwolfach Report 55/2016. Available in: `https://www.mfo.de/occasion/1648`.

[12] N. Kajino, The Laplacian on some round Sierpiński carpets and Weyl's asymptotics for its eigenvalues (in Japanese), *RIMS Kôkyûroku*, **2116** (2019), 47–56.

[13] N. Kajino, The Laplacian on some self-conformal fractals and Weyl's asymptotics for its eigenvalues: A survey of the ergodic-theoretic aspects, *RIMS Kôkyûroku*, **2176** (2021), 111–119, `arXiv:2001.11354`.

[14] N. Kajino, *The Laplacian on the Apollonian gasket and Weyl's asymptotics for its eigenvalues*, 2021, in preparation.

[15] N. Kajino, *Weyl's eigenvalue asymptotics for the Laplacian on circle packing limit sets of certain Kleinian groups*, 2021, in preparation.

[16] N. Kajino, *The Laplacian on some round Sierpiński carpets and Weyl's asymptotics for its eigenvalues*, 2021, in preparation.

[17] L. Keen, B. Maskit and C. Series, Geometric finiteness and uniqueness for Kleinian groups with circle packing limit sets, *J. Reine Angew. Math.*, **436** (1993), 209–219.

[18] L. Keen and C. Series, Pleating coordinates for the Maskit embedding of the Teichmüller space of punctured tori, *Topology*, **32** (1993), 719–749.

[19] H. Kesten, Renewal theory for functionals of a Markov chain with general state space, *Ann. Probab.*, **2** (1974), 355–386.

[20] J. Kigami, Harmonic metric and Dirichlet form on the Sierpinski gasket, In: K. D. Elworthy and N. Ikeda (eds.), *Asymptotic Problems in Probability Theory: Stochastic Models and Diffusions on Fractals (Sanda/Kyoto, 1990)*, Pitman Research Notes in Math., **283**, Longman Sci. Tech., Harlow, 1993, pp. 201–218.

[21] J. Kigami, *Analysis on Fractals*, Cambridge Tracts in Math., **143**, Cambridge Univ. Press, Cambridge, 2001.

[22] J. Kigami, Measurable Riemannian geometry on the Sierpinski gasket: the Kusuoka measure and the Gaussian heat kernel estimate, *Math. Ann.*, **340** (2008), 781–804.

[23] P. Koskela and Y. Zhou, Geometry and analysis of Dirichlet forms, *Adv. Math.*, **231** (2012), 2755–2801.

[24] S. Kusuoka, Dirichlet forms on fractals and products of random matrices, *Publ. Res. Inst. Math. Sci.*, **25** (1989), 659–680.

[25] P. Mattila, *Geometry of Sets and Measures in Euclidean Spaces: Fractals and Rectifiability*, Cambridge Stud. Adv. Math., **44**, Cambridge Univ. Press, Cambridge, 1995.

[26] R. D. Mauldin and M. Urbański, Dimension and measures for a curvilinear Sierpinski gasket or Apollonian packing, *Adv. Math.*, **136** (1998), 26–38.
[27] C. T. McMullen, Hausdorff dimension and conformal dynamics, III: computation of dimension, *Amer. J. Math.*, **120** (1998), 691–721.
[28] R. Meyers, R. Strichartz and A. Teplyaev, Dirichlet forms on the Sierpiński gasket, *Pacific J. Math.*, **217** (2004), 149–174.
[29] D. Mumford, C. Series and D. Wright, *Indra's Pearls: The Vision of Felix Klein*, Cambridge Univ. Press, Cambridge, 2002.
[30] H. Oh and N. Shah, The asymptotic distribution of circles in the orbits of Kleinian groups, *Invent. Math.*, **187** (2012), 1–35.
[31] J. G. Ratcliffe, *Foundations of Hyperbolic Manifolds*, 3rd ed., Grad. Texts in Math., **149**, Springer, Cham, 2019.
[32] D. Sullivan, The density at infinity of a discrete group of hyperbolic motions, *Inst. Hautes Études Sci. Publ. Math.*, **50** (1979), 171–202.
[33] D. Sullivan, Entropy, Hausdorff measures old and new, and limit sets of geometrically finite Kleinian groups, *Acta Math.*, **153** (1984), 259–277.
[34] A. Teplyaev, Energy and Laplacian on the Sierpiński gasket, In: M. L. Lapidus and M. van Frankenhuijsen (eds.), *Fractal Geometry and Applications: A Jubilee of Benoît Mandelbrot*, Proc. Sympos. Pure Math., **72**, Part 1, Amer. Math. Soc., Providence, RI, 2004, pp. 131–154.
[35] G. T. Whyburn, Topological characterization of the Sierpiński curve, *Fund. Math.*, **45** (1958), 320–324.
[36] D. Wright, Searching for the cusp, In: Y. Minsky, M. Sakuma and C. Series (eds.), *Spaces of Kleinian Groups*, London Math. Soc. Lecture Note Ser., **329**, Cambridge Univ. Press, Cambridge, 2005, pp. 301–336.

Department of Mathematics, Graduate School of Science, Kobe University
Rokkodai-cho 1-1, Nada-ku, Kobe 657-8501, Japan

Current Address:
Research Institute for Mathematical Sciences, Kyoto University
Kitashirakawa-Oiwake-cho, Sakyo-ku, Kyoto 606-8502, Japan
E-mail address: nkajino@kurims.kyoto-u.ac.jp

Advanced Studies in Pure Mathematics 87, 2021
Stochastic Analysis, Random Fields and Integrable Probability — Fukuoka 2019
pp. 315–340

Gaussian free fields coupled with multiple SLEs driven by stochastic log-gases

Makoto Katori and Shinji Koshida

Abstract.

Miller and Sheffield introduced the notion of an imaginary surface as an equivalence class of pairs of simply connected proper subdomains of $\mathbb{C}$ and Gaussian free fields (GFFs) on them under the conformal equivalence. They considered the situation in which the conformal maps are given by a chordal Schramm–Loewner evolution (SLE). In the present paper, we construct GFF-valued processes on $\mathbb{H}$ (the upper half-plane) and $\mathbb{O}$ (the first orthant of $\mathbb{C}$) by coupling a GFF with a multiple SLE evolving in time on each domain. We prove that a GFF on $\mathbb{H}$ and $\mathbb{O}$ is locally coupled with a multiple SLE if the multiple SLE is driven by the stochastic log-gas called the Dyson model defined on $\mathbb{R}$ and the Bru–Wishart process defined on $\mathbb{R}_+$, respectively. We obtain pairs of time-evolutionary domains and GFF-valued processes.

§1. Introduction

The present study is motivated by the recent work by Sheffield on the quantum gravity zipper and the AC geometry [32] and a series of papers by Miller and Sheffield on the imaginary geometry [25, 26, 27, 28]. In both of them, a *Gaussian free field* (GFF) on a simply connected proper subdomain D of the complex plane $\mathbb{C}$ (see, for instance, [31]) is coupled with a *Schramm–Loewner evolution* (SLE) [30, 24, 23] driven by a Brownian motion moving on the boundary ∂D, or its variant called an SLE$(\kappa, \underline{\rho})$.

Consider a simply connected domain $D \subsetneq \mathbb{C}$ and write $\mathcal{C}_{\mathrm{c}}^{\infty}(D)$ for the space of real smooth functions on D with compact support. Assume $h \in$

Received January 17, 2020.
Revised August 12, 2020.
2010 *Mathematics Subject Classification.* 60D05, 60J67, 82C22, 60B20.
Key words and phrases. Gaussian free fields, Imaginary surface and imaginary geometry, Schramm–Loewner evolution, Multiple SLE, Stochastic log-gases, Dyson model, Bru–Wishart process.

$C_c^\infty(D)$ and consider a smooth vector field $e^{\sqrt{-1}(h/\chi+\theta)}$ with parameters $\chi, \theta \in \mathbb{R}$. Then, the *flow line* along this vector field, $\eta : (0,\infty) \ni t \mapsto \eta(t) \in D$, starting from $\lim_{t\to 0} \eta(t) =: \eta(0) = x \in \partial D$ is defined (if exists) as the solution of the ordinary differential equation (ODE)

$$\text{(1.1)} \qquad \frac{d\eta(t)}{dt} = e^{\sqrt{-1}\{h(\eta(t))/\chi+\theta\}}, \quad t \geq 0, \quad \eta(0) = x.$$

Let $\widetilde{D} \subsetneq \mathbb{C}$ be another simply connected domain and consider a conformal map $\varphi : \widetilde{D} \to D$. Then, we define the pull-back of the flow line η by φ as $\widetilde{\eta}(t) = (\varphi^{-1} \circ \eta)(t)$. That is, $\varphi(\widetilde{\eta}(t)) = \eta(t)$, and the derivatives with respect to t of both sides of this equation gives $\varphi'(\widetilde{\eta}(t))d\widetilde{\eta}(t)/dt = d\eta(t)/dt$ with $\varphi'(z) := d\varphi(z)/dz$. We use the polar coordinates $\varphi'(\cdot) = |\varphi'(\cdot)|e^{\sqrt{-1}\arg\varphi'(\cdot)}$, where $\arg\zeta$ of $\zeta \in \mathbb{C}$ is a priori defined up to additive multiples of 2π, and hence, we have $d\widetilde{\eta}(t)/dt = e^{\sqrt{-1}\{(h\circ\varphi-\chi\arg\varphi')(\widetilde{\eta}(t))/\chi+\theta\}}/|\varphi'(\widetilde{\eta}(t))|$, $t \geq 0$. If we perform a time change $t \to \tau = \tau(t)$ by putting $t = \int_0^\tau ds/|\varphi'(\widetilde{\eta}(s))|$ and $\widehat{\eta}(t) := \widetilde{\eta}(\tau(t))$, then the above equation becomes

$$\frac{d\widehat{\eta}(t)}{dt} = e^{\sqrt{-1}\{(h\circ\varphi-\chi\arg\varphi')(\widehat{\eta}(t))/\chi+\theta\}}, \quad t \geq 0.$$

Since a time change preserves the image of a flow line, we can identify h on D and $h \circ \varphi - \chi\arg\varphi'$ on $\widetilde{D} = \varphi^{-1}(D)$. In [32, 25, 26, 27, 28], such a flow line is considered also in the case that h is given by an instance of a GFF defined as follows.

Definition 1.1. *Let $D \subsetneq \mathbb{C}$ be a simply connected domain and H be a GFF on D with zero boundary condition (constructed in Section 4). A GFF on D is a random distribution h of the form $h = H + u$, where u is a deterministic harmonic function on D.*

Since a GFF is not function-valued, but it is a *distribution-valued random field* (see Remark 4.1 in Section 4), the ODE in the form (1.1) no longer makes sense mathematically in the classical sense. Using the theory of SLE, however, the notion of flow lines was generalized as follows.

Consider the collection

$$\mathsf{S} := \left\{ (D,h) \,\middle|\, \begin{array}{c} D\subsetneq\mathbb{C}\text{: simply connected} \\ h\text{: GFF on } D \end{array} \right\}.$$

Fixing a parameter $\chi \in \mathbb{R}$, we define the following equivalence relation in S.

Definition 1.2. *Two pairs* (D, h) *and* $(\widetilde{D}, \widetilde{h}) \in \mathsf{S}$ *are equivalent if there exists a conformal map* $\varphi : \widetilde{D} \to D$ *and* $\widetilde{h} \overset{\text{(law)}}{=} h \circ \varphi - \chi \arg \varphi'$. *In this case, we write* $(D, h) \sim (\widetilde{D}, \widetilde{h})$.

We call each element belonging to $\mathsf{S}/\sim$ an *imaginary surface* [25] (or an *AC surface* [32]). That is, in this equivalence class, a conformal map φ causes not only a coordinate change of a GFF as $h \mapsto h \circ \varphi$ associated with changing the domain of definition of the field as $D \mapsto \varphi^{-1}(D)$, but also an addition of a deterministic harmonic function $-\chi \arg \varphi'$ to the field. Notice that this definition depends on one parameter $\chi \in \mathbb{R}$.

As will be explained in Section 4, each instance H of a GFF with zero boundary condition depends on the choice of a complete orthonormal system (CONS) of a Hilbert space starting from which a GFF is constructed. The probability law of a zero-boundary GFF is, however, independent of such construction and uniquely determined.

Consider the case in which D is the upper half-plane $\mathbb{H} := \{z \in \mathbb{C} : \operatorname{Im} z > 0\}$ with $\partial \mathbb{H} = \mathbb{R} \cup \{\infty\}$. Let $(B(t))_{t \geq 0}$ be a one-dimensional standard Brownian motion starting from the origin defined on a probability space $(\Omega^{1\text{SLE}}, \mathcal{F}^{1\text{SLE}}, \mathbb{P}^{1\text{SLE}})$ and adapted to a filtration $(\mathcal{F}_t^{1\text{SLE}})_{t \geq 0}$. We consider the chordal SLE(κ) driven by $(\sqrt{\kappa} B(t))_{t \geq 0}$ on $S := \mathbb{R}$ with $\kappa > 0$ [30, 24, 23], associated to which we obtain a random curve (called a *chordal SLE(*κ*) curve*) parameterized by time, $\eta : (0, \infty) \ni t \mapsto \eta(t) \in \mathbb{H}$, such that $\lim_{t \to 0} \eta(t) =: \eta(0) = 0$, $\lim_{t \to \infty} \eta(t) = \infty$. At each time $t > 0$, let $\eta(0, t] := \{\eta(s) : s \in (0, t]\}$ and we write $\mathbb{H}_t^\eta$ for the unbounded component of $\mathbb{H} \setminus \eta(0, t]$. Then, the chordal SLE($\kappa$) gives a conformal map from $\mathbb{H}_t^\eta$ to $\mathbb{H}$. It is also known that, if $\kappa \in (0, 4]$, then $\eta(0, t]$ is almost surely a simple curve at each $t > 0$ and, hence, $\mathbb{H}_t^\eta = \mathbb{H} \setminus \eta(0, t]$. In this paper, we will write the chordal SLE(κ) as $(g_{\mathbb{H}_t^\eta})_{t \geq 0}$. Let $H(\cdot)$ be a GFF on $\mathbb{H}$ with zero boundary condition on $\mathbb{R}$ that is defined on a probability space $(\Omega^{\text{GFF}}, \mathcal{F}^{\text{GFF}}, \mathbb{P}^{\text{GFF}})$. To couple the SLE and the GFF, we introduce a probability space $(\Omega, \mathcal{F}, \mathbb{P}) = (\Omega^{\text{GFF}} \times \Omega^{1\text{SLE}}, \mathcal{F}^{\text{GFF}} \vee \mathcal{F}^{1\text{SLE}}, \mathbb{P}^{\text{GFF}} \otimes \mathbb{P}^{1\text{SLE}})$ and extend the SLE and the GFF onto this probability space. Then, the SLE is adapted to the filtration $(\mathcal{F}_t)_{t \geq 0}$ defined by $\mathcal{F}_t = \{\emptyset, \Omega^{\text{GFF}}\} \vee \mathcal{F}_t^{1\text{SLE}}$. Instead of $H(\cdot)$ itself, we consider the following GFF on $\mathbb{H}$ by adding a deterministic harmonic function,

$$h(\cdot) := H(\cdot) - \frac{2}{\sqrt{\kappa}} \arg(\cdot). \tag{1.2}$$

Notice that $\arg(\cdot) = \operatorname{Im} \log(\cdot)$ and the real and imaginary parts of a complex analytic function are harmonic. Hence, the random distribution (1.2) is in fact a GFF in the sense of Definition 1.1. Given $\kappa > 0$ for the

SLE(κ), we fix the parameter χ as $\chi = 2/\sqrt{\kappa} - \sqrt{\kappa}/2$. Note that the well-known relation between κ and the *central charge* c of conformal field theory is simply expressed using the present parameter χ as $c = 1 - 6\chi^2$ (see, for instance, [3, Eq.(6)]). Let

$$f_{\mathbb{H}_t^\eta} := g_{\mathbb{H}_t^\eta} - \sqrt{\kappa}B(t) = \sigma_{-\sqrt{\kappa}B(t)} \circ g_{\mathbb{H}_t^\eta},$$

where σ_s denotes the translation by $s \in \mathbb{R}$; $\sigma_s(z) = z + s$, $z \in \mathbb{H}$. Let $A \subset \mathbb{H}$ be an open set and take an $(\mathcal{F}_t)_{t\geq 0}$-stopping time

$$\tau_A := \inf \left\{ t \geq 0 \,\middle|\, \eta(0,t] \cap A \neq \emptyset \right\}.$$

Let τ be any $(\mathcal{F}_t)_{t\geq 0}$-stopping time such that $\tau \leq \tau_A$ a.s. Then, we can prove the following equality in probability [25, Theorem 1.1, Lemma 3.11] (see also [10, Lemma 6.1]); for any $f \in \mathcal{C}_c^\infty(\mathbb{H})$ such that $\mathrm{supp}(f) \subset A$,

$$(1.3) \qquad (h, f) \overset{\text{(law)}}{=} (h \circ f_{\mathbb{H}_\tau^\eta} - \chi \arg f'_{\mathbb{H}_\tau^\eta}, f) \quad \text{under } \mathbb{P},$$

where the pairing $(\cdot,\cdot)$ is defined by (4.3) below. We comment that, due to the conformal invariance of a zero-boundary GFF (see Section 4.2 below), for an instance of the SLE(κ), the random distribution $h \circ f_{\mathbb{H}_t^\eta} - \chi \arg f'_{\mathbb{H}_t^\eta}$ is a GFF on $\mathbb{H}_t^\eta$ in the sense of Definition 1.1. Notice that pairs $(\mathbb{H}, h)$ and $(\mathbb{H}_t^\eta, h \circ f_{\mathbb{H}_t^\eta} - \chi \arg f'_{\mathbb{H}_t^\eta})$ with (1.2) are equivalent in the sense of Definition 1.2. In other words, an imaginary surface whose representative is given by $(\mathbb{H}, h)$ is constructed as a pair of a time-evolutionary domain, $f_{\mathbb{H}_t^\eta}^{-1}(\mathbb{H}) = \mathbb{H}_t^\eta$, $t \geq 0$, and a GFF-valued process, $h \circ f_{\mathbb{H}_t^\eta} - \chi \arg f'_{\mathbb{H}_t^\eta}, t \geq 0$ defined on it. With the establishment of the equality (1.3) we say that the *local coupling between a GFF and an SLE* is constructed (see [10, 32, 25] for lifting the local coupling to the 'global' one). It was proved [10, 25] that, under the coupling between a GFF and an SLE, the SLE-curve is a deterministic functional of the GFF. By virtue of it, in [25], the authors referred to an SLE(κ) curve as a flow line of the GFF h.

Here, first we consider the case in which the conformal maps are generated by a multiple Loewner equation associated with a multi-slit. Let $N \in \mathbb{N} := \{1, 2, \dots\}$ and suppose that we have N slits $\eta_i = \{\eta_i(t) : t \in (0,\infty)\} \subset \mathbb{H}$, $1 \leq i \leq N$, which are simple curves, disjoint with each other, $\eta_i \cap \eta_j = \emptyset$, $i \neq j$, starting from N distinct points $\lim_{t\to 0} \eta_i(t) =: \eta_i(0)$ on $\mathbb{R}$; $\eta_1(0) < \cdots < \eta_N(0)$, and all going to infinity; $\lim_{t\to\infty} \eta_i(t) = \infty$, $1 \leq i \leq N$. A *multi-slit* is defined as the union of them, $\bigcup_{i=1}^N \eta_i$, and

$$\mathbb{H}_t^\eta := \mathbb{H} \setminus \bigcup_{i=1}^N \eta_i(0,t] \quad \text{for each } t > 0 \text{ with } \mathbb{H}_0^\eta := \mathbb{H}.$$

We write the time evolution of the conformal map which transforms $\mathbb{H}_t^\eta$ to $\mathbb{H}$ at each time $t \geq 0$ under the hydrodynamic normalization as $(g_{\mathbb{H}_t^\eta})_{t\geq 0}$ and call it a *multiple SLE*. The images of the tips of the multi-slit $g_{\mathbb{H}_t^\eta}(\eta_i(t))$, $1 \leq i \leq N$ exist as points on $\mathbb{R}$ for $t \geq 0$ and if we put $X_i^{\mathbb{R}}(t) := g_{\mathbb{H}_t^\eta}(\eta_i(t))$, the multiple SLE $(g_{\mathbb{H}_t^\eta})_{t\geq 0}$ is given as a unique solution of the following equation,

$$\begin{aligned} \frac{dg_{\mathbb{H}_t^\eta}(z)}{dt} &= \sum_{i=1}^N \frac{2}{g_{\mathbb{H}_t^\eta}(z) - X_i^{\mathbb{R}}(t)}, \quad t \geq 0, \\ g_{\mathbb{H}_0^\eta}(z) &= z \in \mathbb{H}, \end{aligned} \tag{1.4}$$

under a proper parameterizatoin of the multi-slit. Here

$$\boldsymbol{X}^{\mathbb{R}}(t) = \left(X_1^{\mathbb{R}}(t), \ldots, X_N^{\mathbb{R}}(t)\right) \in \mathbb{R}^N, \quad t \geq 0$$

is called the *driving process* of the multiple SLE.

In the sequel, we will consider the case when $(\boldsymbol{X}^{\mathbb{R}}(t))_{t\geq 0}$ is a stochastic process defined on a probability space $(\Omega^{N\mathrm{SLE}}, \mathcal{F}^{N\mathrm{SLE}}, \mathbb{P}^{N\mathrm{SLE}})$ and adapted to a filtration $(\mathcal{F}_t^{N\mathrm{SLE}})_{t\geq 0}$. In this case, although it is not ensured that the solution generates a multi-slit depending on κ, we can still find a family of domains $\mathbb{H}_t^\eta \subseteq \mathbb{H}$, $t \geq 0$ so that $g_{\mathbb{H}_t^\eta} : \mathbb{H}_t^\eta \to \mathbb{H}$ is a conformal map at each $t \geq 0$. (See Remark 1.4 (a) in [19].)

We again consider a zero-boundary GFF H on $\mathbb{H}$ defined on a probability space $(\Omega^{\mathrm{GFF}}, \mathcal{F}^{\mathrm{GFF}}, \mathbb{P}^{\mathrm{GFF}})$ and introduce a coupled probability space

$$(\Omega, \mathcal{F}, \mathbb{P}) = \left(\Omega^{\mathrm{GFF}} \times \Omega^{N\mathrm{SLE}}, \mathcal{F}^{\mathrm{GFF}} \vee \mathcal{F}^{N\mathrm{SLE}}, \mathbb{P}^{\mathrm{GFF}} \otimes \mathbb{P}^{N\mathrm{SLE}}\right).$$

Then, the multiple SLE and the GFF are naturally extended to $(\Omega, \mathcal{F}, \mathbb{P})$, and the multiple SLE is adapted to the filtration $(\mathcal{F}_t)_{t\geq 0}$ defined by $\mathcal{F}_t = \{\emptyset, \Omega^{\mathrm{GFF}}\} \vee \mathcal{F}_t^{N\mathrm{SLE}}$.

Regarding (1.2) and (1.3), we see that $h \circ f_{\mathbb{H}_t^\eta}(\cdot) - \chi \arg f'_{\mathbb{H}_t^\eta}(\cdot)$ is equal to

$$\begin{aligned} &(H \circ \sigma_{-\sqrt{\kappa}B(t)}) \circ g_{\mathbb{H}_t^\eta}(\cdot) - \frac{2}{\sqrt{\kappa}} \arg\left(g_{\mathbb{H}_t^\eta}(\cdot) - \sqrt{\kappa}B(t)\right) - \chi \arg g'_{\mathbb{H}_t^\eta}(\cdot) \\ &\quad = H \circ g_{\mathbb{H}_t^\eta}(\cdot) - \frac{2}{\sqrt{\kappa}} \arg\left(g_{\mathbb{H}_t^\eta}(\cdot) - g_{\mathbb{H}_t^\eta}(\eta(t))\right) - \chi \arg g'_{\mathbb{H}_t^\eta}(\cdot) \\ &\qquad\qquad \text{under } \mathbb{P}, \end{aligned}$$

$t \geq 0$, where the translation invariance of H was used. Motivated by this observation, we study the GFF-valued process defined by

$$\begin{aligned}(1.5)\quad H_{\mathbb{H}}(\cdot,t) &:= H \circ g_{\mathbb{H}_t^\eta}(\cdot) \\ &\quad - \frac{2}{\sqrt{\kappa}} \sum_{i=1}^{N} \arg\left(g_{\mathbb{H}_t^\eta}(\cdot) - g_{\mathbb{H}_t^\eta}(\eta_i(t))\right) - \chi \arg g'_{\mathbb{H}_t^\eta}(\cdot) \\ &= H \circ g_{\mathbb{H}_t^\eta}(\cdot) \\ &\quad - \frac{2}{\sqrt{\kappa}} \sum_{i=1}^{N} \arg\left(g_{\mathbb{H}_t^\eta}(\cdot) - X_i^{\mathbb{R}}(t)\right) - \chi \arg g'_{\mathbb{H}_t^\eta}(\cdot)\end{aligned}$$

on $\mathbb{H}_t^\eta, t \geq 0$. This process starts from

$$H_{\mathbb{H}}(\cdot,0) = H(\cdot) - \frac{2}{\sqrt{\kappa}} \sum_{i=1}^{N} \arg(\cdot - x_i^{\mathbb{R}}),$$

where we assume that $x_1^{\mathbb{R}} < \cdots < x_N^{\mathbb{R}}$. We let the boundary points evolve according to the stochastic process $(\boldsymbol{X}^{\mathbb{R}}(t))_{t\geq 0}$ starting from $\boldsymbol{x}^{\mathbb{R}} := (x_i^{\mathbb{R}})_{i=1}^N$. At each time $t > 0$, we consider the GFF $H + u_t$ on $\mathbb{H}$ where $u_t(\cdot) = -(2/\sqrt{\kappa}) \sum_{i=1}^N \arg(\cdot - X_i^{\mathbb{R}}(t))$. Then, the GFF $H_{\mathbb{H}}(\cdot,t)$ on $\mathbb{H}_t^\eta$ is defined by the property that $(\mathbb{H}, H + u_t) \sim (\mathbb{H}_t^\eta, H_{\mathbb{H}}(\cdot,t))$ in the sense of Definition 1.2.

A part of the main theorem in this paper (Theorem 5.4) is stated as follows.

Theorem 1.3. *Let $A \subset \mathbb{H}$ be an open subset and take an $(\mathcal{F}_t)_{t\geq 0}$-stopping time*

$$(1.6)\qquad \tau_A := \inf\left\{t \geq 0 \,\middle|\, A \not\subset \mathbb{H}_t^\eta\right\}.$$

Let τ be any $(\mathcal{F}_t)_{t\geq 0}$-stopping time such that $\tau \leq \tau_A$ a.s. Then, for any $f \in C_c^\infty(\mathbb{H})$ such that $\mathrm{supp}(f) \subset A$,

$$(1.7)\qquad (H_{\mathbb{H}}(\cdot,0), f) \overset{\text{(law)}}{=} (H_{\mathbb{H}}(\cdot,\tau), f) \quad \text{under } \mathbb{P},$$

if the driving process $(\boldsymbol{X}^{\mathbb{R}}(t))_{t\geq 0}$ is equal to the time changed version $\boldsymbol{Y}^{\mathbb{R}}(t) = (Y_1^{\mathbb{R}}(t), \ldots, Y_N^{\mathbb{R}}(t))$, $t \geq 0$ of the Dyson model on $\mathbb{R}$ which solves the following system of stochastic differential equations (SDEs) with $\kappa > 0$,

$$(1.8)\qquad dY_i^{\mathbb{R}}(t) = \sqrt{\kappa} dB_i(t) + 4 \sum_{1\leq j\leq N, j\neq i} \frac{dt}{Y_i^{\mathbb{R}}(t) - Y_j^{\mathbb{R}}(t)},$$

$t \geq 0, 1 \leq i \leq N$, where $(B_i(t))_{t\geq 0}$ are mutually independent one-dimensional standard Brownian motions starting from $B_i(0) = Y_i^{\mathbb{R}}(0) =: y_i^{\mathbb{R}} = \eta_i(0)$, $1 \leq i \leq N$, satisfying $y_1^{\mathbb{R}} < \cdots < y_N^{\mathbb{R}}$.

The Dyson model [12] is one of the most studied stochastic log-gases in one dimension, which is a dynamical version of the one-parameter ($\beta = 8/\kappa$) extension of the *Gaussian unitary ensemble* (*GUE*) of point processes studied in random matrix theory [13, 17]. It is also known that the multiple SLE driven by $(\boldsymbol{Y}^{\mathbb{R}}(t))_{t\geq 0}$ is an example of multiple SLEs that is defined in terms of an SLE partition function [4, 15].

It is possible that, for a certain choice of the open set $A \subset \mathbb{H}$ in Theorem 1.3, the stopping time τ has to be $\tau = 0$. Let us see that we can take $\tau > 0$ a.s. for a generic choice of A. For each $1 \leq i \leq N$, let U_i be a neighborhood of $y_i^{\mathbb{R}}$ in $\overline{\mathbb{H}}$ and suppose that $U_i \cap U_j = \emptyset$ if $i \neq j$. We call such U_i, $1 \leq i \leq N$ *localization neighborhoods.* Then, we define the first exit time of the SLE from the union of localization neighborhoods, $\boldsymbol{U} := \bigcup_{i=1}^N U_i$ by

$$\tau_{\boldsymbol{U}} = \inf \left\{ t \geq 0 \,\middle|\, (\mathbb{H} \setminus \boldsymbol{U}) \not\subseteq \mathbb{H}_t^\eta \right\}.$$

Note that we have $\tau_{\boldsymbol{U}} > 0$ a.s. since, otherwise, it contradicts the fact that $(\boldsymbol{Y}^{\mathbb{R}}(t))_{t\geq 0}$ is continuous in $t \geq 0$ [8, 14]. Notice that we can take the stopping time τ in Theorem 1.3 in such a way that $\tau \geq \tau_{\boldsymbol{U}}$ for some localization neighborhoods that are disjoint from A, if the closure of A in $\overline{\mathbb{H}}$ does not contain any of $y_i^{\mathbb{R}}$, $1 \leq i \leq N$. Therefore, we can take it so that $\tau > 0$ a.s. in such a generic case.

As we have already pointed out, it is not clear if a solution of the multiple Loewner equation generates a multi-slit. In our subsequent paper [19], we will prove under the assumption $\kappa \in (0, 8]$ that, if the driving process is the time changed version $\boldsymbol{Y}^{\mathbb{R}}(t) = (Y_1^{\mathbb{R}}(t), \ldots, Y_N^{\mathbb{R}}(t))$, $t \geq 0$ of the Dyson model as in Theorem 1.3, the solution generates a set of curves. We will also prove that the resulting curves $\{\eta_i\}_{i=1}^N$ are simple disjoint curves if $\kappa \in (0, 4]$, self-intersecting if $\kappa \in (4, 8)$, and space-filling if $\kappa = 8$.

The process $(H_{\mathbb{H}}(\cdot, t))_{t\geq 0}$ is a generalization of $(h \circ f_{\mathbb{H}_t^\eta} - \chi \arg f'_{\mathbb{H}_t^\eta})_{t\geq 0}$ considered by Miller and Sheffield [32, 25] as explained above. The equality (1.3) [10, 32, 25] has been extended to the equality (1.7) in Theorem 1.3, which we think of as the local coupling between a GFF and a multiple SLE.

We will also construct another GFF-valued process in the first orthant in $\mathbb{C}$; $\mathbb{O} := \{z \in \mathbb{C} : \operatorname{Re} z > 0, \operatorname{Im} z > 0\}$. There, a GFF on $\mathbb{O}$, denoted as $H_{\mathbb{O}}(\cdot, 0)$ is locally coupled with a multiple version of the *quadrant SLE* [33] defined on $\mathbb{O}$, which is driven by a stochastic log-gas defined on $S = \mathbb{R}_+ := \{x \in \mathbb{R} : x \geq 0\}$. This driving process is a dynamical version of the one-parameter ($\beta = 8/\kappa$) extension of the *chiral GUE* of point processes with parameter $\nu \in [0, \infty)$ studied in random matrix theory [20, 13], and we call it the *Bru–Wishart process* in this paper [34, 6]. We note that $(\mathbb{H}, H_{\mathbb{H}}(\cdot, 0)) \sim (\mathbb{O}, H_{\mathbb{O}}(\cdot, 0))$ in the sense of Definition 1.2.

Construction of such GFF-valued processes will be meaningful for the study of multiple SLEs. The main problem in defining a multiple SLE correctly in $D \subsetneq \mathbb{C}$ may be how to find a correct principle to choose a driving process $(\boldsymbol{X}^S(t))_{t \geq 0}$ defined on a part of the boundary $S \subset \partial D$ (e.g., conformal invariance, statistical mechanics consideration, reparameterization invariance, absolute continuity to the SLE with a single slit, commutation relations) [7, 4, 22, 15, 9]. In the present paper, we simply assume the form of SDEs for $(\boldsymbol{X}^S(t))_{t \geq 0}$ as

$$dX_i^S(t) = \sqrt{\kappa} dB_i(t) + F_i^S(\boldsymbol{X}^S(t))dt, \quad t \geq 0, \quad 1 \leq i \leq N, \tag{1.9}$$

where $(B_i(t))_{t \geq 0}, 1 \leq i \leq N$ are mutually independent one-dimensional standard Brownian motions, $\kappa > 0$, and $F_i^S(\boldsymbol{x}) \in \mathcal{C}^\infty(S^N \setminus \bigcup_{j \neq k}\{x_j = x_k\}), 1 \leq i \leq N$, which do not explicitly depend on t. Then, the equality (1.7) for a GFF-valued process on $D = \mathbb{H}$ determines the driving process $(\boldsymbol{X}^{\mathbb{R}}(t))_{t \geq 0}$ as (a time change of) the Dyson model $(\boldsymbol{Y}^{\mathbb{R}}(t))_{t \geq 0}$. That is, the local coupling between a GFF and a multiple SLE provides a new scheme to choose a driving process for a multiple SLE.

Notice again that $\arg z$ in (1.2) is the imaginary part of the complex analytic function $\log z$. Sheffield studied another type of distribution-valued random field on $\mathbb{H}$ given by [32]

$$\widetilde{h}(\cdot) := \widetilde{H}(\cdot) + \frac{2}{\sqrt{\kappa}} \mathrm{Re} \log(\cdot) = \widetilde{H}(\cdot) + \frac{2}{\sqrt{\kappa}} \log|\cdot|,$$

where $\widetilde{H}(\cdot)$ is a free boundary GFF on $\mathbb{H}$ and found that $\widetilde{h}(\cdot)$ is coupled with a backward SLE in the context of quantum gravity [11]. This coupling was later generalized in [18, 21] to the situations where backward multiple SLEs driven by stochastic log-gases play analogous roles as multiple SLEs did in the present work.

The present paper is organized as follows. We give brief reviews of stochastic log-gases in one dimension in Section 2 and the SLE both for a single-slit and a multi-slit in Section 3. In Section 4, we define a GFF with zero boundary condition on $D \subsetneq \mathbb{C}$ based on the Bochner–Minlos theorem. The construction of GFF-valued processes by locally coupling

GFFs with multiple SLEs driven by specified stochastic log-gases on S are given in Section 5 for $(D, S) = (\mathbb{H}, \mathbb{R})$ and $(\mathbb{O}, \mathbb{R}_+)$.

§2. One-dimensional Stochastic Log-Gases

2.1. Eigenvalue and singular-value processes

For $N \in \mathbb{N}$, let H_N and U_N be the space of $N \times N$ Hermitian matrices and the group of $N \times N$ unitary matrices, respectively. Consider complex-valued processes $(M_{ij}(t))_{t \geq 0}, 1 \leq i, j \leq N$ with the condition $\overline{M_{ji}(t)} = M_{ij}(t)$, where $\overline{z}$ denotes the complex conjugate of $z \in \mathbb{C}$. We consider an H_N-valued process by $M(t) = (M_{ij}(t))_{1 \leq i,j \leq N}$. For $S = \mathbb{R}$ and $\mathbb{R}_+$, define the Weyl chambers as $\mathbb{W}_N(S) := \{\boldsymbol{x} = (x_1, \dots, x_N) \in S^N : x_1 < \cdots < x_N\}$, and write their closures as $\overline{\mathbb{W}_N(S)} = \{\boldsymbol{x} \in S^N : x_1 \leq \cdots \leq x_N\}$. For each $t \geq 0$, there exists $U(t) = (U_{ij}(t))_{1 \leq i,j \leq N} \in \mathsf{U}_N$ such that it diagonalizes $M(t)$ as $U^\dagger(t) M(t) U(t) = \mathrm{diag}(\Lambda_1(t), \dots, \Lambda_N(t))$ with the eigenvalues $\{\Lambda_i(t)\}_{i=1}^N$ of $M(t)$, where $U^\dagger(t)$ is the Hermitian conjugate of $U(t)$; $U^\dagger_{ij}(t) = \overline{U_{ji}(t)}, 1 \leq i, j \leq N$, and we assume $\boldsymbol{\Lambda}(t) := (\Lambda_1(t), \dots, \Lambda_N(t)) \in \overline{\mathbb{W}_N(\mathbb{R})}$, $t \geq 0$. For $dM(t) := (dM_{ij}(t))_{1 \leq i,j \leq N}$, define a set of quadratic variations,

$$\Gamma_{ij,k\ell}(t) := \Big\langle (U^\dagger dMU)_{ij}, (U^\dagger dMU)_{k\ell} \Big\rangle_t, \quad 1 \leq i, j, k, \ell \leq N, \quad t \geq 0.$$

We write $\mathbf{1}_E$ for the indicator function of an event E; $\mathbf{1}_E = 1$ if E occurs, and $\mathbf{1}_E = 0$ otherwise. The following is proved [5, 20, 17]. See Section 4.3 of [1] for details of proof.

Proposition 2.1. *Assume that $(M_{ij}(t))_{t \geq 0}$, $1 \leq i$, $j \leq N$ are continuous semi-martingales. The eigenvalue process $(\boldsymbol{\Lambda}(t))_{t \geq 0}$ satisfies the following system of SDEs,*

$$d\Lambda_i(t) = d\mathcal{M}_i(t) + dJ_i(t), \quad t \geq 0, \quad 1 \leq i \leq N,$$

where $(\mathcal{M}_i(t))_{t \geq 0}, 1 \leq i \leq N$ are martingales with quadratic variations $\langle \mathcal{M}_i, \mathcal{M}_j \rangle_t = \int_0^t \Gamma_{ii,jj}(s) ds$, and $(J_i(t))_{t \geq 0}$, $1 \leq i \leq N$ are the processes with finite variations given by

$$dJ_i(t) = \sum_{j=1}^N \frac{\mathbf{1}_{\Lambda_i(t) \neq \Lambda_j(t)}}{\Lambda_i(t) - \Lambda_j(t)} \Gamma_{ij,ji}(t) dt + d\Upsilon_i(t).$$

Here $d\Upsilon_i(t)$ denotes the finite-variation part of $(U^\dagger(t) dM(t) U(t))_{ii}$, $t \geq 0$, $1 \leq i \leq N$.

We will show two basic examples of $M(t) \in \mathsf{H}_N, t \geq 0$ and applications of Proposition 2.1 [20]. Let $\nu \in \mathbb{N}_0 := \mathbb{N} \cup \{0\}$ and $(B_{ij}(t))_{t\geq 0}$, $(\widetilde{B}_{ij}(t))_{t\geq 0}$, $1 \leq i \leq N+\nu$, $1 \leq j \leq N$ be independent one-dimensional standard Brownian motions. For $1 \leq i \leq j \leq N$, put

$$S_{ij}(t) = \begin{cases} B_{ij}(t)/\sqrt{2}, & (i<j), \\ B_{ii}(t), & (i=j), \end{cases} \quad A_{ij}(t) = \begin{cases} \widetilde{B}_{ij}(t)/\sqrt{2}, & (i<j), \\ 0, & (i=j), \end{cases}$$

and let $S_{ij}(t) = S_{ji}(t)$ and $A_{ij}(t) = -A_{ji}(t)$, $t \geq 0$ for $1 \leq j < i \leq N$.

Example 2.1. Put $M_{ij}(t) = S_{ij}(t) + \sqrt{-1}A_{ij}(t)$, $t \geq 0$, $1 \leq i, j \leq N$. By definition, $\langle dM_{ij}, dM_{k\ell}\rangle_t = \delta_{i\ell}\delta_{jk}dt$, $t \geq 0$, $1 \leq i,j,k,\ell \leq N$. Hence, by unitarity of $U(t), t \geq 0$, we see that $\Gamma_{ij,k\ell}(t) = \delta_{i\ell}\delta_{jk}$, which gives $\langle d\mathcal{M}_i, d\mathcal{M}_j\rangle_t = \Gamma_{ii,jj}(t)dt = \delta_{ij}dt$ and $\Gamma_{ij,ji}(t) \equiv 1$, $t \geq 0$, $1 \leq i,j \leq N$. Then, Proposition 2.1 proves that the eigenvalue process $(\mathbf{\Lambda}(t))_{t\geq 0}$, satisfies the following system of SDEs with $\beta = 2$,

$$d\Lambda_i(t) = dB_i(t) + \frac{\beta}{2} \sum_{1\leq j\leq N, j\neq i} \frac{dt}{\Lambda_i(t) - \Lambda_j(t)}, \tag{2.1}$$

$t \geq 0, 1 \leq i \leq N$. Here, $(B_i(t))_{t\geq 0}, 1 \leq i \leq N$ are independent one-dimensional standard Brownian motions, which are different from $(B_{ij}(t))_{t\geq 0}$ and $(\widetilde{B}_{ij}(t))_{t\geq 0}$ used to define $(S_{ij}(t))_{t\geq 0}$ and $(A_{ij}(t))_{t\geq 0}$, $1 \leq i,j \leq N$.

Example 2.2. Consider an $(N+\nu) \times N$ rectangular-matrix-valued process given by $K(t) = (B_{ij}(t) + \sqrt{-1}\widetilde{B}_{ij}(t))_{1\leq i\leq N+\nu, 1\leq j\leq N}$, $t \geq 0$, and define an H_N-valued process by $M(t) = K^\dagger(t)K(t), t \geq 0$. The matrix $M(t)$ is positive semi-definite and hence the eigenvalues are non-negative; $\Lambda_i(t) \in \mathbb{R}_+$, $t \geq 0, 1 \leq i \leq N$. We see that the finite-variation part of $dM_{ij}(t)$ is equal to $2(N+\nu)\delta_{ij}dt$, $t \geq 0$, and $\langle dM_{ij}, dM_{k\ell}\rangle_t = 2(M_{i\ell}(t)\delta_{jk} + M_{kj}(t)\delta_{i\ell})dt$, $t \geq 0$, $1 \leq i,j,k,\ell \leq N$, which implies that $d\Upsilon_i(t) = 2(N+\nu)dt$, $\Gamma_{ij,ji}(t) = 2(\Lambda_i(t) + \Lambda_j(t))$, and $\langle d\mathcal{M}_i, d\mathcal{M}_j\rangle_t = \Gamma_{ii,jj}(t)dt = 4\Lambda_i(t)\delta_{ij}dt$, $t \geq 0$, $1 \leq i,j \leq N$. Then, we have the SDEs for eigenvalue processes,

$$\begin{aligned} d\Lambda_i(t) &= 2\sqrt{\Lambda_i(t)}d\widetilde{B}_i(t) \\ &\quad + \beta\left[(\nu+1) + 2\Lambda_i(t) \sum_{1\leq j\leq N, j\neq i} \frac{1}{\Lambda_i(t) - \Lambda_j(t)}\right]dt, \end{aligned} \tag{2.2}$$

$t \geq 0, 1 \leq i \leq N$ with $\beta = 2$, where $(\widetilde{B}_i(t))_{t\geq 0}, 1 \leq i \leq N$ are independent one-dimensional standard Brownian motions, which are different from $(B_{ij}(t))_{t\geq 0}$ and $(\widetilde{B}_{ij}(t))_{t\geq 0}$, $1 \leq i,j \leq N$, used above to define the rectangular-matrix-valued process $(K(t))_{t\geq 0}$. The positive roots of eigenvalues of $M(t)$ give the *singular values* of the rectangular matrix $K(t)$, which are denoted by $\mathcal{S}_i(t) = \sqrt{\Lambda_i(t)}, t \geq 0, 1 \leq i \leq N$. The system of SDEs for them is readily obtained from (2.2) as

$$
\begin{aligned}
(2.3) \qquad d\mathcal{S}_i(t) &= d\widetilde{B}_i(t) + \frac{\beta(\nu+1)-1}{2\mathcal{S}_i(t)} dt \\
&\quad + \frac{\beta}{2} \sum_{1\leq j\leq N, j\neq i} \left(\frac{1}{\mathcal{S}_i(t) - \mathcal{S}_j(t)} + \frac{1}{\mathcal{S}_i(t) + \mathcal{S}_j(t)} \right) dt,
\end{aligned}
$$

$t \geq 0, 1 \leq i \leq N$ with $\beta = 2$ and $\nu \in \mathbb{N}_0$.

Other examples of H_N-valued processes $(M(t))_{t\geq 0}$ are shown in [20], in which the eigenvalue processes following the SDEs (2.1), (2.2), and (2.3) with $\beta = 1$ and 4 are also shown.

2.2. 2D-Coulomb gases confined in 1D

In the next section, we will consider the SLE. Schramm used a parameter $\kappa > 0$ in order to parameterize time changes of a Brownian motion [30]. Accordingly, we relate the parameter β to κ by setting $\beta = 8/\kappa$, and perform a time change $t \to \kappa t$. Since $(B(\kappa t))_{t\geq 0} \overset{\text{(law)}}{=} (\sqrt{\kappa}B(t))_{t\geq 0}$, if we put $Y_i^{\mathbb{R}}(t) := \Lambda_i(\kappa t)$, $Y_i^{\mathbb{R}_+}(t) := \mathcal{S}_i(\kappa t)$, $t \geq 0$, $1 \leq i \leq N$, the system of SDEs (2.1) gives (1.8) and that of (2.3) gives

$$
\begin{aligned}
(2.4) \quad dY_i^{\mathbb{R}_+}(t) &= \sqrt{\kappa} d\widetilde{B}_i(t) + \frac{8(\nu+1)-\kappa}{2Y_i^{\mathbb{R}_+}(t)} dt \\
&\quad + 4 \sum_{1\leq j\leq N, j\neq i} \left(\frac{1}{Y_i^{\mathbb{R}_+}(t) - Y_j^{\mathbb{R}_+}(t)} + \frac{1}{Y_i^{\mathbb{R}_+}(t) + Y_j^{\mathbb{R}_+}(t)} \right) dt,
\end{aligned}
$$

$t \geq 0, 1 \leq i \leq N$, where $\nu \geq 0$. In the present paper, we call $(\boldsymbol{Y}^{\mathbb{R}}(t))_{t\geq 0}$ the $(8/\kappa)$-*Dyson model* and $(\boldsymbol{Y}^{\mathbb{R}_+}(t))_{t\geq 0}$ the $(8/\kappa, \nu)$-*Bru–Wishart process*, respectively. The above systems of SDEs for $(\boldsymbol{Y}^S(t))_{t\geq 0}$ can be written as

$$
dY_i^S(t) = \sqrt{\kappa} dB_i(t) + \left. \frac{\partial \phi^S(\boldsymbol{x})}{\partial x_i} \right|_{\boldsymbol{x}=\boldsymbol{Y}^S(t)} dt, \quad t \geq 0, \quad 1 \leq i \leq N,
$$

$S = \mathbb{R}$ or $\mathbb{R}_+$, when we introduce the following logarithmic potentials,

$$(2.5)\qquad \phi^S(\boldsymbol{x}) := \begin{cases} 4 \displaystyle\sum_{1 \le i < j \le N} \log(x_j - x_i), & \text{for } S = \mathbb{R}, \\ 4 \displaystyle\sum_{1 \le i < j \le N} \Big[\log(x_j - x_i) + \log(x_j + x_i)\Big] & \\ \qquad + \dfrac{8(\nu+1) - \kappa}{2} \displaystyle\sum_{1 \le i \le N} \log x_i, & \text{for } S = \mathbb{R}_+. \end{cases}$$

In this sense, the $(8/\kappa)$-Dyson model and the $(8/\kappa, \nu)$-Bru–Wishart process are regarded as *stochastic log-gases* in one dimension [13]. Since the logarithmic potential describes the two-dimensional Coulomb law in electrostatics, the present processes are also considered as stochastic models of *2D-Coulomb gases confined in 1D*.

§3. Multiple Schramm–Loewner Evolution

3.1. Loewner equations for a single-slit and a multi-slit

Let D be a simply connected domain $D \subsetneq \mathbb{C}$ with boundary ∂D. We consider a slit in D, which is defined as a simple curve $\eta = \{\eta(t) : t \in (0, \infty)\} \subset D$; $\eta(s) \neq \eta(t)$ for $s \neq t$ and suppose that $\lim_{t \to 0} \eta(t) =: \eta(0) \in \partial D$. Let $\eta(0, t] := \{\eta(s) : s \in (0, t]\}$ and $D_t^\eta := D \setminus \eta(0, t], t \in (0, \infty)$ with $D_0^\eta := D$. The Loewner theory describes the slit η by encoding it into a time-dependent analytic function $(g_{D_t^\eta})_{t \ge 0}$ such that

$$g_{D_t^\eta} : \text{conformal map } D_t^\eta \to D, \quad t \in [0, \infty).$$

Let us apply the Loewner theory to the case of $D = \mathbb{H}$, in which $\eta(0) \in \mathbb{R}$ and $\eta \subset \mathbb{H}$. Let $\mathbb{H}_t^\eta := \mathbb{H} \setminus \eta(0, t]$, $t > 0$ and $\mathbb{H}_0^\eta := \mathbb{H}$. Then, for each time $t \ge 0$, $\mathbb{H}_t^\eta$ is a simply connected domain in $\mathbb{C}$ and there exists a unique conformal map $\mathbb{H}_t^\eta \to \mathbb{H}$ satisfying the condition $g_{\mathbb{H}_t^\eta}(z) = z + \mathrm{hcap}(\eta(0, t])/z + \mathrm{O}(|z|^{-2})$ as $z \to \infty$, $t > 0$, in which the coefficient of z is unity and no constant term appears. This is called the *hydrodynamic normalization* and $\mathrm{hcap}(\eta(0, t])$ gives the *half-plane capacity* of $\eta(0, t]$. The following can be proved (see, for instance, [23, Proposition 4.4]).

Theorem 3.1. *Let η be a slit in $\mathbb{H}$ such that* $\mathrm{hcap}(\eta(0, t]) = 2t$, $t > 0$. *Then, the solution $(g_t)_{t \ge 0}$ of the differential equation (chordal Loewner equation)*

$$(3.1)\qquad \frac{dg_t(z)}{dt} = \frac{2}{g_t(z) - V(t)}, \quad t \ge 0, \quad g_0(z) = z,$$

where

$$V(t) = g_{\mathbb{H}_t^\eta}(\eta(t)) := \lim_{z \to \eta(t), z \in \mathbb{H}_t^\eta} g_{\mathbb{H}_t^\eta}(z), \quad t \geq 0,$$

coincides with $(g_{\mathbb{H}_t^\eta})_{t \geq 0}$.

Theorem 3.1 can be extended to the situation such that η is given by a multi-slit $\bigcup_{i=1}^N \eta_i \subset \mathbb{H}$ and $\mathbb{H}_t^\eta := \mathbb{H} \setminus \bigcup_{i=1}^N \eta_i(0,t]$, $t > 0$ with $\mathbb{H}_0^\eta := \mathbb{H}$.

Theorem 3.2. *For* $N \in \mathbb{N}$, *let* $\bigcup_{i=1}^N \eta_i$ *be a multi-slit in* $\mathbb{H}$ *such that* $\mathrm{hcap}\big(\bigcup_{i=1}^N \eta(0,t]\big) = 2t$, $t > 0$. *Then, there exists a set of weight functions* $(\lambda_i(t))_{t \geq 0}$, $1 \leq i \leq N$ *satisfying* $\lambda_i(t) \geq 0, 1 \leq i \leq N$, $\sum_{i=1}^N \lambda_i(t) = 1$, $t \geq 0$ *such that the solution* $(g_t)_{t \geq 0}$ *of the differential equation (multiple chordal Loewner equation)*

$$\frac{dg_t(z)}{dt} = \sum_{i=1}^N \frac{2\lambda_i(t)}{g_t(z) - V_i(t)}, \quad t \geq 0, \quad g_0(z) = z, \tag{3.2}$$

where

$$V_i(t) = g_{\mathbb{H}_t^\eta}(\eta_i(t)) := \lim_{z \to \eta_i(t), z \in \mathbb{H}_t^\eta} g_{\mathbb{H}_t^\eta}(z), \quad t \geq 0, \quad 1 \leq i \leq N,$$

coincides with $(g_{\mathbb{H}_t^\eta})_{t \geq 0}$.

Proof. We parameterize each curve η_i separately by $\eta_i : (0, \infty) \to \mathbb{H}; s_i \mapsto \eta_i(s_i)$ so that $\mathrm{hcap}\big(\bigcup_{j=1}^N \eta_j(0, s_j]\big)$ is differentiable with respect to s_i, $1 \leq i \leq N$. For each $\underline{s} = (s_1, \dots, s_N) \in [0, \infty)^N$, we set $\mathbb{H}_{\underline{s}}^\eta = \mathbb{H} \setminus \bigcup_{i=1}^N \eta_i(0, s_i]$. A similar argument as in [23, Proposition 4.4] shows that the family of conformal maps $g_{\underline{s}} = g_{\mathbb{H}_{\underline{s}}^\eta}$, $\underline{s} \in [0, \infty)^N$ satisfies partial differential equations

$$\frac{\partial g_{\underline{s}}(z)}{\partial s_i} = \frac{1}{g_{\underline{s}}(z) - V_i(\underline{s})} \frac{\partial}{\partial s_i} \mathrm{hcap}\left(\bigcup_{j=1}^N \eta_j(0, s_j] \right), \quad 1 \leq i \leq N,$$

where

$$V_i(\underline{s}) = g_{\mathbb{H}_{\underline{s}}^\eta}(\eta_i(s_i)) := \lim_{z \to \eta_i(s_i), z \in \mathbb{H}_{\underline{s}}^\eta} g_{\mathbb{H}_{\underline{s}}^\eta}(z).$$

We have these parameters dependent on a single parameter so that $s_i = s_i(t)$, $t \geq 0$, $1 \leq i \leq N$ are increasing and differentiable in t, and write $\underline{s} = \underline{s}(t)$, $t \geq 0$. Then, we may understand $\mathbb{H}_t^\eta = \mathbb{H}_{\underline{s}(t)}^\eta$, $g_t = g_{\underline{s}(t)}$, and $V_i(t) = V_i(\underline{s}(t))$, $1 \leq i \leq N$, $t \geq 0$. Furthermore, we impose a condition that $\mathrm{hcap}\big(\bigcup_{j=1}^N \eta_j(0, s_j(t)]\big) = 2t$. Then, the family of conformal maps $(g_t)_{t\geq 0}$ satisfies the desired differential equation (3.2), where

$$\lambda_i(t) = \frac{1}{2}\frac{\partial}{\partial s_i}\mathrm{hcap}\left(\bigcup_{j=1}^N \eta_j(0, s_j]\right)\Bigg|_{\underline{s}=\underline{s}(t)} \frac{ds_i(t)}{dt}, \quad 1 \leq i \leq N, \quad t \geq 0$$

are subject to the constraint $\sum_{i=1}^N \lambda_i(t) = 1$, $t \geq 0$. Q.E.D.

The multiple chordal Loewner equation (3.2) for $D = \mathbb{H}$ can be mapped to other simply connected proper subdomains of $\mathbb{C}$ by conformal maps. Here, we consider a conformal map $\widehat{\varphi}(z) = z^2 : \mathbb{O} \to \mathbb{H}$. We set $\widehat{g}_t(z) = \sqrt{g_t(z^2) + c(t)}$, $t \geq 0$ with a function of time $c(t), t \geq 0$. Then, we can see that (3.2) is transformed to

$$\text{(3.3)} \quad \frac{d\widehat{g}_t(z)}{dt} = \sum_{i=1}^N \left(\frac{2\widehat{\lambda}_i(t)}{\widehat{g}_t(z) - \widehat{V}_i(t)} + \frac{2\widehat{\lambda}_i(t)}{\widehat{g}_t(z) + \widehat{V}_i(t)}\right) + \frac{2\widehat{\lambda}_0(t)}{\widehat{g}_t(z)}, \ t \geq 0,$$

$\widehat{g}_0(z) = z \in \mathbb{O}$, where $\widehat{V}_i(t) = \sqrt{V_i(t) + c(t)}$, $t \geq 0, 1 \leq i \leq N$ and $2\sum_{i=1}^N \widehat{\lambda}_i(t) + \widehat{\lambda}_0(t) = (1/4)dc(t)/dt, t \geq 0$. Here, we can assume that $\widehat{V}_i(t) \in \mathbb{R}_+$ by a proper choice of the function $c(t)$, $t \geq 0$. The equation (3.3) can be regarded as a multi-slit version of the *quadrant Loewner equation* studied in [33]. The solution of (3.3) gives a conformal map $\widehat{g}_t = g_{\mathbb{O}_t^\eta} : \mathbb{O}_t^\eta \to \mathbb{O}$, where $\mathbb{O}_t^\eta := \mathbb{O} \setminus \bigcup_{i=1}^N \eta_i(0, t]$, $t > 0$, $\mathbb{O}_0^\eta := \mathbb{O}$, and $g_{\mathbb{O}_t^\eta}(\eta_i(t)) = \widehat{V}_i(t) \in \mathbb{R}_+$, $t \geq 0$, $1 \leq i \leq N$.

3.2. SLE

So far we have considered the problem in which given a single slit $\eta(0, t], t > 0$ or a multi-slit $\bigcup_{i=1}^N \eta(0, t], t > 0$ in $\mathbb{H}$, the time-evolution of the conformal map from $\mathbb{H}_t^\eta$ to $\mathbb{H}$, $t \geq 0$ is asked. The answers are given by Theorem 3.1 and Theorem 3.2. For $\mathbb{H}$ with a single slit, Schramm considered the inverse problem in a probabilistic setting [30]. He first asked a suitable family of driving stochastic processes $(X(t))_{t\geq 0}$ on $\mathbb{R}$. Then, he asked the probability law of the random slit η in $\mathbb{H}$ that is determined by the solution $g_t = g_{\mathbb{H}_t^\eta}, t \geq 0$ of the Loewner equation (3.1) via $X(t) = g_{\mathbb{H}_t^\eta}(\eta(t)), t \geq 0$. Schramm argued that the conformal invariance and the domain Markov property of the law of the curve imply that the driving process $(X(t))_{t\geq 0}$ should be $(B(\kappa t))_{t\geq 0} \overset{\text{(law)}}{=} (\sqrt{\kappa}B(t))_{t\geq 0}$ with a parameter $\kappa > 0$. The solution of the chordal Loewner equation (3.1) driven by $X(t) = \sqrt{\kappa}B(t), t \geq 0$ is called the *chordal Schramm–Loewner evolution* with parameter $\kappa > 0$ and is written as chordal SLE(κ) for short.

The following was proved by Lawler, Schramm, and Werner [24] for $\kappa = 8$ and by Rohde and Schramm [29] for $\kappa \neq 8$.

Proposition 3.3. *A chordal SLE(κ) $(g_{\mathbb{H}_t^\eta})_{t\geq 0}$ determines a continuous curve $\eta = \{\eta(t) : t \in [0,\infty)\} \subset \overline{\mathbb{H}}$ a.s.*

In this inverse problem, the domain $\mathbb{H}_t^\eta$ is defined as the unbounded component of $\mathbb{H}\backslash\eta(0,t]$ so that the solution g_t is a conformal map from $\mathbb{H}_t^\eta$ to $\mathbb{H}$ at each $t > 0$, which verifies writing the solution as $g_{\mathbb{H}_t^\eta} = g_t$, $t \geq 0$.

The continuous curve η determined by an SLE(κ) is called an *SLE(κ) curve.* The probability law of an SLE(κ) curve qualitatively depends on κ. When $\kappa \in (0,4]$, the SLE(κ) curve is a simple curve in $\mathbb{H}$. It becomes self-intersecting and can touch the real axis $\mathbb{R}$ when $\kappa > 4$, and becomes a space-filling curve when $\kappa \geq 8$ (see, for instance, [23, 17]).

3.3. Multiple SLE

For simplicity, we assume that $\lambda_i(t) \equiv 1/N, t \geq 0, 1 \leq i \leq N$ in (3.2) in Theorem 3.2. Then, by a simple time change $t/N \to t$ associated with a change of notation, $g_{Nt} \to g_{\mathbb{H}_t^\eta}$, the multiple chordal Loewner equation is written as (1.4). Then, we ask what is a suitable family of driving stochastic processes of N particles $(\boldsymbol{X}^{\mathbb{R}}(t))_{t\geq 0}$ on $\mathbb{R}$.

Bauer, Bernard, and Kytölä [4] and Graham [15] argued that $X_i^{\mathbb{R}}(t)$, $t \geq 0$, $1 \leq i \leq N$ are semi-martingales and the quadratic variations should be given by $\langle dX_i^{\mathbb{R}}, dX_j^{\mathbb{R}}\rangle_t = \kappa\delta_{ij}dt, t \geq 0$, $1 \leq i,j \leq N$ with $\kappa > 0$. Then, we can assume that the system of SDEs for $(\boldsymbol{X}^{\mathbb{R}}(t))_{t\geq 0}$ is in the form (1.9).

In the orthant system (3.3), we put $\widehat{\lambda}_i(t) \equiv r/(2N)$, $t \geq 0, r \in (0,1], 1 \leq i \leq N$, $dc(t)/dt \equiv 4, t \geq 0$, and perform a time change $rt/(2N) \to t$ associated with a change of notation $\widehat{g}_{2Nt/r} \to g_{\mathbb{O}_t^\eta}$. Then, the multiple Loewner equation in $\mathbb{O}$ is written as

$$(3.4)\quad \frac{dg_{\mathbb{O}_t^\eta}(z)}{dt} = \sum_{i=1}^N \left(\frac{2}{g_{\mathbb{O}_t^\eta}(z) - X_i^{\mathbb{R}_+}(t)} + \frac{2}{g_{\mathbb{O}_t^\eta}(z) + X_i^{\mathbb{R}_+}(t)} \right) + \frac{4\delta}{g_{\mathbb{O}_t^\eta}(z)},$$

$t \geq 0$ with $g_{\mathbb{O}_0^\eta}(z) = z \in \mathbb{O}$, where $\delta := N(1-r)/r \geq 0$. We assume that the system of SDEs for $\boldsymbol{X}^{\mathbb{R}_+}(t) \in (\mathbb{R}_+)^N, t \geq 0$ is in the form (1.9).

Analogously to the case of the SLE for a single slit, in both cases of $D = \mathbb{H}$ and $\mathbb{O}$, we find a family of domains $(D_t^\eta \subseteq D)_{t\geq 0}$ such that $g_{D_t^\eta}$ is a conformal map from D_t^η to D at each $t \geq 0$. (See Remark 1.4 (a) in [19].)

§4. Gaussian Free Field with Zero Boundary Condition

4.1. Bochner–Minlos Theorem

Let $D \subsetneq \mathbb{C}$ be a simply connected domain. Consider the real L^2 space with the inner product, $(f,g) := \int_D f(z)g(z)d\mu(z)$, $f,g \in L^2(D)$, where $\mu(z)$ is the Lebesgue measure on $\mathbb{C}$; $d\mu(z) = \sqrt{-1}dzd\overline{z}/2$. Let Δ be the Dirichlet Laplacian acting on $L^2(D)$. In the present Subsection 4.1 we assume that D is bounded. Then $-\Delta$ has positive discrete eigenvalues so that $-\Delta e_n = \lambda_n e_n$, $e_n \in L^2(D)$, $n \in \mathbb{N}$. We assume that the eigenvalues are labeled in the non-decreasing order; $0 < \lambda_1 \leq \lambda_2 \leq \cdots$. The system of eigenfunctions $\{e_n\}_{n\in\mathbb{N}}$ forms a CONS of $L^2(D)$. The asymptotic behavior of eigenvalues obeys *Weyl's formula*; $\lim_{n\to\infty} \lambda_n/n = \mathrm{O}(1)$.

For $f,g \in \mathcal{C}_{\mathrm{c}}^\infty(D)$, the *Dirichlet inner product* is defined by

$$(f,g)_\nabla := \frac{1}{2\pi}\int_D (\nabla f)(z)\cdot(\nabla g)(z)d\mu(z). \tag{4.1}$$

The Hilbert space completion of $\mathcal{C}_{\mathrm{c}}^\infty(D)$ with respect to $(\cdot,\cdot)_\nabla$ will be denoted by $W(D)$. We write $\|f\|_\nabla = \sqrt{(f,f)_\nabla}, f \in W(D)$. If we set $u_n = \sqrt{2\pi/\lambda_n}\, e_n, n \in \mathbb{N}$, then, by integration by parts, we have $(u_n,u_n)_\nabla = (u_n,(-\Delta)u_m)/(2\pi) = \delta_{nm}$, $n,m \in \mathbb{N}$. Therefore, $\{u_n\}_{n\in\mathbb{N}}$ forms a CONS of $W(D)$.

Let $\widehat{\mathcal{H}}(D)$ be the space of formal infinite series in $\{u_n\}_{n\in\mathbb{N}}$, which is obviously isomorphic to $\mathbb{R}^{\mathbb{N}}$ by setting $\widehat{\mathcal{H}}(D) \ni \sum_{n\in\mathbb{N}} f_n u_n \mapsto (f_n)_{n\in\mathbb{N}} \in \mathbb{R}^{\mathbb{N}}$. As a subspace of $\widehat{\mathcal{H}}(D)$, $W(D)$ is isomorphic to $\ell^2(\mathbb{N}) \subset \mathbb{R}^{\mathbb{N}}$. For two formal series $f = \sum_{n\in\mathbb{N}} f_n u_n$, $g = \sum_{n\in\mathbb{N}} g_n u_n \in \widehat{\mathcal{H}}(D)$ such that $\sum_{n\in\mathbb{N}} |f_n g_n| < \infty$, we define their pairing as $(f,g)_\nabla := \sum_{n\in\mathbb{N}} f_n g_n$. In the case when $f,g \in W(D)$, their pairing, of course, coincides with the Dirichlet inner product (4.1).

Notice that, for any $a \in \mathbb{R}$, the operator $(-\Delta)^a$ acts on $\widehat{\mathcal{H}}(D)$ as $(-\Delta)^a \sum_{n\in\mathbb{N}} f_n u_n := \sum_{n\in\mathbb{N}} \lambda_n^a f_n u_n$, $(f_n)_{n\in\mathbb{N}} \in \mathbb{R}^{\mathbb{N}}$. Using this fact, we define $\mathcal{H}_a(D) := (-\Delta)^a W(D)$, $a \in \mathbb{R}$, each of which is a Hilbert space with the inner product $\langle f,g\rangle_a := ((-\Delta)^{-a}f,(-\Delta)^{-a}g)_\nabla$, $f,g \in \mathcal{H}_a(D)$. We can prove that $\mathcal{H}_a(D) \subset \mathcal{H}_b(D)$ for $a < b$ using Weyl's formula for $\{\lambda_n\}_{n\in\mathbb{N}}$, and that the dual Hilbert space of $\mathcal{H}_a(D)$ is given by $\mathcal{H}_{-a}(D)$ (see [2]).

Remark 4.1. Since

$$\langle f,g\rangle_{1/2} = \left((-\Delta)^{-1/2}f,(-\Delta)^{-1/2}g\right)_\nabla = (f,g)/(2\pi), \quad f,g \in \mathcal{H}_{1/2}(D),$$

$\mathcal{H}_{1/2}(D) \simeq L^2(D)$. This implies that the members of $\mathcal{H}_a(D)$ with $a > 1/2$ cannot be functions, but are distributions.

Define $\mathcal{E}(D) := \bigcup_{a>1/2} \mathcal{H}_a(D)$. Then, its dual Hilbert space is identified with $\mathcal{E}(D)^* := \bigcap_{a<-1/2} \mathcal{H}_a(D)$ and $\mathcal{E}(D)^* \subset W(D) \subset \mathcal{E}(D)$ is established. Here $(\mathcal{E}(D)^*, W(D), \mathcal{E}(D))$ is called a *Gel'fand triple*. We set $\Sigma_{\mathcal{E}(D)} = \sigma(\{(\cdot, f)_\nabla : f \in \mathcal{E}(D)^*\})$. On such a setting, the following is proved. This theorem is called the *Bochner–Minlos theorem* [16, 31, 2].

Theorem 4.1 (Bochner–Minlos theorem). *Let ψ be a continuous function of positive type on $W(D)$ such that $\psi(0) = 1$. Then there exists a unique probability measure $\mathbf{P}$ on $(\mathcal{E}(D), \Sigma_{\mathcal{E}(D)})$ such that $\psi(f) = \int_{\mathcal{E}(D)} e^{\sqrt{-1}(h,f)_\nabla} \mathbf{P}(dh)$ for $f \in \mathcal{E}(D)^*$.*

Under certain conditions on ψ, the domain of the random functional h in the above formula can be extended from $\mathcal{E}(D)^*$ to $W(D)$. It is easy to verify that the functional $\Psi(f) := e^{-\|f\|_\nabla^2/2}$ satisfies the conditions. Then, the following is established with the probability measure $\mathbf{P}$ on $(\mathcal{E}(D), \Sigma_{\mathcal{E}(D)})$,

$$\int_{\mathcal{E}(D)} e^{\sqrt{-1}(h,f)_\nabla} \mathbf{P}(dh) = e^{-\|f\|_\nabla^2/2} \quad \text{for } f \in W(D). \tag{4.2}$$

Definition 4.2 (zero-boundary GFF). *A Gaussian free field (GFF) with zero boundary condition (zero-boundary GFF) is defined as a pair $((\Omega^{\mathrm{GFF}}, \mathcal{F}^{\mathrm{GFF}}, \mathbb{P}^{\mathrm{GFF}}), H)$ of a probability space $(\Omega^{\mathrm{GFF}}, \mathcal{F}^{\mathrm{GFF}}, \mathbb{P}^{\mathrm{GFF}})$ and an isometry $H : W(D) \to L^2(\Omega^{\mathrm{GFF}}, \mathcal{F}^{\mathrm{GFF}}, \mathbb{P}^{\mathrm{GFF}})$ such that each $H(f)$, $f \in W(D)$ is a centered Gaussian random variable.*

For each $f \in W(D)$, we write $(H, f)_\nabla \in L^2(\mathcal{E}(D), \Sigma_{\mathcal{E}(D)}, \mathbf{P})$ for the random variable defined by $h \mapsto (h, f)_\nabla$, $h \in \mathcal{E}(D)$. Then (4.2) ensures that the pair of $((\mathcal{E}(D), \Sigma_{\mathcal{E}(D)}, \mathbf{P}), H)$ gives a GFF with zero boundary condition. We often just call H a zero-boundary GFF without referring to the probability space $(\Omega^{\mathrm{GFF}}, \mathcal{F}^{\mathrm{GFF}}, \mathbb{P}^{\mathrm{GFF}}) = (\mathcal{E}(D), \Sigma_{\mathcal{E}(D)}, \mathbf{P})$.

4.2. Conformal invariance of a zero-boundary GFF

Assume that $D, \widetilde{D} \subsetneq \mathbb{C}$ are simply connected domains and let $\varphi : \widetilde{D} \to D$ be a conformal map.

Lemma 4.3. *The Dirichlet inner product* (4.1) *is conformally invariant. That is, for $f, g \in \mathcal{C}_c^\infty(D)$,*

$$\int_D (\nabla f)(z) \cdot (\nabla g)(z) d\mu(z) = \int_{\widetilde{D}} (\nabla(f \circ \varphi))(z) \cdot (\nabla(g \circ \varphi))(z) d\mu(z).$$

From the above lemma, we see that $\varphi^* : W(D) \ni f \mapsto f \circ \varphi \in W(\widetilde{D})$ is an isomorphism. This allows one to consider a GFF on an unbounded domain. Namely, if $\widetilde{D}$ is bounded on which a zero-boundary GFF is defined, but D is unbounded, we can define a family $\{(\varphi_* H, f)_\nabla : f \in W(D)\}$ by $(\varphi_* H, f)_\nabla := (H, \varphi^* f)_\nabla, f \in W(D)$ so that we have the covariance structure,

$$\mathbb{E}^{\mathrm{GFF}}\Big[(\varphi_* H, f)_\nabla (\varphi_* H, g)_\nabla\Big] = (\varphi^* f, \varphi^* g)_\nabla = (f, g)_\nabla, \quad f, g \in W(D),$$

where $\mathbb{E}^{\mathrm{GFF}}$ is the expectation value with respect to $\mathbb{P}^{\mathrm{GFF}}$. Relying on the formal computation,

$$\begin{aligned}(\varphi_* H, f)_\nabla = (H, \varphi^* f)_\nabla &= \frac{1}{2\pi}\int_{\widetilde{D}} (\nabla H)(z) \cdot (\nabla f \circ \varphi)(z) d\mu(z) \\ &= \frac{1}{2\pi}\int_D (\nabla H \circ \varphi^{-1})(z) \cdot (\nabla f)(z) d\mu(z),\end{aligned}$$

we understand the equality $\varphi_* H = H \circ \varphi^{-1}$. By the fact shown above that the covariance structure does not change under a conformal map φ, we say *a zero-boundary GFF is conformally invariant.*

4.3. Green's function of a zero-boundary GFF

Assume that $D \subsetneq \mathbb{C}$ is a simply connected domain. In the previous subsections, we have constructed a family $\{(H, f)_\nabla : f \in W(D)\}$ of random variables whose covariance structure is given by

$$\mathbb{E}^{\mathrm{GFF}}\Big[(H, f)_\nabla (H, g)_\nabla\Big] = (f, g)_\nabla, \quad f, g \in W(D).$$

By a formal integration by parts, we see that

$$\begin{aligned}(H, f)_\nabla &= \frac{1}{2\pi}\int_D (\nabla H)(z) \cdot (\nabla f)(z) d\mu(z) = \frac{1}{2\pi}\int_D H(z)(-\Delta f)(z) d\mu(z) \\ &= \frac{1}{2\pi}(H, (-\Delta) f).\end{aligned}$$

Motivated by this observation, we define

$$(H, f) := 2\pi (H, (-\Delta)^{-1} f)_\nabla \quad \text{for } f \in \mathsf{D}((-\Delta)^{-1}), \tag{4.3}$$

where $\mathsf{D}((-\Delta)^{-1})$ denotes the domain of $(-\Delta)^{-1}$ in $W(D)$. The action of $(-\Delta)^{-1}$ is expressed as an integral operator as

$$((-\Delta)^{-1} f)(z) = \frac{1}{2\pi}\int_D G_D(z, w) f(w) d\mu(w) \quad \text{a.e. } z \in D,$$

$f \in \mathsf{D}((-\Delta)^{-1})$, where the integral kernel G_D is known as *the Green's function* of D under the Dirichlet boundary condition: $G_D(z, w) =$

$0, w \in D$ if $z \in \partial D$. Hence the covariance of (H, f) and (H, g) with $f, g \in \mathsf{D}((-\Delta)^{-1})$ is written as

$$\mathbb{E}^{\mathrm{GFF}}[(H,f)(H,g)] = \int_{D\times D} f(z) G_D(z,w) g(w) d\mu(z) d\mu(w). \tag{4.4}$$

When we symbolically write

$$(H,f) = \int_D H(z) f(z) d\mu(z), \quad f \in \mathsf{D}((-\Delta)^{-1}),$$

the covariance structure can be expressed as

$$\mathbb{E}^{\mathrm{GFF}}[H(z)H(w)] = G_D(z,w), \quad z, w \in D, \quad z \neq w.$$

The conformal invariance of a zero-boundary GFF implies that for a conformal map $\varphi : \widetilde{D} \to D$, we have the equality,

$$G_{\widetilde{D}}(z,w) = G_D(\varphi(z), \varphi(w)), \quad z, w \in \widetilde{D}, \quad z \neq w.$$

Example 4.1. When $D = \mathbb{H}$,

$$G_{\mathbb{H}}(z,w) = \log \left| \frac{z - \overline{w}}{z - w} \right|, \quad z, w \in \mathbb{H}, \quad z \neq w.$$

Example 4.2. When $D = \mathbb{O}$,

$$G_{\mathbb{O}}(z,w) = \log \left| \frac{(z - \overline{w})(z + \overline{w})}{(z - w)(z + w)} \right|, \quad z, w \in \mathbb{O}, \quad z \neq w.$$

From the formula (4.4), we see that $\mathcal{C}_{\mathrm{c}}^{\infty}(D) \subset \mathsf{D}((-\Delta)^{-1})$. In the following, we will consider the family of random variables $\{(H, f) : f \in \mathcal{C}_{\mathrm{c}}^{\infty}(D)\}$ to characterize a GFF H.

§5. Gaussian Free Fields Coupled with Multiple SLEs

In this section, we take a probability space $(\Omega, \mathcal{F}, \mathbb{P})$ on which a zero-boundary GFF H and a multiple SLE $(g_{D_t^\eta})_{t\geq 0}$ are defined in such a way that they are independent, and a filtration $(\mathcal{F}_t)_{t\geq 0}$ to which the multiple SLE is adapted (see Section 1 for a precise setting). We fix an open set $A \subset D$, where $D = \mathbb{H}$ or $\mathbb{O}$. Then the $(\mathcal{F}_t)_{t\geq 0}$-stopping time τ_A is defined in exactly the same expression as (1.6). We also take an $(\mathcal{F}_t)_{t\geq 0}$-stopping time τ such that $\tau \leq \tau_A$ a.s. as in Theorem 1.3.

5.1. Zero-boundary GFF transformed by a multiple SLE

Here, we write the zero-boundary GFF defined on $D = \mathbb{H}$ or $\mathbb{O}$ as H_D. Consider the transformation of H_D by the multiple SLE, $H_{D_t^\eta} := H_D \circ g_{D_t^\eta}, 0 \leq t \leq \tau$ on D_τ^η. By the conformal invariance, the Green's function of $H_{D_t^\eta}, 0 \leq t \leq \tau$ is given by $G_{D_t^\eta}(z,w) = G_D(g_{D_t^\eta}(z), g_{D_t^\eta}(w))$, $z, w \in D_\tau^\eta$, $z \neq w$, $0 \leq t \leq \tau$. The following is obtained.

Lemma 5.1. *For $D = \mathbb{H}$ and $\mathbb{O}$, the increments of $G_{D_t^\eta}(z,w)$, $z, w \in A$ in time $0 \leq t \leq \tau$ are given as*

$$dG_{\mathbb{H}_t^\eta}(z,w) = -\sum_{i=1}^N \operatorname{Im} \frac{2}{g_{\mathbb{H}_t^\eta}(z) - X_i^{\mathbb{R}}(t)} \operatorname{Im} \frac{2}{g_{\mathbb{H}_t^\eta}(w) - X_i^{\mathbb{R}}(t)} dt,$$

$$\begin{aligned} dG_{\mathbb{O}_t^\eta}(z,w) = -\sum_{i=1}^N \operatorname{Im} & \left(\frac{2}{g_{\mathbb{O}_t^\eta}(z) - X_i^{\mathbb{R}_+}(t)} - \frac{2}{g_{\mathbb{O}_t^\eta}(z) + X_i^{\mathbb{R}_+}(t)} \right) \\ & \times \operatorname{Im} \left(\frac{2}{g_{\mathbb{O}_t^\eta}(w) - X_i^{\mathbb{R}_+}(t)} - \frac{2}{g_{\mathbb{O}_t^\eta}(w) + X_i^{\mathbb{R}_+}(t)} \right) dt. \end{aligned}$$

Proof. Using the explicit expressions of the Green's functions given in Examples 4.1 and 4.2 and the multiple Loewner equations (1.4) and (3.4), the increments of $(G_{D_t^\eta})_{0 \leq t \leq \tau}$ with $D = \mathbb{H}$ and $\mathbb{O}$ are calculated. The above expressions are obtained using the equality $\operatorname{Re} \zeta \overline{\omega} - \operatorname{Re} \zeta \omega = 2 \operatorname{Im} \zeta \operatorname{Im} \omega$ for $\zeta, \omega \in \mathbb{C}$. Q.E.D.

5.2. $\mathbb{C}$-valued logarithmic potentials and martingales

We have remarked in Section 2.2 that the Dyson model and the Bru–Wishart process studied in random matrix theory can be regarded as stochastic log-gasses defined on a line $S = \mathbb{R}$ and a half-line $S = \mathbb{R}_+$, respectively. There, the logarithmic potentials are given by (2.5). Here, we consider a *complex-valued logarithmic potentials* between a point z in the two-dimensional domain $D \subsetneq \mathbb{C}$ and N points $\boldsymbol{x} = (x_1, \dots, x_N)$ on the boundary S. For $(D,S) = (\mathbb{H}, \mathbb{R})$ and $(\mathbb{O}, \mathbb{R}_+)$, put

$$\Phi_{\mathbb{H}}(z, \boldsymbol{x}) = \sum_{i=1}^N \log(z - x_i),$$

$$\Phi_{\mathbb{O}}(z, \boldsymbol{x}) = \Phi_{\mathbb{O}}(z, \boldsymbol{x}; q) = \sum_{i=1}^N \Big\{ \log(z - x_i) + \log(z + x_i) \Big\} + q \log z,$$

where $z \in D, \boldsymbol{x} \in S^N$, and $q \in \mathbb{R}$.

Now, we consider a time evolution of the $\mathbb{C}$-valued potential Φ_D by letting $\boldsymbol{x}$ be the driving process $(\boldsymbol{X}^S(t))_{t \geq 0}$ of the multiple SLE $(g_{D_t^\eta})_{t \geq 0}$

and by transforming the function $\Phi_D(\cdot, \boldsymbol{X}^S(t))$ by $(g_{D_t^\eta})_{t\geq 0}$. We obtain the following.

Lemma 5.2. *For $D = \mathbb{H}$ and $\mathbb{O}$, the increments of the $\mathbb{C}$-valued potentials are given as follows. For $z \in A$, $\boldsymbol{X}^S(t) \in \mathbb{W}_N(S)$, $0 \leq t \leq \tau$,*

$$d\Phi_{\mathbb{H}}\big(g_{\mathbb{H}_t^\eta}(z), \boldsymbol{X}^{\mathbb{R}}(t)\big) = -\sum_{i=1}^N \frac{\sqrt{\kappa} dB_i(t)}{g_{\mathbb{H}_t^\eta}(z) - X_i^{\mathbb{R}}(t)} - \left(1 - \frac{\kappa}{4}\right) d\log g'_{\mathbb{H}_t^\eta}(z)$$
$$- \sum_{i=1}^N \left(F_i^{\mathbb{R}}(\boldsymbol{X}^{\mathbb{R}}(t)) - 4 \sum_{\substack{1\leq j\leq N,\\ j\neq i}} \frac{1}{X_i^{\mathbb{R}}(t) - X_j^{\mathbb{R}}(t)} \right) \frac{dt}{g_{\mathbb{H}_t^\eta}(z) - X_i^{\mathbb{R}}(t)}, \tag{5.1}$$

$$d\Phi_{\mathbb{O}}\big(g_{\mathbb{O}_t^\eta}(z), \boldsymbol{X}^{\mathbb{R}_+}(t); q\big) \tag{5.2}$$
$$= -\sum_{i=1}^N \left(\frac{1}{g_{\mathbb{O}_t^\eta}(z) - X_i^{\mathbb{R}_+}(t)} - \frac{1}{g_{\mathbb{O}_t^\eta}(z) + X_i^{\mathbb{R}_+}(t)} \right) \sqrt{\kappa} d\widetilde{B}_i(t)$$
$$- \sum_{i=1}^N \Bigg[F_i^{\mathbb{R}_+}(\boldsymbol{X}^{\mathbb{R}_+}(t))$$
$$- \Bigg\{ 4 \sum_{\substack{1\leq j\leq N,\\ j\neq i}} \left(\frac{1}{X_i^{\mathbb{R}_+}(t) - X_j^{\mathbb{R}_+}(t)} + \frac{1}{X_i^{\mathbb{R}_+}(t) + X_j^{\mathbb{R}_+}(t)} \right)$$
$$+ 2(1 + 2\delta + q) \frac{1}{X_i^{\mathbb{R}_+}(t)} \Bigg\} \Bigg]$$
$$\times \left(\frac{1}{g_{\mathbb{O}_t^\eta}(z) - X_i^{\mathbb{R}_+}(t)} - \frac{1}{g_{\mathbb{O}_t^\eta}(z) + X_i^{\mathbb{R}_+}(t)} \right) dt$$
$$- 4\delta \left(1 - \frac{\kappa}{4} - q\right) \frac{dt}{(g_{\mathbb{O}_t^\eta}(z))^2} - \left(1 - \frac{\kappa}{4}\right) d\log g'_{\mathbb{O}_t^\eta}(z).$$

Proof. Apply Itô's formula and use the equalities such as

$$\sum_{1\leq i\neq j\leq N} \frac{1}{(g - x_i)(g - x_j)} = 2 \sum_{1\leq i\neq j\leq N} \frac{1}{(g - x_i)(x_i - x_j)}.$$

The proof is given by direct calculation. Q.E.D.

If we assume that $(\boldsymbol{X}^{\mathbb{R}}(t))_{t\geq 0}$ is given by the $(8/\kappa)$-Dyson model $(\boldsymbol{Y}^{\mathbb{R}}(t))_{t\geq 0}$ satisfying (1.8), the third term in the RHS of (5.1) vanishes. Regarding (5.2), we first put $q = 1 - \kappa/4$ to make the third term in the RHS become zero. Then, if we assume that $\delta = \nu$ and $(\boldsymbol{X}^{\mathbb{R}_+}(t))_{t\geq 0}$ is given by the $(8/\kappa, \nu)$-Bru–Wishart process $(\boldsymbol{Y}^{\mathbb{R}_+}(t))_{t\geq 0}$ satisfying (2.4), the second term in the RHS of (5.2) vanishes.

Define

$$\mathcal{M}_{\mathbb{H}}(z,t) = -\Phi_{\mathbb{H}}\big(g_{\mathbb{H}_t^\eta}(z), \boldsymbol{Y}^{\mathbb{R}}(t)\big) - \Big(1-\frac{\kappa}{4}\Big)\log g'_{\mathbb{H}_t^\eta}(z),$$
$$\mathcal{M}_{\mathbb{O}}(z,t) = -\Phi_{\mathbb{O}}\big(g_{\mathbb{O}_t^\eta}(z), \boldsymbol{Y}^{\mathbb{R}_+}(t); 1-\kappa/4\big) - \Big(1-\frac{\kappa}{4}\Big)\log g'_{\mathbb{O}_t^\eta}(z).$$

Proposition 5.3. *Let* $\kappa > 0$, $q = 1-\kappa/4$, $\delta = \nu \geq 0$. *Then, for each point* $z \in A$, $(\mathcal{M}_D(z,t))_{0 \leq t \leq \tau}$, $D = \mathbb{H}$ *and* $\mathbb{O}$, *provide local martingales with increments*

$$d\mathcal{M}_{\mathbb{H}}(z,t) = \sum_{i=1}^N \frac{\sqrt{\kappa}dB_i(t)}{g_{\mathbb{H}_t^\eta}(z) - Y_i^{\mathbb{R}}(t)},$$
$$d\mathcal{M}_{\mathbb{O}}(z,t) = \sum_{i=1}^N \left(\frac{1}{g_{\mathbb{O}_t^\eta}(z) - Y_i^{\mathbb{R}_+}(t)} - \frac{1}{g_{\mathbb{O}_t^\eta}(z) + Y_i^{\mathbb{R}_+}(t)}\right)\sqrt{\kappa}d\widetilde{B}_i(t).$$

5.3. GFF-valued Processes

Now, we consider the sum $H_{D_t^\eta}(\cdot) + \mathsf{F}[\mathcal{M}_D(\cdot,t)]$, $0 \leq t \leq \tau$, where $\mathsf{F}[\,\cdot\,]$ denotes a functional. Comparing Lemma 5.1 and Proposition 5.3 we observe that

$$\text{(5.3)} \qquad d\Big\langle \operatorname{Im}\mathcal{M}_D(z,\cdot), \operatorname{Im}\mathcal{M}_D(w,\cdot)\Big\rangle_t = -\frac{\kappa}{4} dG_{D_t^\eta}(z,w),$$

$z, w \in D_\tau^\eta$, $0 \leq t \leq \tau$, for $(D,S) = (\mathbb{H},\mathbb{R})$ and $(\mathbb{O},\mathbb{R}_+)$. Hence, we put $\mathsf{F}[\,\cdot\,] = (2/\sqrt{\kappa})\operatorname{Im}[\,\cdot\,]$, and define the following GFF-valued processes for $(D,S) = (\mathbb{H},\mathbb{R})$ and $(\mathbb{O},\mathbb{R}_+)$,

$$\text{(5.4)} \qquad H_D(\cdot,t) := H_{D_t^\eta}(\cdot) + \frac{2}{\sqrt{\kappa}}\operatorname{Im}\mathcal{M}_D(\cdot,t), \quad 0 \leq t \leq \tau$$

with $\chi = \frac{2}{\sqrt{\kappa}}(1-\kappa/4) = 2/\sqrt{\kappa} - \sqrt{\kappa}/2$. The second term of (5.4) contains an imaginary part of the complex-valued logarithmic potential $-\Phi_D(g_{D_t^\eta}(z), \boldsymbol{Y}^S(t))$, $t \geq 0$. This is the unique harmonic function satisfying the boundary condition

$$\frac{2}{\sqrt{\kappa}}\operatorname{Im}\mathcal{M}_D(x,t) = \begin{cases} -\dfrac{2\pi}{\sqrt{\kappa}}N, & \text{if } x < Y_1^S(t), \\[2ex] -\dfrac{2\pi}{\sqrt{\kappa}}(N-i), & \text{if } x \in \big(Y_i^S(t), Y_{i+1}^S(t)\big),\ 1 \leq i \leq N, \end{cases}$$

with the convention $Y_{N+1}^S(t) \equiv +\infty$. That is, it has discontinuity at $Y_i^S(t)$ by $2\pi/\sqrt{\kappa}$ along S, $1 \leq i \leq N$, $t \geq 0$. We will think that the GFF $H_D(\cdot,t)$ has the same boundary condition as $(2/\sqrt{\kappa})\operatorname{Im}\mathcal{M}_D(\cdot,t)$,

$0 \leq t \leq \tau$. For further arguments concerning the second term of (5.4), see Section V.C in [18].

Theorem 5.4. *Let* $\kappa > 0$, $q = 1 - \kappa/4, \delta = \nu \geq 0$. *Assume that* $(D, S) = (\mathbb{H}, \mathbb{R})$ *or* $(\mathbb{O}, \mathbb{R}_+)$, *and* $(\boldsymbol{Y}^S(t))_{t \geq 0}$ *is the* $(8/\kappa)$*-Dyson model if* $S = \mathbb{R}$ *and the* $(8/\kappa, \nu)$*-Bru–Wishart process if* $S = \mathbb{R}_+$, *starting from a configuration in* $\mathbb{W}_N(S)$. *Then, for each* $f \in C_c^\infty(D)$ *such that* $\operatorname{supp}(f) \subset A$, *we have the following equality:*

$$(H_D(\cdot, 0), f) \stackrel{\text{(law)}}{=} (H_D(\cdot, \tau), f) \quad \textit{under } \mathbb{P}.$$

Proof. From (5.3) we have

$$d\left\langle \left(\frac{2}{\sqrt{\kappa}} \operatorname{Im} \mathcal{M}_D(\cdot, \cdot), f \right) \right\rangle_t = -dE_t(f), \quad 0 \leq t \leq \tau, \tag{5.5}$$

where

$$E_t(f) := \int_{A \times A} f(z) G_{D_t^\eta}(z, w) f(w) d\mu(z) d\mu(w) \tag{5.6}$$

is called the *Dirichlet energy* of f in D_t^η. We have $\operatorname{Var}[(H_{D_t^\eta}, f)] = E_t(f)$ due to the conformal invariance of a zero-boundary GFF. Introducing a parameter $\theta \in \mathbb{R}$, we see

$$\mathbb{E}\left[e^{\sqrt{-1}\theta (H_D(\cdot,\tau),f)}\right] = \mathbb{E}\left[\mathbb{E}\left[e^{\sqrt{-1}\theta (H_{D_\tau^\eta},f)} \middle| \mathcal{F}_\tau\right] e^{\sqrt{-1}\theta \frac{2}{\sqrt{\kappa}} (\operatorname{Im} \mathcal{M}_D(\cdot,\tau),f)}\right],$$

where $\mathbb{E}$ is the expectation value with respect to the probability measure $\mathbb{P}$, since $(\operatorname{Im} \mathcal{M}_D(\cdot, \tau), f)$ is $\mathcal{F}_\tau$-measurable. By definition of a zero-boundary GFF and the Dirichlet energy (5.6), we obtain

$$\mathbb{E}\left[e^{\sqrt{-1}\theta (H_{D_\tau^\eta},f)} \middle| \mathcal{F}_\tau\right] = e^{-\frac{\theta^2}{2} E_\tau(f)}.$$

Hence, by Proposition 5.3 and (5.5) we have

$$\begin{aligned}
\mathbb{E}\left[e^{\sqrt{-1}\theta (H_D(\cdot,\tau),f)}\right] &= \mathbb{E}\left[e^{\sqrt{-1}\theta \frac{2}{\sqrt{\kappa}} (\operatorname{Im} \mathcal{M}_D(\cdot,\tau),f) - \frac{\theta^2}{2} E_\tau(f)}\right] \\
&= \mathbb{E}\left[e^{\sqrt{-1}\theta \frac{2}{\sqrt{\kappa}} (\operatorname{Im} \mathcal{M}_D(\cdot,0),f) - \frac{\theta^2}{2} E_0(f)}\right] \\
&= \mathbb{E}\left[e^{\sqrt{-1}\theta (H_D(\cdot,0),f)}\right].
\end{aligned}$$

This implies the desired coincidence under the probability law $\mathbb{P}$. Q.E.D.

Remark 5.1. There are two local formulations of multiple SLE; one of them is based on commutation relation of Loewner chains each of which generates a single random curve [9] and the other one is a single Loewner chain driven by multiple driving processes [4]. Though it is expected that these two formulations are equivalent, the equivalence has not been proved and even its precise statement is not obvious. In the work by Miller and Sheffield [25], they studied coupling between a GFF and a variant of SLE called SLE($\kappa, \underline{\rho}$), which reduces to a member of commuting Loewner chains at a specific setting of parameters. Hence, it can be said that they also considered coupling between a GFF and a multiple SLE in the former sense. This does not, however, imply that a multiple SLE in the latter sense can be coupled with the same GFF, which is exactly the result we presented in the current article.

Acknowledgements. The present authors would like to thank Kalle Kytölä for useful comments on multiple SLEs. MK was supported by the Grant-in-Aid for Scientific Research (C) (No.19K03674), (B) (No.18H01124), and (S) (No.16H06338) of Japan Society for the Promotion of Science (JSPS). SK was supported by the Grant-in-Aid for JSPS Fellows (No.19J01279).

References

[1] G. W. Anderson, A. Guionnet and O. Zeitouni, An Introduction to Random Matrices, Cambridge University Press, Cambridge, 2010.

[2] A. Arai, Functional Integral Methods in Quantum Mathematical Physics, (in Japanese), Kyoritsu Shuppan, Tokyo, 2010.

[3] M. Bauer and D. Bernard, SLE$_\kappa$ growth processes and conformal field theories, Phys. Lett. B, **543** (2002), 135–138.

[4] M. Bauer, D. Bernard and K. Kytölä, Multiple Schramm–Loewner evolutions and statistical mechanics martingales, J. Stat. Phys., **120** (2005), 1125–1163.

[5] M. F. Bru, Diffusions of perturbed principal component analysis, J. Multivar. Anal., **29** (1989), 127–136.

[6] M. F. Bru, Wishart processes, J. Theor. Probab., **4** (1991), 725–751.

[7] J. Cardy, Stochastic Loewner evolution and Dyson's circular ensembles, J. Phys. A Math. Gen., **36** (2003), L379-L386; Corrigendum, *ibid.*, 12343.

[8] E. Cépa and D. Lépingle, Diffusing particles with electrostatic repulsion, Probab. Theory Related Fields, **107** (1997), 429–449.

[9] J. Dubédat, Commutation relations for Schramm–Loewner evolutions, Commun. Pure Appl. Math., **60** (2007), 1792–1847.

[10] J. Dubédat, SLE and the free field: partition functions and couplings, J. Amer. Math. Soc., **22** (2009), 995–1054.

[11] B. Duplantier and S. Sheffield, Liouville quantum gravity and KPZ, Invent. Math., **185** (2011), 333–393.

[12] F. J. Dyson, A Brownian-motion model for the eigenvalues of a random matrix, J. Math. Phys., **3** (1962), 1191–1198.

[13] P. J. Forrester, Log-Gases and Random Matrices, London Math. Soc. Monographs, Princeton University Press, Princeton, NJ, 2010.

[14] P. Graczyk and J. Małecki, Strong solutions of non-colliding particle systems, Electron. J. Probab., **19** (2014), 1–21.

[15] K. Graham, On multiple Schramm–Loewner evolutions, J. Stat. Mech. Theory Exp., **2007** (2007), P03008.

[16] T. Hida, Brownian Motion, Applications of Mathematics, **11**, Springer, Heidelberg, 1980.

[17] M. Katori, Bessel Processes, Schramm–Loewner Evolution, and the Dyson Model, SpringerBriefs in Mathematical Physics, **11**, Springer, Tokyo, 2015.

[18] M. Katori and S. Koshida, Conformal welding problem, flow line problem, and multiple Schramm–Loewner evolution, J. Math. Phys., **61** (2020), 083301/1–25.

[19] M. Katori and S. Koshida, Three phases of multiple SLE driven by non-colliding Dyson's Brownian motions, J. Phys. A: Math. Theor., **54** (2021), 325002, 19 pages.

[20] M. Katori and H. Tanemura, Symmetry of matrix-valued stochastic processes and noncolliding diffusion particle systems, J. Math. Phys., **45** (2004), 3058–3085.

[21] S. Koshida, Multiple backward Schramm–Loewner evolution and coupling with Gaussian free field, Lett. Math. Phys., **111** (2021), 30, 41 pages.

[22] M. J. Kozdron and G. F. Lawler, The configurational measure on mutually avoiding SLE paths, In: Universality and Renormalization, Fields Inst. Commun., **50**, Amer. Math. Soc., Providence, RI, 2007, pp.199–224.

[23] G. F. Lawler, Conformally Invariant Processes in the Plane, Amer. Math. Soc., Providence, RI, 2005.

[24] G. F. Lawler, O. Schramm and W. Werner, Conformal invariance of planar loop-erased random walks and uniform spanning trees, Ann. Probab., **32** (2004), 939–995.

[25] J. Miller and S. Sheffield, Imaginary geometry I: Interacting SLEs, Probab. Theory Relat. Fields, **164** (2016), 553–705.

[26] J. Miller and S. Sheffield, Imaginary geometry II: reversibility of $\mathrm{SLE}_\kappa(\rho_1;\rho_2)$ for $\kappa \in (0,4)$, Ann. Probab., **44** (2016), 1647–1722.

[27] J. Miller and S. Sheffield, Imaginary geometry III: reversibility of SLE_κ for $\kappa \in (4,8)$, Ann. Math., **184** (2016), 455–486.

[28] J. Miller and S. Sheffield, Imaginary geometry IV: interior rays, whole-plane reversibility, and space-filling trees, Probab. Theory Relat. Fields, **169** (2017), 729–869.

[29] S. Rohde and O. Schramm, Basic properties of SLE, Ann. Math., **161** (2005), 883–924.

[30] O. Schramm, Scaling limits of loop-erased random walks and uniform spanning trees, Israel J. Math., **118** (2000), 221–288.
[31] S. Sheffield, Gaussian free fields for mathematicians, Probab. Theory Relat. Fields, **139** (2007), 521–541.
[32] S. Sheffield, Conformal weldings of random surfaces: SLE and the quantum gravity zipper, Ann. Probab., **44** (2016) 3474–3545.
[33] T. Takebe, Dispersionless BKP hierarchy and quadrant Löwner equation, SIGMA, **10** (2014), 23, 13 pages.
[34] J. Wishart, The generalized product moment distribution in samples from a normal multivariate population, Biometrika, **20A** (1928), 32–52.

Department of Physics,
Faculty of Science and Engineering,
Chuo University,
Kasuga, Bunkyo-ku, Tokyo 112-8551, Japan
E-mail address: katori@phys.chuo-u.ac.jp
E-mail address: koshida@phys.chuo-u.ac.jp

Advanced Studies in Pure Mathematics 87, 2021
Stochastic Analysis, Random Fields and Integrable Probability — Fukuoka 2019
pp. 341–361

Tightness of the solutions to approximating equations of the stochastic quantization equation associated with the weighted exponential quantum field model on the two-dimensional torus

Masato Hoshino, Hiroshi Kawabi and Seiichiro Kusuoka

Abstract.

We consider stochastic quantization associated with the weighted exponential quantum field model on the two-dimensional torus via a method of singular stochastic partial differential equations and show the tightness of the solutions to approximating equations of the stochastic quantization equation. If the model is not weighted, then the drift term of the stochastic quantization equation, which includes a renormalization term, is nonpositive or nonnegative. However, in the weighted case, generally the drift term is neither nonpositive nor nonnegative. We modify the argument in the case without weights and discuss the weighted model.

§1. Introduction and the main theorem

In [5] we studied a stochastic quantization equation associated with the $\exp(\Phi)_2$-model, which is also called the Høegh-Krohn model. In the present paper, we consider the stochastic quantization equation associated with the weighted $\exp(\Phi)_2$-model.

Let $\Lambda = \mathbb{T}^2 = (\mathbb{R}/2\pi\mathbb{Z})^2$ be the two-dimensional torus equipped with the Lebesgue measure dx. Let

$$e_k(x) = \frac{1}{2\pi} e^{\sqrt{-1}k\cdot x}, \quad k \in \mathbb{Z}^2,\ x \in \Lambda,$$

where $k \cdot x$ is the inner product given by $k \cdot x := k_1x_1 + k_2x_2$ for $k = (k_1, k_2)$ and $x = (x_1, x_2)$. For $s \in \mathbb{R}$, we define the Sobolev space of

---

Received January 11, 2020.
Revised May 19, 2020.
2010 *Mathematics Subject Classification.* 81S20, 60H15, 35Q40, 35R60.
Key words and phrases. stochastic quantization, quantum fields theory, singular SPDE, Høegh-Krohn model.

order s with periodic boundary condition by

$$H^s = H^s(\Lambda) = \left\{ u \in \mathcal{D}'(\Lambda;\mathbb{R}); \sum_{k\in\mathbb{Z}^2} (1+|k|^2)^s |\langle u, e_k\rangle|^2 < \infty \right\}.$$

We define the Gaussian free field measure μ_0 by the centered Gaussian measure on $\mathcal{D}'(\Lambda)$ with covariance $(1-\triangle)^{-1}$. Note that μ_0 has a full support on $H^{-\varepsilon}(\Lambda)$ for any $\varepsilon > 0$. Let $\alpha_0 \in [0, \sqrt{4\pi})$ and ν be a finite Borel measure on $[-\alpha_0, \alpha_0]$, and we define

$$\begin{aligned} (1)\quad & \mu^{(\nu)}(d\phi) \\ & := \frac{1}{Z^{(\nu)}} \exp\left(- \int_{[-\alpha_0,\alpha_0]} \left(\int_\Lambda \exp^\diamond(\alpha\phi)(x)dx \right) \nu(d\alpha) \right) \mu_0(d\phi), \end{aligned}$$

where $Z^{(\nu)} > 0$ is the normalizing constant and $\exp^\diamond(\alpha\cdot)$ is the Wick exponential. See Section 2 in [5] for the detail of the Gaussian free field measure and the Wick exponentials. We call $\mu^{(\nu)}$ a weighted $\exp(\Phi)_2$-measure. Such a quantum field measure was introduced by Høegh-Krohn [4] and has been studied for a long time. For a physical background and related works on this model, see e.g. Section 2 in [1] and Section V.6 in [7]. Since $\int_{[-\alpha_0,\alpha_0]} \left(\int_\Lambda \exp^\diamond(\alpha\phi)(x)dx \right)\nu(d\alpha)$ is nonnegative, the construction of $\mu^{(\nu)}$ is not difficult. See Section 2 for detail.

We consider a stochastic quantization equation associated with $\mu^{(\nu)}$ given by

$$(2)\qquad \partial_t \Phi_t = \frac{1}{2}(\triangle - 1)\Phi_t - \frac{1}{2}\int_{[-\alpha_0,\alpha_0]} \alpha \exp^\diamond(\alpha\phi)\nu(d\alpha) + \dot{W}_t, \quad t > 0,$$

where $W = \{W_t(x); t \geq 0, x \in \Lambda\}$ is an $L^2(\Lambda)$-cylindrical Brownian motion defined on a filtered probability space $(\Omega, \mathcal{F}, (\mathcal{F}_t)_{t\geq 0}, \mathbb{P})$. If ν is a Dirac measure δ_α, then the model coincides with the $\exp(\Phi)_2$-model without weights, which is concerned in [5]. Hence, weighted $\exp(\Phi)_2$-models are generalization of $\exp(\Phi)_2$-models. The difficulty of solving (2) compared with the stochastic quantization equation associated with $\exp(\Phi)_2$-models is that the drift term $\int_{[-\alpha_0,\alpha_0]} \alpha \exp^\diamond(\alpha\phi)\nu(d\alpha)$ can be both positive and negative. For example, in the case that $\nu = (\delta_\alpha + \delta_{-\alpha})/2$, $\int_{[-\alpha_0,\alpha_0]} \alpha \exp^\diamond(\alpha\phi)\nu(d\alpha) = \sinh^\diamond(\alpha\phi)$ where $\sinh^\diamond(\alpha\phi)$ is renormalization of $\sinh(\alpha\phi)$.

Since the Wick exponential $\exp^\diamond(\alpha\phi)$ is renormalized, we consider approximating equations given by regularizing the white noise $\dot{W}_t$. Let ψ be a Borel function on $\mathbb{R}^2$ with the following properties.

- $0 \le \psi(x) \le 1$ for any $x \in \mathbb{R}^2$.
- $\psi(x) = \psi(-x)$ for any $x \in \mathbb{R}^2$.
- $\sup_{x\in\mathbb{R}^2\setminus\{0\}} |x|^{-\theta}|\psi(x) - 1| < \infty$ for some $\theta \in (0, 1)$.
- $\sup_{x\in\mathbb{R}^2} |x|^m|\psi(x)| < \infty$ for some $m \ge 4$.

For a fixed ψ, we define an operator P_N on $\mathcal{D}'(\Lambda)$ by

$$P_N f(x) = \sum_{k\in\mathbb{Z}^2} \psi(2^{-N}k)\langle f, e_k\rangle e_k(x), \quad N \in \mathbb{N},\ x \in \Lambda.$$

This class of approximation operators has appeared in [5]. For the main theorem, we prepare a function space

$$\mathcal{Y}_T = \left\{ \Upsilon \in L^2([0,T]; H^1) \cap C([0,T]; L^2)\ ;\ e^{\alpha_0|\Upsilon|} \in L^1([0,T]; C(\Lambda)) \right\}.$$

We remark that this space is different from that appeared in Section 3.2 in [5]. We denote the Wick exponential, which is given by the renormalization of $\exp(\alpha\phi)$, by $\exp^\diamond(\alpha\phi)$ (see Section 2 in [5] for the precise definition of the Wick exponentials). The main result in the present paper is the following.

Theorem 1.1. *Assume that $\alpha_0 \in [0, \sqrt{2\pi})$ and the support of ν is included in $A_0 := [-\alpha_0, -\alpha_0/2] \cup [\alpha_0/2, \alpha_0]$. For each $N \in \mathbb{N}$, consider the initial value problem*

$$(3)\qquad \begin{cases} \partial_t \Phi_t^N = \dfrac{1}{2}(\triangle - 1)\Phi_t^N - \dfrac{1}{2}\displaystyle\int_{A_0} \alpha \exp\left(\alpha\Phi_t^N - \frac{\alpha^2}{2}C_N\right)\nu(d\alpha) \\ \qquad\qquad + P_N\dot{W}_t, \\ \Phi_0^N = P_N\xi, \end{cases}$$

where ξ is a random variable with the law μ_0 and

$$C_N := \frac{1}{4\pi^2}\sum_{k\in\mathbb{Z}^2}\frac{\psi(2^{-N}k)^2}{1+|k|^2}.$$

Then, the laws of the solutions $\{\Phi^N\}$ to (3) are tight as probability measures on $C([0,T]; H^{-\varepsilon})$ for any $\varepsilon \in (0,\infty)$.

Moreover, if Φ is a limit of a subsequence of $\{\Phi^N\}$ in law, then Φ is decomposed as $\Phi = X + Y$ where X is the stationary Ornstein-Uhlenbeck process and $Y \in \mathcal{Y}_T$ almost surely.

Proof. It is easy to see that for each N, the unique solution Φ^N to (3) is given by the sum of the solutions X^N and Y^N to (6) and (7) below with $\eta = 0$, respectively. On the other hand, in view of Theorem 3.6

below, the laws of $\{Y^N\}$ are tight in $C([0,T];L^2)$. Moreover, the laws of $\{X^N\}$ are tight in $C([0,T];H^{-\varepsilon})$. Hence, in view of Corollary A.5 in [2], we have the tightness of the laws of $\{\Phi^N\}$.

Assume that Φ is a limit of a subsequence of $\{\Phi^N\}$ in law. Since $\{X^N\}$ converges to the stationary Ornstein-Uhlenbeck process X almost surely in $C([0,T];H^{-\varepsilon})$ (see Section 2 in [5]), the associated subsequence of $\{Y^N\}$ also converges to an $H^{-\varepsilon}$-valued continuous process Y in law. Hence, from Theorem 3.6 we have $Y \in \mathcal{Y}_T$ almost surely. Q.E.D.

The tightness obtained in Theorem 1.1 is associated with the existence of the time-global solution to the stochastic quantization equation (2). In the case without weights, uniqueness (in a suitable sense) is also obtained in [5]. However, the uniqueness is not obtained here, because we do not have the characterization of Y. We discuss the uniqueness shortly in the end of this paper (see Section 4).

In [6], the case of stationary approximation with $\nu = (\delta_\alpha + \delta_{-\alpha})/2$, i.e., $\int_{\mathbb{R}} \alpha \exp^\diamond(\alpha\phi)\nu(d\alpha) = \sinh^\diamond(\alpha\phi)$, is concerned and obtained the existence and uniqueness in a suitable sense under the condition $\alpha \in \left(0, \sqrt{8\pi/(3+2\sqrt{2})}\right)$ (see Remark 1.15 in [6]). For the difference of the approximations see Section 1.2 in [5]. In view of $\sqrt{8\pi/(3+2\sqrt{2})} < \sqrt{2\pi}$, our range of α is larger than their range.

The present paper is organized as follows. In Section 2 we see the definition and properties of weighted Wick exponentials of the Ornstein-Uhlenbeck process. Approximation of weighted Wick exponentials will appear in the shifted equations of (3) (see Section 3). In Section 3, by using results in Section 2, we will show some uniform estimates in N of the solutions to the shifted equations and finally show the tightness of the solutions. Section 3 is the main part of the present paper. In Section 4, as a remark, we discuss the uniqueness of the limit equation of the shifted equations of (3) by following the argument in [5]. However, unfortunately the obtained uniqueness is not applicable to Y appeared in Theorem 1.1.

§2. Weighted Wick exponentials of the Ornstein-Uhlenbeck process

In this section, we introduce the definition and properties of weighted Wick exponentials of the Ornstein-Uhlenbeck process.

Let $X = X(\phi)$ be the Ornstein-Uhlenbeck process starting at ϕ, which is defined by the unique solution of the initial value problem

$$\begin{cases} \partial_t X_t = \frac{1}{2}(\triangle - 1)X_t + \dot{W}_t, \\ X_0 = \phi \end{cases} \tag{4}$$

for $\phi \in \mathcal{D}'(\Lambda)$. We remark that μ_0 is the invariant measure of X.

For $x \in \mathbb{R}$ and $\sigma \geq 0$, let $\{H_n(x;\sigma)\}_{n=0}^{\infty}$ be the Hermite polynomials defined via the generating function

$$e^{\alpha x - \frac{\alpha^2}{2}\sigma} = \sum_{n=0}^{\infty} \frac{\alpha^n}{n!} H_n(x;\sigma), \quad \alpha \in \mathbb{R}.$$

It is well-known that, if X and Y are jointly Gaussian random variables with means 0 and covariances σ_X and σ_Y respectively, then one has

$$\mathbb{E}\big[H_n(X;\sigma_X)H_m(Y;\sigma_Y)\big] = \delta_{nm} n! \mathbb{E}[XY]^n. \tag{5}$$

Let ϕ be an element of the probability space $(\mathcal{D}'(\Lambda), \mathcal{B}(\mathcal{D}'(\Lambda)), \mu_0)$, where $\mathcal{B}(\mathcal{D}'(\Lambda))$ is the Borel σ-algebra of $\mathcal{D}'(\Lambda)$. For simplicity, denote $\psi(2^{-N}\cdot)$ by ψ_N. We define the approximating Wick exponential $\exp_N^{\diamond}(\alpha\phi)$ by

$$\exp_N^{\diamond}(\alpha\phi) := \sum_{n=0}^{\infty} \frac{\alpha^n}{n!} H_n(P_N\phi; C_N), \quad \alpha \in (-\sqrt{4\pi}, \sqrt{4\pi}),$$

where

$$C_N = \int_{\mathcal{D}'(\Lambda)} (P_N\phi(x))^2 \mu_0(d\phi) = \frac{1}{4\pi^2} \sum_{k \in \mathbb{Z}^2} \frac{\psi_N(k)^2}{1 + |k|^2}.$$

Note that $\exp_N^{\diamond}(\alpha\phi) \geq 0$ for $\alpha \in (-\sqrt{4\pi}, \sqrt{4\pi})$. To define the weighted Wick exponentials we give a finite Borel measure ν on $[-\alpha_0, \alpha_0]$ with $\alpha_0 \in [0, \sqrt{4\pi})$, and let $\beta \in (\alpha_0^2/(4\pi), 1)$.

Theorem 2.1. *The sequence of functions*

$$\left\{ \int_{[-\alpha_0,\alpha_0]} \exp_N^{\diamond}(\alpha\phi)\nu(d\alpha) \right\}$$

converges in $H^{-\beta}$, μ_0*-almost every* ϕ, *and in* $L^2(\mu_0; H^{-\beta})$. *Moreover, the limit is independent of the choice of* ψ.

Proof. As in the proof of Theorem 2.2 in [5], for $\lambda \in (0,1)$ it holds that

$$\sum_{N=1}^{\infty} \left(\int_{\mathcal{D}'(\Lambda)} \left\| \exp_{N+1}^{\diamond}(\alpha\phi) - \exp_N^{\diamond}(\alpha\phi) \right\|_{H^{-\beta}}^2 \mu_0(d\phi) \right)^{1/2} \leq \sum_{n=1}^{\infty} \frac{(C\alpha^2)^n}{(2\pi)^n (n-1)!} + C \sum_{n=1}^{\infty} \left(\frac{\alpha^2(1+\lambda)}{4\pi(\beta-\lambda)} \right)^n$$

for $\alpha \in [-\alpha_0, \alpha_0]$, where C is a constant depending on λ. Therefore, by choosing $\lambda \in (0,1)$ sufficiently small we obtain

$$\sum_{N=1}^{\infty} \left(\int_{\mathcal{D}'(\Lambda)} \left\| \int_{[-\alpha_0,\alpha_0]} \exp_{N+1}^{\diamond}(\alpha\phi)\nu(d\alpha) - \int_{[-\alpha_0,\alpha_0]} \exp_N^{\diamond}(\alpha\phi)\nu(d\alpha) \right\|_{H^{-\beta}}^2 \mu_0(d\phi) \right)^{1/2} < \infty.$$

This yields the convergence in $H^{-\beta}$, μ_0-almost every ϕ, and in $L^2(\mu_0; H^{-\beta})$.

The proof of the uniqueness is similar to the proof of Theorem 2.2 in [5]. So, we omit it. Q.E.D.

In the present paper, we call the limit obtained in Theorem 2.1, which is formally denoted by

$$\int_{[-\alpha_0,\alpha_0]} \exp^{\diamond}(\alpha\phi)\nu(d\alpha),$$

the weighted Wick exponential.

Corollary 2.2. *The weighted* $\exp(\Phi)_2$*-measure* $\mu^{(\nu)}$ *defined by* (1) *is well-defined as the limit of the approximating measures* $\{\mu_N^{(\nu)}\}$ *defined by*

$$\mu_N^{(\nu)}(d\phi) := \frac{1}{Z_N^{(\nu)}} \exp\left\{ -\int_{\Lambda} \left(\int_{[-\alpha_0,\alpha_0]} \exp_N^{\diamond}(\alpha\phi)(x)\nu(d\alpha) \right) dx \right\} \mu_0(d\phi)$$

in weak topology, and absolutely continuous with respect to μ_0. *Here,* $Z_N^{(\nu)}$ *is the normalizing constant. In particular, the support of* $\mu^{(\nu)}$ *is in* $H^{-\varepsilon}$ *for any* $\varepsilon > 0$. *Moreover, the Radon-Nikodym derivatives* $\{\frac{d\mu_N^{(\nu)}}{d\mu_0}\}$ *are uniformly bounded.*

Proof of Corollary 2.2 is the same as that of Corollary 2.3 in [5]. So, we omit it.

For the Ornstein-Uhlenbeck process $X = X(\phi)$, we also define the approximating weighted Wick exponential $\mathcal{X}^{N,\nu}$ by

$$\mathcal{X}_t^{N,\nu}(\phi) := \int_{[-\alpha_0,\alpha_0]} \exp_N^{\diamond}(\alpha X_t(\phi))\nu(d\alpha).$$

Then, similarly to Theorem 2.4 and Lemma 2.5 in [5], we obtain the following.

Theorem 2.3. *For any $T > 0$, $\{\mathcal{X}^{N,\nu}\}$ converges in $L^2([0,T]; H^{-\beta})$, $\mathbb{P} \otimes \mu_0$-almost surely, and in $L^2(\mathbb{P} \otimes \mu_0)$. Moreover, the limit $\mathcal{X}^\nu$ is independent of the choice of ψ.*

Proposition 2.4. *Let ξ_N and ξ_∞ be H^{-2}-valued random variables independent of W. Assume that the laws of ξ_N and ξ are absolutely continuous with respect to μ_0, and their Radon-Nikodym derivatives are uniformly bounded over N. If ξ_N converges to ξ_∞ in H^{-2} almost surely, then we have*

$$\mathcal{X}^\nu(\xi_N) \to \mathcal{X}^\nu(\xi_\infty)$$

in $L^2([0,T]; H^{-\beta})$ for any $T > 0$, in probability.

The proofs of Theorem 2.3 and Proposition 2.4 are similar to those of Theorem 2.4 and Lemma 2.5 in [5], respectively. So, we omit them.

§3. Tightness of the approximation sequence

First we introduce the shifted equations. Precisely, we decompose $\Phi^N = X^N + Y^N$, where X^N and Y^N solve

$$\begin{cases} \partial_t X_t^N = \frac{1}{2}(\triangle - 1)X_t^N + P_N \dot{W}_t, \\ X_0^N = P_N \xi, \end{cases} \tag{6}$$

$$\begin{cases} \partial_t Y_t^N = \frac{1}{2}(\triangle - 1)Y_t^N - \frac{1}{2}\int_{[-\alpha_0,\alpha_0]} \alpha \exp(\alpha Y_t^N)\mathcal{X}_t^{N,\alpha}\nu(d\alpha), \\ Y_0^N = \eta \in C(\Lambda) \cap H^{2\varepsilon}, \end{cases} \tag{7}$$

and ξ is a random variable with the law μ_0 and independent of $\dot{W}$, η is nonrandom, and

$$\mathcal{X}_t^{N,\alpha} := \exp_N^{\diamond}(\alpha X_t), \quad \alpha \in (-\sqrt{4\pi}, \sqrt{4\pi}).$$

Note that $X^N = P_N X(\xi)$, where $X(\xi)$ is the solution of (4) with the initial value ξ. We remark that (6) and (7) have the pathwise-unique mild solutions $X^N, Y^N \in C^\infty(\Lambda)$, which follows from the continuity of $\mathcal{X}_t^{N,\alpha}(x)$ in (t,x) and a priori estimate of the solutions. The approach to stochastic partial differential equations via shifted equations is called the Da Prato-Debussche argument (see [3]).

To show the tightness of the solutions Φ^N to (3), we show that of the solutions Y^N of the shifted equations (7). In this section we often write C as a constant independent of N, and C can be different from line to line. Let $\alpha_0 \in (0, \sqrt{4\pi})$ and assume that $\mathrm{supp}(\nu) \subset A_0 := [-\alpha_0, -\alpha_0/2] \cup [\alpha_0/2, \alpha_0]$. Let $\beta \in (\alpha_0^2/(4\pi), 1)$. Later, we will restrict the range of α_0 in order to take sufficiently small β. We remark that $\mathcal{X}^{N,\nu}$ is a stationary process, because $X_0^N = P_N \xi$ and ξ has μ_0 as its law.

Now we are going to derive uniform estimates for $\cosh(\alpha_0 Y_t^N)$ with respect to N. Since

$$\partial_t \cosh(\alpha_0 Y_t^N) = \alpha_0 \sinh(\alpha_0 Y_t^N) \partial_t Y_t^N,$$
$$\triangle \cosh(\alpha_0 Y_t^N) = \alpha_0^2 \cosh(\alpha_0 Y_t^N) |\nabla Y_t^N|^2 + \alpha_0 \sinh(\alpha_0 Y_t^N) \triangle Y_t^N,$$

by (7) we have

$$\begin{aligned}
&\left(\partial_t - \frac{1}{2}\triangle\right) \cosh(\alpha_0 Y_t^N) \\
&= -\frac{\alpha_0^2}{2} \cosh(\alpha_0 Y_t^N) |\nabla Y_t^N|^2 - \frac{\alpha_0}{2} Y_t^N \sinh(\alpha_0 Y_t^N) \\
&\quad - \frac{\alpha_0}{2} \sinh(\alpha_0 Y_t^N) \int_{A_0} \alpha \exp(\alpha Y_t^N) \mathcal{X}_t^{N,\alpha} \nu(d\alpha) \\
&\le -\frac{\alpha_0^2}{2} \cosh(\alpha_0 Y_t^N) |\nabla Y_t^N|^2 - \frac{\alpha_0}{2} Y_t^N \sinh(\alpha_0 Y_t^N) \\
&\quad + \frac{\alpha_0}{4} \exp(-\alpha_0 Y_t^N) \int_{[\alpha_0/2, \alpha_0]} \alpha \exp(\alpha Y_t^N) \mathcal{X}_t^{N,\alpha} \nu(d\alpha) \\
&\quad - \frac{\alpha_0}{4} \exp(\alpha_0 Y_t^N) \int_{[-\alpha_0, -\alpha_0/2]} \alpha \exp(\alpha Y_t^N) \mathcal{X}_t^{N,\alpha} \nu(d\alpha).
\end{aligned}$$

In the last inequality, we decompose $\sinh(\alpha_0 Y_t^N) = \frac{1}{2}(e^{\alpha_0 Y_t^N} - e^{-\alpha_0 Y_t^N})$ and we use the inequalities

$$\frac{\alpha_0}{4} \exp(\alpha_0 Y_t^N) \int_{[\alpha_0/2, \alpha_0]} \alpha \exp(\alpha Y_t^N) \mathcal{X}_t^{N,\alpha} \nu(d\alpha) \ge 0,$$
$$\frac{\alpha_0}{4} \exp(-\alpha_0 Y_t^N) \int_{[-\alpha_0, -\alpha_0/2]} \alpha \exp(\alpha Y_t^N) \mathcal{X}_t^{N,\alpha} \nu(d\alpha) \le 0.$$

This inequality implies

$$
\begin{aligned}
&\left(\partial_t - \frac{1}{2}\triangle\right)\cosh(\alpha_0 Y_t^N) + \frac{\alpha_0^2}{2}\cosh(\alpha_0 Y_t^N)|\nabla Y_t^N|^2 \\
&\qquad\qquad\qquad\qquad + \frac{\alpha_0}{2} Y_t^N \sinh(\alpha_0 Y_t^N) \\
&\quad \leq \frac{\alpha_0}{4}\int_{[\alpha_0/2,\alpha_0]} \alpha \exp\big((\alpha-\alpha_0)Y_t^N\big)\mathcal{X}_t^{N,\alpha}\nu(d\alpha) \\
&\qquad + \frac{\alpha_0}{4}\int_{[-\alpha_0,-\alpha_0/2]} (-\alpha)\exp\big((\alpha+\alpha_0)Y_t^N\big)\mathcal{X}_t^{N,\alpha}\nu(d\alpha) \\
&\quad \leq \frac{\alpha_0^2}{4}\exp\left(\frac{\alpha_0}{2}|Y_t^N|\right)\int_{[\alpha_0/2,\alpha_0]} \mathcal{X}_t^{N,\alpha}\nu(d\alpha) \\
&\qquad + \frac{\alpha_0^2}{4}\exp\left(\frac{\alpha_0}{2}|Y_t^N|\right)\int_{[-\alpha_0,-\alpha_0/2]} \mathcal{X}_t^{N,\alpha}\nu(d\alpha) \\
&\quad \leq \frac{\alpha_0^2}{4}\exp\left(\frac{\alpha_0}{2}|Y_t^N|\right)\mathcal{X}_t^{N,\nu}.
\end{aligned}
$$

Hence, by using the mild form, we have

$$
\begin{aligned}
(8)\quad & \cosh(\alpha_0 Y_t^N) + \frac{\alpha_0^2}{2}\int_0^t e^{(t-s)\triangle/2}\left[\cosh(\alpha_0 Y_s^N)|\nabla Y_s^N|^2\right]ds \\
&\qquad\qquad\qquad + \frac{\alpha_0}{2}\int_0^t e^{(t-s)\triangle/2}\left[Y_s^N \sinh(\alpha_0 Y_s^N)\right]ds \\
&\quad \leq e^{t\triangle/2}\cosh(\alpha_0\eta) + \frac{\alpha_0^2}{4}\int_0^t e^{(t-s)\triangle/2}\left[\exp\left(\frac{\alpha_0}{2}|Y_s^N|\right)\mathcal{X}_s^{N,\nu}\right]ds.
\end{aligned}
$$

Let $\varepsilon \in (0,(1-\beta)/2)$. From (8), the Sobolev embedding theorem and the smoothing property of the heat semigroup it follows that

$$
\begin{aligned}
&\|\cosh(\alpha_0 Y_t^N)\|_{C(\Lambda)} \\
&\quad \leq \left\|e^{t\triangle/2}\cosh(\alpha_0\eta)\right\|_{C(\Lambda)} \\
&\qquad + \frac{\alpha_0^2}{4}\int_0^t \left\|e^{(t-s)\triangle/2}\left(\exp\left(\frac{\alpha_0}{2}|Y_s^N|\right)\mathcal{X}_s^{N,\nu}\right)\right\|_{C(\Lambda)} ds \\
&\quad \leq \|\cosh(\alpha_0\eta)\|_{C(\Lambda)} \\
&\qquad + C\int_0^t (t-s)^{-(1+\beta+\varepsilon)/2}\left\|\exp\left(\frac{\alpha_0}{2}|Y_s^N|\right)\mathcal{X}_s^{N,\nu}\right\|_{H^{-\beta}} ds.
\end{aligned}
$$

Applying Theorem 3.4 in [5], we have

$$\begin{aligned}(9)\quad &\| \cosh(\alpha_0 Y_t^N)\|_{C(\Lambda)} \\ &\leq C\| \cosh(\alpha_0 \eta)\|_{C(\Lambda)} \\ &\quad + C\int_0^t (t-s)^{-(1+\beta+\varepsilon)/2} \left\| \exp\left(\frac{\alpha_0}{2}|Y_s^N|\right)\right\|_{C(\Lambda)} \|\mathcal{X}_s^{N,\nu}\|_{H^{-\beta}}\, ds.\end{aligned}$$

This inequality, Hölder's inequality, the stationarity of $\mathcal{X}^{N,\nu}$ and Theorem 2.3 imply

$$\begin{aligned}&\mathbb{E}\left[\| \cosh(\alpha_0 Y_t^N)\|_{C(\Lambda)}\right] \\ &\leq \| \cosh(\alpha_0 \eta)\|_{C(\Lambda)} \\ &\quad + C\int_0^t (t-s)^{-(1+\beta+\varepsilon)/2}\, \mathbb{E}\left[\|\mathcal{X}_s^{N,\nu}\|_{H^{-\beta}}^2\right]^{1/2} \\ &\qquad\qquad \times \mathbb{E}\left[\left\| \exp\left(\frac{\alpha_0}{2}|Y_s^N|\right)\right\|_{C(\Lambda)}^2\right]^{1/2} ds \\ &\leq \| \cosh(\alpha_0 \eta)\|_{C(\Lambda)} \\ &\quad + C\int_0^t (t-s)^{-(1+\beta+\varepsilon)/2}\, \mathbb{E}\left[\|\exp(\alpha_0|Y_s^N|)\|_{C(\Lambda)}\right]^{1/2} ds.\end{aligned}$$

Hence, by Young's inequality for $p \in [1,\infty)$ we have

$$\begin{aligned}&\left(\int_0^T \mathbb{E}\left[\| \cosh(\alpha_0 Y_t^N)\|_{C(\Lambda)}\right]^p dt\right)^{1/p} \\ &\leq C\| \cosh(\alpha_0 \eta)\|_{C(\Lambda)} + C\left(\int_0^T \mathbb{E}\left[\|\exp(\alpha_0|Y_s^N|)\|_{C(\Lambda)}\right]^{p/2} ds\right)^{1/p}.\end{aligned}$$

Since

$$\begin{aligned}&\left(\int_0^T \mathbb{E}\left[\|\exp(\alpha_0|Y_s^N|)\|_{C(\Lambda)}\right]^{p/2} ds\right)^{1/p} \\ &\leq \lambda\left(\int_0^T \mathbb{E}\left[\|\exp(\alpha_0|Y_s^N|)\|_{C(\Lambda)}\right]^{p} ds\right)^{1/p} + C \\ &\leq 2\lambda\left(\int_0^T \mathbb{E}\left[\|\cosh(\alpha_0 Y_s^N)\|_{C(\Lambda)}\right]^{p} ds\right)^{1/p} + C\end{aligned}$$

for any $\lambda \in (0,1)$, by applying this inequality with sufficiently small λ we obtain

$$\int_0^T \mathbb{E}\left[\|\cosh(\alpha_0 Y_t^N)\|_{C(\Lambda)}\right]^p dt \leq C\|\cosh(\alpha_0\eta)\|_{C(\Lambda)}^p \tag{10}$$

for $p \in [1,\infty)$.

Lemma 3.1. *Assume that* $\alpha_0 \in [0,\sqrt{2\pi})$ *and take* $\beta \in (\alpha_0^2/(4\pi), 1/2)$ *and* $\varepsilon \in (0,(1-2\beta)/4)$. *For* $N \in \mathbb{N}$ *we have*

$$\mathbb{E}\left[\sup_{t\in[0,T]}\int_0^t (t-\tau)^{(1/2)-2\beta-(5\varepsilon/2)}\left\|\exp(\alpha_0|Y_\tau^N|)\right\|_{C(\Lambda)} d\tau\right] \leq C\big(1+\|\cosh(\alpha_0\eta)\|_{C(\Lambda)}\big).$$

Proof. If $(1/2)-2\beta-(5\varepsilon/2) \geq 0$, then (10) yields the assertion. Consider the case that $(1/2)-2\beta-(5\varepsilon/2) < 0$. Let $\gamma \in (0,1)$. From (9), Hölder's inequality and Lemma 2.3 in [2], we have

$$\begin{aligned}
&\int_0^t (t-\tau)^{-\gamma}\left\|\exp(\alpha_0|Y_\tau^N|)\right\|_{C(\Lambda)} d\tau \\
&\leq C\|\cosh(\alpha_0\eta)\|_{C(\Lambda)} \\
&\quad + C\int_0^t (t-\tau)^{-\gamma}\bigg(\int_0^\tau (\tau-s)^{-(1+\beta+\varepsilon)/2} \\
&\qquad\qquad \times \left\|\exp\left(\frac{\alpha_0}{2}|Y_s^N|\right)\right\|_{C(\Lambda)}\left\|\mathcal{X}_s^{N,\nu}\right\|_{H^{-\beta}} ds\bigg) d\tau \\
&\leq C\|\cosh(\alpha_0\eta)\|_{C(\Lambda)} \\
&\quad + C_\gamma\int_0^t (t-s)^{-\gamma+(1-\beta-\varepsilon)/2}\left\|\exp\left(\frac{\alpha_0}{2}|Y_s^N|\right)\right\|_{C(\Lambda)}\left\|\mathcal{X}_s^{N,\nu}\right\|_{H^{-\beta}} ds \\
&\leq C\|\cosh(\alpha_0\eta)\|_{C(\Lambda)} + \int_0^t \left\|\mathcal{X}_s^{N,\nu}\right\|_{H^{-\beta}}^2 ds \\
&\quad + C_\gamma\int_0^t (t-s)^{-2\gamma+1-\beta-\varepsilon}\left\|\exp(\alpha_0|Y_s^N|)\right\|_{C(\Lambda)} ds.
\end{aligned}$$

Hence, it holds that

$$\begin{aligned}
&\mathbb{E}\left[\sup_{t\in[0,T]}\int_0^t (t-\tau)^{-\gamma}\left\|\exp(\alpha_0|Y_\tau^N|)\right\|_{C(\Lambda)} d\tau\right] \\
&\leq C\big(1+\|\cosh(\alpha_0\eta)\|_{C(\Lambda)}\big) \\
&\quad + C_\gamma\mathbb{E}\left[\sup_{t\in[0,T]}\int_0^t (t-s)^{-2\gamma+1-\beta-\varepsilon}\left\|\exp(\alpha_0|Y_s^N|)\right\|_{C(\Lambda)} ds\right]
\end{aligned} \tag{11}$$

for $\gamma \in (0,1)$. We repeat this calculation. Let $f(\gamma) := 2\gamma - (1-\beta-\varepsilon)$ for $\gamma \in \mathbb{R}$, and denote the n-composition of f by f^n. Then, $\gamma = f(\gamma)$ if and only if $\gamma = 1-\beta-\varepsilon$. Moreover, for $\gamma < 1-\beta-\varepsilon$, $\gamma > f(\gamma)$. These facts imply that $\lim_{n\to\infty} f^n(\gamma) = -\infty$ for $\gamma < 1-\beta-\varepsilon$. Since we are considering the case that $0 < -[(1/2) - 2\beta - (5\varepsilon/2)] < 1-\beta-\varepsilon$, there exists $n \in \mathbb{N}$ such that

$$f^n\left(-\left(\frac{1}{2} - 2\beta - \frac{5}{2}\varepsilon\right)\right) < 0 < f^{n-1}\left(-\left(\frac{1}{2} - 2\beta - \frac{5}{2}\varepsilon\right)\right).$$

Applying (11) recursively with $\gamma = f^{k-1}(-[(1/2) - 2\beta - (5\varepsilon/2)])$ for $k = 1, 2, \dots, n$, we obtain

$$\begin{aligned}
&\mathbb{E}\left[\sup_{t\in[0,T]} \int_0^t (t-\tau)^{(1/2)-2\beta-(5\varepsilon/2)} \left\|\exp(\alpha_0|Y_\tau^N|)\right\|_{C(\Lambda)} d\tau\right] \\
&\quad \le C\big(1 + \|\cosh(\alpha_0\eta)\|_{C(\Lambda)}\big) \\
&\qquad + C\mathbb{E}\left[\sup_{t\in[0,T]} \int_0^t (t-s)^{-f^n((1/2)-2\beta-(5\varepsilon/2))} \left\|\exp(\alpha_0|Y_s^N|)\right\|_{C(\Lambda)} ds\right].
\end{aligned}$$

Since (10) implies

$$\begin{aligned}
&\mathbb{E}\left[\sup_{t\in[0,T]} \int_0^t (t-s)^{-f^n(-[(1/2)-2\beta-(5\varepsilon/2)])} \left\|\exp(\alpha_0|Y_s^N|)\right\|_{C(\Lambda)} ds\right] \\
&\quad \le C\mathbb{E}\left[\int_0^T \left\|\exp(\alpha_0|Y_s^N|)\right\|_{C(\Lambda)} ds\right] \\
&\quad \le C\|\cosh(\alpha_0\eta)\|_{C(\Lambda)},
\end{aligned}$$

we have the assertion. Q.E.D.

Lemma 3.2. *Assume that $\alpha_0 \in [0, \sqrt{2\pi})$, and take $\beta \in (\alpha_0^2/(4\pi), 1/2)$ and $\varepsilon \in (0, (1-2\beta)/4)$. For $N \in \mathbb{N}$ we have*

$$\begin{aligned}
&\mathbb{E}\left[\sup_{t\in[0,T]} \int_0^t (t-\tau)^{-\beta-2\varepsilon} \left\|\exp(\alpha_0|Y_\tau^N|)\right\|_{C(\Lambda)}^2 d\tau\right] \\
&\quad \le C\big(1 + \|\cosh(\alpha_0\eta)\|_{C(\Lambda)}\big).
\end{aligned}$$

Proof. From (9) and the smoothing property of the heat semigroup and Hölder's inequality, for $\tau \in (0,t)$ we have

$$
\begin{aligned}
&\left\|\exp(\alpha_0|Y_\tau^N|)\right\|_{C(\Lambda)}^2 \\
&\quad \le C\|\cosh(\alpha_0\eta)\|_{C(\Lambda)}^2 \\
&\qquad + C\left(\int_0^\tau (\tau-s)^{-(1+\beta+\varepsilon)/2}\left\|\exp\left(\frac{\alpha_0}{2}|Y_s^N|\right)\right\|_{C(\Lambda)}\left\|\mathcal{X}_s^{N,\nu}\right\|_{H^{-\beta}} ds\right)^2 \\
&\quad \le C\|\cosh(\alpha_0\eta)\|_{C(\Lambda)}^2 \\
&\qquad + C(t-\tau)^{-\beta/2}\int_0^\tau (\tau-s)^{-(1+\beta+\varepsilon)/2}\left\|\exp\left(\frac{\alpha_0}{2}|Y_s^N|\right)\right\|_{C(\Lambda)}^2 ds \\
&\qquad + C(t-\tau)^{\beta/2}\int_0^\tau (\tau-s)^{-(1+\beta+\varepsilon)/2}\left\|\mathcal{X}_s^{N,\nu}\right\|_{H^{-\beta}}^2 ds.
\end{aligned}
$$

Hence, Lemma 2.3 in [2] implies

$$
\begin{aligned}
&\int_0^t (t-\tau)^{-\beta-2\varepsilon}\left\|\exp(\alpha_0|Y_\tau^N|)\right\|_{C(\Lambda)}^2 d\tau \\
&\quad \le C\|\cosh(\alpha_0\eta)\|_{C(\Lambda)}^2 \\
&\qquad + C\int_0^t (t-s)^{(1/2)-2\beta-(5\varepsilon/2)}\left\|\exp(\alpha_0|Y_s^N|)\right\|_{C(\Lambda)} ds \\
&\qquad + C\int_0^t (t-s)^{(1/2)-\beta-(5\varepsilon/2)}\left\|\mathcal{X}_s^{N,\nu}\right\|_{H^{-\beta}}^2 ds.
\end{aligned}
$$

Since $(1/2)-\beta-(5\varepsilon/2)>0$, we obtain

$$
\begin{aligned}
&\sup_{t\in[0,T]}\int_0^t (t-\tau)^{-\beta-2\varepsilon}\left\|\exp(\alpha_0|Y_\tau^N|)\right\|_{C(\Lambda)}^2 d\tau \\
&\quad \le C\|\cosh(\alpha_0\eta)\|_{C(\Lambda)}^2 + C\int_0^T \left\|\mathcal{X}_s^{N,\nu}\right\|_{H^{-\beta}}^2 ds \\
&\qquad + C\sup_{t\in[0,T]}\int_0^t (t-s)^{(1/2)-2\beta-(5\varepsilon/2)}\left\|\exp(\alpha_0|Y_s^N|)\right\|_{C(\Lambda)} ds.
\end{aligned}
$$

Taking expectation and applying Lemma 3.1 yield the conclusion. Q.E.D.

Lemma 3.3. *Assume that* $\alpha_0\in[0,\sqrt{2\pi})$*, and take* $\beta\in(\alpha_0^2/(4\pi),1/2)$ *and* $\varepsilon\in(0,(1-2\beta)/4)$*. For* $N\in\mathbb{N}$ *we have*

$$
\begin{aligned}
&\mathbb{E}\left[\sup_{s,t\in[0,T];s<t}\frac{\left\|Y_t^N-Y_s^N\right\|_{L^2}}{(t-s)^\varepsilon}\right] \\
&\quad \le CE\left[\sup_{r\in[0,T]}\left\|Y_r^N\right\|_{H^{2\varepsilon}}\right] + C\|\exp(\alpha_0\eta)\|_{C(\Lambda)}^2 + C.
\end{aligned}
$$

Proof. The mild form of Y_t^N implies

$$Y_t^N = e^{-(t-s)(1-\triangle)/2} Y_s^N - \int_s^t e^{-(t-\tau)(1-\triangle)/2} \left[\int_{A_0} \alpha \exp(\alpha Y_\tau^N) \mathcal{X}_\tau^{N,\alpha} \nu(d\alpha) \right] d\tau \tag{12}$$

for $s, t \in [0, T]$ such that $s < t$. From this equality, the nonnegativity and smoothing property of heat semigroup, we have

$$\begin{aligned}
&\|Y_t^N - Y_s^N\|_{L^2} \\
&\leq \left\| \left(e^{-(t-s)(1-\triangle)/2} - I \right) Y_s^N \right\|_{L^2} \\
&\quad + \alpha_0 \int_s^t \left\| e^{-(t-\tau)(1-\triangle)/2} \int_{A_0} \exp(\alpha Y_\tau^N) \mathcal{X}_\tau^{N,\alpha} \nu(d\alpha) \right\|_{L^2} d\tau \\
&\leq C(t-s)^\varepsilon \|Y_s^N\|_{H^{2\varepsilon}} \\
&\quad + C \int_s^t \left\| e^{-(t-\tau)(1-\triangle)/2} \left[\exp(\alpha_0 |Y_\tau^N|) \mathcal{X}_\tau^{N,\nu} \right] \right\|_{L^2} d\tau \\
&\leq C(t-s)^\varepsilon \|Y_s^N\|_{H^{2\varepsilon}} + C \int_s^t (t-\tau)^{-\beta/2} \|\exp(\alpha_0 |Y_\tau^N|) \mathcal{X}_\tau^{N,\nu}\|_{H^{-\beta}} d\tau.
\end{aligned}$$

Hence, by Theorem 3.4 in [5] and Young's inequality for products

$$\begin{aligned}
&\|Y_t^N - Y_s^N\|_{L^2} \\
&\leq C(t-s)^\varepsilon \|Y_s^N\|_{H^{2\varepsilon}} \\
&\quad + C \int_s^t (t-\tau)^{-\beta/2} \|\mathcal{X}_\tau^{N,\nu}\|_{H^{-\beta}} \|\exp(\alpha_0 |Y_\tau^N|)\|_{C(\Lambda)} d\tau \\
&\leq C(t-s)^\varepsilon \|Y_s^N\|_{H^{2\varepsilon}} + C(t-s)^\varepsilon \int_s^t \|\mathcal{X}_\tau^{N,\nu}\|_{H^{-\beta}}^2 d\tau \\
&\quad + (t-s)^{-\varepsilon} \int_s^t (t-\tau)^{-\beta} \|\exp(\alpha_0 |Y_\tau^N|)\|_{C(\Lambda)}^2 d\tau.
\end{aligned}$$

From this inequality we have

$$\begin{aligned}
&\mathbb{E}\left[\sup_{s,t \in [0,T]; s<t} \frac{\|Y_t^N - Y_s^N\|_{L^2}}{(t-s)^\varepsilon} \right] \\
&\leq C\mathbb{E}\left[\sup_{r \in [0,T]} \|Y_r^N\|_{H^{2\varepsilon}} \right] + \int_0^T \mathbb{E}\left[\|\mathcal{X}_\tau^{N,\nu}\|_{H^{-\beta}}^2 \right] d\tau \\
&\quad + C\mathbb{E}\left[\sup_{t \in [0,T]} \int_0^t (t-\tau)^{-\beta-2\varepsilon} \|\exp(\alpha_0 |Y_\tau^N|)\|_{C(\Lambda)}^2 d\tau \right].
\end{aligned}$$

This inequality, Theorem 2.3 and Lemma 3.2 yield the conclusion. Q.E.D.

Lemma 3.4. *Assume that* $\alpha_0 \in [0, \sqrt{2\pi})$, *and take* $\beta \in (\alpha_0^2/(4\pi), 1/2)$ *and* $\varepsilon \in (0, (1-2\beta)/4)$. *We have*

$$\mathbb{E}\left[\sup_{t\in[0,T]} \|Y_t^N\|_{H^{2\varepsilon}}\right] \le C\big(1 + \|\cosh(\alpha_0\eta)\|_{C(\Lambda)}\big)$$

for $N \in \mathbb{N}$.

Proof. Similarly to above, from (12) and the smoothing property of the heat semigroup we have

$$\begin{aligned}
&\|Y_t^N\|_{H^{2\varepsilon}} \\
&\quad \le \left\|e^{-t(1-\triangle)/2}\eta\right\|_{H^{2\varepsilon}} \\
&\qquad + \alpha_0 \int_0^t \left\| e^{-(t-\tau)(1-\triangle)/2} \int_{A_0} \exp(\alpha Y_\tau^N)\mathcal{X}_\tau^{N,\alpha}\nu(d\alpha)\right\|_{H^{2\varepsilon}} d\tau \\
&\quad \le \left\|e^{-t(1-\triangle)/2}\eta\right\|_{H^{2\varepsilon}} \\
&\qquad + C\int_0^t (t-\tau)^{-\varepsilon} \left\| e^{-(t-\tau)(1-\triangle)/4} \int_{A_0} \exp(\alpha Y_\tau^N)\mathcal{X}_\tau^{N,\alpha}\nu(d\alpha)\right\|_{L^2} d\tau.
\end{aligned}$$

Hence, the nonnegativity, contraction and smoothing property of the heat semigroup imply

$$\begin{aligned}
&\|Y_t^N\|_{H^{2\varepsilon}} \\
&\le C\|\eta\|_{H^{2\varepsilon}} + C\int_0^t (t-\tau)^{-\varepsilon} \left\| e^{-(t-\tau)(1-\triangle)/4}\big[\exp(\alpha_0|Y_\tau^N|)\mathcal{X}_\tau^{N,\nu}\big]\right\|_{L^2} d\tau \\
&\le C\|\eta\|_{H^{2\varepsilon}} + C\int_0^t (t-\tau)^{-\beta/2-\varepsilon} \left\|\exp(\alpha_0|Y_\tau^N|)\mathcal{X}_\tau^{N,\nu}\right\|_{H^{-\beta}} d\tau.
\end{aligned}$$

By applying Theorem 3.4 in [5] and Hölder's inequality we have

$$\begin{aligned}
&\|Y_t^N\|_{H^{2\varepsilon}} \\
&\le C\|\eta\|_{H^{2\varepsilon}} + C\int_0^t (t-\tau)^{-\beta/2-\varepsilon} \left\|\mathcal{X}_\tau^{N,\nu}\right\|_{H^{-\beta}} \left\|\exp(\alpha_0|Y_\tau^N|)\right\|_{C(\Lambda)} d\tau \\
&\le C\|\eta\|_{H^{2\varepsilon}} + C\int_0^t \left\|\mathcal{X}_\tau^{N,\nu}\right\|_{H^{-\beta}}^2 d\tau \\
&\qquad\qquad + \int_0^t (t-\tau)^{-\beta-2\varepsilon} \left\|\exp(\alpha_0|Y_\tau^N|)\right\|_{C(\Lambda)}^2 d\tau.
\end{aligned}$$

This inequality implies

$$\mathbb{E}\left[\sup_{t\in[0,T]}\|Y_t^N\|_{H^{2\varepsilon}}\right] \leq C\|\eta\|_{H^{2\varepsilon}} + C\int_0^T \mathbb{E}\left[\|\mathcal{X}_\tau^{N,\nu}\|_{H^{-\beta}}^2\right]d\tau + \mathbb{E}\left[\sup_{t\in[0,T]}\int_0^t (t-\tau)^{-\beta-2\varepsilon}\left\|\exp(\alpha_0|Y_\tau^N|)\right\|_{C(\Lambda)}^2 d\tau\right].$$

By applying Theorem 2.3 and Lemma 3.2, we obtain the assertion.

Q.E.D.

Theorem 3.5. *Assume that* $\alpha_0 \in [0, \sqrt{2\pi})$, *and take* $\beta \in (\alpha_0^2/(4\pi), 1/2)$ *and* $\varepsilon \in (0, (1-2\beta)/4)$. *For* $N \in \mathbb{N}$

$$\mathbb{E}\left[\sup_{t\in[0,T]}\|Y_t^N\|_{H^{2\varepsilon}}\right] + \mathbb{E}\left[\sup_{s,t\in[0,T];s<t}\frac{\|Y_t^N - Y_s^N\|_{L^2}}{(t-s)^\varepsilon}\right] \leq C\big(1 + \|\cosh(\alpha_0\eta)\|_{C(\Lambda)}\big).$$

Proof. Lemmas 3.3 and 3.4 yield the assertion. Q.E.D.

Theorem 3.6. *Assume that* $\alpha_0 \in [0, \sqrt{2\pi})$, *and take* $\beta \in (\alpha_0^2/(4\pi), 1/2)$ *and* $\varepsilon \in (0, (1-2\beta)/4)$. *Then, the laws of* $\{Y^N\}$ *are tight on* $C([0,T]; L^2)$.

Moreover, if Y *is a limit of a subsequence of* $\{Y^N\}$ *in law, then* $Y \in \mathcal{Y}_T$ *almost surely.*

Proof. For $h \in (0, 1]$ and $\varepsilon' \in (0, 1]$, Chebyshev's inequality implies that

$$\sup_{N\in\mathbb{N}}\mathbb{P}\left(\sup_{s,t\in[0,T];|s-t|<h}\|Y_t^N - Y_s^N\|_{L^2} > \varepsilon'\right) \leq \frac{h^\varepsilon}{\varepsilon'}\mathbb{E}\left[\sup_{s,t\in[0,T];s<t,t-s<h}\frac{\|Y_t^N - Y_s^N\|_{L^2}}{(t-s)^\varepsilon}\right].$$

Hence, from Theorem 3.5 we obtain

$$\lim_{h\downarrow 0}\sup_{N\in\mathbb{N}}\mathbb{P}\left(\sup_{s,t\in[0,T];|s-t|<h}\|Y_t^N - Y_s^N\|_{L^2} > \varepsilon'\right) = 0$$

for $\varepsilon' \in (0, 1]$. In view that the initial value of Y^N is a common value $\eta \in L^2$ in N, the tightness of the laws of $\{Y^N\}$ on $C([0,T]; L^2)$ follows.

Let Y be a limit of a subsequence of $\{Y^N\}$ in law. Then, $Y \in C([0,T];L^2)$. Taking limit in (10), we have $\exp(\alpha_0|Y|) \in L^1([0,T];C(\Lambda))$ almost surely. Integrating both sides of (8) by dx on Λ, we have

$$\begin{aligned}
&\int_0^T \int_\Lambda \cosh(\alpha_0 Y_s^N)|\nabla Y_s^N|^2 dxds \\
&\quad \le C\int_\Lambda \cosh(\alpha_0\eta)dx + C\int_0^T \int_\Lambda \cosh\left(\frac{\alpha_0}{2}Y_s^N\right)\mathcal{X}_s^{N,\nu} dxds \\
&\quad \le C\|\cosh(\alpha_0\eta)\|_{L^1} + C\int_0^T \left\|\cosh\left(\frac{\alpha_0}{2}Y_s^N\right)\right\|_{H^1} \|\mathcal{X}_s^{N,\nu}\|_{H^{-1}} ds.
\end{aligned}$$

Hence, for $\lambda \in (0,1)$ it holds that

$$\begin{aligned}
&\int_0^T \int_\Lambda \cosh(\alpha_0 Y_s^N)|\nabla Y_s^N|^2 dxds \\
&\quad \le C\|\cosh(\alpha_0\eta)\|_{L^1} + \lambda\int_0^T \left\|\cosh\left(\frac{\alpha_0}{2}Y_s^N\right)\right\|_{H^1}^2 ds \\
&\qquad + C\lambda^{-1}\int_0^T \|\mathcal{X}_s^{N,\nu}\|_{H^{-1}}^2 ds.
\end{aligned}$$

This inequality and the fact that

$$\begin{aligned}
&\left\|\cosh\left(\frac{\alpha_0}{2}Y_s^N\right)\right\|_{H^1}^2 \\
&\quad \le 2\left\|\cosh\left(\frac{\alpha_0}{2}Y_s^N\right)\right\|_{L^2}^2 + \frac{\alpha_0^2}{2}\left\|\sinh\left(\frac{\alpha_0}{2}Y_s^N\right)\nabla Y_s^N\right\|_{L^2}^2 \\
&\quad \le 2\left\|\cosh(\alpha_0 Y_s^N)\right\|_{L^1} + \frac{\alpha_0^2}{2}\left\|\cosh\left(\frac{\alpha_0}{2}Y_s^N\right)|\nabla Y_s^N|\right\|_{L^2}^2 \\
&\quad \le 2\left\|\cosh(\alpha_0 Y_s^N)\right\|_{L^1} + \frac{\alpha_0^2}{2}\left\|\cosh(\alpha_0 Y_s^N)|\nabla Y_s^N|^2\right\|_{L^1},
\end{aligned}$$

imply

$$\begin{aligned}
&\int_0^T \int_\Lambda \cosh(\alpha_0 Y_s^N)|\nabla Y_s^N|^2 dxds \\
&\quad \le C\|\cosh(\alpha_0\eta)\|_{L^1} + 2\lambda\int_0^T \int_\Lambda \cosh(\alpha_0 Y_s^N)dxds \\
&\qquad + C\lambda\int_0^T \int_\Lambda \cosh(\alpha_0 Y_s^N)|\nabla Y_s^N|^2 dxds + C\lambda^{-1}\int_0^T \|\mathcal{X}_s^{N,\nu}\|_{H^{-1}}^2 ds.
\end{aligned}$$

Taking λ sufficiently small and applying (10) and Theorem 2.3, we have

$$\mathbb{E}\left[\int_0^T\int_\Lambda \cosh(\alpha_0 Y_s^N)|\nabla Y_s^N|^2 dx ds\right] \le C\|\cosh(\alpha_0\eta)\|_{L^1} + C.$$

Thus, by noting that $1 \le \cosh(\alpha_0 Y_s^N)$, and taking limit as $N \to \infty$, we have $Y \in L^2([0,T];H^1)$ almost surely. Q.E.D.

Remark 3.7. *We do not know whether Y obtained in Theorem 3.6 satisfies the limit equation of (7) as $N \to \infty$, because the uniform estimates obtained in this section are not sufficient to take the limits of each terms in (7). Indeed, we do not have the strong convergence of* $\exp(\alpha Y^N)$ *in $L^1([0,T];C(\Lambda))$ (see Theorem 3.6 of [5]), while we obtained in the proof of Theorem 3.6 that the limit satisfies* $\exp(\alpha Y) \in L^1([0,T];C(\Lambda))$.

As in Section 1, by applying Theorem 3.6, we obtain the main theorem (Theorem 1.1).

§4. Remark on the uniqueness of the shifted equation

Let $\beta \in (\alpha_0^2/(4\pi), 1)$. As a special case of Theorem 2.3, we have the convergence of $\mathcal{X}^{N,\alpha}$ to an $L^2([0,T];H^{-\beta})$-valued nonnegative random variable $\mathcal{X}^\alpha$ for $\alpha \in [-\alpha_0, \alpha_0]$. In view of this fact, in this section we consider the deterministic equation

$$\partial_t \Upsilon_t = \frac{1}{2}(\triangle - 1)\Upsilon_t - \frac{1}{2}\int_{[-\alpha_0,\alpha_0]} \alpha e^{\alpha\Upsilon_t}\mathcal{X}_t^\alpha \nu(d\alpha)$$

for any generic nonnegative $\mathcal{X}^\alpha \in L^2([0,T];H^{-\beta})$ such that $\alpha \mapsto \mathcal{X}^\alpha$ is Borel measurable and integrable with respect to ν i.e.

$$\int_{[-\alpha_0,\alpha_0]} \|\mathcal{X}_\cdot^\alpha\|_{L^2([0,T];H^{-\beta})}\, \nu(d\alpha) < \infty.$$

This equation is regarded as the limit equation of the shifted equation (7). In this section, we consider the uniqueness of this equation by following the argument in Section 3.2 in [5].

Consider the initial value problem

$$\text{(13)}\qquad \begin{cases} \partial_t \Upsilon_t = \frac{1}{2}(\triangle - 1)\Upsilon_t - \frac{1}{2}\int_{[-\alpha_0,\alpha_0]} \alpha\mathcal{M}(e^{\alpha\Upsilon_t}, \mathcal{X}_t^\alpha)\nu(d\alpha), \\ \Upsilon_0 = \upsilon, \end{cases}$$

for any given $\mathcal{X} \in L^2([0,T];H_+^{-\beta})$ and $\upsilon \in H^{2-\beta}$. Here, $\mathcal{M}(f,\xi)$ is the nonnegative distribution given by the multiplication of ξ by f for

$f \in C(\Lambda)$ and a nonnegative distribution ξ on Λ, which is defined via the one-to-one correspondence between Borel measures and nonnegative distributions (see Section 3.1 in [5]). We introduce the space

$$\widetilde{\mathcal{Y}}_T = \{\Upsilon \in L^2([0,T]; C(\Lambda) \cap H^1) \cap C([0,T]; L^2) \ ; \\ e^{\alpha_0|\Upsilon|} \in L^\infty([0,T]; C(\Lambda))\}.$$

We remark that $\widetilde{\mathcal{Y}}_T \subsetneq \mathcal{Y}_T$. When we choose $\widetilde{\mathcal{Y}}_T$ as the solution space, we have the uniqueness of the solution to (13) as follows.

Proposition 4.1. *For any nonnegative $\mathcal{X}^\alpha \in L^2([0,T]; H^{-\beta})$ such that $\alpha \mapsto \mathcal{X}^\alpha$ is Borel measurable and integrable with respect to ν, and $\upsilon \in H^{2-\beta}$, there is at most one mild solution $\Upsilon \in \widetilde{\mathcal{Y}}_T$ of the equation* (13).

Proof. Let $\Upsilon, \Upsilon' \in \widetilde{\mathcal{Y}}_T$ be two solutions of (13) with the same $\mathcal{X}^\alpha$ and υ. Then $Z = \Upsilon - \Upsilon'$ solves the equation

$$\left\{\partial_t - \frac{1}{2}(\triangle - 1)\right\} Z_t = -\frac{1}{2}\int_{[-\alpha_0,\alpha_0]} \alpha\mathcal{M}(e^{\alpha\Upsilon_t} - e^{\alpha\Upsilon'_t}, \mathcal{X}_t^\alpha)\nu(d\alpha) =: D_t.$$

Since $e^{\alpha\Upsilon}, e^{\alpha\Upsilon'} \in L^\infty([0,T]; C(\Lambda))$ and $\mathcal{X}^\alpha \in L^2([0,T]; H_+^{-\beta})$, we have that $D \in L^2([0,T]; H^{-\beta})$ by Theorem 3.4 in [5]. Similarly to the proof of Lemma 3.7 in [5], we have

$$\int_\Lambda |Z_t(x)|^2 dx = -\int_0^t \int_\Lambda |\nabla Z_s(x)|^2 dxds \\ -\int_0^t \int_\Lambda |Z_s(x)|^2 dxds + 2\int_0^t \int_\Lambda Z_s(x) D_s(x) dxds.$$

Since

$$\begin{aligned} &2\int_\Lambda Z_s(x) D_s(x) dx \\ &= -\int_\Lambda \left(\int_{[-\alpha_0,\alpha_0]} \alpha\mathcal{M}\left((e^{\alpha\Upsilon_s(x)} - e^{\alpha\Upsilon'_s(x)})Z_s(x), \mathcal{X}_s^\alpha\right)\nu(d\alpha)\right) dx \\ &= -\int_\Lambda \left(\int_{[-\alpha_0,\alpha_0]} \alpha^2\mathcal{M}\left(e^{A(\alpha\Upsilon_s(x),\alpha\Upsilon'_s(x))}|Z_s(x)|^2, \mathcal{X}_s^\alpha\right)\nu(d\alpha)\right) dx \\ &\le 0, \end{aligned}$$

where $A(x,y)$ is a continuous function on $\mathbb{R}^2$ defined by

$$A(x,y)=\begin{cases}\log\dfrac{e^x-e^y}{x-y}, & x\neq y,\\ x, & x=y,\end{cases}$$

we have $\|Z_t\|_{L^2}=0$ for any $t\in(0,T]$. Hence, $\Upsilon=\Upsilon'$ in $\widetilde{\mathcal{Y}}_T$. Q.E.D.

Unfortunately we are not able to apply Proposition 4.1 to Y appeared in Theorem 1.1, because $\widetilde{\mathcal{Y}}_T\subsetneq\mathcal{Y}_T$ and we do not know whether Y obtained in Theorem 3.6 satisfies (13).

Acknowledgements. This work was partially supported by JSPS KAKENHI Grant Numbers 17K05300, 17K14204 and 19K14556.

References

[1] S. Albeverio and R. Høegh-Krohn, Uniqueness of the physical vacuum and the Wightman functions in the infinite volume limit for some non polynomial interactions, Comm. Math. Phys., **30** (1973), 171–200.
[2] S. Albeverio and S. Kusuoka, The invariant measure and the flow associated to the Φ^4_3-quantum field model, Ann. Sc. Norm. Super. Pisa Cl. Sci. (5), **20** (2020), 1359–1427.
[3] G. Da Prato and A. Debussche, Strong solutions to the stochastic quantization equations, Ann. Probab., **31** (2003), 1900–1916.
[4] R. Høegh-Krohn, A general class of quantum fields without cut-offs in two space-time dimensions, Comm. Math. Phys., **21** (1971), 244–255.
[5] M. Hoshino, H. Kawabi and S. Kusuoka, Stochastic quantization associated with the $\exp(\Phi)_2$-quantum field model driven by space-time white noise on the torus, J. Evol. Equ., **21** (2021), 339–375.
[6] T. Oh, T. Robert and Y. Wang, On the parabolic and hyperbolic Liouville equations, arXiv:1908.03944.
[7] B. Simon, The $P(\phi)_2$ Euclidean (Quantum) Field Theory, Princeton Series in Physics, Princeton University Press, Princeton, N.J., 1974.

Masato Hoshino:
Graduate School of Engineering Science, Osaka University
1-3 Machikaneyama, Toyonaka,
Osaka 560-8531, Japan
E-mail address: hoshino@sigmath.es.osaka-u.ac.jp

Hiroshi Kawabi:
Department of Mathematics, Hiyoshi Campus, Keio University,
4-1-1 Hiyoshi, Kohoku-ku,
Yokohama 223-8521, Japan
E-mail address: kawabi@keio.jp

Seiichiro Kusuoka:
Department of Mathematics,Graduate School of Science, Kyoto University,
Kitashirakawa Oiwakecho, Sakyo-ku,
Kyoto 606-8502, Japan
E-mail address: kusuoka@math.kyoto-u.ac.jp

Advanced Studies in Pure Mathematics 87, 2021
Stochastic Analysis, Random Fields and Integrable Probability — Fukuoka 2019
pp. 363–379

Divergence of non-random fluctuation for Euclidean first-passage percolation

Shuta Nakajima

Abstract.

The non-random fluctuation is one of the central objects in first passage percolation. It was proved in [11] that for a particular asymptotic direction, it diverges in a lattice first passage percolation with an explicit lower bound. In this paper, we discuss the non-random fluctuation in Euclidean first passage percolations and show that it diverges in dimension $d \geq 2$ in this model also. Compared with the result in [11], the present result is proved for any direction and improves the lower bound.

§1. Introduction

First-passage percolation (FPP) was introduced by Hammersley and Welsh as a dynamical model of infection. One of the motivations of the studies on FPP is to understand the general behavior of subadditive processes. To do this, a number of techniques and phenomena, such as Kingman's subadditive ergodic theorem and a sublinear variance, have been discovered and they have born fruitful results. See [2] on the backgrounds and related topics.

We consider an Euclidean FPP on $\mathbb{R}^d$ with $d \geq 2$, which is a variant of classical FPP and introduced in [9]. The model is defined as follows. We consider a Poisson point process Ξ with Lebesgue intensity on $\mathbb{R}^d$. We regard Ξ as a subset of $\mathbb{R}^d$. For any $x \in \mathbb{R}^d$, we denote by $D(x)$ the closest point of Ξ to x with respect to the Euclidean norm $|\cdot|$. If there are multiple choices, we take one of them with a deterministic rule to break ties, though it does not happen almost surely.

Received March 19, 2020.
Revised June 20, 2020.
2010 *Mathematics Subject Classification.* Primary 60K37; Secondary 60K35, 82A51, 82D30.
Key words and phrases. random environment, First-passage percolation.

A path γ is a finite sequence of points $(x_0, \dots, x_\ell) \subset \Xi$. Then we write $\gamma : x_0 \to x_\ell$. We fix $\alpha > 1$. Given a path γ, we define the passage time of $\gamma = (x_i)_{i=0}^{\ell}$ as

$$\mathrm{T}(\gamma) = \sum_{i=1}^{\ell} |x_i - x_{i-1}|^\alpha, \tag{1.1}$$

where $|\cdot|$ is the Euclidean norm. For $x, y \in \mathbb{R}^d$, we define the *first passage time* between x and y as

$$\mathrm{T}(x,y) = \inf_{\gamma : D(x) \to D(y)} \mathrm{T}(\gamma),$$

where the infimum is taken over all finite paths γ starting at $D(x)$ and ending at $D(y)$. It should be noted that if $\alpha \leq 1$, $\mathrm{T}(x,y) = |D(x) - D(y)|^\alpha$, which is a rather trivial model. Hence we suppose $\alpha > 1$. A path γ from $D(x)$ to $D(y)$ is said to be *optimal* if it attains the first passage time between x and y, i.e. $\mathrm{T}(\gamma) = \mathrm{T}(x,y)$. Note that for $x, y \in \mathbb{R}^d$, the optimal path between x and y is uniquely determined almost surely.

One of the important property in our model is the so-called rotational invariance [6, p.20]. Indeed, for any rotation matrix, say A, $A\Xi = \{Ax \mid x \in \Xi\}$ has the same distribution as Ξ. Hence, $(\mathrm{T}(Ax, Ay))_{x,y \in \mathbb{R}^d}$ also has the same distribution as $(\mathrm{T}(x,y))_{x,y \in \mathbb{R}^d}$.

By Kingman's subadditive ergodic theorem, for any $x \in \mathbb{R}^d \backslash \{0\}$, there exists a non-random constant $\mathrm{g} \geq 0$ such that

$$\mathrm{g} = \lim_{t \to \infty} (t|x|)^{-1} \mathrm{T}(0, tx) = \lim_{t \to \infty} (t|x|)^{-1} \mathbb{E}[\mathrm{T}(0, tx)] \quad \text{a.s.} \tag{1.2}$$

This g, called the *time constant*, is independent of the choice of x because of the rotational invariance. Moreover it is known from [9, Theorem 1] that g is positive. Note that, since T is subadditive (i.e. $\mathrm{T}(x,z) \leq \mathrm{T}(x,y) + \mathrm{T}(y,z)$), we have for $x \in \mathbb{R}^d$,

$$\mathrm{g}|x| \leq \mathbb{E}\mathrm{T}(0, x). \tag{1.3}$$

1.1. Main results

We define

$$\psi(t) = \mathrm{Var}(\mathrm{T}(0, t\mathbf{e}_1)),\ \phi(t) = \sqrt{\frac{t}{\psi(t)}},$$

where $(\mathbf{e}_i)_{i=1}^d$ is the canonical basis of $\mathbb{R}^d$. It was proved in [4] that $\psi(t) \leq \frac{C\,t}{\log t}$ with some constant $C > 0$, and thus $\phi(t) \geq c\sqrt{\log t}$ with $c = C^{-\frac{1}{2}} > 0$. Moreover it is expected that $\psi(t) = O(t^{\beta})$ with some $\beta < 1/2$. It is also known that $\psi(t) \geq c$, and hence $\phi(t) \leq C\sqrt{t}$ with some c, $C > 0$ (see [10, (1.13)] for a lattice FPP, and Appendix for the detailed proof).

The following is our main result, which gives an explicit lower bound coming from the variance $\psi(t)$. It should be noted that there have been no results concerning with the lower bound of the non-random fluctuation in Euclidean FPP. The related work in a lattice FPP is introduced in Section 1.2.

Theorem 1. *There exists $c > 0$ such that for any $x \in \mathbb{R}^d$ satisfying $|x| > 1$,*

$$\mathbb{E}\mathrm{T}(0,x) - \mathrm{g}|x| \geq c \log \phi(|x|).$$

In particular, by Jensen's inequality,

$$\mathbb{E}\big|\mathrm{T}(0,x) - \mathrm{g}|x|\big| \geq c \log \phi(|x|).$$

1.2. Related work

The main issue of FPP is to understand the behavior of $\mathrm{T}(0,x)$ as $|x| \to \infty$. Since Kingman proved a kind of law of large numbers as in (1.2), the next question was the asymptotic of $\mathrm{T}(0,x) - \mathrm{g}(x)$ with a natural scaling. Thus, the typical order of $\mathrm{T}(0,x) - \mathrm{g}(x)$ was of great interest in search of scaling. To study this, Kesten considered the following decomposition:

$$\mathrm{T}(0,x) - \mathrm{g}(x) = [\mathrm{T}(0,x) - \mathbb{E}\mathrm{T}(0,x)] + [\mathbb{E}\mathrm{T}(0,x) - \mathrm{g}(x)]. \tag{1.4}$$

The first term $\mathrm{T}(0,x) - \mathbb{E}\mathrm{T}(0,x)$ is called the random fluctuation, while the second term $\mathrm{T}(0,x) - \mathbb{E}\mathrm{T}(0,x)$ is called the non-random fluctuation. Kesten's idea is the following. First, we study the variance of $\mathrm{T}(0,x)$ and estimate the random fluctuation from it. Second, using the estimate of the random fluctuation, we estimate the non-random fluctuation.

Following the idea, there have been several attempts to study them [10, 1, 3, 7, 8, 4]. In particular, Alexander [1] developed a new and strong method to derive the upper bound of the non-random fluctuation from a concentration inequality for the random fluctuation. Nevertheless, there are few results on the lower bounds of the non-random fluctuations due to the lack of understanding and techniques.

In the classical FPP, the author proved the divergence of the non-random fluctuation [11]. However, there are at least two drawbacks. First, the result was not stated for a fixed direction. Second, the estimate is anything but sharp, where the lower bound is given by $(\log\log n)^{\frac{1}{d}}$. In this paper, changing the model, we overcome these problems. Indeed, by the rotational invariance of our model, we not only prove the result for any fixed direction, but improve the bound, though we are not sure if this is sharp. Moreover, the argument may be transparent because some of the cumbersome terms disappear in our argument.

1.3. Notation and terminology

This subsection collects some notations and terminologies for the proof.

- Let us define the Euclidean ball $\mathrm{B}(x,r)$ for $x \in \mathbb{R}^d$ and $r > 0$ as
$$\mathrm{B}(x,r) = \left\{y \in \mathbb{R}^d \mid |x-y| \leq r\right\}.$$
For $x = 0$, we simply write $\mathrm{B}(r)$ instead of $\mathrm{B}(x,r)$.
- For $a \in \mathbb{R}$, $\lfloor a \rfloor$ is the greatest integer less than or equal to a. Given $x = (x_i)_{i=1}^d \in \mathbb{R}^d$, we define $\lfloor x \rfloor = (\lfloor x_i \rfloor)_{i=1}^d$.
- Given $a, b, y \in \mathbb{R}^d$, we define $\mathrm{T}(a,y,b) = \mathrm{T}(a,y) + \mathrm{T}(y,b)$, which is the first passage time from $D(a)$ to $D(b)$ passing through $D(y)$.
- We denote by $\Gamma(x,y)$ and $\Gamma(x,y,z)$ the optimal paths of $\mathrm{T}(x,y)$ and $\mathrm{T}(x,y,z)$, respectively.
- Given a Borel set $A \subset \mathbb{R}^d$, let denote $\mathrm{Vol}(A)$ the d-dimensional volume of A.

§2. Proof of the main theorem

We only consider the $\mathbf{e}_1$-direction, i.e. $x = \mathbf{e}_1$, since another direction is the same by the rotational invariance. We write $\mathrm{T}_n = \mathrm{T}(0, n\mathbf{e}_1)$. Let us denote $\mathbb{L} = \{(x_i) \in \mathbb{R}^d \mid x_1 = 0\}$. Given sufficiently large $n > 0$, one can find a finite subset $\mathbb{L}_n$ of $\mathbb{L}$ such that

$$\begin{cases} \sharp\mathbb{L}_n = \lfloor \phi(n)^{\frac{1}{2}} \rfloor, \\ \text{if } a \neq b \in \mathbb{L}_n,\ |a-b| \geq \sqrt{n}\,\phi(n)^{-\frac{1}{2}}, \\ \text{for any } a \in \mathbb{L}_n,\ \sqrt{n}\phi(n)^{-\frac{1}{2}} \leq |a| \leq \sqrt{n}. \end{cases} \tag{2.1}$$

Given $y \in \mathbb{L}_n$, let us define

(2.2)
$$\mathcal{A}^y_{(2.2)} = \left\{\forall z \in \mathbb{L}_n \text{ with } z \neq y,\ \mathrm{T}(-n\mathbf{e}_1, y, n\mathbf{e}_1) < \mathrm{T}(-n\mathbf{e}_1, z, n\mathbf{e}_1)\right\}.$$

Proposition 1. *For any $K > 0$,*

$$2(\mathbb{E}\mathrm{T}_n - \mathrm{g}n) \geq K \sum_{y \in \mathbb{L}_n} \mathbb{P}\big(\big\{\mathrm{T}(-n\mathbf{e}_1, 0, n\mathbf{e}_1) - \mathrm{T}(-n\mathbf{e}_1, y, n\mathbf{e}_1) > K\big\} \cap \mathcal{A}^y_{(2.2)}\big). \tag{2.3}$$

Proof. For any $n > 1$, observe that by (1.3),

$$\begin{aligned} &2(\mathbb{E}\mathrm{T}_n - \mathrm{g}n) \\ &= \mathbb{E}[\mathrm{T}(-n\mathbf{e}_1, 0, n\mathbf{e}_1) - \mathrm{T}(-n\mathbf{e}_1, n\mathbf{e}_1)] + (\mathbb{E}[\mathrm{T}(-n\mathbf{e}_1, n\mathbf{e}_1)] - 2\mathrm{g}n) \\ &\geq \mathbb{E}[\mathrm{T}(-n\mathbf{e}_1, 0, n\mathbf{e}_1) - \mathrm{T}(-n\mathbf{e}_1, n\mathbf{e}_1)]. \end{aligned}$$

Since $\mathrm{T}(x, y, z) \geq \mathrm{T}(x, z)$ for any $x, y, z \in \mathbb{R}^d$ and $\{\mathcal{A}^y_{(2.2)}\}_{y \in \mathbb{L}_n}$ are disjoint, we have

$$\begin{aligned} &\mathbb{E}[\mathrm{T}(-n\mathbf{e}_1, 0, n\mathbf{e}_1) - \mathrm{T}(-n\mathbf{e}_1, n\mathbf{e}_1)] \\ &\geq \sum_{y \in \mathbb{L}_n} \mathbb{E}[\mathrm{T}(-n\mathbf{e}_1, 0, n\mathbf{e}_1) - \mathrm{T}(-n\mathbf{e}_1, n\mathbf{e}_1);\ \mathcal{A}^y_{(2.2)}] \\ &\geq \sum_{y \in \mathbb{L}_n} \mathbb{E}[\mathrm{T}(-n\mathbf{e}_1, 0, n\mathbf{e}_1) - \mathrm{T}(-n\mathbf{e}_1, y, n\mathbf{e}_1);\ \mathcal{A}^y_{(2.2)}]. \end{aligned}$$

By the first moment mothod, this is further bounded from below by the right hand side of (2.3). □

We take $K = K_n(\theta) = \theta \log \phi(n)$ for a fixed θ to be chosen later. The next proposition is useful to estimate the right hand side of (2.3) from below.

Proposition 2. *There exists $\theta > 0$ such that for sufficiently large $n > 1$ and $y \in \mathbb{L}_n$,*

$$\begin{aligned} &\mathbb{P}\big(\big\{\mathrm{T}(-n\mathbf{e}_1, 0, n\mathbf{e}_1) - \mathrm{T}(-n\mathbf{e}_1, y, n\mathbf{e}_1) > K_n\big\} \cap \mathcal{A}^y_{(2.2)}\big) \\ &\geq \exp\left(-\frac{1}{4}\log\phi(n)\right)\left(\frac{3}{4} - K_n^{-1}\big(\mathbb{E}[\mathrm{T}(-n\mathbf{e}_1, y, n\mathbf{e}_1)] - 2\mathrm{g}n\big)\right). \end{aligned} \tag{2.4}$$

Before proving Propositions 2, we shall complete the proof of Theorem 1. First, suppose that n is large enough and there exists $y \in \mathbb{L}_n$ such that $\mathbb{E}[\mathrm{T}(-n\mathbf{e}_1, y)] - \mathrm{g}n \geq K_n/4$. By $n \leq |y + n\mathbf{e}_1| \leq \sqrt{n^2 + n} \leq n + 1$

and the rotational invariance of T,

$$\begin{aligned}
&\mathbb{E}[\mathrm{T}_n] - \mathrm{g}n \\
&= \mathbb{E}\left[\mathrm{T}\left(0, n\frac{y+n\mathbf{e}_1}{|y+n\mathbf{e}_1|}\right)\right] - \mathrm{g}n \\
&= \mathbb{E}[\mathrm{T}(0, y+n\mathbf{e}_1)] - \mathrm{g}n + \mathbb{E}\left[\mathrm{T}\left(0, n\frac{y+n\mathbf{e}_1}{|y+n\mathbf{e}_1|}\right) - \mathrm{T}(0, y+n\mathbf{e}_1)\right] \\
&\geq \mathbb{E}[\mathrm{T}(0, y+n\mathbf{e}_1)] - \mathrm{g}n - \mathbb{E}\left[\mathrm{T}\left(n\frac{y+n\mathbf{e}_1}{|y+n\mathbf{e}_1|}, |y+n\mathbf{e}_1|\frac{y+n\mathbf{e}_1}{|y+n\mathbf{e}_1|}\right)\right] \\
&= \mathbb{E}[\mathrm{T}(-n\mathbf{e}_1, y)] - \mathrm{g}n - \mathbb{E}\mathrm{T}_{|y+n\mathbf{e}_1|-n} \geq \frac{K_n}{8} = \frac{\theta}{8}\log\phi(n),
\end{aligned}$$

as desired. Otherwise, if for any $y \in \mathbb{L}_n$, $\mathbb{E}[\mathrm{T}(-n\mathbf{e}_1, y) - \mathrm{g}n] \leq K_n/4$, then

$$\begin{aligned}
\mathbb{E}[\mathrm{T}(-n\mathbf{e}_1, y, n\mathbf{e}_1)] - 2\mathrm{g}n &= 2\mathbb{E}[\mathrm{T}(-n\mathbf{e}_1, y) - \mathrm{g}n] \\
&\leq \frac{1}{2}K_n = \frac{\theta}{2}\log\phi(n).
\end{aligned}$$

This, combined with Propositions 1 and 2, implies that

$$\begin{aligned}
\mathbb{E}[\mathrm{T}_n] - \mathrm{g}n &\geq \frac{1}{4}K_n \sum_{y\in\mathbb{L}_n} \exp\left(-\frac{1}{4}\log\phi(n)\right) \\
&= \frac{1}{4}K_n \lfloor \phi(n)^{\frac{1}{2}} \rfloor \exp\left(-\frac{1}{4}\log\phi(n)\right) > \frac{K_n}{4} = \frac{\theta}{4}\log\phi(n).
\end{aligned}$$

Therefore, the proof of Theorem 1 is completed. Thus, it remains to prove Proposition 2. We prepare some notations for the proof.

Definition 1. *We define the events* $\mathcal{A}_{(2.5)}(\mathbb{L}_n)$, $\mathcal{A}_{(2.6)}(\mathbb{L}_n)$ *and* $\mathcal{A}_{(2.7)}(\mathbb{L}_n)$ *as*

(2.5)
$$\mathcal{A}_{(2.5)}(\mathbb{L}_n) = \left\{\forall a, b \in \mathbb{L}_n \cup \{0\} \text{ with } a \neq b,\ \mathrm{T}(a,b) \geq \frac{\sqrt{n}}{\phi(n)^{\frac{3}{5}}}\right\},$$

(2.6)
$$\mathcal{A}_{(2.6)}(\mathbb{L}_n) = \left\{\forall y \in \mathbb{L}_n \cup \{0\},\ \max_{z=\pm n\mathbf{e}_1} \left|\mathrm{T}(z,y) - \mathbb{E}[\mathrm{T}(z,y)]\right| \leq \frac{\sqrt{n}}{\phi(n)^{\frac{2}{3}}}\right\},$$

(2.7)
$$\mathcal{A}_{(2.7)}(\mathbb{L}_n) = \mathcal{A}_{(2.5)}(\mathbb{L}_n) \cap \mathcal{A}_{(2.6)}(\mathbb{L}_n).$$

Let δ be a small positive number to be specified later, and set $C_\delta = 4(1+\delta^{-1})$.

Definition 2. *Recall the notation α from* (1.1). *We define*

(2.8)
$$\mathrm{V}_{(2.8)}(\mathbb{L}_n) = \left\{ y \in \mathbb{L}_n \;\middle|\; \begin{array}{c} \forall \ell \in \mathbb{Z} \text{ with } \ell \geq (K_n)^{\frac{1}{2\alpha}}, \\ \forall x \in \mathrm{B}(y, C_\delta K_n + \ell) \cap \mathbb{Z}^d, \\ \Xi \cap \mathbb{B}(x, \ell^{\frac{1}{2}}) \neq \emptyset \end{array} \right\},$$

(2.9)
$$\mathrm{W}_{(2.9)}(\mathbb{L}_n) = \left\{ y \in \mathbb{L}_n \;\middle|\; \begin{array}{c} \forall a, b \in \mathrm{B}(y, 2C_\delta K_n) \text{ with } |a-b| \geq K_n, \\ \mathrm{T}(a,b) \geq \delta |a-b| \end{array} \right\},$$

(2.10)
$$\mathrm{X}_{(2.10)}(\mathbb{L}_n) = \{ y \in \mathbb{L}_n \mid \mathrm{T}(-n\mathbf{e}_1, y, n\mathbf{e}_1) - \mathrm{T}(-n\mathbf{e}_1, n\mathbf{e}_1) < K_n \},$$

(2.11)
$$\mathrm{Y}_{(2.11)}(\mathbb{L}_n) = \mathrm{V}_{(2.8)}(\mathbb{L}_n) \cap \mathrm{W}_{(2.9)}(\mathbb{L}_n) \cap \mathrm{X}_{(2.10)}(\mathbb{L}_n).$$

If it is clear from the context, we simply write $\mathcal{A}_{(2.5)}, \mathcal{A}_{(2.6)}, \mathcal{A}_{(2.7)}$ instead of $\mathcal{A}_{(2.5)}(\mathbb{L}_n)$, $\mathcal{A}_{(2.6)}(\mathbb{L}_n)$, $\mathcal{A}_{(2.7)}(\mathbb{L}_n)$, respectively. Similarly, we write $\mathrm{V}_{(2.8)} = \mathrm{V}_{(2.8)}(\mathbb{L}_n)$ etc.

Proposition 3.

$$\lim_{n\to\infty} \inf_{\mathbb{L}_n} \mathbb{P}(\mathcal{A}_{(2.5)}(\mathbb{L}_n)) = 1, \tag{2.12}$$

$$\lim_{n\to\infty} \inf_{\mathbb{L}_n} \mathbb{P}(\mathcal{A}_{(2.6)}(\mathbb{L}_n)) = 1, \tag{2.13}$$

$$\lim_{n\to\infty} \inf_{\mathbb{L}_n} \min_{y \in \mathbb{L}_n} \mathbb{P}(y \in \mathrm{V}_{(2.8)}(\mathbb{L}_n) \cap \mathrm{W}_{(2.9)}(\mathbb{L}_n)) = 1, \tag{2.14}$$

where $\mathbb{L}_n$ runs over all subset of $\mathbb{L}$ satisfying (2.1).

We pospone the proof until Appendix, but we give some words on the proof here. In fact, (2.12) comes from the linearity of the first passage time, i.e, $\mathrm{T}(a,b) = O(|a-b|)$ with high probability, and (2.13) comes from the variance $\psi(t) = \mathrm{Var}(\mathrm{T}(0, t\mathbf{e}_1))$ and $\sqrt{n}\,\phi(n)^{-\frac{2}{3}} \gg \sqrt{n}\,\phi(n)^{-1} = \sqrt{\psi(n)}$. On the other hand, (2.14) comes from basic computations of the Poisson point process and the linearity of the first passage time.

Given $y \in \mathbb{L}_n$, for the optimal path $(\gamma_y(i))_{i=1}^l = \Gamma(-n\mathbf{e}_1, y, n\mathbf{e}_1)$, we set

$$s_y = \min\{ i \in \{1, \ldots, l\} \mid \gamma_y(i) \in \mathrm{B}(y, C_\delta K_n) \},$$
$$t_y = \max\{ i \in \{1, \ldots, l\} \mid \gamma_y(i) \in \mathrm{B}(y, C_\delta K_n) \}.$$

Proposition 4. *On the event* $\{y \in \mathrm{V}_{(2.8)}\}$,

$$\max\{ |\gamma_y(s_y) - \gamma_y(s_y - 1)|, |\gamma_y(t_y) - \gamma_y(t_y + 1)| \} \leq (K_n)^{\frac{1}{2\alpha}} + 1.$$

Proof. For simplicity of notation, we drop subscripts y in the proof such as $s = s_y$, $\gamma_i = \gamma_y(i)$. Let $\ell = \lfloor|\gamma_s - \gamma_{s-1}|\rfloor$. Suppose $\ell \geq (K_n)^{\frac{1}{2\alpha}}$ and we shall derive a contradiction.

Since $\lfloor\frac{\gamma_s+\gamma_{s-1}}{2}\rfloor \in \mathrm{B}(y, C_\delta K_n + \ell)$ and $y \in \mathrm{V}_{(2.8)}$,

$$\mathrm{B}\left(\left\lfloor\frac{\gamma_s + \gamma_{s-1}}{2}\right\rfloor, \ell^{\frac{1}{2}}\right) \cap \Xi \neq \emptyset.$$

Let us take $x \in \mathrm{B}\big(\lfloor\frac{\gamma_s+\gamma_{s-1}}{2}\rfloor, \ell^{\frac{1}{2}}\big) \cap \Xi$. Since the jump $\{\gamma_{s-1}, \gamma_s\}$ is itself optimal,

$$\begin{aligned}
\ell^\alpha &\leq |\gamma_{s-1} - \gamma_s|^\alpha = \mathrm{T}(\gamma_{s-1}, \gamma_s) \\
&\leq |\gamma_{s-1} - x|^\alpha + |\gamma_s - x|^\alpha \\
&= \left|\gamma_{s-1} - \frac{\gamma_s + \gamma_{s-1}}{2} + \frac{\gamma_s + \gamma_{s-1}}{2} - x\right|^\alpha \\
&\qquad + \left|\gamma_s - \frac{\gamma_s + \gamma_{s-1}}{2} + \frac{\gamma_s + \gamma_{s-1}}{2} - x\right|^\alpha \\
&\leq 2\left(\left|\frac{\gamma_s + \gamma_{s-1}}{2}\right| + \left|\frac{\gamma_s + \gamma_{s-1}}{2} - x\right|\right)^\alpha.
\end{aligned}$$

Since $\left|\frac{\gamma_s-\gamma_{s-1}}{2}\right| \leq \frac{\ell}{2}+1$ and $\left|\frac{\gamma_s+\gamma_{s-1}}{2}-x\right| \leq \ell^{\frac{1}{2}}+d \leq 2\ell^{\frac{1}{2}}$ and $\ell \geq (K_n)^{\frac{1}{2\alpha}}$, for sufficiently large n, this is bounded from above by

$$2\left(\frac{1}{2} + 3\ell^{-\frac{1}{2}}\right)^\alpha \ell^\alpha < \ell^\alpha.$$

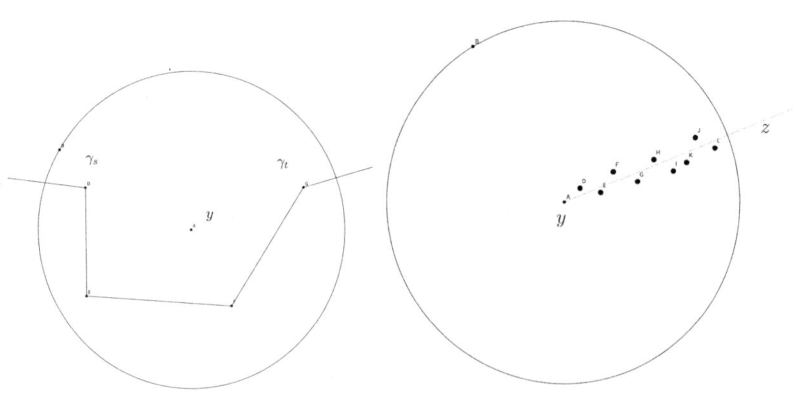

Fig. 1

Left: γ_s and γ_t, Right: Schematic picture of $\mathcal{C}_{c,y}(z)$

Therefore $\ell^\alpha < \ell^\alpha$, which is a contradiction. Thus $\ell < (K_n)^{\frac{1}{2\alpha}}$ and $|\gamma_y(s_y) - \gamma_y(s_y - 1)| \le (K_n)^{\frac{1}{2\alpha}} + 1$. Similarly, we obtain $|\gamma_y(t_y) - \gamma_y(t_y + 1)| \le (K_n)^{\frac{1}{2\alpha}} + 1$. □

Given $z_1, z_2 \in \mathrm{B}(2C_\delta K_n)$, we define the event

$$\mathcal{B}_{(2.15)}(z_1, z_2) = \big\{|\gamma_y(s_y) - (y + z_1)| \le d,\ |\gamma_y(t_y) - (y + z_2)| \le d\big\}, \tag{2.15}$$

where d is the dimension. Given $x \in \mathbb{R}^d$ and $c, K > 0$, we define

$$\mathbb{Z}_{c,K}(x) = \big\{k \in \mathbb{Z}_{\ge 0} \bigm| 2c\,k|x| \le K - 1\big\}.$$

Given $y \in \mathbb{L}_n$, $z \in \mathbb{R}^d \backslash \{0\}$ and $c > 0$, we define

$$\mathcal{C}_{c,y}(z) = \left\{\forall k \in \mathbb{Z}_{c, C_\delta K_n}\left(\frac{z}{|z|}\right),\ \Xi \cap \mathrm{B}\left(y + 2c\,k\frac{z}{|z|}, c\right) \neq \emptyset\right\}.$$

Roughly speaking, $\mathcal{C}_{c,y}(z)$ implies that there are ubiquitous points of Ξ around the line segment $\{y + tz \mid t \ge 0\} \cap \mathrm{B}(y, C_\delta K_n)$ (see Figure 2). Note that, for $c < 1/4$, $\mathcal{C}_{c,y}(z)$ depends only on $\Xi \cap \mathrm{B}(y, C_\delta K_t)$. Independently of Ξ, we take independent random variables Z_1, Z_2 with uniform distributions on $\mathrm{B}(2C_\delta K_n) \cap (\mathbb{Z}^d \backslash \{0\})$.

Lemma 1. *If we take $c > 0$ sufficiently small such that $4^\alpha c^{\alpha-1} C_\delta < \frac{1}{2}$, then for sufficiently large $n > 1$ and $y \in \mathbb{L}_n$,*

$$\begin{aligned} &\mathbb{P}\big(\{\mathrm{T}(-n\mathbf{e}_1, 0, n\mathbf{e}_1) - \mathrm{T}(-n\mathbf{e}_1, y, n\mathbf{e}_1) > K_n\} \cap \mathcal{A}^y_{(2.2)}\big) \\ &\ge \min_{z_1, z_2 \in \mathrm{B}(2C_\delta K_n) \backslash \{0\}} \mathbb{P}\big(\mathcal{C}_{c,y}(z_1) \cap \mathcal{C}_{c,y}(z_2)\big) \\ &\qquad \times \mathbb{P}\big(\mathcal{A}_{(2.7)} \cap \{y \in \mathrm{Y}_{(2.11)}\} \cap \mathcal{B}_{(2.15)}(Z_1, Z_2)\big). \end{aligned} \tag{2.16}$$

Proof. We first explain the idea of the proof. We start with the event $\mathcal{A}_{(2.7)} \cap \{y \in \mathrm{Y}_{(2.11)}\}$. Then we resample all the configurations in $\mathrm{B}(y, C_\delta K_n)$ and suppose $\mathcal{C}_{c,y}(Z_1) \cap \mathcal{C}_{c,y}(Z_2) \cap \mathcal{B}_{(2.15)}(Z_1, Z_2)$ after resampling. Then we will check that $\mathrm{T}(-n\mathbf{e}_1, y, n\mathbf{e}_1)$ decreases by at least $2K_n$. On the other hand, since y and 0 are far away from each other, $\mathrm{T}(-n\mathbf{e}_1, 0, n\mathbf{e}_1)$ is unchanged. Similarly, we have the same thing for $\{\mathrm{T}(-n\mathbf{e}_1, z, n\mathbf{e}_1)\}_{z \neq y \in \mathbb{L}_n}$. Thus after resampling, one may get

$$\big\{\mathrm{T}(-n\mathbf{e}_1, 0, n\mathbf{e}_1) - \mathrm{T}(-n\mathbf{e}_1, y, n\mathbf{e}_1) > K_n\big\} \cap \mathcal{A}^y_{(2.2)}.$$

To make the above rigorous, we use the resampling argument introduced in [5].

Let Ξ^* be an independent copy of the Poisson point process Ξ. We assume that (Ξ, Ξ^*, Z_1, Z_2) are all independent. We enlarge the probability space so that we can measure the event depending on them and we still denote the joint probability measure by $\mathbb{P}$. We define the resampled Poisson point process as

$$\widetilde{\Xi} = \left(\Xi \cap (\mathrm{B}(y, C_\delta K_n))^c\right) \cup (\Xi^* \cap \mathrm{B}(y, C_\delta K_n)).$$

We write $\widetilde{\mathrm{T}}(a,b)$ for the first passage time from a to b with respect to $\widetilde{\Xi}$. Similarly, we define $\widetilde{\mathrm{T}}(a,y,b)$, $\widetilde{\mathcal{C}}_{c,y}(z)$ etc. Note that the distributions of Ξ and $\widetilde{\Xi}$ are the same under $\mathbb{P}$ since Ξ and Ξ^* are independent. Thus the left hand side of (2.16) is equal to

$$\mathbb{P}\left(\widetilde{\mathcal{A}}^y_{(2.2)} \cap \left\{\widetilde{\mathrm{T}}(-n\mathbf{e}_1, 0, n\mathbf{e}_1) - \widetilde{\mathrm{T}}(-n\mathbf{e}_1, y, n\mathbf{e}_1) > K_n\right\}\right),$$

where

$$\widetilde{\mathcal{A}}^y_{(2.2)} = \left\{\forall z \in \mathbb{L}_n \text{ with } z \neq y,\ \widetilde{\mathrm{T}}(-n\mathbf{e}_1, y, n\mathbf{e}_1) < \widetilde{\mathrm{T}}(-n\mathbf{e}_1, z, n\mathbf{e}_1)\right\}.$$

By independence of Ξ and Ξ^*, the right hand side of (2.16) is bounded from above by

$$\begin{aligned} &\sum_{z_1,z_2} \mathbb{P}(Z_1 = z_1, Z_2 = z_2)\mathbb{P}\big(\widetilde{\mathcal{C}}_{c,y}(z_1) \cap \widetilde{\mathcal{C}}_{c,y}(z_2)\big) \\ &\qquad\qquad \times \mathbb{P}\big(\mathcal{A}_{(2.7)} \cap \{y \in \mathrm{Y}_{(2.11)}\} \cap \mathcal{B}_{(2.15)}(z_1, z_2)\big) \\ (2.17) &= \mathbb{P}\Big(\widetilde{\mathcal{C}}_{c,y}(Z_1) \cap \widetilde{\mathcal{C}}_{c,y}(Z_2) \cap \mathcal{A}_{(2.7)} \cap \{y \in \mathrm{Y}_{(2.11)}\} \cap \mathcal{B}_{(2.15)}(Z_1, Z_2)\Big). \end{aligned}$$

Thus, it suffices to show that the event inside the probability in (2.17) implies $\widetilde{\mathcal{A}}^y_{(2.2)}$ and $\widetilde{\mathrm{T}}(-n\mathbf{e}_1, 0, n\mathbf{e}_1) - \widetilde{\mathrm{T}}(-n\mathbf{e}_1, y, n\mathbf{e}_1) > K_n$. To do this, we suppose that (Ξ, Ξ^*, Z_1, Z_2) belongs to the event in (2.17).

Step 1 We prove that $\widetilde{\mathrm{T}}(-n\mathbf{e}_1, y, n\mathbf{e}_1) + 2K_n < \mathrm{T}(-n\mathbf{e}_1, y, n\mathbf{e}_1)$. Take the optimal path $(\gamma_i)_{i=1}^{\ell} = \Gamma(-n\mathbf{e}_1, y, n\mathbf{e}_1)$ and let

$$\begin{aligned} s &= \min\{i \in \{1, \dots, \ell\} \mid \gamma_i \in \mathrm{B}(y, C_\delta K_n)\}, \\ t &= \max\{i \in \{1, \dots, \ell\} \mid \gamma_i \in \mathrm{B}(y, C_\delta K_n)\}. \end{aligned}$$

Since $|\gamma_s - (y+Z_1)| \leq d$ where d is the dimension, taking $k = \lfloor (2c)^{-1}(|Z_1| - 2d)\rfloor \vee 0$, one has $2ck \leq C_\delta K_n - 1$. On the event $\widetilde{\mathcal{C}}_{c,y}(Z_1)$, for any $0 \leq k' \leq k$, there exists $q_{k'} \in \widetilde{\Xi} \cap \mathrm{B}\big(y + 2c\,k' \frac{Z_1}{|Z_1|}, c\big)$. Then, by Proposition 4,

$$
\begin{aligned}
|\gamma_{s-1}-q_k| &\le |\gamma_{s-1}-\gamma_s| + |(y+Z_1)-\gamma_s| + |(y+Z_1)-q_k| \\
&\le \left((K_n)^{\frac{1}{2\alpha}}+1\right)+d \\
&\quad + \left|\left(y+2c\,k\frac{Z_1}{|Z_1|}\right)-(y+Z_1)\right| + \left|q_k-\left(y+2c\,k\frac{Z_1}{|Z_1|}\right)\right| \\
&\le \left((K_n)^{\frac{1}{2\alpha}}+1\right)+d+3d+c \le 2(K_n)^{\frac{1}{2\alpha}}. \qquad (2.18)
\end{aligned}
$$

Since $\gamma_{s-1}\in \mathrm{B}(y,2C_\delta K_t)\backslash \mathrm{B}(y,C_\delta K_n)$ and $C_\delta = 4(1+\delta^{-1})$, on the event $\mathcal{A}_{(2.7)}$,

$$
\mathrm{T}(\gamma_{s-1},y) \ge \delta|\gamma_{s-1}-y| \ge \delta C_\delta K_n \ge 2K_n. \qquad (2.19)
$$

Furthermore, on the event $\widetilde{\mathcal{C}}_{c,y}(Z_1)$,

$$
\begin{aligned}
\widetilde{\mathrm{T}}(q_k,y) &\le \sum_{i=1}^{k} |q_i-q_{i-1}|^\alpha \qquad (2.20)\\
&\le k(4c)^\alpha \le 4^\alpha c^{\alpha-1} C_\delta K_n. \qquad (\text{by } k\le c^{-1}C_\delta K_n)
\end{aligned}
$$

Thus, we have

$$
\begin{aligned}
\widetilde{\mathrm{T}}(-n\mathbf{e}_1,y) &\le \widetilde{\mathrm{T}}(-n\mathbf{e}_1,\gamma_{s-1}) + \widetilde{\mathrm{T}}(\gamma_{s-1},y) \\
&\le \mathrm{T}(-n\mathbf{e}_1,\gamma_{s-1}) + |\gamma_{s-1}-q_k|^\alpha + \widetilde{\mathrm{T}}(q_k,y) \\
(\text{by (2.18), (2.20)}) \quad &\le \mathrm{T}(-n\mathbf{e}_1,\gamma_{s-1}) + 2^\alpha K_n^{\frac{1}{2}} + 4^\alpha c^{\alpha-1} C_\delta K_n \\
(\text{by } 4^\alpha c^{\alpha-1} C_\delta < 2^{-1}) \quad &\le \mathrm{T}(-n\mathbf{e}_1,\gamma_{s-1}) + K_n \\
&= \mathrm{T}(-n\mathbf{e}_1,y) - \mathrm{T}(\gamma_{s-1},y) + K_n \\
(\text{by (2.19)}) \quad &< \mathrm{T}(-n\mathbf{e}_1,y) - K_n.
\end{aligned}
$$

Similarly, $\widetilde{\mathrm{T}}(y,n\mathbf{e}_1) \le \mathrm{T}(y,n\mathbf{e}_1) - K_n$ holds. Consequently, we obtain

$$
\widetilde{\mathrm{T}}(-n\mathbf{e}_1,y,n\mathbf{e}_1) < \mathrm{T}(-n\mathbf{e}_1,y,n\mathbf{e}_1) - 2K_n.
$$

Step 2 We prove that $\widetilde{\mathrm{T}}(-n\mathbf{e}_1,y,n\mathbf{e}_1)+K_n < \widetilde{\mathrm{T}}(-n\mathbf{e}_1,z,n\mathbf{e}_1)$ for any $z\in \mathbb{L}_n\cup\{0\}$ with $z\neq y$. Let $z\in \mathbb{L}_n\cup\{0\}$ with $z\neq y$. If $\widetilde{\Gamma}(-n\mathbf{e}_1,z,n\mathbf{e}_1)$ does not touch with $\mathrm{B}(y,C_\delta K_n)$, then one has $\mathrm{T}(-n\mathbf{e}_1,z,n\mathbf{e}_1) \le \widetilde{\mathrm{T}}(-n\mathbf{e}_1, z,n\mathbf{e}_1)$ and thus

$$
\begin{aligned}
\widetilde{\mathrm{T}}(-n\mathbf{e}_1,y,n\mathbf{e}_1) &\le \mathrm{T}(-n\mathbf{e}_1,y,n\mathbf{e}_1) - 2K_n \qquad (2.21)\\
(\text{by } y\in \mathrm{X}_{(2.10)}) \quad &\le \mathrm{T}(-n\mathbf{e}_1,z,n\mathbf{e}_1) - K_n \\
&\le \widetilde{\mathrm{T}}(-n\mathbf{e}_1,z,n\mathbf{e}_1) - K_n,
\end{aligned}
$$

which is the desired conclusion. Hereafter, we suppose that $\mathrm{B}(y, C_\delta K_n) \cap \widetilde{\Gamma}(-n\mathbf{e}_1, z, n\mathbf{e}_1) \neq \emptyset$. For the optimal path $(\widetilde{\gamma}_i)_{i=1}^{\widetilde{\ell}} = \widetilde{\Gamma}(-n\mathbf{e}_1, z, n\mathbf{e}_1)$, we define

$$\widetilde{s} = \min\bigl\{i \in \{1, \ldots, \widetilde{\ell}\} \mid \widetilde{\gamma}_i \in \mathrm{B}(y, K_n)\bigr\},$$
$$\widetilde{t} = \max\bigl\{i \in \{1, \ldots, \widetilde{\ell}\} \mid \widetilde{\gamma}_i \in \mathrm{B}(y, C_\delta K_n)\bigr\}.$$

Then,

$$\begin{aligned}
\widetilde{\mathrm{T}}(-n\mathbf{e}_1, z) &= \widetilde{\mathrm{T}}(-n\mathbf{e}_1, \widetilde{\gamma}_{\widetilde{s}-1}) + \widetilde{\mathrm{T}}(\widetilde{\gamma}_{\widetilde{s}-1}, \widetilde{\gamma}_{\widetilde{t}+1}) + \widetilde{\mathrm{T}}(\widetilde{\gamma}_{\widetilde{t}+1}, z) \\
&\geq \mathrm{T}(-n\mathbf{e}_1, \widetilde{\gamma}_{\widetilde{s}-1}) + \mathrm{T}(\widetilde{\gamma}_{\widetilde{t}+1}, z) \\
&\geq \mathrm{T}(-n\mathbf{e}_1, y) + \mathrm{T}(y, z) - \mathrm{T}(\widetilde{\gamma}_{\widetilde{s}-1}, y) - \mathrm{T}(y, \widetilde{\gamma}_{\widetilde{t}+1}).
\end{aligned} \tag{2.22}$$

By $y \in \mathrm{V}_{(2.8)}$ and the same proof as in Proposition 4,

$$\begin{aligned}
\mathrm{T}(\widetilde{\gamma}_{\widetilde{s}-1}, y) &\leq |\widetilde{\gamma}_{\widetilde{s}-1} - D(y)|^\alpha \\
&\leq (|\widetilde{\gamma}_{\widetilde{s}-1} - y| + |D(y) - y|)^\alpha \leq (4C_\delta K_n)^\alpha.
\end{aligned}$$

Similarly $\mathrm{T}(y, \widetilde{\gamma}_{\widetilde{t}+1}) \leq (4C_\delta K_n)^\alpha$ holds. Furthermore, on the event $\mathcal{A}_{(2.7)}$, $\mathrm{T}(y, z) \geq \sqrt{n}\phi(n)^{-3/5}$ holds. Thus, (2.22) is further bounded from below by

$$\begin{aligned}
&\mathbb{E}\mathrm{T}(-n\mathbf{e}_1, y) - \sqrt{n}\phi(n)^{-\frac{2}{3}} + \sqrt{n}\phi(n)^{-\frac{3}{5}} - 2(4C_\delta K_n)^\alpha \\
&\geq \mathbb{E}\mathrm{T}(-n\mathbf{e}_1, z) + \frac{1}{2}\sqrt{n}\phi(n)^{-3/5} \\
&\geq \mathrm{T}(-n\mathbf{e}_1, z). \qquad (\text{by } \mathrm{T}(-n\mathbf{e}_1, z) \leq \mathbb{E}\mathrm{T}(-n\mathbf{e}_1, z) + \sqrt{n}\phi(n)^{-\frac{2}{3}}).
\end{aligned}$$

Similarly, we get $\widetilde{\mathrm{T}}(z, n\mathbf{e}_1) \geq \mathrm{T}(z, n\mathbf{e}_1)$, which implies $\widetilde{\mathrm{T}}(-n\mathbf{e}_1, z, n\mathbf{e}_1) \geq \mathrm{T}(-n\mathbf{e}_1, z, n\mathbf{e}_1)$. Then, as in (2.21), we have

$$\widetilde{\mathrm{T}}(-n\mathbf{e}_1, y, n\mathbf{e}_1) \leq \widetilde{\mathrm{T}}(-n\mathbf{e}_1, z, n\mathbf{e}_1) - K_n.$$

The proof is completed combining these two steps. □

Lemma 2. *If $\theta < 2^{-8d} c^d C_\delta^{-1}$, then*

$$\min_z \mathbb{P}(\mathcal{C}_{c,y}(z)) \geq \exp\left(-\frac{1}{16}\log\phi(n)\right).$$

Proof. We simply calculate

$$\begin{aligned}
\mathbb{P}(\mathcal{C}_{c,y}(z)) &\geq \bigl(\mathbb{P}(\mathrm{B}(c) \cap \Xi \neq \emptyset)\bigr)^{2C_\delta K_t} \\
&= \exp\left(-2C_\delta K_t \operatorname{Vol}(\mathrm{B}(c))\right) \\
&\geq \exp\left(-\frac{1}{16}\log\phi(n)\right). \qquad (\text{by } \operatorname{Vol}(\mathrm{B}(c)) \leq (2c)^d)
\end{aligned}$$ □

Proof of Proposition 2. For ease of notation, we write $\mathrm{Y}_{(2.11)}(y) = \{y \in \mathrm{Y}_{(2.11)}\}$ and $\mathcal{B}_n(y) = \mathrm{B}(2C_\delta K_n) \cap (\mathbb{Z}^d \setminus \{0\})$. By FKG inequality, we will compute the right hand side of (2.16) as

(2.23)

$$\begin{aligned}
&\min_{z_1,z_2} \mathbb{P}\big(\mathcal{C}_{c,y}(z_1) \cap \mathcal{C}_{c,y}(z_2)\big)\mathbb{P}\big(\mathcal{A}_{(2.7)} \cap \mathrm{Y}_{(2.11)}(y) \cap \mathcal{B}_{(2.15)}(Z_1, Z_2)\big) \\
&\geq \min_{z} \mathbb{P}(\mathcal{C}_{c,y}(z))^2 \sum_{z_1,z_2} \mathbb{P}(Z_1 = z_1, Z_2 = z_2) \\
&\qquad \times \mathbb{P}\big(\mathcal{A}_{(2.7)} \cap \mathrm{Y}_{(2.11)}(y) \cap \mathcal{B}_{(2.15)}(z_1, z_2)\big).
\end{aligned}$$

Under $\mathcal{A}_{(2.7)} \cap \mathrm{Y}_{(2.11)}(y)$, taking $z_1 = \lfloor \gamma_y(s_y) - y \rfloor$ and $z_2 = \lfloor \gamma_y(t_y) - y \rfloor$, $\mathcal{B}_{(2.15)}(z_1, z_2)$ holds and $z_1, z_2 \in \mathcal{B}_n(y)$. Thus

$$\begin{aligned}
&\sum_{z_1,z_2 \in \mathcal{B}_n(y)} \mathbb{P}\big(\mathcal{A}_{(2.7)} \cap \mathrm{Y}_{(2.11)}(y) \cap \mathcal{B}_{(2.15)}(z_1, z_2)\big) \\
&= \mathbb{E}[\sharp\{(z_1, z_2) \in \mathcal{B}_n(y)^{\otimes 2} \mid \mathcal{B}_{(2.15)}(z_1, z_2)\};\ \mathcal{A}_{(2.7)} \cap \mathrm{Y}_{(2.11)}(y)] \\
&\geq \mathbb{P}(\mathcal{A}_{(2.7)} \cap \mathrm{Y}_{(2.11)}(y)).
\end{aligned}$$

Since $\mathbb{P}(Z_1 = z_1, Z_2 = z_2) \geq \big(\sharp[\mathrm{B}(2C_\delta K_n) \cap \mathbb{Z}^d]\big)^{-2} \geq (4C_\delta K_n)^{-2d}$, (2.12), (2.13) and $K_n = \theta \log \phi(n)$, for sufficiently large n, (2.23) is further bounded from below by

$$\begin{aligned}
&\min_{z} \mathbb{P}(\mathcal{C}_{c,y}(z))^2 (4C_\delta K_n)^{-2d}\, \mathbb{P}\big(\mathcal{A}_{(2.7)} \cap \mathrm{Y}_{(2.11)}(y)\big) \\
&\geq \exp\left(-\frac{1}{8}\log \phi(n)\right)(4C_\delta \theta \log \phi(n))^{-2d} \\
&\qquad \times \Big(\mathbb{P}(y \in \mathrm{X}_{(2.10)}) - \mathbb{P}(\mathcal{A}_{(2.7)}(\mathbb{L}_n)^c) - \mathbb{P}(y \notin \mathrm{V}_{(2.8)} \cup \mathrm{W}_{(2.9)})\Big) \\
&\geq \exp\left(-\frac{1}{4}\log \phi(n)\right) \\
&\qquad \times \Big(\mathbb{P}\big(\mathrm{T}(-n\mathbf{e}_1, y, n\mathbf{e}_1) - \mathrm{T}(-n\mathbf{e}_1, n\mathbf{e}_1) \leq K_n\big) - 1/4\Big).
\end{aligned}$$

By the first moment method, $\mathrm{T}(-n\mathbf{e}_1, y, n\mathbf{e}_1) \geq \mathrm{T}(-n\mathbf{e}_1, n\mathbf{e}_1)$ and (1.3),

$$\begin{aligned}
&\mathbb{P}\big(\mathrm{T}(-n\mathbf{e}_1, y, n\mathbf{e}_1) - \mathrm{T}(-n\mathbf{e}_1, n\mathbf{e}_1) \leq K_n\big) \\
&= 1 - \mathbb{P}\big(\mathrm{T}(-n\mathbf{e}_1, y, n\mathbf{e}_1) - \mathrm{T}(-n\mathbf{e}_1, n\mathbf{e}_1) > K_n\big) \\
&\geq 1 - K_n^{-1}\mathbb{E}\left[\mathrm{T}(-n\mathbf{e}_1, y, n\mathbf{e}_1) - \mathrm{T}(-n\mathbf{e}_1, n\mathbf{e}_1)\right] \\
&\geq 1 - K_n^{-1}\mathbb{E}\left[\mathrm{T}(-n\mathbf{e}_1, y, n\mathbf{e}_1) - 2\mathrm{g}n\right].
\end{aligned}$$

With these observations, the proof is completed. □

§3. Appendix

3.1. Proof of $\liminf_{n\geq 0} \psi(n) > 0$

In this section, we give a proof of

$$\mathrm{Var}(\mathrm{T}_n) > c, \tag{3.1}$$

with some $c > 0$ independent of n, where we recall that $\mathrm{T}_n = \mathrm{T}(0, n\mathbf{e}_1)$. Let Ξ^* be an independent copy of Ξ and set $\widetilde{\Xi} = (\Xi \backslash \mathrm{B}(2)) \cup (\Xi^* \cap \mathrm{B}(2))$. Let us denote by $\mathcal{F}$ the σ-field generated by $\Xi \backslash \mathrm{B}(2)$. We denote by $\widetilde{\mathrm{T}}_n$ the first passage time from 0 to $n\,\mathbf{e}_1$ with respect to $\widetilde{\Xi}$. Then, since martingale differences are uncorrelated, we have

$$\begin{aligned}\mathrm{Var}(\mathrm{T}_n) &= \mathbb{E}[(\mathbb{E}[\mathrm{T}_\mathrm{n}|\mathcal{F}] - \mathbb{E}[\mathrm{T}_\mathrm{n}])^2] + \mathbb{E}[(\mathrm{T}_\mathrm{n} - \mathbb{E}[\mathrm{T}_\mathrm{n}|\mathcal{F}])^2] \\ &\geq \mathbb{E}[\mathbb{E}[(\mathrm{T}_\mathrm{n} - \mathbb{E}[\mathrm{T}_\mathrm{n}|\mathcal{F}])^2|\mathcal{F}]] = \mathbb{E}[\mathbb{E}[(\mathrm{T}_\mathrm{n} - \widetilde{\mathrm{T}}_\mathrm{n})^2|\mathcal{F}]].\end{aligned}$$

Hence, it suffice to show that there exists a non-random constant $c > 0$ such that for any $n > 2$

$$\mathbb{E}[(\mathrm{T}_\mathrm{n} - \widetilde{\mathrm{T}}_\mathrm{n})^2|\mathcal{F}] > \mathrm{c} \qquad \text{a.s.}$$

To this end, we consider the event:

$$\mathcal{E}_{(3.2)} = \left\{\Xi \cap \mathrm{B}(2) = \emptyset,\ \Xi^* \cap \mathrm{B}(1) \neq \emptyset,\ \Xi^* \cap (\mathrm{B}(2) \backslash \mathrm{B}(1)) = \emptyset\right\}. \tag{3.2}$$

Note that $\mathcal{E}_{(3.2)}$ is independent of $\mathcal{F}$. Since, on the event $\mathcal{E}_{(3.2)}$, $D(0)$ (the closest point of Ξ to 0) is located in $\mathrm{B}(2)^\mathrm{c}$, $\widetilde{D}(0)$ (the closest point of $\widetilde{\Xi}$ to 0) is located in $\mathrm{B}(1)$ and $\Xi \backslash \mathrm{B}(2)^\mathrm{c} = \widetilde{\Xi} \backslash \mathrm{B}(2)^\mathrm{c}$, we have $\widetilde{\mathrm{T}}_n - \mathrm{T}_\mathrm{n} \geq 1$. Hence,

$$\begin{aligned}\mathbb{E}[(\mathrm{T}_\mathrm{n} - \widetilde{\mathrm{T}}_\mathrm{n})^2|\mathcal{F}] &\geq \mathbb{E}[(\mathrm{T}_\mathrm{n} - \widetilde{\mathrm{T}}_\mathrm{n})^2 \mathbf{1}(\mathcal{E}_{(3.2)})|\mathcal{F}] \\ &\geq \mathbb{E}[\mathbf{1}(\mathcal{E}_{(3.2)})|\mathcal{F}] = \mathbb{P}(\mathcal{E}_{(3.2)}),\end{aligned}$$

and since $\mathcal{E}_{(3.2)}$ is independent of n, the proof of (3.1) is completed.

3.2. Proof of Proposition 3

Proof of (2.12). Note that for $a \neq b \in \mathbb{L}_n \cup \{0\}$,

$$|a - b| \geq n^{\frac{1}{2}} \phi(n)^{-\frac{1}{2}} \gg \sqrt{n}\phi(n)^{-\frac{2}{3}}.$$

By using [9, Lemma 1] and $\phi(n) \le C\, n^{\frac{1}{2}}$ with some $C > 0$,

$$\begin{aligned}\mathbb{P}((\mathcal{A}_{(2.5)})^c) &\le \sum_{a,b \in \mathbb{L}_n \cup \{0\}} \mathbb{P}\big(\mathrm{T}(a,b) < \sqrt{n}\phi(n)^{-\frac{2}{3}}\big)\mathbf{1}_{\{a \neq b\}} \\ &\le 2\phi(n) \exp\left(-\epsilon\left(\sqrt{n}\phi(n)^{-\frac{1}{2}}\right)^{\kappa}\right) \\ &\le \exp\left(-\epsilon'\, n^{\kappa/4}\right),\end{aligned}$$

with some $\kappa, \epsilon, \epsilon' > 0$. □

Proof of (2.13). Since for $y \in \mathbb{L}_n \cup \{0\}$, $n \le |y| \le n+1$, by Chebyshev's inequality,

$$\begin{aligned}&\mathbb{P}\left(|\mathrm{T}(0,y) - \mathbb{E}\,\mathrm{T}(0,y)| \ge \frac{\sqrt{n}}{\phi(n)^{\frac{2}{3}}}\right) \\ &= \mathbb{P}\left(|\mathrm{T}_{|y|} - \mathbb{E}\mathrm{T}_{|y|}| \ge \frac{\sqrt{n}}{\phi(n)^{\frac{2}{3}}}\right) \\ &\le \mathbb{P}\left(|\mathrm{T}_n - \mathbb{E}\mathrm{T}_n| + \mathrm{T}(n\mathbf{e}_1, |y|\mathbf{e}_1) + \mathbb{E}\mathrm{T}(n\mathbf{e}_1, |y|\mathbf{e}_1) \ge \frac{\sqrt{n}}{\phi(n)^{\frac{2}{3}}}\right) \\ &\le \mathbb{P}\left(|\mathrm{T}_n - \mathbb{E}\mathrm{T}_n| \ge \frac{1}{3}\frac{\sqrt{n}}{\phi(n)^{\frac{2}{3}}}\right) + \mathbb{P}\left(\mathrm{T}_{|y|-n} \ge \frac{1}{3}\frac{\sqrt{n}}{\phi(n)^{\frac{2}{3}}}\right) \\ &\le \frac{9\phi(n)^{4/3}}{n}\left(\mathbb{E}\mathrm{T}^2_{|y|-n} + \mathrm{Var}(\mathrm{T}_n)\right) \le C\phi(n)^{-\frac{2}{3}},\end{aligned}$$

with some constant $C > 0$ independent of y and n. Then by the union bound, we have

$$\begin{aligned}&\mathbb{P}(\mathcal{A}^c_{(2.6)}) \\ &= \mathbb{P}\left(\exists y \in \mathbb{L}_n \cup \{0\} \text{ s.t. } \max_{z=\pm n\mathbf{e}_1} \{|\mathrm{T}(z,y) - \mathbb{E}\,\mathrm{T}(z,y)|\} \ge \frac{\sqrt{n}}{\phi(n)^{\frac{2}{3}}}\right) \\ &\le 2\sharp\mathbb{L}_n \sup_{y \in \mathbb{L}_n \cup \{0\}} \mathbb{P}\big(|\mathrm{T}(0,y) - \mathbb{E}\,\mathrm{T}(0,y)| \ge \sqrt{n}\,\phi(n)^{-\frac{2}{3}}\big) \\ &\le 2C\phi(n)^{\frac{1}{2}}\,\phi(n)^{-\frac{2}{3}} = 2C\phi(n)^{-\frac{1}{6}}.\end{aligned}$$

□

Proof of (2.14). We first prove that

$$\lim_{n\to\infty} \inf_{\mathbb{L}_n} \mathbb{P}(y \in \mathrm{V}_{(2.8)}) = 1.$$

Indeed, by the union bound,

$$\begin{aligned}
&\mathbb{P}(\{y \in \mathrm{V}_{(2.8)}\}^c) \\
\leq &\sum_{\ell \geq (K_n)^{\frac{1}{2\alpha}}} \sum_{x \in \mathrm{B}(y, C_\delta K_n + \ell) \cap \mathbb{Z}^d} \mathbb{P}\big(\Xi \cap \mathrm{B}(x, \ell^{\frac{1}{2}}) = \emptyset\big) \\
= &\sum_{\ell \geq (K_n)^{\frac{1}{2\alpha}}} \sharp\big(\mathrm{B}(y, C_\delta K_n + \ell) \cap \mathbb{Z}^d\big)\, \mathbb{P}\big(\Xi \cap \mathrm{B}(\ell^{\frac{1}{2}}) = \emptyset\big) \\
\leq &\, 2^d \sum_{\ell \geq (K_n)^{\frac{1}{2\alpha}}} (K_n + \ell)^d \exp\big(-\mathrm{Vol}(\mathrm{B}(\ell^{\frac{1}{2}}))\big) \to 0, \qquad \text{as } n \to \infty.
\end{aligned}$$

It remains to prove

$$\mathbb{P}\big(\{y \in \mathrm{W}_{(2.9)}(\mathbb{L}_n)\}^c \cap \{y \in \mathrm{V}_{(2.8)}\}\big) \to 0.$$

First we note that by [9, Lemma 1], for sufficiently small δ,

$$\begin{aligned}
&\mathbb{P}\big(\exists a, b \in \mathrm{B}(y, 4C_\delta K_n) \cap \mathbb{Z}^d \text{ s.t. } |a-b| \geq K_n/2,\ \mathrm{T}(a,b) \leq 4\delta|a-b|\big) \\
\leq &\sum_{a,b \in \mathrm{B}(y, 4C_\delta K_n) \cap \mathbb{Z}^d} \mathbb{P}\big(\mathrm{T}(a,b) \leq 4\delta|a-b|\big) \mathbf{1}_{\{|a-b| \geq K_n/2\}} \\
\leq &\,(8C_\delta K_n)^{2d} \exp(-\epsilon K_n^\kappa) \leq \exp\Big(-\frac{\epsilon}{2} K_n^\kappa\Big),
\end{aligned}$$

with some $\epsilon, \kappa > 0$. Hereafter, we suppose that $y \in \mathrm{V}_{(2.8)}$ and for any $a, b \in \mathrm{B}(y, 4C_\delta K_n) \cap \mathbb{Z}^d$ with $|a-b| \geq K_n/2$, $\mathrm{T}(a,b) > 4\delta|a-b|$. Let $a, b \in \mathrm{B}(y, 2C_\delta K_n)$ with $|a-b| \geq K_n$. Then by $y \in \mathrm{V}_{(2.8)}$,

$$\max\big\{|D(a) - a|, |D(b) - b|\big\} \leq 2(K_n)^{\frac{1}{2\alpha}}.$$

Similarly, $\max\{|D(\lfloor a \rfloor) - \lfloor a \rfloor|, |D(\lfloor b \rfloor) - \lfloor b \rfloor|\} \leq 2(K_n)^{\frac{1}{2\alpha}}$. Hence,

$$\max\big\{|D(a) - D(\lfloor a \rfloor)|, |D(b) - D(\lfloor b \rfloor)|\big\} \leq 6(K_n)^{\frac{1}{2\alpha}}. \tag{3.3}$$

Since $\lfloor a \rfloor, \lfloor b \rfloor \in \mathrm{B}(y, 4CK_n) \cap \mathbb{Z}^d$ and $|\lfloor a \rfloor - \lfloor b \rfloor| \geq |a-b|/2 \geq K_n/2$,

$$\begin{aligned}
&\mathrm{T}(a,b) = \mathrm{T}(D(a), D(b)) \\
&\geq \mathrm{T}(D(\lfloor a \rfloor), D(\lfloor b \rfloor)) - \mathrm{T}(D(a), D(\lfloor a \rfloor)) - \mathrm{T}(D(b), D(\lfloor b \rfloor)) \\
&\geq \mathrm{T}(D(\lfloor a \rfloor), D(\lfloor b \rfloor)) - |D(a) - D(\lfloor a \rfloor)|^\alpha - |D(b) - D(\lfloor b \rfloor)|^\alpha \\
&\geq 2\delta|a-b| - 12^\alpha (K_n)^{\frac{1}{2}} \geq \delta|a-b|. \qquad \text{(by (3.3))}
\end{aligned}$$

Therefore, the proof is completed. □

§ Acknowledgments

I am indebted to an anonymous reviewer of an earlier paper for providing insightful comments. The work is partially supported by JSPS KAKENHI 19J00660 and SNSF grant 176918.

References

[1] K. S. Alexander, Approximation of subadditive functions and convergence rates in limiting-shape results, *Ann. Probab.*, 25, 30–55, 1997, MR 1428498.

[2] A. Auffinger, M. Damron and J. Hanson, 50 years of first-passage percolation, *University Lecture Series*, 68, *American Mathematical Society, Providence, RI*, 2017, MR 3729447.

[3] A. Auffinger, M. Damron and J. Hanson, Rate of convergence of the mean for sub-additive ergodic sequences, Adv. Math., 285 (2015), 138–181, MR 3406498.

[4] M. Bernstein, M. Damron and T. Greenwood, Sublinear variance in Euclidean first-passage percolation, *preprint.*

[5] J. van den Berg and H. Kesten, Inequalities for the time constant in first-passage percolation, *Ann. Appl. Probab.*, 56–80, 1993, MR 1202515.

[6] F. Comets and C. Cosco, Brownian Polymers in Poissonian Environment: a survey, arXiv:1805.10899, (2018).

[7] M. Damron and N. Kubota, Rate of convergence in first-passage percolation under low moments, *Stochastic Process. Appl.*, 126 (10), 3065–3076, 2016.

[8] M. Damron and X. Wang, Entropy reduction in Euclidean first-passage percolation, *Electron. J. Probab.*, 21 (65), 1–23, 2016, MR 3580031.

[9] C. D. Howard and C. M. Newman, Euclidean models of first-passage percolation, *Probab. Theory Related Fields*, 108, 153–170, 1997, MR 1452554.

[10] H. Kesten, On the speed of convergence in first-passage percolation, *Ann. Appl. Probab.*, 3, 296–338, 1993, MR 1221154.

[11] S. Nakajima, Divergence of non-random fluctuation in First Passage Percolation, *Electron. Commun. Probab.*, 24 (65), 1–13. 2019, MR 4029434.

University of Basel,
Basel, Switzerland
E-mail address: shuta.nakajima@unibas.ch

Advanced Studies in Pure Mathematics 87, 2021
Stochastic Analysis, Random Fields and Integrable Probability — Fukuoka 2019
pp. 381–402

Discrete integrable systems and Pitman's transformation

David A. Croydon and Makiko Sasada

Abstract.

We survey recent work that relates Pitman's transformation to a variety of classical integrable systems, including the box-ball system, the ultra-discrete and discrete KdV equations, and the ultra-discrete and discrete Toda lattice equations. It is explained how this connection enables the dynamics of the integrable systems to be initiated from infinite configurations, which is important in the study of invariant measures. In the special case of spatially independent and identically distributed configurations, progress on the latter topic is also reported.

§1. Introduction

Let us recall the well-known Pitman's transformation, which for discrete-time paths $S : \mathbb{Z}_+ \to \mathbb{R}$ and continuous-time paths $S : \mathbb{R}_+ \to \mathbb{R}$ is defined by the mapping

$$S \mapsto 2M. - S. \tag{1}$$

where $M_x = \sup_{0 \le y \le x} S_y$. (In the continuous-time case, one needs to make an assumption on S to ensure that M is finite.) We will denote this transform by $T^{\mathbb{Z}}S := 2M - S$ and $T^{\mathbb{R}}S := 2M - S$ respectively.

Since its introduction in [32], where it was shown to relate the one-dimensional Brownian motion and the three-dimensional Bessel process, Pitman's transformation and variations thereof have been extensively studied. Following Pitman's original work, much of this interest has been in the context of stochastic processes (see [1, 14, 17, 24, 25, 26,

---

Received January 30, 2020.
Revised May 15, 2020.
2010 *Mathematics Subject Classification.* Primary 37K60; Secondary 37B15, 37K10, 37L40, 60G50, 82B99.
Key words and phrases. Box-ball system, integrable systems, invariant measures, KdV equation, Pitman's transformation, random walks, Toda lattice.

27, 33, 34, 35], for example). Moreover, there has been lively activity involving Pitman-type transformations in the areas of queuing theory and stochastic integrable systems, where it has links to models such as random polymers, random matrices, the quantum Toda lattice and the KPZ equation (see [2, 28, 29, 31], for example), and such transformations have also been used to define a continuous model of crystals [3]. We highlight that one of the most important and well-studied adaptations of Pitman's transformation arising within this research is its exponential version, which is defined for a continuous-time path $S : \mathbb{R}_+ \to \mathbb{R}$ by the mapping

$$S \mapsto 2M^{\int}_{\cdot} - S_{\cdot},$$

where

$$M^{\int}_{\cdot} := \frac{1}{2}\log\left(\int_0^{\cdot} \exp(2S_y)dy\right).$$

(Again, it is necessary to make suitable assumptions on S to ensure that the above transformation is well-defined.) We will denote this transform by $T^{\int}S := 2M^{\int} - S$.

The purpose of the present paper is to briefly survey recent work involving the authors of the present article and their collaborators that relates Pitman's transformation to classical integrable systems, such as the box-ball system, the ultra-discrete and discrete KdV equations, and the ultra-discrete and discrete Toda lattice equations [4, 5, 6, 7, 8, 9, 10]. Importantly, the framework that is developed across the latter articles enables the dynamics of the various systems to be initiated from infinite configurations, a novelty which is crucial for the study of invariant measures. In the special case of spatially independent and identically distributed (i.i.d.) configurations, the authors' progress upon the latter topic is reported here. Whilst the study of classical integrable systems from a probabilistic perspective has hitherto been somewhat limited, we note that beyond the results discussed, there have been increasing efforts in this direction, see [11, 12, 13, 20, 21, 22, 23] for some of the interesting developments.

Although in Pitman's original study the discrete-time version of Pitman's transformation was central to the proof of the main result concerning Brownian motion, the majority of the stochastic processes literature has focussed on the continuous-time setting. Part of the reason for this is that perhaps the most canonical discretisation of $T^{\int}$, whereby the integral is replaced by a sum, has somewhat different properties to the continuous-time version, cf. [36]. One of new aspects of our work is that through the connection to discrete integrable systems, we find an alternative, but still natural, discretisation of the operator $T^{\int}$ that has the appealing property of admitting an i.i.d. invariant measure for which the marginals of the process corresponding to $M - S$ precisely match those of the related continuous-time invariant measure (see Remark 6.2). In [8], we will discuss further advantages of this discretisation.

The remainder of the article is organised as follows. In Section 2 we present versions of Pitman's transformation that will appear later in the article. The corresponding discrete integrable systems are introduced in Section 3, and their relation with Pitman-type transformations is outlined in Section 4. Section 5 incorporates a presentation of results on invariant measures for the systems under study, which are then summarised in Section 6.

§2. Pitman-type transformations

Similarly to the original version of Pitman's transformation that appears at (1), the Pitman-type transformations that we consider will be of the form

$$T^*S = 2M^* - S - 2M_0^*$$

for some path-valued functional M^*. Note that, in contrast to the introduction, we will typically focus on two-sided paths $S : \mathbb{Z} \to \mathbb{R}$. Moreover, we will typically assume that $S_0 = 0$, and it is for this reason that we include the constant shift by $2M_0^*$, which ensures that we also have $(T^*S)_0 = 0$. Specifically, in the two-sided discrete-time case, namely for $S : \mathbb{Z} \to \mathbb{R}$, we will consider the following choices for M^*:

(a)

$$M(S)_n := \sup_{m \leq n} S_m;$$

(b)

$$M^{\vee}(S)_n := \sup_{m \leq n} \left(\frac{S_m + S_{m-1}}{2} \right);$$

(c)

$$M^{\Sigma}(S)_n := \log \left(\sum_{m \leq n} \exp \left(\frac{S_m + S_{m-1}}{2} \right) \right);$$

(d)

$$M^{\vee^*}(S)_n := \begin{cases} \sup_{m \leq \frac{n-1}{2}} S_{2m}, & n \text{ odd}, \\ \dfrac{M^{\vee^*}(S)_{n+1} + M^{\vee^*}(S)_{n-1}}{2}, & n \text{ even}; \end{cases}$$

(e)

$$M^{\Sigma^*}(S)_n := \begin{cases} \log\left(\sum_{m\leq \frac{n-1}{2}} \exp(S_{2m})\right), & n \text{ odd}, \\ \dfrac{M^{\Sigma^*}(S)_{n+1} + M^{\Sigma^*}(S)_{n-1}}{2}, & n \text{ even}; \end{cases}$$

writing the corresponding path transformations as $T^{\mathbb{Z}}$, $T^{\vee}$, T^{Σ}, $T^{\vee^*}$, T^{Σ^*}, respectively. Clearly, we need the right-hand sides in the definitions of M^* to converge for these operators to be well-defined, and we note that this is always the case on the space of asymptotically linear functions, as given by

(2)

$$\mathcal{S}^{lin} :=$$

$$\left\{(S_n)_{n\in\mathbb{Z}} : \lim_{n\to-\infty} \frac{S_n}{n} \text{ and } \lim_{n\to+\infty} \frac{S_n}{n} \text{ exist and are strictly positive}\right\}.$$

It will further be convenient to introduce shifted versions of $T^{\vee^*}$ and T^{Σ^*}, and so we define

$$\mathcal{T}^{\vee^*}(S) := \theta \circ T^{\vee^*}(S) - \theta \circ T^{\vee^*}(S)_0,$$
$$\mathcal{T}^{\Sigma^*}(S) := \theta \circ T^{\Sigma^*}(S) - \theta \circ T^{\Sigma^*}(S)_0,$$

where θ is the usual left-shift, i.e. if $x \in \mathbb{R}^{\mathbb{Z}}$, then $\theta(x) \in \mathbb{R}^{\mathbb{Z}}$ is given by $\theta(x)_n = x_{n+1}$. Again, we include a constant shift to ensure the new path passes through the origin.

§3. Discrete integrable systems

The discrete integrable systems that we consider in this article are based on discretisations (or ultra-discretisations) of the KdV and Toda lattice equations, as introduced in [19] and [39], respectively. In each case, the variables in the model include those representing the configuration, $(z_n^t)_{n,t\in\mathbb{Z}}$ say, and those representing an auxiliary 'carrier', $(w_n^t)_{n,t\in\mathbb{Z}}$ say, where the index n refers to the spatial location of the variable, and the index t refers to the temporal location. Moreover, the dynamics of the systems are locally-defined, in the sense that we have maps $(F_n)_{n\in\mathbb{Z}}$ and $m \in \mathbb{Z}$ such that $(z_{n-m}^{t+1}, w_n^t) = F_n(z_n^t, w_{n-1}^t)$, or graphically, we have the following lattice structure:

(3)

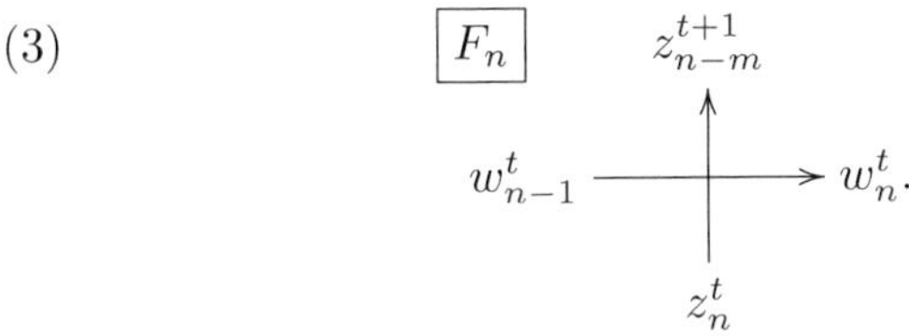

We now describe a number of particular examples of such locally-defined dynamics, which, as we will explain in the next section, relate to the Pitman-type transformations of Section 2. To keep the presentation concise, we refer the reader to [10, 16, 37, 40, 41, 42] for further background concerning the various models discussed.

(a) *Box-ball system (BBS).* The most basic model in our discussion is the BBS, which was introduced in [38] as a simple example of a discrete model exhibiting the solitonic behaviour of the KdV equation. For this model, the configuration variables will be written $(\eta^t_n)_{n,t\in\mathbb{Z}}$. These take values in $\{0,1\}$, with $\eta^t_n = 1$ representing the presence of a ball in the box at spatial location n at time t, and $\eta^t_n = 0$ representing the absence of such a ball. The carrier variables $(W^t_n)_{n,t\in\mathbb{Z}}$ are $\mathbb{Z}_+$-valued, with W^t_n representing the number of balls transported from location n to $n+1$ by the carrier between time t and $t+1$. The evolution of the system is described as follows:

$$\begin{cases} \eta^{t+1}_n = \min\{1-\eta^t_n, W^t_{n-1}\}, \\ W^t_n = \eta^t_n + W^t_{n-1} - \eta^{t+1}_n, \end{cases} \tag{4}$$

which, roughly speaking, means that on each time step the carrier moves from left to right, picking up each ball it passes, and putting down a ball when it is carrying at least one and sees an empty box. There are many variations of the BBS, including the case of BBS(J,K), where the boxes have capacity $J \in \mathbb{N}\cup\{\infty\}$ and the carrier has capacity $K \in \mathbb{N}\cup\{\infty\}$. Whilst we focus on the original model here (which can be thought of as BBS(1,∞)), it is in fact possible to describe the evolution of BBS(J,K) in terms of a Pitman-type transformation whenever $J < K$ (see [7]). Other variations of the BBS incorporate balls of multiple colours. Again, such models can be described using Pitman-type transformations, but require higher dimensional versions to encode the relevant information [18].

(b) *Ultra-discrete KdV (udKdV) equation.* Generalising the BBS is the udKdV equation. In this model, both the configuration variables $(\eta_n^t)_{n,t\in\mathbb{Z}}$ and carrier variables $(U_n^t)_{n,t\in\mathbb{Z}}$ take values in $\mathbb{R}$, though one can also consider specialisations (including the BBS). Given a parameter $L \in \mathbb{R}$, the udKdV equation is given as follows:

$$\begin{cases} \eta_n^{t+1} = \min\{L - \eta_n^t, U_{n-1}^t\}, \\ U_n^t = \eta_n^t + U_{n-1}^t - \eta_n^{t+1}, \end{cases} \tag{5}$$

which clearly matches (4) when $L = 1$. We can summarise the above equations by writing $(\eta_n^{t+1}, U_n^t) = F_{udK}^{(L)}(\eta_n^t, U_{n-1}^t)$, which can be understood as locally-defined dynamics with a lattice structure for the variables as at (3) (with $m = 0$). As with the BBS, there are other versions of the udKdV system that can be described by Pitman-type transformations; the L parameter represents the box capacity, and we can also vary the carrier capacity, see [8].

(c) *Discrete KdV (dKdV) equation.* Obtained from the KdV equation by a natural discretisation procedure, and yielding the udKdV equation through an ultra-discretisation procedure, is the dKdV equation. For this system, the configuration variables $(\omega_n^t)_{n,t\in\mathbb{Z}}$ and carrier variables $(U_n^t)_{n,t\in\mathbb{Z}}$ take values in $(0, \infty)$. Fixing $\delta \in (0, \infty)$, the dKdV equation is given by

$$\begin{cases} \omega_n^{t+1} = \left(\delta\omega_n^t + (U_{n-1}^t)^{-1}\right)^{-1}, \\ U_n^t = U_{n-1}^t \omega_n^t (\omega_n^{t+1})^{-1}. \end{cases}$$

Again, by writing $(\omega_n^{t+1}, U_n^t) = F_{dK}^{(\delta)}(\omega_n^t, U_{n-1}^t)$, we can be understand the system being given by locally-defined dynamics with lattice structure as at (3) (with $m = 0$). Moreover, variations of the dKdV equation with an additional parameter to above can also be handled within our framework, see [8].

(d) *Ultra-discrete Toda lattice (udToda) equation.* Parallel to KdV-type equations, we consider Toda-type equations. The ultra-discrete Toda equation in particular has configuration variables $(Q_n^t, E_n^t)_{n,t\in\mathbb{Z}}$ and carrier variables $(U_n^t)_{n,t\in\mathbb{Z}}$ that take values in $\mathbb{R}$. The evolution of the system is described as follows:

$$\begin{cases} Q_n^{t+1} = \min\{U_n^t, E_n^t\}, \\ E_n^{t+1} = Q_{n+1}^t + E_n^t - Q_n^{t+1}, \\ U_{n+1}^t = U_n^t + Q_{n+1}^t - Q_n^{t+1}, \end{cases}$$

and summarised as $(Q_n^{t+1}, E_n^{t+1}, U_{n+1}^t) = F_{udT}(Q_{n+1}^t, E_n^t, U_n^t)$. We note that the udToda system also connects with the BBS in that we can view Q_n^t as the length of the nth interval containing balls, and E_n^t representing the length of the nth empty interval (at time t); of course, in the case of infinite balls, there is an issue of how to enumerate the intervals. For the udToda equation, the lattice structure is of the form

(6)

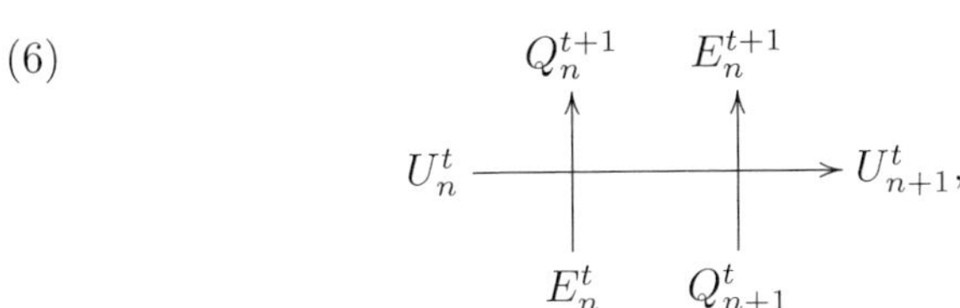

and so does not match (3). However, it is possible to decompose the single map F_{udT} with three inputs and three outputs into two maps F_{udT^*} and $F_{udT^*}^{-1}$, each with two inputs and two outputs:

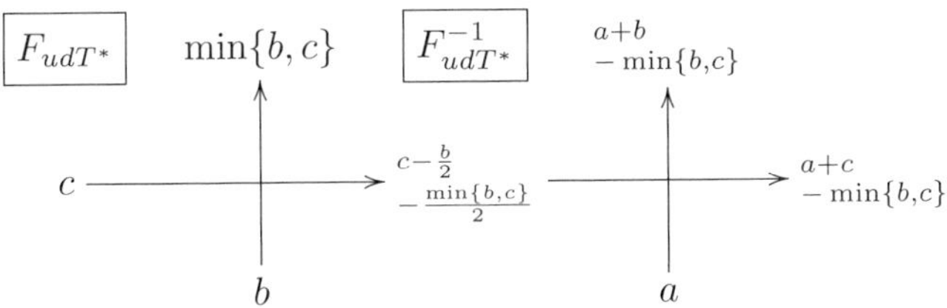

where we generically take $(a, b, c) = (Q_{n+1}^t, E_n^t, U_n^t)$. Including the additional lattice variables, we can thus view the system as locally-defined dynamics as at (3), with F_n alternating between F_{udT^*} for n even and $F_{udT^*}^{-1}$ for n odd, and $m = 1$.

(e) *Discrete Toda lattice (dToda) equation.* Sitting between the original Toda lattice equation and its ultra-discrete version is the discrete Toda lattice equation, as given by:

$$\begin{cases} I_n^{t+1} = J_n^t + U_n^t, \\ J_n^{t+1} = I_{n+1}^t J_n^t (I_n^{t+1})^{-1}, \\ U_{n+1}^t = I_{n+1}^t U_n^t (I_n^{t+1})^{-1}. \end{cases}$$

Here, the configuration variables $(I_n^t, J_n^t)_{n,t\in\mathbb{Z}}$ and carrier variables $(U_n^t)_{n,t\in\mathbb{Z}}$ take values in $(0,\infty)$, and we can summarise the above dynamics by $(I_n^{t+1}, J_n^{t+1}, U_{n+1}^t) = F_{dT}(I_{n+1}^t, J_n^t, U_n^t)$.

Similarly to (6), in this case we have a lattice structure

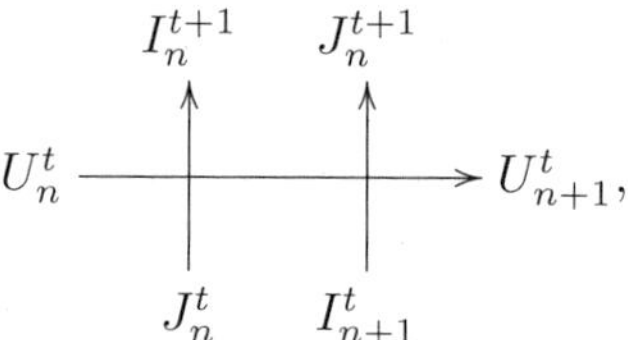

which can be decomposed into two maps, F_{dT^*} and $F_{dT^*}^{-1}$, as follows:

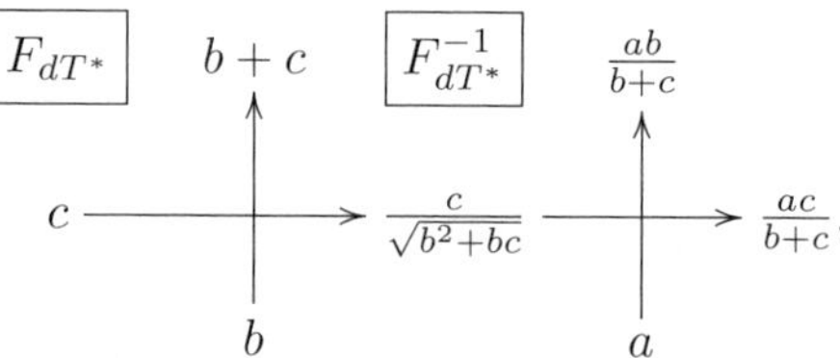

where we generically take $(a, b, c) = (I_{n+1}^t, J_n^t, U_n^t)$. So, again including the additional lattice variables, we can view the system as locally-defined dynamics as at (3), with F_n now alternating between F_{dT^*} for n even and $F_{dT^*}^{-1}$ for n odd, and $m = 1$.

§4. Relation between locally-defined dynamics and Pitman-type transformations

For a system of locally-defined dynamics, as introduced at the start of the previous section, it is natural to consider an initial value problem of the following form: given initial condition $(z_n^0)_{n\in\mathbb{Z}}$, is it possible to find $(z_n^t)_{n,t\in\mathbb{Z}}$ and $(w_n^t)_{n,t\in\mathbb{Z}}$ such that $(z_{n-m}^{t+1}, w_n^t) = F_n(z_n^t, w_{n-1}^t)$ holds for all $n, t \in \mathbb{Z}$? Moreover, if a solution exists, then is it unique?

Given the form of the equation $(z_{n-m}^{t+1}, w_n^t) = F_n(z_n^t, w_{n-1}^t)$, it is clear that if both $(z_n^0)_{n\in\mathbb{Z}}$ and $(w_n^0)_{n\in\mathbb{Z}}$ are given, then we can compute $(z_n^1)_{n\in\mathbb{Z}}$ by setting $z_{n-m}^1 = F_n^{(1)}(z_n^0, w_{n-1}^0)$, where the superscript (1) refers to the first component of F_n. Hence, to solve the forward problem (where we only consider $t \geq 0$), it is enough to find a carrier $(w_n^0)_{n\in\mathbb{Z}}$ for $(z_n^0)_{n\in\mathbb{Z}}$ such that: $w_n^0 = F_n^{(2)}(z_n^0, w_{n-1}^0)$ for all $n \in \mathbb{Z}$, and for which it is possible to repeat the procedure starting from the resulting $(z_n^1)_{n\in\mathbb{Z}}$. Moreover, in all the cases considered in this article, the locally-defined dynamics have a symmetry that means the backward problem can be solved in the same way.

Now, for any bijections $\mathcal{A}_n$ and $\mathcal{B}_n$ (defined on appropriate subsets of $\mathbb{R}$), we can rephrase the above problem in terms of the variables $x_n^t := \mathcal{A}_n(z_n^t)$, $u_n^t := \mathcal{B}_n(w_n^t)$. Indeed, for these variables, the equation of interest becomes

$$\left(x_{n-m}^{t+1}, u_n^t\right) = K_n\left(x_n^t, u_{n-1}^t\right),$$

where $K_n := (\mathcal{A}_n \times \mathcal{B}_n) \circ F_n \circ (\mathcal{A}_n \times \mathcal{B}_{n-1})^{-1}$. (The product map $\mathcal{A}_n \times \mathcal{B}_n$ is defined by setting $\mathcal{A}_n \times \mathcal{B}_n(a,b) := (\mathcal{A}_n(a), \mathcal{B}_n(b))$.) Of course, this is nothing but a change of variables. However, as a key assumption that enables a link to be made to a Pitman-type transformation, we will suppose that the variables have been changed in such a way that K_n satisfies the conservation law:

$$K_n^{(1)}(a,b) - 2K_n^{(2)}(a,b) = a - 2b. \tag{7}$$

In our work, we do not attempt to classify for which integrable systems such a change of variables exists, but we note that it is possible to do so for the systems considered here, i.e. BBS, udKdV, dKdV, udToda, dToda.

To relate to Pitman-type transformations, we need to introduce the path encoding of a configuration. In particular, given $(x_n)_{n\in\mathbb{Z}}$, we define the associated path $S : \mathbb{Z} \to \mathbb{R}$ by setting $S_0 = 0$ and

$$S_n - S_{n-1} = x_n, \qquad \forall n \in \mathbb{Z}.$$

Note that the mapping $x \mapsto S$ is one-to-one. By construction, it is easy to see that the existence of a carrier $(u_n)_{n\in\mathbb{Z}}$ for $(x_n)_{n\in\mathbb{Z}}$ (i.e. a solution to $u_n = K_n^{(2)}(x_n, u_{n-1})$) is equivalent to the existence of a path $M : \mathbb{Z} \to \mathbb{R}$ satisfying:

$$M_n = K_n^{(2)}(S_n - S_{n-1}, M_{n-1} - S_{n-1}) + S_n. \tag{8}$$

Indeed, if such an M exists, then defining $(u_n)_{n\in\mathbb{Z}}$ by setting $u_n = M_n - S_n$, $n \in \mathbb{Z}$, gives a carrier for $(x_n)_{n\in\mathbb{Z}}$, and vice versa. Moreover, if such an M exists and we define $(x_n^1)_n \in \mathbb{Z}$ to be the corresponding updated configuration (i.e. $x_{n-m}^1 := K_n^{(1)}(x_n, u_{n-1})$ for $u_n = M_n - S_n$), then the conservation law (7) yields that

$$\begin{aligned}
x_{n-m}^1 &= 2K_n^{(2)}(x_n, u_{n-1}) + x_n - 2u_{n-1} \\
&= 2K^{(2)}(S_n - S_{n-1}, M_{n-1} - S_{n-1}) + (S_n - S_{n-1}) - 2(M_{n-1} - S_{n-1}) \\
&= 2M_n - S_n - (2M_{n-1} - S_{n-1}).
\end{aligned}$$

Thus if we define $TS : \mathbb{Z} \to \mathbb{R}$ by setting $(TS)_n := 2M_n - S_n - 2M_0$, then we see that the path encoding of $(x^1_n)_{n\in\mathbb{Z}}$ is precisely given by $\theta^m(TS) - \theta^m(TS)_0$, which is a (spatially-shifted) Pitman-type transformation of S with respect to M.

To summarise, for any locally-defined dynamics $(K_n)_{n\in\mathbb{Z}}$ satisfying the conservation law at (7), we can associate a Pitman-type transformation of the corresponding path encoding. NB. By the change of variables, the conservation law can alternatively be expressed as

$$\mathcal{A}_n(F^{(1)}(a,b)) - 2\mathcal{B}_n(F^{(2)}(a,b)) = \mathcal{A}_n(a) - 2\mathcal{B}_n(b).$$

So far, this is simply a change of language. Importantly, however, in many examples, the equation (8) can be solved explicitly. Moreover, it is often possible to determine uniquely a choice of M for which the procedure can be iterated. In particular, in such cases, one obtains the existence and uniqueness of solutions to the initial value problem of interest.

4.1. Examples

It transpires that it is possible to follow the procedure described above in all of the examples of discrete integrable systems presented in Section 3. In particular, we are able to solve initial value problems for these systems whenever the path encoding of the initial configuration is an element of $\mathcal{S}^{lin}$, where the latter set was defined at (2). These results are presented in detail in [10]. Rather than exhaustively repeat the exact statements here, however, we simply summarise how the framework applies in each case within the following table, and present one indicative example – the udKdV equation – in more detail (see Theorem 4.1 below). The maps $K^{\mathbb{Z}}$, $K^{\vee}$, K^{Σ}, $K^{\vee^*}$ and K^{Σ^*} that appear in the table are given as follows: for $a, b \in \mathbb{R}$,

$$K^{\mathbb{Z}}(a,b) := \left(-\min\{a,2b-1\}, b - \frac{a}{2} - \frac{\min\{a,2b-1\}}{2}\right),$$

$$K^{\vee}(a,b) := \left(-\min\{a,2b\}, b - \frac{a}{2} - \frac{\min\{a,2b\}}{2}\right),$$

$$K^{\Sigma}(a,b) := \left(2\log(e^{-\frac{a}{2}} + e^{-b}), b - \frac{a}{2} + \log(e^{-\frac{a}{2}} + e^{-b})\right),$$

$$K^{\vee^*}(a,b) := \left(-\min\{a,b\}, b - \frac{a}{2} - \frac{\min\{a,b\}}{2}\right),$$

$$K^{\Sigma^*}(a,b) := \left(\log(e^{-a} + e^{-b}), b - \frac{a}{2} + \frac{\log(e^{-a} + e^{-b})}{2}\right).$$

Integrable system	z_n^t	w_n^t	$\mathcal{A}_n(a)$	$\mathcal{B}_n(b)$	m	K_n	*Pitman-type transformation*
BBS	η_n^t	W_n^t	$1-2a$	b	0	$K^{\mathbb{Z}}$	$T^{\mathbb{Z}}$
udKdV	η_n^t	U_n^t	$L-2a$	$b-\frac{L}{2}$	0	$K^{\vee}$	$T^{\vee}$
dKdV	ω_n^t	U_n^t	$-\log\delta-2\log a$	$\log b+\frac{\log\delta}{2}$	0	K^{Σ}	T^{Σ}
udToda	$z_{2n-1}^t=Q_n^t$ $z_{2n}^t=E_n^t$	$w_{2n-1}^t=U_n^t$ $w_{2n}^t=F_{udT^*}^{(2)}(E_n^t,U_n^t)$	$(-1)^n a$	b	1	$K_{2n-1}=(K^{\vee^*})^{-1}$ $K_{2n}=K^{\vee^*}$	$\mathcal{T}^{\vee^*}$
dToda	$z_{2n-1}^t=I_n^t$ $z_{2n}^t=J_n^t$	$w_{2n-1}^t=U_n^t$ $w_{2n}^t=F_{dT^*}^{(2)}(J_n^t,U_n^t)$	$(-1)^{n+1}\log a$	$-\log b$	1	$K_{2n-1}=(K^{\Sigma^*})^{-1}$ $K_{2n}=K^{\Sigma^*}$	$\mathcal{T}^{\Sigma^*}$

Note that since we only apply $K^{\mathbb{Z}}$ to elements of $\{-1,1\} \times \{0,1,2,\dots\}$, it would be possible to consider other expressions that are equivalent on this subset of $\mathbb{R}^2$. The latter four of the maps are obtained by a change of variables from $F^{(L)}_{udK}$, $F^{(\delta)}_{dK}$, F_{udT^*} and F_{dT^*}, respectively. Observe that, unlike $F^{(L)}_{udK}$ and $F^{(\delta)}_{dK}$, the maps $K^{\vee}$ and K^{Σ} are parameter free.

As an illustrative example, in the next theorem, we present an application of the Pitman-type transformation approach to the udKdV equation.

Theorem 4.1. *Given $\eta = (\eta_n)_{n\in\mathbb{Z}} \in \mathbb{R}^{\mathbb{Z}}$, let S be the path given by setting $S_0 = 0$ and $S_n - S_{n-1} = L - 2\eta_n$ for $n \in \mathbb{Z}$. If $S \in \mathcal{S}^{lin}$, then there is a unique solution $(\eta^t_n, U^t_n)_{n,t\in\mathbb{Z}}$ to (5) that satisfies the initial condition $\eta^0 = \eta$. This solution is given by*

$$\eta^t_n := \frac{L - S^t_n + S^t_{n-1}}{2}, \qquad U^t_n := M^{\vee}(S^t)_n - S^t_n + \frac{L}{2}, \qquad \forall n, t \in \mathbb{Z},$$

where $S^t := (T^{\vee})^t(S)$ for all $t \in \mathbb{Z}$.

Remark 4.2. *Each of the operators $T^{\vee}$, T^{Σ}, $\mathcal{T}^{\vee^*}$ and $\mathcal{T}^{\Sigma^*}$ is a bijection from $\mathcal{S}^{lin}$ to itself. In particular, if T^* is any of these operators and $S \in \mathcal{S}^{lin}$, then $(T^*)^t(S)$ is well-defined for all $t \in \mathbb{Z}$. Similarly, $T^{\mathbb{Z}}$ is a bijection on $\mathcal{S}^{lin} \cap \{S : \mathbb{Z} \to \mathbb{Z} : S_0 = 0, |S_n - S_{n-1}| = 1\}$.*

Remark 4.3. *The statement of Theorem 4.1 is given in a slightly different form to the corresponding result in [10]. This is because in the latter article, the definition of the transformation T^* does not include the constant shift downwards by $2M^*_0$.*

Remark 4.4. *As is discussed in detail in [10], the assumption that $S \in \mathcal{S}^{lin}$ includes various known solutions to the udKdV equation, such that as those based on finite or periodic initial configurations. The main novelty of our framework is that it also allows us to consider infinite initial configurations. As a particularly important example, one might consider $\eta = (\eta_n)_{n\in\mathbb{Z}}$ to be a random sequence that is stationary and ergodic under spatial shifts. If the density condition $\mathbf{E}\eta_0 < \frac{L}{2}$ holds, then one almost-surely has that $S \in \mathcal{S}^{lin}$, and so the system can be started from the initial configuration η.*

Remark 4.5. *In the case of the udKdV equation, writing the conservation law at (7) in terms of the original lattice variables gives*

$$\eta^t_n + U^t_{n-1} = \eta^{t+1}_n + U^t_n,$$

which can be interpreted transparently as conservation of mass by the system. The corresponding conservation law for the dKdV equation gives

$$\log \omega_n^t + \log U_{n-1}^t = \log \omega_n^{t+1} + \log U_n^t,$$

which can again be understood as conservation of mass. In the case of udToda, putting together the conservation laws for $K^{\vee *}$ and $(K^{\vee *})^{-1}$ in a way that eliminates the variable that was not part of the original lattice yields

$$E_n^t - Q_{n+1}^t - 2U_n^t = E_n^{t+1} - Q_n^{t+1} - 2U_{n+1}^t,$$

which is a combination of the conservation laws for the mass and length of the interval to which the local dynamics applies. An essentially similar argument yields the corresponding conservation rule for the dToda system. See [10] *for further discussion of this point.*

Remark 4.6. *Using a different path encoding, it is shown in* [9] *that the dynamics of the ultra-discrete Toda equation can also be expressed in terms of $T^{\mathbb{R}}$.*

§5. Invariant measures

For the discrete integral systems of Section 3, it is a natural question to ask for which random initial configurations do the dynamics of the model leave the distribution of the configuration unchanged. Or, given that the mappings from the configurations of the discrete integrable systems to their path encodings are one-to-one, one might equivalently ask what distributions on paths are invariant under the associated Pitman-type transformations. In this section, we describe two approaches for verifying the invariance of measures, one of which is based on the path encoding viewpoint, and the other involves working directly with the configurations. Once we have done this, we describe some of our results concerning invariant measures for the discrete integrable systems of Section 3 and Pitman-type transformations of Section 2.

5.1. Approaches for establishing invariance

We call the first approach the 'three conditions theorem', since it establishes a connection between the invariance of S under a Pitman-type transformation and two natural symmetry properties. To describe these, the precise statement involves two reflection operators: a map $S \mapsto R(S)$, where

$$R(S)_n = -S_{-n}$$

(which corresponds to reflection of the configuration); and a map $W \mapsto \tilde{R}(W)$, where

$$\tilde{R}W_n := W_{-n}$$

(which will be applied to the carrier $M(S) - S$). The following result initially appeared as [4, Theorem 1.8] in the context of the BBS, and has been generalised in [8]. Since the latter version is stated in an abstract framework, one needs to make some further assumptions to capture the kind of operator to which the result applies. We will not detail these here, but simply note that these include a 'reversibility' condition (REV), which ensures an appropriate interplay between the operators R and T, and a 'local time' condition (LOC), which enables the function $M(S)-M(S)_0$ to be recovered from $M(S)-S$. Both (REV) and (LOC) hold for all the discrete integrable systems of Section 3.

Theorem 5.1 (Three conditions theorem). *Suppose that $T(S) = 2M(S)-S-2M(S)_0$ is a Pitman-type transformation operator satisfying (REV) and (LOC). It is then the case that, for any probability measure supported on the domain of T, any two of the following conditions imply the third*:

$$R(S) \stackrel{d}{=} S; \qquad \tilde{R}(W) \stackrel{d}{=} W; \qquad T(S) \stackrel{d}{=} S,$$

where $W = M(S) - S$.

Remark 5.2. *If S has i.i.d. increments, then $R(S) \stackrel{d}{=} S$ automatically holds, and so the invariance of S under T follows from the invariance of W under $\tilde{R}$.*

For the second approach, which was developed for BBS(J, K) in [7] and generalised in [6], we directly consider the evolution under a system of locally-defined dynamics, as introduced at the start of Section 3. We call the result a 'detailed balance condition', since it is reminiscent of the corresponding result for determining invariant measures of Markov chains. Note that, for a measurable function F and measure μ on the same space, we define $F(\mu) := \mu \circ F^{-1}$.

Theorem 5.3 (Detailed balance condition). *Suppose $(F_n)_{n\in\mathbb{Z}}$ are all bijections. Moreover, let $\mathcal{C} \subseteq \mathbb{R}^{\mathbb{Z}}$ be such that for each $(z_n)_{n\in\mathbb{Z}} \in \mathcal{C}$, the system of locally-defined dynamics $(F_n)_{n\in\mathbb{Z}}$ has a unique solution $(z_n^t, w_n^t)_{n,t\in\mathbb{Z}}$ with $z^0 = z$.*
(a) *Suppose $F_n \equiv F$, and $z = (z_n)_{n\in\mathbb{Z}}$ is an i.i.d. sequence supported on $\mathcal{C}$. It is then the case that z is invariant under the dynamics if and only if*

$$F(\mu \times \nu) = \mu \times \nu,$$

where μ is the distribution of z_0^0 and ν is the distribution of w_0^0.
(b) *Suppose $F_{2n} \equiv F$ and $F_{2n+1} \equiv F^{-1}$, and $z = (z_n)_{n\in\mathbb{Z}}$ is an alternating i.i.d. sequence (i.e. the terms in the sequence are independent, with the odd terms and the even terms each being identically distributed) supported on $\mathcal{C}$. It is then the case that z is invariant under the dynamics if and only if*

$$F(\mu \times \nu) = \tilde{\mu} \times \tilde{\nu},$$

where $\mu, \tilde{\mu}$ are the distributions of z_0^0, z_1^0, and $\nu, \tilde{\nu}$ are the distributions of w_{-1}^0, w_0^0, respectively.

Remark 5.4. *Each of the two approaches has particular advantages and disadvantages, some of which we now briefly discuss.*
(a) *For the original Pitman transform $T^{\mathbb{Z}}$, if we restrict the state space to $\{S : \mathbb{Z} \to \mathbb{Z} \,:\, S_0 = 0,\, |S_n - S_{n-1}| = 1\}$ (i.e. the original BBS), then both Theorems 5.1 and 5.3 can be applied. On the other hand, if we consider more general increments for S, then the associated locally-defined dynamics are not given by bijections, and so we cannot apply the detailed balance condition.*
(b) *The three conditions theorem is applicable to both discrete-time random paths $S : \mathbb{Z} \to \mathbb{R}$ and continuous-time random paths $S : \mathbb{R} \to \mathbb{R}$. However, the detailed balance condition works only in the discrete-time case.*
(c) *The three conditions theorem only gives a sufficient condition for invariance, whereas the detailed balance condition gives an equivalent condition when it applies. Hence the latter result can be used to characterise all invariant measures of a Pitman-type transformations with i.i.d., or alternating i.i.d., increments.*
(d) *Unlike the detailed balance condition, the three conditions theorem does not depend on z being an i.i.d. or alternating i.i.d. sequence. In fact, in [4], we apply the three conditions theorem to establish the invariance of some random configurations that are not i.i.d.*

5.2. Invariant measures for discrete integrable systems

Since the discrete integrable systems of Section 3 have many conserved quantities, each of the models will admit a rich array of invariant measures (cf. [5, 11, 13]). In the following four theorems, we give examples of invariant configurations based on i.i.d. or alternating i.i.d. sequences. Verification of the results via the detailed balance condition appears in [6]. The latter article incorporates further discussion concerning the complete characterisation of such invariant measures. In the statements of the results, Z represents a normalising constant, which will in general be different on each appearance.

Theorem 5.5. *Suppose $\eta = (\eta_n)_{n\in\mathbb{Z}}$ is an i.i.d. sequence with marginals given by one of the following distributions:*
(a) *for some $\lambda > 0$ and $c \in \mathbb{R}$ such that $c < \frac{L}{2}$,*

$$\mathbf{P}(\eta_n \in dx) = \frac{1}{Z} e^{-\lambda x} \mathbf{1}_{[c,L-c]}(x)dx;$$

(b) *for $h > 0$ such that $L = \ell h$ for some $\ell \in \mathbb{Z}$, $\lambda > 0$ and $k \in \mathbb{Z}$ such that $k < \frac{\ell}{2}$,*

$$\mathbf{P}(\eta_n = hm) = \frac{1}{Z} e^{-\lambda m} \mathbf{1}_{\{k,k+1,\dots,\ell-k\}}(m).$$

It is then the case that the path encoding of η takes values in $\mathcal{S}^{lin}$ almost-surely, and the distribution of η is invariant under the dynamics of the udKdV equation.

Theorem 5.6. *Suppose $\omega = (\omega_n)_{n\in\mathbb{Z}}$ is an i.i.d. sequence with marginals given by the distribution: for some $\lambda > 0$ and $c > 0$,*

$$\mathbf{P}(\omega_n \in dx) = \frac{1}{Z} e^{-cx^{-1}-c\delta x} x^{-\lambda-1} \mathbf{1}_{(0,\infty)}(x)dx.$$

It is then the case that the path encoding of ω takes values in $\mathcal{S}^{lin}$ almost-surely, and the distribution of ω is invariant under the dynamics of the dKdV equation.

Theorem 5.7. *Suppose $(Q,E) = (Q_n,E_n)_{n\in\mathbb{Z}}$ is an i.i.d. sequence (with Q_n independent of E_n and) with marginals given by one of the following distributions:*
(a) *for some $0 < \lambda_2 < \lambda_1$ and $c \in \mathbb{R}$,*

$$\mathbf{P}(Q_n \in dx) = \frac{1}{Z} e^{-\lambda_1 x} \mathbf{1}_{[c,\infty)}(x)dx,$$

$$\mathbf{P}(E_n \in dx) = \frac{1}{Z} e^{-\lambda_2 x} \mathbf{1}_{[c,\infty)}(x)dx;$$

(b) *for some $0 < \lambda_2 < \lambda_1$, $h > 0$ and $k \in \mathbb{Z}$,*

$$\mathbf{P}(Q_n = mh) = \frac{1}{Z} e^{-\lambda_1 m} \mathbf{1}_{\{k,k+1,k+2,\dots\}}(m),$$

$$\mathbf{P}(E_n = mh) = \frac{1}{Z} \exp^{-\lambda_2 m} \mathbf{1}_{\{k,k+1,k+2,\dots\}}(m).$$

It is then the case that the path encoding of (Q,E) takes values in $\mathcal{S}^{lin}$ almost-surely, and the distribution of (Q,E) is invariant under the dynamics of the udToda equation.

Theorem 5.8. *Suppose $(I, V) = (I_n, V_n)$ is an i.i.d. sequence (with I_n independent of V_n and) with marginals given by the following distribution: for some $0 < \lambda_2 < \lambda_1$ and $c > 0$,*

$$\mathbf{P}(I_n \in dx) = \frac{1}{Z} e^{-cx} x^{\lambda_1 - 1} \mathbf{1}_{(0,\infty)}(x) dx,$$

$$\mathbf{P}(V_n \in dx) = \frac{1}{Z} e^{-cx} x^{\lambda_2 - 1} \mathbf{1}_{(0,\infty)}(x) dx.$$

It is then the case that the path encoding of (I, J) takes values in $\mathcal{S}^{lin}$ almost-surely, and the distribution of (I, J) is invariant under the dynamics of the dToda equation.

5.3. Invariant measures for Pitman's type operators

The following theorem transfers the results of the previous section to the setting of path encodings. In the statement, S is either a simple random walk path (i.e. $S_0 = 0$, and $(S_n - S_{n-1})_{n\in\mathbb{Z}}$ is an i.i.d. sequence) or an alternating random walk path (i.e. $S_0 = 0$, and $(S_n - S_{n-1})_{n\in\mathbb{Z}}$ are independent, with the odd terms and the even terms each being identically distributed).

Theorem 5.9. (a) *If S is a simple random walk such that*

$$\mathbf{P}(S_n - S_{n-1} = x) = \begin{cases} p, & \text{if } x = a, \\ 1 - p - q, & \text{if } x = 0, \\ q, & \text{if } x = -a, \end{cases}$$

for some $a > 0$ and $0 \le q < p \le 1$ satisfying $p + q \le 1$, then S is invariant under $T^{\mathbb{Z}}$.
(b) *If S is a simple random walk such that*

$$\mathbf{P}(S_n - S_{n-1} \in dx) = \frac{1}{Z} e^{\lambda x} \mathbf{1}_{[-a,a]}(x) dx$$

for some $a > 0$ and $\lambda > 0$, then S is invariant under $T^{\vee}$. Moreover, if S is a simple random walk such that either

$$\mathbf{P}(S_n - S_{n-1} = mh) = \frac{1}{Z} e^{\lambda m} \mathbf{1}_{\{-k,-k+1,\dots,k-1,k\}}(m),$$

or

$$\mathbf{P}(S_n - S_{n-1} = mh) = \frac{1}{Z} e^{\lambda m} \mathbf{1}_{\{-k+\frac{1}{2},-k+\frac{3}{2},\dots,k-\frac{1}{2}\}}(m),$$

for some $h > 0$, $\lambda > 0$ and $k \in \mathbb{N}$, then S is invariant under $T^{\vee}$.
(c) *If S is a simple random walk such that*

$$\mathbf{P}(S_n - S_{n-1} \in dx) = \frac{1}{Z} e^{\lambda x - a\cosh\left(\frac{x}{2}\right)}$$

for some $a > 0$ and $\lambda > 0$, then S is invariant under T^{Σ}.
(d) *If S is an alternating random walk such that*

$$\mathbf{P}(S_{2n} - S_{2n-1} \in dx) = \frac{1}{Z} e^{-\lambda_1 x} \mathbf{1}_{(a,\infty)}(x) dx,$$

$$\mathbf{P}(S_{2n+1} - S_{2n} \in dx) = \frac{1}{Z} e^{\lambda_2 x} \mathbf{1}_{(-\infty,-a)}(x) dx,$$

for some $0 < \lambda_1 < \lambda_2$ and $a \in \mathbb{R}$, then S is invariant under $\mathcal{T}^{\vee^*}$. Moreover, if S is an alternating random walk such that

$$\mathbf{P}(S_{2n} - S_{2n-1} = mh) = \frac{1}{Z} e^{-\lambda_1 m} \mathbf{1}_{\{k,k+1,k+2\dots,\}}(m),$$

$$\mathbf{P}(S_{2n+1} - S_{2n} = mh) = \frac{1}{Z} e^{\lambda_2 m} \mathbf{1}_{\{\dots,-k-1,-k\}}(m),$$

for some $h > 0$, $0 < \lambda_1 < \lambda_2$ and $k \in \mathbb{Z}$, then S is invariant under $\mathcal{T}^{\vee^*}$.
(e) *Let S be an alternating random walk such that*

$$\mathbf{P}(S_{2n} - S_{2n-1} \in dx) = \frac{1}{Z} e^{\lambda_1 x - ae^x} dx,$$

$$\mathbf{P}(S_{2n+1} - S_{2n} \in dx) = \frac{1}{Z} e^{-\lambda_2 x - ae^{-x}} dx,$$

for some $0 < \lambda_1 < \lambda_2$ and $a > 0$, then S is invariant under $\mathcal{T}^{\Sigma^*}$.

Remark 5.10. *For* (b)–(e), *there are essentially two parameters that uniquely determine the drift and the variance of the increment.* (*Here, for* (d) *and* (e), *by increment we mean $S_{2n} - S_{2n-2}$, the distribution of which does not depend on the parameters a or k.*) *For* (a), *since adding the delta measure at 0 to the increment distribution does not affect to the invariance of S under $T^{\mathbb{Z}}$, we have an extra parameter.*

Remark 5.11. *For* (a), (d) *and* (e), *all invariant measures for the relevant Pitman-type transformations with i.i.d.* (*in the case of* (a)) *or alternating i.i.d.* (*in the case of* (d) *and* (e)) *increments are given by the previous theorem* [6].

§6. Summary

Summarising the above, the following table outlines increment distributions of simple random walks/alternating random walks that are invariant under the Pitman-type transformations of Section 2. In all cases, the process $W = M(S) - S$ is a reversible Markov chain, and it is possible to compute the distribution of W_0; this is also included in the table. For further details, see [6, 8].

Operator	$S_n - S_{n-1}$	W_0	*Integrable system*
$T^{\mathbb{Z}}$	Bernoulli (on $\{-1,1\}$), plus δ_0	Geometric	BBS
$T^{\vee}$	Truncated exponential or truncated geometric	Exponential or geometric	udKdV
T^{Σ}	Log of generalised inverse Gaussian	Log of inverse gamma	dKdV
$\mathcal{T}^{\vee^*}$	Exponential or geometric, with alternating signs and parameters (NB. $S_{2n} - S_{2n-2}$: asymmetric Laplace)	Exponential or geometric	udToda
$\mathcal{T}^{\Sigma^*}$	Log of gamma, with alternating signs and parameters (NB. $S_{2n} - S_{2n-2}$: log of beta)	Log of inverse gamma	dToda

Remark 6.1. *For any $c > 0$ and $v \geq 0$, if $S : \mathbb{R} \to \mathbb{R}$ is defined by setting $S_x = vB_x + cx$, where $B = (B_x)_{x \in \mathbb{R}}$ is a two-sided standard Brownian motion, then the distribution of S is invariant under both $T^{\mathbb{R}}$ and $T^{\int}$, see* [15, 30]. *In the former case, $M(S) - S$ is a reversible reflected Brownian motion with drift, which has exponential stationary distribution. In the latter, $M^{\int}(S) - S$ is a reversible diffusion, with stationary distribution being log of inverse gamma.*

Remark 6.2. *The parameters and distributions of the various models can be connected by ultra-discretisation, a correspondence between the KdV- and Toda-type systems, and continuous scaling limits* [6, 8]. *Indeed, taking an appropriate choice of parametrisation, it is possible to scale the invariant measures of T^{Σ} to arrive at the Brownian invariant measure for $T^{\int}$ of Remark* 6.1. *Moreover, this can be done in such a way that both the discrete and continuous models admit the same distribution for W_0.*

§ Acknowledgements

The research of DC was supported by JSPS Grant-in-Aid for Scientific Research (C), 19K03540. The research of MS was supported by JSPS Grant-in-Aid for Scientific Research (B), 19H01792. We thank the referee for their careful reading of the paper, which resulted in a number of improvements being made.

References

[1] J. Bertoin, *An extension of Pitman's theorem for spectrally positive Lévy processes*, Ann. Probab., **20** (1992), no. 3, 1464–1483.

[2] P. Biane, P. Bougerol and N. O'Connell, *Littelmann paths and Brownian paths*, Duke Math. J., **130** (2005), no. 1, 127–167.

[3] ———, *Continuous crystal and Duistermaat-Heckman measure for Coxeter groups*, Adv. Math., **221** (2009), no. 5, 1522–1583.

[4] D. A. Croydon, T. Kato, M. Sasada and S. Tsujimoto, *Dynamics of the box-ball system with random initial conditions via Pitman's transformation*, to appear in Mem. Amer. Math. Soc., preprint appears at arXiv:1806.02147, 2018.

[5] D. A. Croydon and M. Sasada, *Invariant measures for the box-ball system based on stationary Markov chains and periodic Gibbs measures*, J. Math. Phys., **60** (2019), no. 8, 083301, 25.

[6] ———, *Detailed balance and invariant measures for systems of locally-defined dynamics*, preprint appears at arXiv:2007.06203, 2020.

[7] ———, *Duality between box-ball systems of finite box and/or carrier capacity*, RIMS Kôkyûroku Bessatsu, **B79** (2020), 63–107.

[8] ———, *Pitman-type transformations and their invariant measures*, forthcoming, 2021.

[9] D. A. Croydon, M. Sasada and S. Tsujimoto, *Dynamics of the ultra-discrete Toda lattice via Pitman's transformation*, RIMS Kôkyûroku Bessatsu, **B78** (2020), 235–250.

[10] ———, *General solutions for KdV- and Toda-type discrete integrable systems based on path encodings*, preprint appears at arXiv:2011.00690, 2020.

[11] P. A. Ferrari and D. Gabrielli, *BBS invariant measures with independent soliton components*, Electron. J. Probab., **25** (2020), paper no. 78, 26.

[12] ———, *Box-ball system: soliton and tree decomposition of excursions*, XIII Symposium on Probability and Stochastic Processes, (S. I. López, V. M. Rivero, A. Rocha-Arteaga and A. Siri-Jégousse, eds.), Springer International Publishing, 2020.

[13] P. A. Ferrari, C. Nguyen, L. Rolla and M. Wang, *Soliton decomposition of the box-ball system*, preprint appears at arXiv:1806.02798, 2018.

[14] B. M. Hambly, J. B. Martin and N. O'Connell, *Pitman's $2M - X$ theorem for skip-free random walks with Markovian increments*, Electron. Comm. Probab., **6** (2001), 73–77.

[15] J. M. Harrison and R. J. Williams, *On the quasireversibility of a multiclass Brownian service station*, Ann. Probab., **18** (1990), no. 3, 1249–1268.

[16] R. Inoue, A. Kuniba and T. Takagi, *Integrable structure of box-ball systems: crystal, Bethe ansatz, ultradiscretization and tropical geometry*, J. Phys. A, **45** (2012), no. 7, 073001, 64.

[17] T. Jeulin, *Un théorème de J. W. Pitman*, Séminaire de Probabilités, XIII (Univ. Strasbourg, Strasbourg, 1977/78), Lecture Notes in Math., vol. 721, Springer, Berlin, 1979, With an appendix by M. Yor, pp. 521–532,

[18] K. Kondo, *Dynamics of the multicolor box-ball system with random initial conditions via Pitman's transformation*, preprint appears at arXiv:2003.12974, 2020.

[19] D. J. Korteweg and G. de Vries, *On the change of form of long waves advancing in a rectangular canal, and on a new type of long stationary waves*, Philos. Mag. (5), **39** (1895), no. 240, 422–443.

[20] A. Kuniba and H. Lyu, *Large deviations and one-sided scaling limit of randomized multicolor box-ball system*, J. Stat. Phys., **178** (2020), no. 1, 38–74.

[21] A. Kuniba, H. Lyu and M. Okado, *Randomized box-ball systems, limit shape of rigged configurations and thermodynamic Bethe ansatz*, Nuclear Phys. B, **937** (2018), 240–271.

[22] L. Levine, H. Lyu and J. Pike, *Double jump phase transition in a soliton cellular automaton*, preprint appears at arXiv:1706.05621, 2017.

[23] J. Lewis, H. Lyu, P. Pylvavskyy and A. Sen, *Scaling limit of soliton lengths in a multicolor box-ball system*, preprint appears at arXiv:1911.04458, 2019.

[24] H. Matsumoto and M. Yor, *Some changes of probabilities related to a geometric Brownian motion version of Pitman's $2M - X$ theorem*, Electron. Comm. Probab., **4** (1999), 15–23.

[25] ———, *A version of Pitman's $2M - X$ theorem for geometric Brownian motions*, C. R. Acad. Sci. Paris Sér. I Math., **328** (1999), no. 11, 1067–1074.

[26] ———, *An analogue of Pitman's $2M - X$ theorem for exponential Wiener functionals. I. A time-inversion approach*, Nagoya Math. J., **159** (2000), 125–166.

[27] ———, *An analogue of Pitman's $2M - X$ theorem for exponential Wiener functionals. II. The role of the generalized inverse Gaussian laws*, Nagoya Math. J., **162** (2001), 65–86.

[28] N. O'Connell, *Random matrices, non-colliding processes and queues*, Séminaire de Probabilités, XXXVI, Lecture Notes in Math., vol. 1801, Springer, Berlin, 2003, pp. 165–182.

[29] ———, *Directed polymers and the quantum Toda lattice*, Ann. Probab., **40** (2012), no. 2, 437–458.

[30] N. O'Connell and M. Yor, *Brownian analogues of Burke's theorem*, Stochastic Process. Appl., **96** (2001), no. 2, 285–304.

[31] ———, *A representation for non-colliding random walks*, Electron. Comm. Probab., **7** (2002), 1–12.

[32] J. W. Pitman, *One-dimensional Brownian motion and the three-dimensional Bessel process*, Advances in Appl. Probability, **7** (1975), no. 3, 511–526.

[33] L. C. G. Rogers, *Characterizing all diffusions with the $2M - X$ property*, Ann. Probab., **9** (1981), no. 4, 561–572.

[34] L. C. G. Rogers and J. W. Pitman, *Markov functions*, Ann. Probab., **9** (1981), no. 4, 573–582.

[35] Y. Saisho and H. Tanemura, *Pitman type theorem for one-dimensional diffusion processes*, Tokyo J. Math., **13** (1990), no. 2, 429–440.

[36] T. Szabados and B. Székely, *An exponential functional of random walks*, J. Appl. Probab., **40** (2003), no. 2, 413–426.

[37] D. Takahashi and J. Matsukidaira, *Box and ball system with a carrier and ultradiscrete modified KdV equation*, J. Phys. A, **30** (1997), no. 21, L733–L739.

[38] D. Takahashi and J. Satsuma, *A soliton cellular automaton*, J. Phys. Soc. Japan, **59** (1990), no. 10, 3514–3519.

[39] M. Toda, *Vibration of a chain with nonlinear interaction*, J. Phys. Soc. Japan, **22** (1967), no. 2, 431–436.

[40] T. Tokihiro, *The mathematics of box-ball systems*, Asakura Shoten, 2010.

[41] T. Tokihiro, D. Takahashi, J. Matsukidaira and J. Satsuma, *From soliton equations to integrable cellular automata through a limiting procedure*, Phys. Rev. Lett., **76** (1996), no. 18, 3247–3250.

[42] S. Tsujimoto and R. Hirota, *Ultradiscrete KdV equation*, J. Phys. Soc. Japan, **67** (1998), no. 6, 1809–1810.

Research Institute for Mathematical Sciences,
Kyoto University,
Kyoto 606-8502, Japan
E-mail address: `croydon@kurims.kyoto-u.ac.jp`

Graduate School of Mathematical Sciences,
University of Tokyo,
3-8-1, Komaba, Meguro-ku,
Tokyo, 153-8914, Japan
E-mail address: `sasada@ms.u-tokyo.ac.jp`

Advanced Studies in Pure Mathematics 87, 2021
Stochastic Analysis, Random Fields and Integrable Probability — Fukuoka 2019
pp. 403–414

The number of spanning clusters of the uniform spanning tree in three dimensions

Omer Angel, David A. Croydon, Sarai Hernandez-Torres and Daisuke Shiraishi

Abstract.

Let $\mathcal{U}_\delta$ be the uniform spanning tree on $\delta\mathbb{Z}^3$. A spanning cluster of $\mathcal{U}_\delta$ is a connected component of the restriction of $\mathcal{U}_\delta$ to the unit cube $[0,1]^3$ that connects the left face $\{0\} \times [0,1]^2$ to the right face $\{1\} \times [0,1]^2$. In this note, we will prove that the number of the spanning clusters is tight as $\delta \to 0$, which resolves an open question raised by Benjamini in [4].

§1. Introduction

Given a finite connected graph $G = (V, E)$, a spanning tree T of G is a subgraph of G that is a tree (i.e. is connected and contains no cycles) with vertex set V. A uniform spanning tree (UST) of G is obtained by choosing a spanning tree of G uniformly at random. This is an important model in probability and statistical physics, with beautiful connections to other subjects, such as electrical potential theory, loop-erased random walk and Schramm-Loewner evolution. See [10] for an introduction to various aspects of USTs.

Fix $\delta \in (0,1)$ and $d \in \mathbb{N}$. In [11] it was shown that, by taking the local limit of the uniform spanning trees on an exhaustive sequence of finite subgraphs of $\delta\mathbb{Z}^d$, it is possible to construct a random subgraph $\mathcal{U}_\delta$ of $\delta\mathbb{Z}^d$. Whilst the resulting graph $\mathcal{U}_\delta$ is almost-surely a forest consisting on an infinite number of disjoint components that are trees when $d \geq 5$, it is also the case that $\mathcal{U}_\delta$ is almost-surely a spanning tree of $\delta\mathbb{Z}^d$ with one topological end for $d \leq 4$, see [11]. In the latter low-dimensional case, $\mathcal{U}_\delta$ is commonly referred to as the UST on $\delta\mathbb{Z}^d$.

Received January 11, 2020.
Revised May 19, 2020.
2010 *Mathematics Subject Classification.* Primary 60D05; Secondary 05C80.
Key words and phrases. uniform spanning tree, spanning clusters.

In this note, we study a macroscopic scale property of $\mathcal{U}_\delta$, namely the number of its spanning clusters, as previously studied by Benjamini in [4]. To be more precise, let us proceed to introduce some notation. Write

$$\mathbb{B} = [0,1]^d = \{(x_1, x_2, \ldots, x_d) \in \mathbb{R}^d : 0 \le x_i \le 1,\ i = 1, 2, \ldots, d\} \tag{1}$$

for the unit hypercube in $\mathbb{R}^d$. Also, set

$$F = \{(x_1, x_2, \ldots, x_d) \in \mathbb{R}^d : x_1 = 0\} \tag{2}$$

and

$$G = \{(x_1, x_2, \ldots, x_d) \in \mathbb{R}^d : x_1 = 1\} \tag{3}$$

for the hyperplanes intersecting the 'left' and 'right' sides of the hypercube $\mathbb{B}$. Given a subgraph $U = (V, E)$ of $\delta\mathbb{Z}^d$, we write $U' = (V', E')$ for the restriction of U to the cube $\mathbb{B}$, i.e. we set $V' = V \cap \mathbb{B}$ and $E' = \{\{x, y\} \in E : x, y \in V'\}$. A connected component of U' is called a cluster of U. Moreover, following [4], a spanning cluster of U is a cluster of U containing vertices x and y such that $\mathrm{dist}(x, F) < \delta$ and $\mathrm{dist}(y, G) < \delta$, where $\mathrm{dist}(z, A) := \inf_{w \in A} |z - w|$ is the Euclidean distance between a point $z \in \mathbb{R}^d$ and subset $A \subseteq \mathbb{R}^d$. That is, a cluster of U is called spanning when it connects F to G (at the level of discretization being considered).

Concerning the number of spanning clusters of $\mathcal{U}_\delta$, it was proved in [4] that:

- for $d \ge 4$, the expected number of spanning clusters of $\mathcal{U}_\delta$ grows to infinity as $\delta \to 0$;
- for $d = 2$, the number of spanning clusters of $\mathcal{U}_\delta^+$ is tight as $\delta \to 0$, where $\mathcal{U}_\delta^+$ denotes the uniform spanning tree of the square $\mathbb{B} \cap \delta\mathbb{Z}^2$ when all the vertices on the right side of the square are identified to a single point (which is called the right wired uniform spanning tree in [4]). Figure 1 shows the spanning cluster of a realisation of (an approximation to) $\mathcal{U}_\delta$ on $\delta\mathbb{Z}^2$.

The case $d = 3$ was left as an open question in [4]. The main purpose of this note is to resolve it by showing the following theorem.

Theorem 1.1. *Let $d = 3$. It holds that the number of spanning clusters of $\mathcal{U}_\delta$ is tight as $\delta \to 0$.*

Remark 1.2. *The proof for Theorem 1.1 can be adapted to show tightness of the number of spanning clusters of $\mathcal{U}_\delta$ on $\delta\mathbb{Z}^2$. This is an improvement over the result in [4], which required right-wired boundary conditions.*

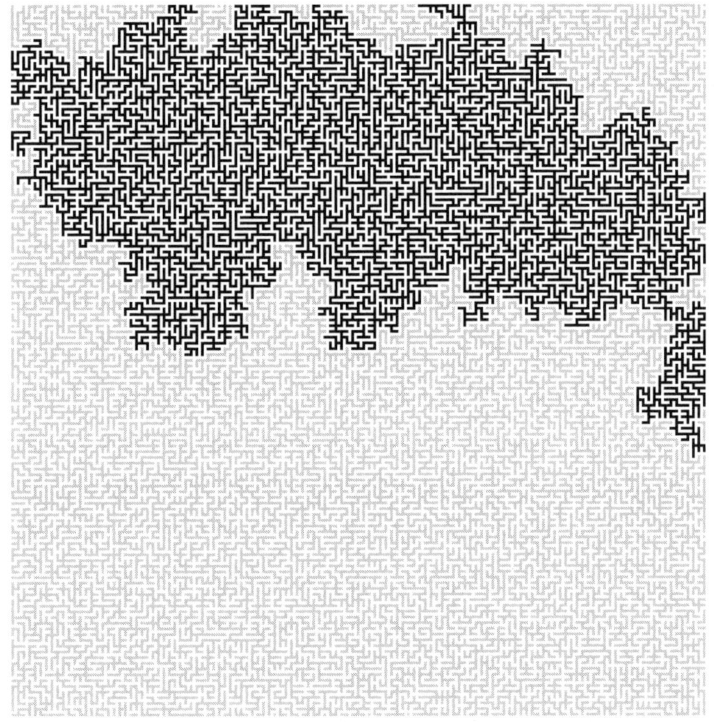

Fig. 1. Part of a UST in a two-dimensional box; the part shown is the central 115×115 section of a UST on a 229×229 box. The single cluster spanning the two sides of the box is highlighted.

Remark 1.3. *Part of Benjamini's motivation for studying the number of spanning clusters came from percolation. Indeed, for critical Bernoulli percolation in $\mathbb{Z}^d$, it is conjectured that the number of spanning clusters is tight when $d \leq 6$, while the expected number of spanning clusters grows to infinity as the mesh size goes to zero for $d > 6$, see for instance* [1, 5, 6]. *Putting our main conclusion together with the results obtained by Benjamini in* [4], *the corresponding qualitative picture is proved for the uniform spanning tree.*

Remark 1.4. *In the article* [2], *we establish a scaling limit for the three-dimensional UST in a version of the Gromov-Hausdorff topology, at least along the subsequence $\widehat{\delta}_n := 2^{-n}$. The corresponding two-dimensional result is also known (along an arbitrary sequence $\delta \to 0$),* *see* [3] *and* [7, Remark 1.2]. *In both cases, we expect that the techniques used to prove such a scaling limit can be used to show that the number of spanning clusters of $\mathcal{U}_\delta$ actually converges in distribution. We plan to pursue this in a subsequent work that focusses on the topological properties of the three-dimensional UST.*

The organization of the remainder of the paper is as follows. In Section 2, we introduce some notation that will be used in the paper. The proof of Theorem 1.1 is then given in Section 3.

§2. Notation

In this section, we introduce the main notation needed for the proof of Theorem 1.1. We write $|\cdot|$ for the Euclidean norm on $\mathbb{R}^3$ and, as in the introduction, $\mathrm{dist}(\cdot,\cdot)$ for the Euclidean distance between a point and a subset of $\mathbb{R}^3$. Given $\delta \in (0,1)$, if $x \in \delta\mathbb{Z}^3$ and $r > 0$, then we write

$$B(x,r) = \left\{ y \in \delta\mathbb{Z}^3 : \, |x-y| < r \right\}$$

for the lattice ball of centre x and radius r (we will commonly omit dependence on δ for brevity). Let $\mathbb{B}$, F and G be defined as at (1), (2) and (3) in the case $d = 3$.

For $\delta \in (0,1)$, a sequence $\lambda = (\lambda(0), \lambda(1), \dots, \lambda(m))$ is said to be a path of length m if $\lambda(i) \in \delta\mathbb{Z}^3$ and $|\lambda(i) - \lambda(i+1)| = \delta$ for every i. A path λ is simple if $\lambda(i) \neq \lambda(j)$ for all $i \neq j$. For a path $\lambda = (\lambda(0), \lambda(1), \dots, \lambda(m))$, we define its loop-erasure $\mathrm{LE}(\lambda)$ as follows. Firstly, let

$$s_0 = \max\left\{ j \leq m : \, \lambda(j) = \lambda(0) \right\},$$

and for $i \geq 1$, set

$$s_i = \max\left\{ j \leq m : \, \lambda(j) = \lambda(s_{i-1}+1) \right\}.$$

Moreover, write $n = \min\{i : s_i = m\}$. The loop-erasure of λ is then given by

$$\mathrm{LE}(\lambda) = \left(\lambda(s_0), \lambda(s_1), \dots, \lambda(s_n) \right).$$

We write $\mathrm{LE}(\lambda)(k) = \lambda(s_k)$ for each $0 \leq k \leq n$. Note that the vertices hit by $\mathrm{LE}(\lambda)$ are a subset of those hit by λ, and that $\mathrm{LE}(\lambda)$ is a simple path such that $\mathrm{LE}(\lambda)(0) = \lambda(0)$ and $\mathrm{LE}(\lambda)(n) = \lambda(m)$. Although the loop-erasure of λ has so far only been defined in the case that λ has a finite length, it is clear that we can define $\mathrm{LE}(\lambda)$ similarly for an infinite path λ if the set $\{k \geq 0 : \lambda(j) = \lambda(k)\}$ is finite for each $j \geq 0$. Additionally, when the path λ is given by a simple random walk, we call $\mathrm{LE}(\lambda)$ a loop-erased random walk (see [9] for an introduction to loop-erased random walks).

Again given $\delta \in (0,1)$, write $\mathcal{U}_\delta$ for the uniform spanning tree on $\delta\mathbb{Z}^3$. As noted in the introduction, this object was constructed in [11], and shown to be a tree with a single end, almost-surely. The graph $\mathcal{U}_\delta$ can be generated from loop-erased random walks by a procedure now referred to as Wilson's algorithm (after [13]), which is described as follows.

- Let $(x_i)_{i\geq 1}$ be an arbitrary, but fixed, ordering of $\delta\mathbb{Z}^3$.
- Write R^{x_1} for a simple random walk on $\delta\mathbb{Z}^3$ started at x_1. Let $\gamma_{x_1} = \mathrm{LE}(R^{x_1})$ be the loop-erasure of R^{x_1} – this is well-defined since R^{x_1} is transient. Set $\mathcal{U}^1 = \gamma_{x_1}$.
- Given $\mathcal{U}^i$ for $i \geq 1$, let $R^{x_{i+1}}$ be a simple random walk (independent of $\mathcal{U}^i$) started at x_{i+1} and stopped on hitting $\mathcal{U}^i$. We let $\mathcal{U}^{i+1} = \mathcal{U}^i \cup \mathrm{LE}(R^{x_{i+1}})$.

It is then the case that the output random tree $\bigcup_{i=1}^\infty \mathcal{U}^i$ has the same distribution as $\mathcal{U}_\delta$. In particular, the distribution of the output tree does not depend on the ordering of points $(x_i)_{i\geq 1}$.

Similarly to above, for $z \in \delta\mathbb{Z}^3$, we will write γ_z for the infinite simple path in $\mathcal{U}_\delta$ starting from z. Given a point $z \in \delta\mathbb{Z}^3$, it follows from the construction of $\mathcal{U}_\delta$ explained hitherto that the distribution of γ_z coincides with that of $\mathrm{LE}(R^z)$, where R^z is a simple random walk on $\delta\mathbb{Z}^3$ started at z.

Furthermore, as we explained in the introduction, we will write $\mathcal{U}'_\delta$ for the restriction of $\mathcal{U}_\delta$ to the cube $\mathbb{B}$. A connected component of $\mathcal{U}'_\delta$ is called a cluster. Also, as we defined previously, a spanning cluster is a cluster connecting F to G. We let N_δ be the number of spanning clusters of $\mathcal{U}_\delta$.

Finally, we will use c, C, c_0, etc. to denote universal positive constants which may change from line to line.

§3. Proof of the main result

In this section, we will prove the following theorem, which incorporates Theorem 1.1.

Theorem 3.1. *There exists a universal constant C such that: for all $M < \infty$ and $\delta > 0$,*

$$\mathbf{P}(N_\delta \geq M) \leq CM^{-1}. \tag{4}$$

In particular, the laws of $(N_\delta)_{\delta\in(0,1)}$ form a tight sequence of probability measures on $\mathbb{Z}_+$.

Remark 3.2. *In [1], Aizenman proved that for critical percolation in two dimensions, the probability of seeing M distinct spanning clusters is bounded above by Ce^{-cM^2}. We do not expect that the polynomial bound in (4) is sharp, but leave it as an open problem to determine the correct tail behaviour for number of spanning clusters of the UST in three dimensions, and, in particular, ascertain whether it also exhibits Gaussian decay.*

Proof. Let $\delta \in (0,1)$, and suppose $M \geq 1$ is such that $\delta < M^{-1}$. For $r \in [0,1]$, we let

$$A(r) = \{(x_1, x_2, x_3) \in \mathbb{B} : x_1 = r\}.$$

We also define

$$A = [-1,2]^3, \quad \mathbb{B}' = \{(x_1, x_2, x_3) \in \mathbb{B} : x_1 \leq 2/3\}.$$

Moreover, let $(z_i)_{i=1}^L$ be a sequence of points in $A \cap \delta\mathbb{Z}^3$ such that $A \subseteq \bigcup_{i=1}^L B(z_i, 1/M)$ and $L \leq 10^5 M^3$.

To construct $\mathcal{U}_\delta$, we first perform Wilson's algorithm for $(z_i)_{i=1}^L$ (see Section 2). Namely, we consider

$$\mathcal{U}^1 := \bigcup_{i=1}^L \gamma_{z_i},$$

which is the subtree of $\mathcal{U}_\delta$ spanned by $(z_i)_{i=1}^L$. (Recall that for $z \in \delta\mathbb{Z}^3$ we denote the infinite simple path in $\mathcal{U}_\delta$ starting from z by γ_z.) The idea of the proof is then as follows. Crucially, each branch of $\mathcal{U}^1$ is a 'hittable' set, in the sense that for a simple random walk R whose starting point is close to $\mathcal{U}^1$, it is likely that R hits $\mathcal{U}^1$ before moving far away. As a result, Wilson's algorithm guarantees that, with high probability, the spanning clusters of $\mathcal{U}_\delta$ correspond to those of $\mathcal{U}^1$ when M is sufficiently large. So, the problem boils down to the tightness of the number of spanning clusters of $\mathcal{U}^1$, which is not difficult to prove.

To make the above argument rigorous, we introduce the following two "good" events for $\mathcal{U}^1$:

$$H_i = H_i(\xi) := \left\{ \begin{array}{c} \text{For any } x \in B(0,4) \cap \delta\mathbb{Z}^3 \text{ with } \mathrm{dist}(x, \gamma_{z_i}) \leq 1/M, \\ P_R^x\left(R[0,T] \cap \gamma_{z_i} = \emptyset\right) \leq M^{-\xi} \end{array} \right\},$$

$$I_i := \left\{ \begin{array}{c} \text{The number of crossings of } \gamma_{z_i} \text{ between } A(0) \text{ and } A(2/3) \\ \text{in } \mathbb{B}' \text{ is smaller than } M \end{array} \right\},$$

for $1 \leq i \leq L$, where

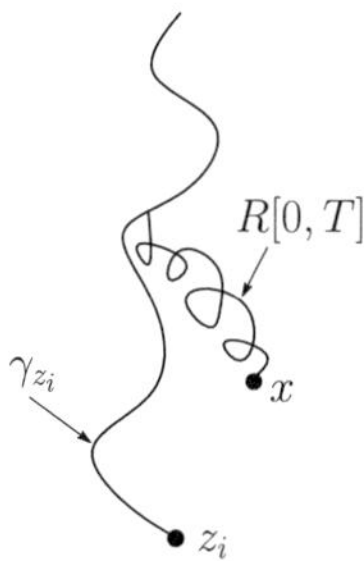

Fig. 2. Conditional on the event H_i, for any $x \in B(0,4) \cap \delta\mathbb{Z}^3$ with $\mathrm{dist}(x, \gamma_{z_i}) \le 1/M$, the above configuration occurs with probability at least $1 - M^{-\xi}$.

- R is a simple random walk which is independent of γ_{z_i}, the law of which is denoted by P_R^x when we assume $R(0) = x$;
- T is the first time that R exits $B(x, 1/\sqrt{M})$;
- a crossing of γ_{z_i} between $A(0)$ and $A(2/3)$ in $\mathbb{B}'$ is a connected component of the restriction of γ_{z_i} to $\mathbb{B}'$ that connects $A(0)$ to $A(2/3)$.

Namely, the event H_i guarantees that the branch γ_{z_i} is a hittable set (see Figure 2), and the event I_i controls the number of crossings of γ_{z_i}.

Now, [12, Theorem 3.1] ensures that there exist universal constants $\xi_0, C > 0$ such that

$$\mathbf{P}\left(\bigcap_{i=1}^{L} H_i(\xi_0) \right) \ge 1 - CM^{-10}.$$

Thus, with high probability (for $\mathcal{U}^1$), each branch of $\mathcal{U}^1$ is a hittable set.

The probability of the event I_i is easy to estimate. Indeed, suppose that the event I_i does not occur. This implies that the number of "traversals" of S^{z_i} from $A(0)$ to $A(2/3)$ or vice versa must be bigger than M, where S^{z_i} stands for a simple random walk starting from z_i. Notice that there exists a universal constant $c_0 > 0$ such that for any point $w \in A(0)$ (respectively $w \in A(2/3)$), the probability that S^w hits $A(2/3)$ (respectively $A(0)$) is smaller than $1 - c_0$ (see [8, Proposition 1.5.10], for example). Thus, the probability of the event I_i is bounded below by $1 - (1 - c_0)^M =: 1 - e^{-aM}$, where $a > 0$. Taking sum over $1 \le i \le L$, we find that

$$\mathbf{P}\left(\bigcap_{i=1}^{L} I_i\right) \geq 1 - Le^{-aM}.$$

To put the above together, let

$$J = \bigcap_{i=1}^{L} H_i(\xi_0) \cap I_i.$$

For $1 \leq i \leq L$, set $\mathcal{U}_i^1 = \bigcup_{j=1}^{i} \gamma_{z_j}$ so that $\mathcal{U}^1 = \mathcal{U}_L^1$. As above, by a spanning cluster of $\mathcal{U}_i^1$ between $A(0)$ and $A(2/3)$ in $\mathbb{B}'$ we mean a connected component of the restriction of $\mathcal{U}_i^1$ to $\mathbb{B}'$ which connects $A(0)$ to $A(2/3)$. We write n_i for the number of spanning clusters of $\mathcal{U}_i^1$ between $A(0)$ and $A(2/3)$ in $\mathbb{B}'$. On the event J, we have that

$$n_i \leq iM + i - 1,$$

for all $1 \leq i \leq L$, since $n_{i+1} - n_i$ is at most $M + 1$ for each $i \geq 1$. In particular, we see that the number of spanning clusters of $\mathcal{U}^1$ between $A(0)$ and $A(2/3)$ in $\mathbb{B}'$ is bounded above by $L(M+1)$, which is comparable to M^4.

We next consider a sequence of subsets of A as follows. Let $a^* > 0$ be the positive constant such that

$$a^* \sum_{k=1}^{\infty} k^{-2} = 10^{-1}. \tag{5}$$

Set $\eta_1 = 0$, and $\eta_k = a^* \sum_{j=1}^{k-1} j^{-2}$ for $k \geq 2$. Finally, for $k \geq 1$, let

$$A_k = [-1 + \eta_k, 2 - \eta_k]^3.$$

Notice that $A_{k+1} \subseteq A_k$ and $[-1/2, 3/2]^3 \subseteq A_k$ for all $k \geq 1$, and moreover $\mathrm{dist}(\partial A_k, \partial A_{k+1}) = a^* k^{-2}$. We further introduce sequences $(z_i^k)_{i=1}^{L_k}$ consisting of points in $A_k \cap \delta\mathbb{Z}^3$ such that

$$A_k \subseteq \bigcup_{i=1}^{L_k} B\left(z_i^k, \delta_k\right),$$

and

$$L_k \leq 10^5 \delta_k^{-3}, \text{ where } \delta_k := M^{-1} 2^{-(k-1)}. \tag{6}$$

Note that we may assume that $L_1 = L$ and $(z_i^1)_{i=1}^{L_1} = (z_i)_{i=1}^{L}$.

For $\xi > 0$, we set

$$H_i^k = H_i^k(\xi) := \left\{ \begin{array}{c} \text{For any } x \in B(0,4) \cap \delta\mathbb{Z}^3 \text{ with } \operatorname{dist}\left(x, \gamma_{z_i^k}\right) \leq \delta_k, \\ P_R^x\left(R[0,T^k] \cap \gamma_{z_i^k} = \emptyset\right) \leq \delta_k^{\xi} \end{array} \right\},$$

where R is a simple random walk that is independent of $\gamma_{z_i^k}$, with law denoted by P_R^x when we assume $R(0) = x$, and T^k is the first time that R exits $B(x, \sqrt{\delta_k})$. By [12, Theorem 3.1] again, there exist universal constants $\xi_1, C > 0$ (which do not depend on k) such that

$$\mathbf{P}\left(\bigcap_{i=1}^{L_k} H_i^k(\xi_1)\right) \geq 1 - C\delta_k^{10},$$

for all $k = 1, 2, \dots, k_0$, where k_0 is the smallest integer k such that $\delta_k < \delta$. Thus if we write

$$H^k = \bigcap_{i=1}^{L_k} H_i^k(\xi_1)$$

and

$$J' = J \cap \bigcap_{k=1}^{k_0} H^k,$$

we have

$$\mathbf{P}(J') \geq 1 - CM^{-10}.$$

Given the above setup, we perform Wilson's algorithm as follows:

- recall that $\mathcal{U}^1$ is the tree spanned by $(z_i^1)_{i=1}^{L_1} = (z_i)_{i=1}^{L}$;
- next perform Wilson's algorithm for $(z_i^2)_{i=1}^{L_2}$ – for each z_i^2, run a simple random walk $R^{z_i^2}$ from z_i^2 until it hits the part of the tree that has already been constructed, and adding its loop-erasure as a new branch – the output tree is denoted by $\mathcal{U}^2$;
- repeat the previous step for $(z_i^k)_{i=1}^{L_k}$ to construct $\mathcal{U}^k$ for $k = 1, 2, \dots, k_0$.

Now, condition $\mathcal{U}^1$ on the event J above. We will show that, with high (conditional) probability, every new branch in $\mathcal{U}^2 \setminus \mathcal{U}^1$ has diameter smaller than $M^{-1/4}$. To this end, for $1 \leq i \leq L_2$, we write d_i^2 for the Euclidean diameter of the path from z_i^2 to $\mathcal{U}^1$ in $\mathcal{U}^2$, and define the event W_i^2 by setting

$$W_i^2 = \left\{d_i^2 \geq M^{-1/4}\right\}.$$

Suppose that the event W_i^2 occurs. By Wilson's algorithm, the simple random walk $R^{z_i^2}$ must not hit $\mathcal{U}^1$ until it exits $B(z_i^2, M^{-1/4})$. Since $\operatorname{dist}(z_i^2, \partial A) \geq a^*$ (for the constant a^* defined at (5)), it holds that $B(z_i^2, M^{-1/4}) \subseteq A$. With this in mind, we set $u_0 = 0$, and

$$u_m = \inf\left\{ j \geq u_{m-1} : \left| R^{z_i^2}(j) - R^{z_i^2}(u_{m-1}) \right| \geq M^{-1/2} \right\}$$

for $m \geq 1$. We then have that

$$R^{z_i^2}[u_{m-1}, u_m] \cap \mathcal{U}^1 = \emptyset$$

for all $1 \leq m \leq M^{1/4}$. Since $A \subseteq \bigcup_{i=1}^{L} B(z_i, 1/M)$, it follows that for each $1 \leq m \leq M^{1/4}$, there exists a z_i such that $R^{z_i^2}(u_{m-1}) \in B(z_i, 1/M)$. Thus the event $H_i(\xi_0)$ guarantees that

$$\mathbf{P}\left(R^{z_i^2}[u_{m-1}, u_m] \cap \mathcal{U}^1 = \emptyset \text{ for all } 1 \leq m \leq M^{1/4} \right) \leq M^{-\xi_0 M^{1/4}}.$$

Consequently, the conditional probability of $\bigcup_{i=1}^{L_2} W_i^2$ is bounded above by $L_2 M^{-\xi_0 M^{1/4}}$, which is smaller than $CM^{-\xi_0 M^{1/4}+3}$ for some universal constant $C \in (0, \infty)$ (see (6), for the definition of L_2). Replacing constants if necessary, this implies that, with probability at least $1 - Ce^{-cM^{1/4}}$, every new branch in $\mathcal{U}^2 \setminus \mathcal{U}^1$ has diameter smaller than $M^{-1/4}$. Notice that once each new branch has such a small diameter, the event J guarantees that the number of spanning clusters of $\mathcal{U}^2$ between $A(0)$ and $A(2/3 + M^{-1/4})$ in $\mathbb{B}$ is bounded above by $L(M+1) \leq 10^6 M^4$.

Essentially the same argument is valid for $\mathcal{U}^k$. Indeed, conditioning $\mathcal{U}^k$ on the good event $J \cap \bigcap_{l=1}^{k} H^l$ as above, it holds that, with probability at least $1 - Ce^{-c\delta_k^{-1/4}}$ every new branch in $\mathcal{U}^{k+1} \setminus \mathcal{U}^k$ has diameter smaller than $\delta_k^{1/4}$. Notice that $\sum_k \delta_k^{1/4} \leq 10M^{-1/4} < 10^{-2}$ when M is large. Therefore, with probability at least $1 - CM^{-10}$, the number of spanning clusters of $\mathcal{U}^{k_0}$ between $A(0)$ and $A(3/4)$ in $\mathbb{B}$ is bounded above by $L(M+1) \leq CM^4$ for some universal constant C.

Finally, we perform Wilson's algorithm for all of the remaining points in $\delta\mathbb{Z}^3$ to construct $\mathcal{U}_\delta$. Since k_0 is the smallest integer k such that $\delta_k < \delta$, it follows that the restriction of $\mathcal{U}_\delta$ to $\mathbb{B}$ coincides with that of $\mathcal{U}^{k_0}$. Thus we conclude that there exists a universal constant C such that: for all $M < \infty$ and $\delta \in (0, M^{-1})$,

$$\mathbf{P}(N_\delta \geq M) \leq CM^{-2}.$$

For the case that $\delta > M^{-1}$, it is clear that $N_\delta < 100M^2$. Combining these two bounds, we readily obtain the bound at (4). Q.E.D.

§ Acknowledgements

DC would like to acknowledge the support of a JSPS Grant-in-Aid for Research Activity Start-up, 18H05832 and a JSPS Grant-in-Aid for Scientific Research (C), 19K03540. SHT is supported by a fellowship from the Mexican National Council for Science and Technology (CONACYT). DS is supported by a JSPS Grant-in-Aid for Early-Career Scientists, 18K13425 and JSPS KAKENHI Grant Number 17H02849 and 18H01123.

References

[1] M. Aizenman, *On the number of incipient spanning clusters*, Nuclear Phys. B, **485** (1997), no. 3, 551–582.

[2] O. Angel, D. A. Croydon, S. Hernandez-Torres and D. Shiraishi, *Scaling limit of the three-dimensional uniform spanning tree and the associated random walk*, preprint, see https://arxiv.org/abs/2003.09055.

[3] M. T. Barlow, D. A. Croydon and T. Kumagai, *Subsequential scaling limits of simple random walk on the two-dimensional uniform spanning tree*, Ann. Probab., **45** (2017), no. 1, 4–55.

[4] I. Benjamini, *Large scale degrees and the number of spanning clusters for the uniform spanning tree*, Perplexing problems in probability, Progr. Probab., **44**, Birkhäuser Boston, Boston, MA, 1999, pp. 175–183.

[5] C. Borgs, J. Chayes, H. Kesten and J. Spencer, *Uniform boundedness of critical crossing probabilities implies hyperscaling*, Random Structures and Algorithms., **15** (1999), no. 3-4, 368–413.

[6] J. Cardy, *The number of incipient spanning clusters in two-dimensional percolation*, Journal of Physics A: Mathematical and General, **31** (1998), no 5, L105.

[7] N. Holden and X. Sun, *SLE as a mating of trees in Euclidean geometry*, Comm. Math. Phys., **364** (2018), no. 1, 171–201.

[8] G. F. Lawler, *Intersections of random walks*, Probability and its Applications, Birkhäuser Boston, Inc., Boston, MA, 1991.

[9] ———, *Loop-erased random walk*, Perplexing problems in probability, Progr. Probab., **44**, Birkhäuser Boston, Boston, MA, 1999, pp. 197–217.

[10] R. Lyons and Y. Peres, *Probability on trees and networks*, Cambridge Series in Statistical and Probabilistic Mathematics, **42**, Cambridge University Press, New York, 2016.

[11] R. Pemantle, *Choosing a spanning tree for the integer lattice uniformly*, Ann. Probab., **19** (1991), no. 4, 1559–1574.

[12] A. Sapozhnikov and D. Shiraishi, *On Brownian motion, simple paths, and loops*, Probab. Theory Related Fields, **172** (2018), no. 3-4, 615–662.

[13] D. B. Wilson, *Generating random spanning trees more quickly than the cover time*, Proceedings of the Twenty-eighth Annual ACM Symposium on the Theory of Computing (Philadelphia, PA, 1996), ACM, New York, 1996, pp. 296–303.

Department of Mathematics,
University of British Columbia,
Vancouver, BC, V6T 1Z2, Canada
E-mail address: angel@math.ubc.ca

Research Institute for Mathematical Sciences,
Kyoto University,
Kyoto 606-8502, Japan
E-mail address: croydon@kurims.kyoto-u.ac.jp

Department of Mathematics,
University of British Columbia,
Vancouver, BC, V6T 1Z2, Canada
E-mail address: saraiht@math.ubc.ca

Department of Advanced Mathematical Sciences,
Graduate School of Informatics,
Kyoto University,
Kyoto 606-8501, Japan
E-mail address: shiraishi@acs.i.kyoto-u.ac.jp

Advanced Studies in Pure Mathematics 87, 2021
Stochastic Analysis, Random Fields and Integrable Probability — Fukuoka 2019
pp. 415–429

Large deviations of the KPZ equation via the Stochastic Airy operator

Li-Cheng Tsai

Abstract.

In this article we review the ideas in [Tsa18] toward proving the one-point, lower-tail large deviation principle for the Kardar–Parisi–Zhang equation.

§1. Introduction

The Kardar–Parisi–Zhang (KPZ) equation was introduced in [KPZ86] as a model of random surface growth. In one spatial dimensional the equation reads

$$\partial_t h = \frac{1}{2}\partial_{xx} h + \frac{1}{2}(\partial_x h)^2 + \xi, \tag{1}$$

where $\xi = \xi(t,x)$ denotes the spacetime white noise, and the solution $h = h(t,x)$ is a random function that describes the height at time $t \in \mathbb{R}_+$ and position $x \in \mathbb{R}$. Together with a host of related models, the KPZ equation has been intensively studied, due to its rich connections to other physical phenomena and mathematical structures. We refer to [FS11, Qua11, Cor12, QS15, CW17, CS19] for reviews on studies related to the KPZ equation.

Due to the roughness of ξ, the solution h is only a-Hölder continuous in x for $a < \frac{1}{2}$. This fact together with the presence of the nonlinear term $(\partial_x h)^2$ makes the KPZ equation (1) ill-posed. New theories have been built toward making sense of the KPZ equation and constructing the corresponding solution. We refer to [Hai14, GIP15, GJ14, GP18] and the references therein for related developments. An alternative formulation to these theories is the **Hopf–Cole solution**. That is, a *formal*

Received December 26, 2019.
Revised September 25, 2020.
2010 *Mathematics Subject Classification.* Primary 60F10; Secondary 60H25.
Key words and phrases. Kardar–Parisi–Zhang equation, large deviations, Airy point process, random operators, stochastic Airy operator.

exponentiation $Z(t,x) := e^{h(t,x)}$ brings (1) to the Stochastic Heat Equation (SHE)

$$\partial_t Z = \frac{1}{2}\partial_{xx} Z + \xi Z. \tag{2}$$

This equation is well-posed [Wal86, BC95] and the solution $Z(t,x)$ is strictly positive for $t > 0$ and for generic nonnegative and nonzero initial data [Mue91, Flo14]. These facts allow us to define the Hopf–Cole solution $h(t,x) := \log Z(t,x)$. The Hopf–Cole formulation arises in several discrete or regularized versions of the KPZ equation, and other notions of solutions from the aforementioned theories have been shown to coincide with the Hopf–Cole solution within the relevant class of initial data.

In this article we are concerned with the one-point, lower-tail Large Deviation Principle (LDP) for the KPZ equation. Consider the Hopf–Cole solution $h(t,x) := \log Z(t,x)$ with the initial data $Z(0,x) = \delta(x)$, a Dirac delta at the origin. It is known that, for large time $t \gg 1$, the height $h(2t,0)$ concentrates around $-\frac{t}{12}$. The question of interest here is to estimate the probability of $h(2t,0)$ being much smaller than this typical value $-\frac{t}{12}$. This question has been much studied recently in the physics and mathematics communities. In particular, the physics works [SMP17, CGKLDT18, KLDP18] each employed a different method to derive the explicit rate function

$$\mathbf{P}\Big[h(2t,0) \leq -\frac{t}{12} + tz\Big] \approx e^{-t^2\Phi_-(z)}, \qquad z < 0,$$

where

$$\Phi_-(z) := \frac{4}{15\pi^6}(1-\pi^2 z)^{\frac{5}{2}} - \frac{4}{15\pi^6} + \frac{2}{3\pi^4}z - \frac{1}{2\pi^2}z^2. \tag{3}$$

The first rigorous proof came soon later, via yet another method:

Theorem 1 ([Tsa18]). *Consider the Hopf–Cole solution $h(t,x) := \log Z(t,x)$ of the KPZ equation with the initial data $Z(0,x) = \delta(x)$. For any $z < 0$,*

$$\lim_{t\to\infty} \frac{1}{t^2} \log \mathbf{P}\Big[h(2t,0) \leq -\frac{t}{12} + tz\Big] = -\Phi_-(z).$$

The four different methods [SMP17, CGKLDT18, KLDP18, Tsa18] were later shown to be closely related in [KLD19]. Two new methods have been recently obtained, in the mathematically rigorous work [CC19] and the physics work [LD19]. This article focuses on reviewing the method used in [Tsa18].

It is known that the large deviation rate function depends on the initial data, and there has been many recent results for general initial data or special initial data other than $Z(0,x) = \delta(x)$. In the mathematics literature, the work [CG20] proved upper- and lower-tail probability bounds for general initial data; the work [Kim19] proved lower-tail probability bounds for the narrow-wedge initial data in the half-space geometry; the work [GL20] proved the one-point, upper-tail LDP for general initial data; the work [Lin20] proved the one-point, upper-tail LDP for the narrow-wedge initial data in the half-space geometry. For the physics literature we refer to [Kra19] and the references therein.

Acknowledgements.
Research partially supported by the NSF through DMS-1712575.

§2. Exact formulas and the stochastic Airy operator

Hereafter $Z(t,x)$ denotes the solution of the SHE with $Z(0,x) = \delta(x)$ and $h(t,x) := \log Z(t,x)$. We recall the formula that expresses the Laplace transform of $Z(2t,0)$ in terms of a Fredholm determinant:

$$\mathbf{E}\big[e^{-sZ(2t,0)e^{\frac{t}{12}}}\big] = \det(I - K_{s,t}), \quad s,t > 0, \tag{4}$$

where $K_{s,t}$ is a trace-class operator on $L^2[0,\infty)$ with the integral kernel $K_{s,t}(x,y) := \int_{\mathbb{R}} \frac{dr}{1+s^{-1}e^{-t^{1/3}r}} \mathrm{Ai}(x+r)\mathrm{Ai}(y+r)$, and Ai denotes the Airy function. The formula (or a closely related version of it) was derived simultaneously and independently in the works [CLDR10, ACQ11, Dot10, SS10], and [ACQ11] provided a rigorous proof.

The formula (4) provides access to the distribution of $h(2t,0)$, for example, in deriving the Tracy–Widom fluctuation of $h(2t,0)$ at large time [CLDR10, ACQ11, Dot10, SS10]. Another instance is the upper-tail large deviations. For $s = e^{-zt}$ and $z > 0$, the determinant in (4) behaves perturbatively, and (with some modifications) can be used to derive the upper-tail large deviations of $h(2t,0)$ [LDMS16, DT19].

In the lower-tail regime considered here, the determinant in (4) does not provide a handy access to the LDP. Instead, we appeal to a different expression of the formula from [BG16]:

$$\mathbf{E}\big[e^{-sZ(2t,0)e^{\frac{t}{12}}}\big] = \mathbf{E}\Bigg[\prod_{i=1}^{\infty} \frac{1}{1+se^{t^{1/3}\mathbf{a}_i}}\Bigg], \quad t,s > 0. \tag{5}$$

On the r.h.s., the expectation is taken with respect to the **Airy point process** $-\infty < \cdots < \mathbf{a}_3 < \mathbf{a}_2 < \mathbf{a}_1 < \infty$, which is the determinantal point process on $\mathbb{R}$ with the correlation kernel $K_{\mathrm{Airy\ PP}}(x,y) :=$

$\int_{\mathbb{R}_+} dr \mathrm{Ai}(x+r)\mathrm{Ai}(y+r)$. In (5), substitute in $s = e^{-tz}$ and $Z(2t,0) = e^{h(2t,0)}$. We rewrite the formula as

$$\mathbf{E}\Big[F\Big(h(2t,0) + \frac{t}{12} - tz\Big)\Big] = \mathbf{E}\Big[\exp\Big(-t\int_{\mathbb{R}} \mathrm{d}\mu_{\mathbf{a},t}(a)\,\psi_{t,z}(a)\Big)\Big], \tag{6}$$

where $F(x) := \exp(-e^x)$ and $\psi_{t,z}(a) := \log(1+e^{-t(z+a)})$, and $\mu_{\mathbf{a},t}(a) := t^{-1}\sum_{i=1}^{\infty}\delta_{-t^{-2/3}\mathbf{a}_i}(a)$ denotes the empirical measure of the scaled, spaced-reversed Airy point process.

As noted in [CG20a], the formulas (5)–(6) provide the suitable framework for the lower-tail LDP. To see how, we discuss the left and right hand sides of (6):

(LHS) The function $F(x)$ approaches 0 and 1 respectively as $x \to \infty$ and as $x \to -\infty$. Together with the t scaling, the function $F(x)$ serves as a good proxy for $\mathbf{1}_{x<0}$, thereby

$$(\text{l.h.s. of (6)}) \approx \mathbf{P}\Big[h(2t,0) + \frac{t}{12} - tz \le 0\Big].$$

Note that this approximation holds even in the large deviation regime, because $F(x) \to 0$ *super*-exponentially as $x \to \infty$.

(RHS) For $t \gg 1$, $\psi_{t,z}(a) \approx t(z+a)_-$, where $x_- := \max\{-x, 0\}$ denotes the negative part of x. Hence

$$(\text{r.h.s. of (6)}) \approx \mathbf{E}\Big[\exp\Big(-t^2\int_{\mathbb{R}} \mathrm{d}\mu_{\mathbf{a},t}(a)(a+z)_-\Big)\Big].$$

Assuming that the random measure $\mu_{\mathbf{a},t}$ enjoys an LDP with speed t^2 and a rate function $I_{\mathbf{a}}$, we should have

$$(\text{r.h.s. of (6)}) \approx \exp\Big(-t^2 \inf_{\mu}\Big\{\int_{\mathbb{R}} \mathrm{d}\mu(a)\,(a+z)_- + I_{\mathbf{a}}(\mu)\Big\}\Big),$$

where the infimum is taken over a suitable class of measures μ.

(Var) Combining the preceding two observations, one expects

$$\Phi_-(z) = \inf_{\mu}\Big\{\int_{\mathbb{R}} \mathrm{d}\mu(a)\,(a+z)_- + I_{\mathbf{a}}(\mu)\Big\}. \tag{7}$$

The observations (RHS)–(Var) were first made and noted in [CG20a]. Based on these observations, [CG20a] obtained detailed bounds on the tail probability of $h(2t,0)$. The physics work [CGKLDT18] continued along this path. It is known that the Airy point process $\{\mathbf{a}_i\}_{i=1}^{\infty}$ is the limit near the top edge of the spectrum of the Gaussian Unitary Ensemble (GUE). Employing a non-rigorous limit transition from the known rate function of the GUE [BAG97], the work [CGKLDT18] obtained a conjectural form of $I_{\mathbf{a}}$, and solved the variational problem (7) to obtain the rate function (3).

The work [Tsa18] also proceeds through (7), but, instead of viewing the Airy point process as a limit of the GUE, appeals to the stochastic Airy operator. For $\beta > 0$, consider the random operator

$$\mathcal{A}_\beta = -\frac{\mathrm{d}^2}{\mathrm{d}x^2} + x + \frac{2}{\sqrt{\beta}} B', \quad x \in [0, \infty), \tag{8}$$

where B' denotes the derivative of a Brownian Motion (BM). It is standard to construct $\mathcal{A}_\beta$ as an unbounded, self-adjoint operator on $L^2[0, \infty)$ with the Dirichlet boundary condition at $x = 0$, and the so constructed operator has a pure-point, bounded below spectrum, $-\infty < \boldsymbol{\lambda}_1 < \boldsymbol{\lambda}_2 < \boldsymbol{\lambda}_3 < \cdots$. This spectrum offers an alternative description of the Airy point process:

Theorem 2 ([RRV11]). *The spectrum of $\mathcal{A}_2$ is equal in law to the space-reversed Airy point process, i.e., $\{\boldsymbol{\lambda}_i\}_{i=1}^\infty \stackrel{law}{=} \{-\mathbf{a}_i\}_{i=1}^\infty$.*

The main theorem of [Tsa18] can now be stated as:

Theorem 3 ([Tsa18]). *For any $\beta > 0$ and $z < 0$, let $\boldsymbol{\lambda}_i$, $i = 1, 2, \ldots$, denote the eigenvalues of $\mathcal{A}_\beta$. We have*

$$\lim_{t \to \infty} \frac{1}{t^2} \log\Big(\mathbf{E}\big[e^{-\sum_{i=1}^\infty (\boldsymbol{\lambda}_i t^{1/3} + zt)_-} \big] \Big) = -\Big(\frac{2}{\beta}\Big)^5 \Phi_-\Big(\Big(\frac{\beta}{2}\Big)^2 z \Big). \tag{9}$$

Theorem 1 then follows from Theorem 3 for $\beta = 2$ and the preceding observations (LHS)–(RHS).

We devote the rest of the article to explaining the ideas of the proof of Theorem 3. The proof proceeds in steps, which are designated in the titles of the remaining sections.

§3. Localization via Riccati transform

The stochastic Airy operator (8) has a linear potential x. Such a potential is physically relevant because it ensures that $\mathcal{A}_\beta$, which acts on the unbounded interval $[0, \infty)$, has a pure-point spectrum. For our analysis, however, a varying potential is inconvenient. We hence seek to approximate $\mathcal{A}_\beta$ by a sequence of operators with translation-invariant potentials. Fix a mesoscopic scale $\Xi = t^a$, where the value of a will be specified later in (18). For $j \in \mathbb{Z}_{\geq 0}$, consider the (shifted) Hill's operator

$$\mathcal{H}_j := -\frac{\mathrm{d}^2}{\mathrm{d}y^2} + j\Xi + \frac{2}{\sqrt{\beta}} W', \quad y \in [0, \Xi], \tag{10}$$

with the Dirichlet boundary condition at $y = 0$ and Ξ, where W denotes a BM, and for different j's the operators $\{\mathcal{H}_j\}_{j=0}^{\infty}$ are independent.

The idea is that the term $j\Xi$ in (10) approximates the linear potential x for $x = y + (j-1)\Xi$, on the interval $y \in [0, \Xi]$ or $x \in [(j-1)\Xi, j\Xi]$. We then seek to 'piece together' the operators $\mathcal{H}_1, \mathcal{H}_2, \ldots$ to approximate $\mathcal{A}_\beta$. Doing so requires the **Riccati transform**. The transform begins with the eigenvalue problem $\mathcal{A}_\beta f = -f'' + xf + \frac{2}{\sqrt{\beta}} B' f = \lambda f$, with the boundary condition $f(0) = 0$. Viewing this equation as a second-order ODE, we perform the transformation $g = f'/f$ into a first-order ODE of the Riccati type: $g' = x - \lambda - g^2 + \frac{2}{\sqrt{\beta}} B'$. See Figure 1 for schematic graphs of f and g. With $g = f'/f$, we see that g explodes to $\pm\infty$ whenever f crosses zero. Since the underlying space $[0, \infty)$ is one-dimensional, the k-th eigenvector has exactly k roots, and hence the function g explodes exactly k times, excluding the explosion at $x = 0$.

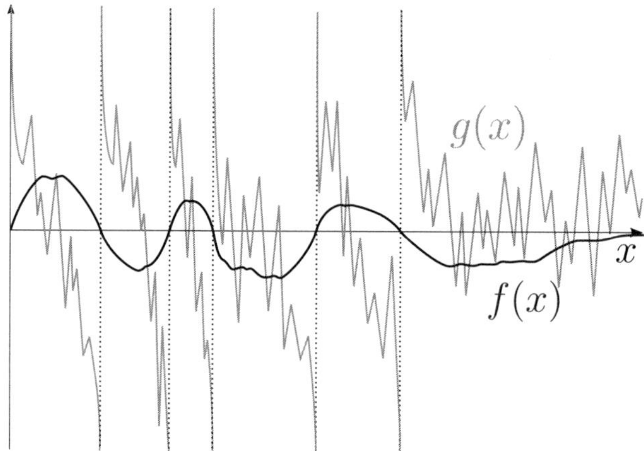

Fig. 1. Schematic graphs of f and g in the Riccati transform.

Now, let us view $\lambda \in \mathbb{R}$ as an *arbitrary* parameter, and solve the following ODE:

$$(11) \quad g_\lambda'(x) = x - \lambda - g_\lambda^2(x) + \frac{2}{\sqrt{\beta}} B'(x), \; x \in (0, \infty), \; g_\lambda(0) = +\infty.$$

The solution g_λ may undergo explosions to $-\infty$, and whenever that happens we immediately reinitiate g_λ from $+\infty$. Let $N(\lambda) := \#\{\boldsymbol{\lambda}_i \leq \lambda\}$ counts the eigenvalues of $\mathcal{A}_\beta$ at most λ.

Proposition 4 (Proposition 3.4. in [RRV11]). *Almost surely for all* $\lambda \in \mathbb{R}$, $\#\{x \in (0,\infty) : |g_\lambda(x)| = \infty\} = N(\lambda)$.

We can also consider the analogous ODE for Hill's operator

$$g'_{\lambda,j}(y) = j\Xi - \lambda - g^2_{\lambda,i}(y) + \frac{2}{\sqrt{\beta}}W'(y),\ g_{\lambda,i}(0) = +\infty,\ y \in (0,\Xi]. \tag{12}$$

Let $\{-\infty < \boldsymbol{\lambda}_1^{(j)} \leq \boldsymbol{\lambda}_2^{(j)} \leq \cdots\}$ denote the spectrum of $\mathcal{H}_j$, and similarly $N_j(\lambda) := \#\{\boldsymbol{\lambda}_i^{(j)} \leq \lambda\}$. Similarly to Proposition 4, $N_j(\lambda)$ is equal to the number of explosions of (12).

Having introduced the Riccati transforms for $\mathcal{A}_\beta$ and $\mathcal{H}_j$, we now apply these transforms to compare the counting functions $N(\lambda)$ and $N_j(\lambda)$. Localize (11) onto the interval $x \in ((j-1)\Xi, j\Xi]$, and set $x = y + (j-1)\Xi$ to get

$$g'_\lambda = (y + (j-1)\Xi)) - \lambda - g^2_\lambda + \frac{2}{\sqrt{\beta}}B'. \tag{11'}$$

We can now couple (11') with (12) by $W(y) = B(y + (j-1)\Xi)$. On the interval $y \in (0,\Xi]$ we see that (12) has a larger potential $j\Xi \geq (y+(j-1)\Xi))$ and a larger entrance value $g_{\lambda,j}(0) = +\infty \geq g_\lambda((j-1)\Xi)$. Comparison arguments then yield

$$N_j(\lambda) \leq \#\big\{x \in ((j-1)\Xi, j\Xi] : |g_\lambda(x)| = \infty\big\}. \tag{13}$$

To get the reverse inequality, we apply the same coupling $W(y) = B(y+(j-1)\Xi)$ for (12) with $j \mapsto j-1$ and for (11') with j. The potential is now reversely ordered $(j-1)\Xi \leq (y + (j-1)\Xi))$, though the entrance values are not. The issue of entrance values can be cured by forgoing the first explosion of (11') on $y \in (0,\Xi]$, which gives

$$N_{j-1}(\lambda) + 1 \geq \#\big\{x \in ((j-1)\Xi, j\Xi] : |g_\lambda(x)| = \infty\big\}. \tag{14}$$

Next, to make use of the comparison results (13)–(14), we rewrite the quantity of interest in Theorem 3 as

$$-\sum_{i=1}^{\infty}(\boldsymbol{\lambda}_i t^{1/3} + zt)_- = -\int_{(-\infty,-zt^{2/3}]} \mathrm{d}N(\lambda)\,(t^{1/3}\lambda + z)_- = -t^{1/3}\int_{(-\infty,-zt^{2/3}]} \mathrm{d}\lambda\, N(\lambda), \tag{15}$$

and similarly

$$-\sum_{i=1}^{\infty}(\boldsymbol{\lambda}_i^{(j)} t^{1/3} + zt)_- = -t^{1/3}\int_{(-\infty,-zt^{2/3}]} \mathrm{d}\lambda\, N_j(\lambda). \tag{16}$$

Combining (13) and (15) gives, for any $n \in \mathbb{Z}_{>0}$,

$$-\sum_{i=1}^{\infty}(\boldsymbol{\lambda}_i t^{1/3} + zt)_- \leq -\sum_{j=1}^{n}\sum_{i=1}^{\infty}(\boldsymbol{\lambda}_i^{(j)} t^{1/3} + zt)_-.$$

Here we forgo the explosions of (11) on $[n\Xi, \infty)$ because they contribute negatively. For the reverse inequality, we need to account for these remaining explosions. To this end, consider the operator $\mathcal{A}_{\beta,n} := (-\frac{\mathrm{d}^2}{\mathrm{d}x^2} + x + \frac{2}{\sqrt{\beta}}B')$ acting on $x \in [n\Xi, \infty)$, with the Dirichlet boundary condition at $x = n\Xi$. We make $(\mathcal{A}_\beta + n\Xi)$ independent of $\{\mathcal{H}_j\}_{j=0}^{n-1}$. Let $\{\boldsymbol{\eta}_i\}_{i=1}^{\infty}$ denote the spectrum of $\mathcal{A}_{\beta,n}$. We have

$$-\sum_{i=1}^{\infty}(\boldsymbol{\lambda}_i t^{1/3} + zt)_- \geq -n - \sum_{j=0}^{n-1}\sum_{i=1}^{\infty}(\boldsymbol{\lambda}_i^{(j)} t^{1/3} + zt)_- - \sum_{i=1}^{\infty}(\boldsymbol{\eta}_i t^{1/3} + zt)_-.$$

Exponentiate the preceding inequalities, take expectation, and utilize the independence of $\{\mathcal{H}_i\}_{i=1}^{n}$ and of $\{\mathcal{H}_i\}_{i=0}^{n-1}, \mathcal{A}_{\beta,n}$ to split the resulting expectations into products. Note that $\mathcal{A}_{\beta,n} \stackrel{\text{law}}{=} \mathcal{A}_\beta + n\Xi$. We have

Proposition 5. *For any* $n \in \mathbb{Z}_{>0}$,

$$\text{(17a)} \quad \prod_{j=0}^{n-1}\mathbf{E}\big[e^{-\sum_{i=1}^{\infty}(\boldsymbol{\lambda}_i^{(j)} t^{1/3}+zt)_-}\big] \cdot e^{-n} \cdot \mathbf{E}\big[e^{-\sum_{i=1}^{\infty}((\boldsymbol{\lambda}_i+n\Xi)t^{1/3}+zt)_-}\big]$$

$$\text{(17b)} \quad \leq \mathbf{E}\big[e^{-\sum_{i=1}^{\infty}(\boldsymbol{\lambda}_i t^{1/3}+zt)_-}\big] \leq \prod_{j=1}^{n}\mathbf{E}\big[e^{-\sum_{i=1}^{\infty}(\boldsymbol{\lambda}_i^{(j)} t^{1/3}+zt)_-}\big].$$

§4. Lower bound

Proposition 5 reduces the problem of analyzing $\mathcal{A}_\beta$ to analyzing each $\mathcal{H}_j$. Based on this reduction, we will explain how to obtain the desired lower and upper bounds on the quantity of interest, i.e., the l.h.s. of (9) in Theorem 3.

Let us begin by fixing the scale $\Xi = t^a$. Referring to the l.h.s. of (9), since $z < 0$ is fixed, we see that the relevant eigenvalues should be of order $t^{2/3}$. Refer to (11); if we also match the order of λ to $t^{2/3}$, then the linear potential should vary at scale $t^{2/3}$. This observation forces us to choose $a < 2/3$, since otherwise we cannot expect x to be approximated by a constant on the interval $((j-1)\Xi, j\Xi]$. On the other hand, we wish $\Xi = t^a$ to be greater than the time scale between explosions of (11). Doing so gives us some room for analysis. Performing scaling in

(11) under the assumptions that $\lambda, x \asymp t^{2/3}$ shows that explosions of the ODE should occur at scale $t^{-1/3}$. Hence we require $a > -\frac{1}{3}$. It turns out that our analysis does not require any further condition on a, and we hereafter fix

$$\Xi = t^a, \quad a \in \Big(-\frac{1}{3}, \frac{2}{3}\Big). \tag{18}$$

We now return to the task of obtaining the lower bound. The expression (17a) contains three terms. Let us focus on the first term and show that, for some suitable n given later in (23),

$$\liminf_{t\to\infty} \frac{1}{t^2} \log\Big(\prod_{j=0}^{n-1} \mathbf{E}\big[e^{-\sum_{i=1}^{\infty}(\boldsymbol{\lambda}_i^{(j)} t^{1/3}+zt)_-}\big]\Big) \geq \Big(\frac{2}{\beta}\Big)^5 \Phi_-\Big(\Big(\frac{\beta}{2}\Big)^2 z\Big). \tag{19}$$

Once this is done, we will argue that the remaining two terms in (17a) are negligible.

Recall from (10) that the randomness of $\mathcal{H}_j$ and hence of $\{\boldsymbol{\lambda}_i^{(j)}\}_{i=1}^{\infty}$ comes solely from W. Our task is to find an 'optimal' deviation of W that realizes the lower bound (19). Namely, we seek a deviation such that, when evaluating the l.h.s. of (19) around this deviation one obtains the desired lower bound. As will be explained in Section 5, such a deviation can be chosen to be a constantly drifted BM. That is, we consider the deviation where $W(y)$ behaves like a constantly drifted BM with a drift $t^{2/3} v_j \mathrm{d}y$. Here $v_j \in \mathbb{R}$ is a parameter, and the scaling $t^{2/3}$ matches the aforementioned scaling of λ and x.

We now evaluate the l.h.s. of (19) around the deviation. Girsanov's theorem asserts that the probability of having such a deviation is

$$\text{Prob.} \approx \exp\Big(-\frac{1}{2} t^{a+4/3} v_j^2\Big). \tag{20}$$

Around such a deviation, Hill's operator behaves like the shifted Laplace operator $-\frac{\mathrm{d}^2}{\mathrm{d}y^2} + j\Xi + \frac{2}{\sqrt{\beta}} t^{2/3} v_j$, acting on $[0, \Xi]$ with the Dirichlet boundary condition. From this we calculate

$$\begin{aligned}
-\sum_{i=1}^{\infty}(\boldsymbol{\lambda}_i^{(j)} t^{1/3} + zt)_- &= -t^{1/3} \int_{(-\infty, -zt^{2/3}]} \mathrm{d}\lambda\, N_j(\lambda) \\
&\approx -\frac{t^{1/3}\Xi}{\pi} \int_{-\infty}^{-zt^{2/3}} \mathrm{d}\lambda \sqrt{\Big(\lambda - \frac{2}{\sqrt{\beta}} t^{2/3} v_j - j\Xi\Big)_+} \\
&= -\frac{2t^{a+4/3}}{3\pi} \Big(\Big(-z - \frac{2}{\sqrt{\beta}} v_j - j t^{a-2/3}\Big)_+\Big)^{3/2}.
\end{aligned} \tag{21}$$

Combining (20)–(21) gives

$$t^{-4/3-a}\log\mathbf{E}\big[e^{-\sum_{i=1}^{\infty}(\boldsymbol{\lambda}_i^{(j)}t^{1/3}+zt)_-}\big] \gtrsim -\frac{1}{2}v_j^2-\frac{2}{3\pi}\Big(-z-\frac{2}{\sqrt{\beta}}v_j-jt^{a-2/3}\Big)_+^{3/2}. \tag{22}$$

Here $\gtrsim$ means $\geq$ with some lower order (in t) error terms.

The approximate inequality (22) holds for all $v_j\in\mathbb{R}$. It is natural to optimize over v_j. Differentiating in v_j shows that the optimum is achieved at

$$v_{j,*}:=4\pi^{-2}\beta^{-3/2}\left(-1+\sqrt{1+\Big(\frac{\beta\pi}{2}\Big)^2(-z-jt^{a-2/3})_+}\right).$$

Note that $v_{j,*}=0$ for all $j>-zt^{2/3-a}$, which suggests that we need only to invoke $j\leq -zt^{2/3-a}$. With this in mind, we set

$$n=-zt^{2/3-a}. \tag{23}$$

Sum (22) over $j=0,1,\dots,n-1$ for $v_j=v_{j,*}$. Within the result, the sum over j can be recognized as a Riemann sum, with $jn^{a-2/3}$ approximating a continuous variable $\nu\in[0,\infty)$. This gives

$$(\text{l.h.s. of (19)})\geq -\int_0^\infty \mathrm{d}\nu\ \frac{1}{2}v_*^2(\nu)+\left(\Big(-z-\frac{2}{\sqrt{\beta}}v_*(\nu)-\nu\Big)_+\right)^{3/2}, \tag{24}$$

where $v_*(\nu):=4\pi^{-2}\beta^{-3/2}(-1+\sqrt{1+(\frac{\beta\pi}{2})^2(-z-\nu)_+})$. The integral in (24) is explicit and can be evaluated to be $(2/\beta)^5\Phi_-((\beta/2)^2z)$.

We have concluded (19) for the n in (23). A careful analysis shows that, for such an n, the last expectation in (17a) is negligible after being taken $t^{-2}\log(\boldsymbol{\cdot})$ and passed to the limit $t\to\infty$. The analysis is too involved for the purpose of this article and hence not presented. The term e^{-n} in (17a) is also negligible because $t^{-2}\log(e^{-n})=-t^{-2}(\log(-z)+t^{2/3-a})\to 0$ by (18). This concludes the discussion of the lower bound.

§5. Upper bound, the WKB condition

In Section 4, we utilized a certain type of deviations of W — namely constantly drifted BM — to produce the desired lower bound. To complete the proof, we need to argue that such deviations are optimal, asymptotically as $t \to \infty$. We refer to this assertion as the **WKB condition**. The terminology is motivated by the fact that our analysis in Sections 3–4 can be interpreted as the WKB approximation of the stochastic Airy operator.

The WKB condition is by no means obvious in the current context. To see why, recall that the BM enters Hill's operator (10) through the derivative W'. Consider the Fourier transform of W' on the interval $[0, \Xi]$, i.e., $W'(y) = \sum_{k\in\mathbb{Z}} W_k e^{2\pi \mathbf{i} y/\Xi}$. A priori, since W' is *very rough*, it seems that the high frequency modes W_k, $k \gg 1$, could have significant impacts on the spectrum of Hill's operator. The WKB condition, however, asserts that only the constant mode W_0 matters for the LDP in question. Let us further emphasize that the WKB condition may be violated for some other cost functions. More precisely, here we are concerned with $\mathbf{E}[\exp(\sum_{i=1}^{\infty} \widetilde{\psi}_t(\boldsymbol{\lambda}_i^{(j)})]$, with the cost function $\widetilde{\psi}_t(\lambda) := -(\lambda t^{1/3} + zt)_-$. As claimed previously and will be verified in the sequel, the major contribution of this expectation comes from configurations with constantly drifted W. On the other hand, we expect that there exists some other cost function $\widehat{\psi}_t$, such that the major contribution of $\mathbf{E}[\exp(\sum_{i=1}^{\infty} \widehat{\psi}_t(\boldsymbol{\lambda}_i^{(j)})]$ arises from deviations where high frequency modes of W contribute. A class of cost functions that should enjoy the WKB condition have been studied in [KLD19].

We now return to the task of verifying the WKB condition. The first step is to argue that, we can replace the Dirichlet boundary condition for $\mathcal{H}_j$ with the periodic boundary condition. This is proven in [Tsa18] by utilizing the interlacing of eigenvalues under different boundary conditions. We do not repeat the technical argument here, and simply *switch* to the periodic boundary condition hereafter.

We will verify the WKB condition at a deterministic level. To set up the notation, for a real $f \in C[0, 1]$, consider

$$H := -\frac{\mathrm{d}^2}{\mathrm{d}y^2} + f', \quad \widetilde{H} := -\frac{\mathrm{d}^2}{\mathrm{d}y^2} + \frac{1}{\Xi}(f(\Xi) - f(0)) \tag{25}$$

acting on $[0, \Xi]$ with the periodic boundary condition. Note that f does *not* have to be periodic. The following Proposition encapsulates the WKB condition. To see how, apply Proposition 6 with $r = t^{2/3}z + j\Xi$ and with $f = \frac{2}{\sqrt{\beta}}W$. One finds that the linear statistics $-\sum_{i=1}^{\infty}(t^{1/3}\boldsymbol{\lambda}_i^{(j)} + tz)_-$ of $\mathcal{H}_j$ is bounded above by the same linear statistics of an operator with W' replaced by the average $\frac{1}{\Xi}(W(\Xi) - W(0))$. This shows that, among all configurations of W with both ends $W(0)$ and $W(\Xi)$ fixed, the constantly drifted configuration performs the best.

Proposition 6. *For any real $f \in C[0,1]$, consider the operators H and $\widetilde{H}$ defined in* (25), *and let $\{\lambda_1 \le \lambda_2 \le \cdots\}$ and $\{\widetilde{\lambda}_1 \le \widetilde{\lambda}_2 \le \cdots\}$ denote their respective spectra. For any $r \in \mathbb{R}$ we have*

$$-\sum_{i=1}^{\infty}(r+\lambda_i)_- \le -\sum_{i=1}^{\infty}(r+\widetilde{\lambda}_i)_-.$$

Proof. The first step is to recognize that, for any sequence of real numbers put in ascending order $-\infty < a_1 \le a_2 \le \cdots$, we have

$$-\sum_{i=1}^{\infty}(r+a_i)_- = \inf_N \left\{ \sum_{i=1}^{N}(a_i+r) \right\}. \tag{26}$$

We will apply this inequality with $a_i = \lambda_i$, so for now let us focus on bounding the sum $\sum_{i=1}^{N}\lambda_i$, for generic $N \in \mathbb{Z}_{\ge 0}$. The operator H is self-adjoint on $L^2[0,\Xi]$ (with the periodic boundary condition), and hence the corresponding eigenvectors $\phi_1, \phi_2, \ldots$ form an orthonormal basis for $L^2[0,\Xi]$. We can thus express the sum of the first N eigenvalues as $\sum_{i=1}^{N}\lambda_i = \sum_{i=1}^{N}\langle\phi_i, H\phi_i\rangle_{L^2}$. In fact, the sum can be characterized as the infimum of the same quantity when tested over orthonormal sets:

$$\sum_{i=1}^{N}\lambda_i = \inf_{\{\psi_1,\ldots,\psi_N\}} \sum_{i=1}^{N}\langle\psi_i, H\psi_i\rangle_{L^2}, \tag{27}$$

where the infimum is taken over orthonormal $\psi_1, \ldots, \psi_N$ in the domain of H. The assertion (27) can be proven by expanding each ψ_i into a linear combination of $\{\phi_j\}_{j=1}^{\infty}$. We do not perform the calculation here.

To bound the r.h.s. of (27), we insert a particular orthonormal set from the Fourier basis. That is, we let $\psi_{1,*}(y), \ldots, \psi_{N,*}(y)$ be the first N among

$$\frac{1}{|\Xi|^{1/2}}, \quad \frac{1}{|\Xi|^{1/2}}e^{2\pi \mathbf{i}y/|\Xi|}, \quad \frac{1}{|\Xi|^{1/2}}e^{-2\pi \mathbf{i}y/|\Xi|}, \quad \frac{1}{|\Xi|^{1/2}}e^{4\pi \mathbf{i}y/|\Xi|}, \ldots.$$

From (27) we obtain

$$\sum_{i=1}^{N}\lambda_i \le \sum_{i=1}^{N}\langle\psi_{i,*}, H\psi_{i,*}\rangle_{L^2} = \sum_{i=1}^{N}\int_0^{\Xi}\big(\mathrm{d}y\,|\psi_{i,*}'|^2 + \mathrm{d}f(y)\,|\psi_{i,*}|^2\big),$$

where the integral against $\mathrm{d}f(y)$ is interpreted in the Riemann–Stieltjes sense. Noting that $|\psi_{i,*}|^2 \equiv \frac{1}{\Xi}$ and referring to (25), we see that the last sum is equal to $\sum_{i=1}^{N}\langle\psi_{i,*}, \widetilde{H}\psi_{i,*}\rangle_{L^2}$. Further, since $\widetilde{H}$ is just a shifted

Laplace operator, the Fourier vectors $\psi_{i,*}$, $i = 1, \ldots, N$, are the first N eigenvectors of $\widetilde{H}$. Consequently, the last sum is equal to $\sum_{i=1}^{N} \widetilde{\lambda}_i$, and therefore

$$\sum_{i=1}^{N} \lambda_i \leq \sum_{i=1}^{N} \widetilde{\lambda}_i. \tag{28}$$

Now, apply (26) with $a_i = \lambda_i$, use (28) to bound the result, and apply (26) with $a_i = \widetilde{\lambda}_i$ in reverse. This concludes the desired result. Q.E.D.

Proposition 6 verifies the WKB condition. The upper bound can now be proven in the same fashion as the lower bound.

References

[ACQ11] G. Amir, I. Corwin and J. Quastel, Probability distribution of the free energy of the continuum directed random polymer in 1 + 1 dimensions, *Comm. Pure Appl. Math.*, 64(4):466–537, 2011.

[BAG97] G. Ben Arous and A. Guionnet, Large deviations for Wigner's law and Voiculescu's non-commutative entropy, *Probab. Theory Related Fields*, 108(4):517–542, 1997.

[BC95] L. Bertini and N. Cancrini, The stochastic heat equation: Feynman–Kac formula and intermittence, *J. Stat. Phys.*, 78(5-6):1377–1401, 1995.

[BG16] A. Borodin and V. Gorin, Moments match between the KPZ equation and the Airy point process, *SIGMA*, 12:102, 2016.

[CC19] M. Cafasso and T. Claeys, A Riemann–Hilbert approach to the lower tail of the KPZ equation, *arXiv*:1910.02493, 2019.

[CLDR10] P. Calabrese, P. Le Doussal and A. Rosso, Free-energy distribution of the directed polymer at high temperature, *Europhys. Lett.*, 90(2):20002, 2010.

[CW17] A. Chandra and H. Weber, Stochastic PDEs, regularity structures, and interacting particle systems, *Ann. Fac. Sci. Toulouse Math.* (6), 26(4):847–909, 2017.

[Cor12] I. Corwin, The Karder–Parisi–Zhang equation and universality class, *Random Matrices: Theory Appl.*, 01(01):1130001, 2012.

[CG20] I. Corwin and P. Ghosal, KPZ equation tails for general initial data, *Electron J. Probab.*, 25, 2020.

[CG20a] I. Corwin and P. Ghosal, Lower tail of the KPZ equation, *Duke Math. J.*, 169(7):1329–1395, 2020.

[CGKLDT18] I. Corwin, P. Ghosal, A. Krajenbrink, P. Le Doussal and L.-C. Tsai, Coulomb-gas electrostatics controls large fluctuations of the Kardar–Parisi–Zhang equation, *Phys. Rev. Lett.*, 121(6):060201, 2018.

[CS19] I. Corwin and H. Shen, Some recent progress in singular stochastic PDEs, *Bulletin of the AMS*, electronically published, 2019.

[DT19] S. Das and L.-C. Tsai, Fractional moments of the Stochastic Heat Equation, to appear in *Ann. Inst. Henri Poincaré (B) Probab. Stat.*, *arXiv*:1910.09271, 2019.

[Dot10] V. Dotsenko, Bethe ansatz derivation of the Tracy–Widom distribution for one-dimensional directed polymers, *Europhys. Lett.*, 90(2):20003, 2010.

[Flo14] G. R. M. Flores, On the (strict) positivity of solutions of the stochastic heat equation, *Ann. Probab.*, 42(4):1635–1643, 2014.

[FS11] P. Ferrari and H. Spohn, Random growth models, In: J. B. G. Akemann and P. D. Francesco, editors, *Oxford Handbook of Random Matrix Theory*, Oxford University Press, 2011.

[GL20] P. Ghosal and Y. Lin, Lyapunov exponents of the SHE for general initial data, *arXiv*:2007.06505, 2020.

[GJ14] P. Gonçalves and M. Jara, Nonlinear fluctuations of weakly asymmetric interacting particle systems, *Arch. Ration. Mech. Anal.*, 212(2):597–644, 2014.

[GIP15] M. Gubinelli, P. Imkeller and N. Perkowski, Paracontrolled distributions and singular PDEs, In: *Forum of Mathematics, Pi*, volume 3, 2015.

[GP18] M. Gubinelli and N. Perkowski, Energy solutions of KPZ are unique, *J. Amer. Math. Soc.*, 31(2):427–471, 2018.

[Hai14] M. Hairer, A theory of regularity structures, *Invent. Math.*, 198(2): 269–504, 2014.

[KPZ86] M. Kardar, G. Parisi and Y.-C. Zhang, Dynamic scaling of growing interfaces, *Phys. Rev. Lett.*, 56(9):889, 1986.

[Kim19] Y. Kim, The lower tail of the half-space KPZ equation, *arXiv*:1905.07703, 2019.

[Kra19] A. Krajenbrink, Beyond the typical fluctuations: a journey to the large deviations in the Kardar–Parisi–Zhang growth model, PhD thesis, PSL Research University, 2019.

[KLDP18] A. Krajenbrink, P. Le Doussal and S. Prolhac, Systematic time expansion for the Kardar–Parisi–Zhang equation, linear statistics of the GUE at the edge and trapped Fermions, *Nuclear Phys. B*, 936:239–305, 2018.

[KLD19] A. Krajenbrink and P. Le Doussal, Linear statistics and pushed Coulomb gas at the edge of β-random matrices: Four paths to large deviations, *Europhys. Lett.*, 125(2):20009, 2019.

[Lin20] Y. Lin, Lyapunov exponents of the half-line SHE, *arXiv*:2007.10212, 2020.

[LDMS16] P. Le Doussal, S. N. Majumdar and G. Schehr, Large deviations for the height in 1D Kardar–Parisi–Zhang growth at late times, *Europhys. Lett.*, 113(6):60004, 2016.

[LD19] P. Le Doussal, Large deviations for the Kardar–Parisi–Zhang equation from the Kadomtsev–Petviashvili equation, *J. Stat. Mech. Theory Exp.*, 2020(4):043201.

[Mue91] C. Mueller, On the support of solutions to the heat equation with noise, *Stochastics: An International Journal of Probability and Stochastic Processes*, 37(4):225–245, 1991.

[Qua11] J. Quastel, Introduction to KPZ, *Current Developments in Mathematics*, 2011(1), 2011.

[QS15] J. Quastel and H. Spohn, The one-dimensional KPZ equation and its universality class, *J. Stat. Phys.*, 160(4):965–984, 2015.

[RRV11] J. Ramirez, B. Rider and B. Virág, Beta ensembles, stochastic Airy spectrum, and a diffusion, *J. Amer. Math. Soc.*, 24(4):919–944, 2011.

[SS10] T. Sasamoto and H. Spohn, One-dimensional Kardar–Parisi–Zhang equation: an exact solution and its universality, *Phys. Rev. Lett.*, 104(23):230602, 2010.

[SMP17] P. Sasorov, B. Meerson and S. Prolhac, Large deviations of surface height in the 1+ 1-dimensional Kardar–Parisi–Zhang equation: exact long-time results for $\lambda H < 0$, *J. Stat. Mech. Theory Exp.*, 2017(6):063203, 2017.

[Tsa18] L.-C. Tsai, Exact lower tail large deviations of the KPZ equation, *arXiv*:1809.03410, 2018.

[Wal86] J. B. Walsh, An introduction to stochastic partial differential equations, In: *École d'Été de Probabilités de Saint Flour* XIV-1984, pages 265–439, Springer, 1986.

Department of Mathematics,
Rutgers University — New Brunswick
110 Frelinghuysen Road, Piscataway, NJ 08854, USA
E-mail address: lctsai.math@gmail.com

Advanced Studies in Pure Mathematics

A SERIES OF UP-TO-DATE GUIDES OF LASTING INTEREST
TO ADVANCED MATHEMATICS

Volume 16	Conformal Field Theory and Solvable Lattice Models. Edited by M. Jimbo, T. Miwa and A. Tsuchiya. June, 1988
Volume 17	Algebraic Number Theory—in honor of K. Iwasawa. Edited by J. Coates, R. Greenberg, B. Mazur and I. Satake. August, 1989
Volume 18	I—Recent Topics in Differential and Analytic Geometry. Edited by T. Ochiai. November, 1990 II—Kähler Metric and Moduli Spaces. Edited by Edited by T. Ochiai. November, 1990
Volume 19	Integrable Systems in Quantum Field Theory and Statistical Mechanics. Edited by M. Jimbo, T. Miwa and A. Tsuchiya. November, 1989
Volume 20	Aspects of Low Dimensional Manifolds. Edited by Y. Matsumoto and S. Morita. December, 1992
Volume 21	Zeta Functions in Geometry. Edited by N. Kurokawa and T. Sunada. December, 1992
Volume 22	Progress in Differential Geometry. Edited by K. Shiohama. September, 1993
Volume 23	Spectral and Scattering Theory and Applications. Edited by K. Yajima. February, 1994
Volume 24	Progress in Algebraic Combinatorics. Edited by E. Bannai and A. Munemasa. May, 1996
Volume 25	CR-Geometry and Overdetermined Systems. Edited by T. Akahori, G. Komatsu, K. Miyajima, M. Namba and K. Yamaguchi. July, 1997
Volume 26	Analysis on Homogeneous Spaces and Representation Theory of Lie Groups, Okayama-Kyoto. Edited by M. Kashiwara, T. Kobayashi, T. Matsuki, K. Nishiyama and T. Oshima. April, 2000
Volume 27	Arrangements – Tokyo 1998. Edited by M. Falk and H. Terao. August, 2000
Volume 28	Combinatorial Methods in Representation Theory. Edited by M. Kashiwara, K. Koike, S. Okada, I. Terada, and H.-F. Yamada. December, 2000

Volume 29 Singularities – Sapporo 1998.
Edited by J.-P. Brasselet and T. Suwa.
December, 2000

Volume 30 Class Field Theory – Its Centenary and Prospect.
Edited by K. Miyake. April, 2001

Volume 31 Taniguchi Conference on Mathematics Nara '98.
Edited by M. Maruyama and T. Sunada. May, 2001

Volume 32 Groups and Combinatorics — in memory of Michio Suzuki.
Edited by E. Bannai, H. Suzuki, H. Yamaki and T. Yoshida. October, 2001

Volume 33 Computational Commutative Algebra and Combinatorics.
Edited by T. Hibi. February, 2002

Volume 34 Minimal Surfaces, Geometric Analysis and Symplectic Geometry.
Edited by K. Fukaya, S. Nishikawa and J. Spruck.
June, 2002

Volume 35 Higher Dimensional Birational Geometry.
Edited by S. Mori and Y. Miyaoka. July, 2002

Volume 36 Algebraic Geometry 2000, Azumino.
Edited by S. Usui, M. Green, L. Illusie, K. Kato, E. Looijenga, S. Mukai and S. Saito. November, 2002

Volume 37 Lie Groups, Geometric Structures and Differential Equations — One Hundred Years after Sophus Lie —.
Edited by T. Morimoto, H. Sato and K. Yamaguchi.
December, 2002

Volume 38 Operator Algebras and Applications.
Edited by H. Kosaki. January, 2004

Volume 39 Stochastic Analysis on Large Scale Interacting Systems.
Edited by T. Funaki and H. Osada. March, 2004

Volume 40 Representation Theory of Algebraic Groups and Quantum Groups.
Edited by T. Shoji, M. Kashiwara, N. Kawanaka, G. Lusztig and K. Shinoda. March, 2004

Volume 41 Stochastic Analysis and Related Topics in Kyoto
In honour of Kiyosi Itô.
Edited by H. Kunita, S. Watanabe and Y. Takahashi.
March, 2004

Volume 42 Complex Analysis in Several Variables
– Memorial Conference of Kiyoshi Oka's Centennial Birthday, Kyoto/Nara 2001.
Edited by K. Miyajima, M. Furushima, H. Kazama, A. Kodama, J. Noguchi, T. Ohsawa, H. Tsuji and T. Ueda. June, 2004

Volume 43 Singularity Theory and Its Applications.
Edited by S. Izumiya, G. Ishikawa, H. Tokunaga, I. Shimada and T. Sano. January, 2007

Volume 44 Potential Theory in Matsue.
Edited by H. Aikawa, T. Kumagai, Y. Mizuta and N. Suzuki. August, 2006

Volume 45 Moduli Spaces and Arithmetic Geometry (Kyoto, 2004).
Edited by S. Mukai, Y. Miyaoka, S. Mori, A. Moriwaki and I. Nakamura. January, 2007

Volume 46 Singularities in Geometry and Topology 2004.
Edited by J.-P. Brassele and T. Suwa. February, 2007

Volume 47 Asymptotic Analysis and Singularities.
1—Hyperbolic and dispersive PDEs and fluid mechanics.
2—Elliptic and parabolic PDEs and related problems.
Edited by H. Kozono, T. Ogawa, K. Tanaka, Y. Tsutsumi and E. Yanagida. October, 2007

Volume 48 Finsler Geometry, Sapporo 2005
— In Memory of Makoto Matsumoto.
Edited by S. V. Sabau and H. Shimada.
November, 2007

Volume 49 Probability and Number Theory — Kanazawa 2005.
Edited by S. Akiyama, K. Matsumoto, L. Murata and H. Sugita. December, 2007

Volume 50 Algebraic Geometry in East Asia — Hanoi 2005.
Edited by K. Konno and V. Nguyen-Khac.
February, 2008

Volume 51 Surveys on Geometry and Integrable Systems.
Edited by M. Guest, R. Miyaoka and Y. Ohnita.
November, 2008

Volume 75 Algebraic Varieties and Automorphism Groups.
Edited by K. Masuda, T. Kishimoto, H. Kojima, M. Miyanishi and M. Zaidenberg. December, 2017

Volume 76 Representation Theory, Special Functions and Painlevé Equations — RIMS 2015.
Edited by H. Konno, H. Sakai, J. Shiraishi, T. Suzuki and Y. Yamada. June, 2018

Volume 77 The 50th Anniversary of Gröbner Bases.
Edited by T. Hibi. August, 2018

Volume 78 Singularities in Generic Geometry.
Edited by S. Izumiya, G. Ishikawa, M. Yamamoto, K. Saji, T. Yamamoto and M. Takahashi. September, 2018

Volume 79 Mathematics of Takebe Katahiro and History of Mathematics in East Asia.
Edited by T. Ogawa and M. Morimoto.
December, 2018

Volume 80 Operator Algebras and Mathematical Physics.
Edited by M. Izumi, Y. Kawahigashi, M. Kotani, H. Matui and N. Ozawa. May, 2019

Volume 81 Asymptotic Analysis for Nonlinear Dispersive and Wave Equations.
Edited by K. Kato, T. Ogawa and T. Ozawa.
October, 2019

Volume 82 Differential Geometry and Tanaka Theory — Differential Systems and Hypersurface Theory—.
Edited by T. Shoda and K. Shibuya. November, 2019

Volume 83 Primitive Forms and Related Subjects — Kavli IPMU 2014.
Edited by K. Hori, C. Li, S. Li and K. Saito.
December, 2019

Volume 84 Various Aspects of Multiple Zeta Functions — in honor of Professor Kohji Matsumoto's 60th birthday.
Edited by H. Mishou, T. Nakamura, M. Suzuki and Y. Umegaki. March, 2020

Volume 85 The Role of Metrics in the Theory of Partial Differential Equations.
Edited by Y. Giga, N. Hamamuki, H. Kubo, H. Kuroda and T. Ozawa. December, 2020

Volume 86	Development of Iwasawa Theory — the Centennial of K. Iwasawa's Birth. Edited by M. Kurihara, K. Bannai, T. Ochiai and T. Tsuji. December, 2020
Volume 87	Stochastic Analysis, Random Fields and Integrable Probability — Fukuoka 2019. Edited by Y. Inahama, H. Osada and T. Shirai. November, 2021

TO BE CONTINUED